SME
Mining
Reference
Handbook

Raymond L. Lowrie, P.E.
Editor

Published by the
Society for Mining, Metallurgy, and Exploration, Inc.

Title Page: Aerial photograph of the Kemmerer Mine in Wyoming. P&M Coal Company. Courtesy of Manley Prim Photography, Inc.

Society for Mining, Metallurgy, and Exploration, Inc. (SME)
8307 Shaffer Parkway
Littleton, Colorado, USA 80127
(303) 973-9550 / (800) 763-3132
www.smenet.org

SME advances the worldwide mining and minerals community through information exchange and professional development. SME is the world's largest association of mining and minerals professionals.

Copyright © 2002 Society for Mining, Metallurgy, and Exploration, Inc.

All Rights Reserved. Printed in the United States of America.

Information contained in this work has been obtained by SME, Inc. from sources believed to be reliable. However, neither SME nor its authors guarantee the accuracy or completeness of any information published herein, and neither SME nor its authors shall be responsible for any errors, omissions, or damages arising out of use of this information. This work is published with the understanding that SME and its authors are supplying information but are not attempting to render engineering or other professional services. If such services are required, the assistance of an appropriate professional should be sought.

No part of this publication may be reproduced, stored in a retrieval system, or transmitted in any form or by any means, electronic, mechanical, photocopying, recording, or otherwise, without the prior written permission of the publisher. Any statement or views presented here are those of the author and are not necessarily those of SME. The mention of trade names for commercial products does not imply the approval or endorsement of SME.

ISBN 0-87335-175-4

Library of Congress Cataloging-in-Publication Data

SME mining reference handbook / edited by Raymond L. Lowrie.
    p. cm.
 Includes bibliographical references and index.
 ISBN 0-87335-175-4
  1. Mines and mineral resources--Handbooks, manuals, etc. I. Lowrie, Raymond L. II. Society for Mining, Metallurgy, and Exploration (U.S.)

TN151 .S64 2002
622--dc21
                                              2002026804

Contents

	PREFACE xi
	ABOUT THE EDITOR xiii
	CHAPTER EDITORS xv
CHAPTER 1	**PROJECT EVALUATION** 1
	Feasibility Studies 1
	Due Diligence Studies 4
	Due Diligence Checklist 4
	References 7
CHAPTER 2	**MATERIAL PROPERTIES** 9
	Soils 9
	Rocks 9
	Minerals 10
	Coal 10
	References 11
CHAPTER 3	**EXPLORATION AND GEOLOGY** 55
	Geologic Time 55
	Geologic Map Symbols 56
	Rock Classification 57
	Structural Geology 60
	Geophysics 65
	Aerial Photography 65
	Drilling 67
	References 67
CHAPTER 4	**PHYSICAL SCIENCE AND ENGINEERING** 69
	Chemistry 69
	Statics 71
	Dynamics 80
	Mechanics of Materials 83
	Fluid Mechanics 90
	Thermodynamics and Heat Transfer 96

Electricity and Magnetism **99**
References **102**

CHAPTER 5 **MATHEMATICS, STATISTICS, AND PROBABILITY 103**
Elementary Analysis **103**
Algebra **105**
Trigonometry **107**
Geometry **109**
Derivatives **112**
Integrals **113**
Probability and Statistics **114**
Miscellaneous **119**
References **125**

CHAPTER 6 **WEIGHTS, MEASURES, CONVERSIONS, CONSTANTS, AND SYMBOLS 127**
International System of Units **127**
Fundamental Physical Constants **140**
Selected Constants, Measures, and Time **140**
Selected Unit Equivalencies and Approximations **141**
SI Prefixes **143**
Greek Alphabet **143**
References **143**

CHAPTER 7 **SAMPLING AND ANALYSIS 145**
Sampling **145**
Analysis **155**
References **159**

CHAPTER 8 **ECONOMICS AND COSTING 161**
Interest Formulas **161**
Discrete Compound Interest Tables **162**
Capitalized Costs **166**
Depreciation **166**
Depletion Allowance **167**
Effective Tax Rate **168**
Benefit Cost Ratio (BCR) **168**
Present Value Ratio (PVR) **168**
Inflation **168**
Consumer Price Index **169**
Capital Structure **169**
Cash Flow **170**
Costing **171**
References **172**

CHAPTER 9 **QUALITY AND SPECIFICATIONS OF PRODUCTS 173**
Abrasives **173**
Aggregate **174**
Aluminum **175**
Barite **176**

Chromium   **176**
Clays   **176**
Coal   **177**
Cobalt   **178**
Copper   **178**
Diamonds, Industrial   **179**
Diatomite and Perlite   **179**
Fluorspar   **179**
Gold   **180**
Graphite   **180**
Iron Ore   **180**
Lead   **180**
Limestone and Dolomite   **181**
Manganese   **182**
Mercury   **182**
Mica   **183**
Molybdenum   **183**
Nickel   **184**
Phosphate Rock   **184**
Platinum Group Metals   **184**
Potash   **185**
Pumice   **185**
Quartz Crystal   **185**
Sand and Gravel   **185**
Silver   **186**
Soda Ash   **186**
Stone, Crushed   **186**
Sulfur   **187**
Talc   **187**
Tin   **188**
Titanium   **189**
Vanadium   **190**
Zeolites   **190**
Zinc   **191**
References   **191**

CHAPTER 10   **HAUL ROADS   195**
Haul Truck Specifications   **195**
Stopping Distance   **195**
Number of Lanes   **196**
Safe Distance Between Trucks   **197**
Road Width   **198**
Sight Distance   **199**
Gradient   **199**
Super Elevation   **199**
Construction Material Factors   **200**
Other Considerations   **200**
References   **200**

CHAPTER 11	**BLASTING AND EXPLOSIVES** 203	
	Characteristics of Explosives 203	
	Blast Design 206	
	Ground Vibrations and Air Concussion from Blasting 211	
	Vibration Attenuation and Peak Particle Velocity Predictions 211	
	References 214	
CHAPTER 12	**EXCAVATION, LOADING, AND MATERIAL TRANSPORT** 215	
	Equipment Performance 215	
	Production Calculations 218	
	Equipment Operating Costs 220	
	Draglines 222	
	Clamshells 222	
	Shovels 222	
	Backhoes 222	
	Wheel Loaders 222	
	Crawler-Mounted Loaders 225	
	Scrapers 227	
	Bulldozers 228	
	Rail Haulage 229	
	Belt Conveyors 232	
	Hoisting 238	
	Selected Underground Equipment 239	
	References 241	
CHAPTER 13	**GROUND CONTROL/SUPPORT** 243	
	Introduction 243	
	Safety Factor 246	
	Rock Bolting 246	
	Pillars 248	
	Subsidence 254	
	Slope Stability 256	
	References 257	
CHAPTER 14	**VENTILATION** 259	
	Air Composition, Density, and Psychrometry 259	
	Gas Laws 263	
	Airflow 263	
	Fans 266	
	Natural Ventilating Pressure 267	
	Parallel Flow 267	
	Natural Splitting 267	
	Controlled Splitting 267	
	Mine Gases 268	
	Regulatory Requirements 270	
	Tubing 270	
	Velocities for Mineral Dusts and Gases 273	
	Explosion-Resistant Seals 273	
	References 274	

CHAPTER 15	**PUMPING** 275
	Introduction 275
	Centrifugal Pumps 275
	Calculating Dynamic Head 275
	Net Positive Suction Head (NPSH) 277
	Pump Characteristic Curves 277
	Pumping Power Formulas 278
	Flow Rate Formula and Other Facts 278
	Affinity Laws 279
	References 279
CHAPTER 16	**POWER: ELECTRICAL AND COMPRESSED AIR** 281
	Electrical Power 281
	Compressed Air 283
	References 286
CHAPTER 17	**MINERAL PROCESSING** 287
	Crushing 287
	Grinding 289
	Classification 293
	Screening 294
	Sorting 298
	Gravity Concentration 300
	Flotation 301
	Solid–Liquid Separations 303
	Solvent Extraction (SX) 312
	Heap Leaching 315
	Carbon Plants 317
	Bacterial Oxidation 319
	Coal Cleaning 323
	Dust Collection 326
	Electrowinning (EW) and Electrorefining (ER) 329
	Pyrometallurgy 332
	Industrial Minerals 341
	Gold Leaching 343
	Vat Leaching 344
	Miscellaneous 344
	References 348
CHAPTER 18	**SITE STRUCTURES AND HYDROLOGY** 351
	Facility Layout 351
	Earthworks 353
	Liner Systems 353
	Pipes 360
	Hydrology and Hydraulics 362
	Inspection and Maintenance 367
	References 367

CHAPTER 19	**PLACER MINING** 369
	Placer Ore Bodies 369
	Placer Evaluation 371
	Placer Operations 373
	Costs 377
	Gravel 377
	References 377
CHAPTER 20	**IN SITU LEACHING** 379
	In Situ Leaching 379
	In Situ Leaching of Uranium 380
CHAPTER 21	**MAINTENANCE AND INVENTORY** 383
	Maintenance 383
	Recordkeeping 389
	Inventory 390
	References 392
CHAPTER 22	**ENVIRONMENT AND RECLAMATION** 393
	Major Federal Laws 393
	Citations and Regulations 395
	Permits and Approvals 397
	Air Quality 398
	Water Quality 400
	Treatment 406
	References 409
CHAPTER 23	**HEALTH AND SAFETY** 411
	Legislative History 411
	Health and Safety Regulations 412
	Mine Safety and Health Administration's Mission 413
	Statutory Functions 414
	Organizational Structure 414
	Enforcement 415
	Technical Assistance 416
	Testing Products Used in Mining 417
	Mandatory Training 417
	Historical Data on Mine Disasters in the United States 418
	References 418
CHAPTER 24	**BONDING AND LIABILITIES** 419
	Bonding (Financial Assurance) 419
	Liabilities 420
	References 422

CHAPTER 25	**WEB SITES RELATED TO MINING** **423**	

Professional Societies, Institutes, Councils, Associations, Foundations, Boards, and Commissions **423**
Reference, Information, and Publications **425**
Associations **426**
Economics **427**
Computing Applications **428**
Machinery and Equipment Trading **428**
Trade Unions **428**
Government **428**
International **430**

INDEX 435

Preface

Engineers in the mining industry often must solve problems while in the field at prospects, projects, or places far from any personal bookshelf, company office, or public or private library. And it isn't always feasible to bring along the voluminous authoritative books on mining topics so familiar to the profession. This handbook, then, is designed to fill the technical reference gap for the mobile professional who is away from the normal workplace with its comprehensive store of technical information and resources. It is a distillation of key technical information from the mining literature.

To keep this handbook a reasonable and portable size, the volume editor and all chapter editors had to strictly budget the number of pages allocated to particular subjects. We assumed that the reader is already knowledgeable about the topics and may just need a reminder "on the fly." For this reason, many of the ideas, data, graphs, tables, equations, constants, and rules of thumb are presented with little if any explanation. Detailed explanation or elaboration can be found in the original source, such as the venerable but large two-volume *SME Mining Engineering Handbook*.

The volume editor and all chapter editors are currently licensed or retired registered professional engineers in one or more states. Although the intended audience for this handbook is primarily mining/mineral engineers who work for mining companies and other mining-oriented firms around the world, we hope that academia, students, and state and federal government agencies will also find it useful.

This is a first effort. We anticipate that technological change, along with experience in using the handbook in the field, will allow improvements in future editions.

About the Editor

Raymond L. Lowrie, P.E. (Texas), is the professional registration coordinator for the Society for Mining, Metallurgy, and Exploration, Inc. (SME) and serves as staff liaison for SME's Professional Registration Committee. The committee prepares the mining/mineral P.E. examination each year for the National Council of Examiners for Engineering and Surveying, which distributes it to the various states for administration.

Lowrie started in the mining industry as a development miner at Climax, Colorado. He then worked as a mining engineer at two underground coal mines in Oklahoma and became a supervisory mining engineer for and later chief of the Intermountain Field Operations Center of the former U.S. Bureau of Mines. He was chief of the Division of Reclamation of the Ohio Department of Natural Resources and also served as an assistant director of the Office of Surface Mining within the U.S. Department of the Interior.

He has a B.S. in mining engineering from Texas Western College (now the University of Texas at El Paso) and an M.S. in mineral economics from Colorado School of Mines.

Lowrie is a member of SME, along with the Mineral Economics and Management Society, the Rocky Mountain Coal Mining Institute, the Colorado Mining Association, and the Denver Coal Club.

Chapter Editors

Brent C. Bailey, P.E. (Colorado, Idaho, Utah)
Consultant
Littleton, Colorado
bcbailey@idcomm.com

Jack W. Burgess, P.E. (New Mexico)
Consulting mining engineer
Corrales, New Mexico
(505) 898-7234

Alan A. Campoli, P.E. (Pennsylvania, Kentucky, West Virginia)
Business Development manager, Fosroc, Inc.
Georgetown, Kentucky
Al_Campoli@burmahcastrol.com

Paul D. Chamberlin, P.E. (Colorado)
President, Chamberlin and Associates
Littleton, Colorado
pdcham@wcox.com

Louis W. Cope, P.E. (Colorado)
Process and placer consultant
Denver, Colorado
processman@aol.com

Keith E. Dyas (retired P.E., Arizona, Nevada)
Mining engineer
Parker, Colorado
keithdyas@msn.com

John E. Feddock, P.E. (Ohio, West Virginia, Pennsylvania, Kentucky, Virginia, Illinois, Utah)
Senior vice president, Marshall Miller & Associates
Lexington, Kentucky
john.feddock@mma1.com

Frank J. Filas, P.E. (Colorado, Utah, Arizona, New Mexico)
Senior project engineer, EnecoTech Inc.
Golden, Colorado
(303) 526-4495

Brett F. Flint, P.E. (Alaska, Nevada, Colorado)
Manager, Engineering Services, Alaska Operations, URS Corporation
Anchorage, Alaska
Brett_Flint@URSCorp.com

William J. Francart, P.E. (Pennsylvania)
Mining engineer, Mine Safety and Health Administration
Pittsburgh, Pennsylvania
francart-william@msha.gov

Daniel F. Kump, P.E. (Nevada)
Senior project engineer, Barrick Bullfrog Inc.
Beatty, Nevada
dankump@aol.com

Raymond L. Lowrie, P.E. (Texas)
Professional registration coordinator, SME
Franktown, Colorado
raylowrie@aol.com

Conrad H. Parrish, P.E. (Colorado, Wyoming)
Principal, C.H. Parrish & Associates, Inc.
chparrish@worldnet.att.net

Russell F. Price, P.E. (New Mexico, Virginia, Colorado)
Consulting engineer
Alamogordo, New Mexico
russprice@zianet.com

Larry C. Schneider, P.E. (Kentucky)
Director, Division of Explosives and Blasting, Kentucky Department of Mines and Minerals
Frankfort, Kentucky
Lschnei276@aol.com

William K. Smith, P.E. (Colorado)
Geologist, U.S. Geological Survey
Golden, Colorado
wksmith@usgs.gov

Marcus A. Wiley, P.E. (Colorado; formerly Oklahoma, New Mexico)
Manager, Wiley Consulting, LLC
Englewood, Colorado
mark@wileyconsulting.net

Kelvin K. Wu, P.E. (Pennsylvania)
Chief, Mine Waste and Geotechnical Engineering Division, Mine Safety and Health Administration
Pittsburgh, Pennsylvania
wu-kelvin@msha.gov

R. Karl Zipf, Jr., P.E. (Colorado)
Chief, Catastrophic Failure Detection & Prevention Branch, Spokane Research Laboratory, National Institute of Occupational Safety & Health
Spokane, Washington
RZipf@cdc.gov

Contributors to Mineral Processing Chapter

Edwin H. Bentzen, III
Senior project manager, Resource Development, Inc.
Wheat Ridge, Colorado
bentzeniii@aol.com

Andy Briggs, P.E. (Northwest Territories, Canada)
Vice president, Signet Technology, Inc.
Evergreen, Colorado
andy.briggs@fluor.com

Paul D. Chamberlin, P.E. (Colorado)
President, Chamberlin and Associates
Littleton, Colorado
pdcham@wcox.com

Louis W. Cope, P.E. (Colorado)
Process and placer consultant
Denver, Colorado
processman@aol.com

David S. Davies, P.E. (Connecticut)
Consulting engineer, Pocock Industrial, Inc.
Salt Lake City, Utah
pocock_industrial@msn.com

John E. Litz
President, J.E. Litz & Associates, LLC
Golden, Colorado
jklitz7@ix.netcom.com

Deepak Malhotra
President, Resource Development, Inc.
Wheat Ridge, Colorado
dmalhotra@aol.com

Douglas K. Maxwell, P.E. (Colorado)
Consultant
Arvada, Colorado
dmaxwell39@home.com

Terry P. McNulty, P.E. (Colorado)
President, T.P. McNulty and Associates, Inc.
Tucson, Arizona
tpmacon1@aol.com

Gary F. Meenan
Senior engineering scientist, Consol Energy, Inc.
South Park, Pennsylvania
garymeenan@consolenergy.com

Douglas J. Robinson
President, Dremco, Inc.
Phoenix, Arizona
doug@dremco1.com

D. Erik Spiller
Outokumpu Technology, Inc.
Jacksonville, Florida
erik.spiller@outokumpu.com

Vernon F. (Fred) Swanson, P.E. (Colorado)
Consultant
Lakewood, Colorado
fswanson@excelonline.com

Contributors to In Situ Leaching Chapter

Ray V. Huff (formerly P.E., Oklahoma)
President, R.V. Huff & Associates, Inc.
Florence, Alabama
janeray@mail.getaway.net

Joseph R. Stano
Senior process engineer, Washington Group International, Inc.
Englewood, Colorado
jstano@ecentral.com

CHAPTER 1

Project Evaluation

Keith E. Dyas (retired P.E.)

FEASIBILITY STUDIES

A "project evaluation" includes all the activities conducted to develop information for determining a project's economic viability. The results are included in a "feasibility study," which integrates the data into a project plan and analyzes the economics. A feasibility study examines all phases of an investment proposal in as much detail as necessary to justify dropping it or continuing expenses through the next stage. A project evaluation progresses through several levels of feasibility as additional data are accumulated. Generally, feasibility levels advance from conceptual (prefeasibility) to preliminary to intermediate to final. After each level, investigators decide, based on economics, environmental considerations, and market timing, whether to advance to the next level and ultimately whether to build. Different companies use different terminology to define levels of feasibility.

Investigators must be careful to include all required elements and to conduct their studies at the appropriate level of investigation. If the investigation is preliminary, additional time and expense should not be spent to secure detail that belongs in the intermediate or final level.

Table 1.1 is a checklist, developed by Pincock Allen & Holt (2000), of minimum reporting requirements for feasibility studies. The process has three steps: conceptual, prefeasibility, and feasibility.

A conceptual study is a preliminary evaluation of a project. Although the level of drilling and sampling must be sufficient to define a resource adequately, flowsheet development, cost estimation, and production scheduling are often based on limited test work and engineering design. This level of study is useful for defining subsequent engineering inputs and further required studies, but is not appropriate for economic decision-making.

The prefeasibility study represents an intermediate step between the conceptual and final feasibility studies. Cost estimates are of the order of ±30% accuracy, but not considered sufficiently accurate for final decision-making.

The feasibility study, considered the "bankable document," is detailed and accurate enough to be used for positive "go" decisions and financing purposes. Cost estimates are ±20% accurate or better. Mine plans show materials movements and ore grades on an annual basis. Flowsheet development is based on extensive test work, material balances, and general arrangement drawings. Economic evaluation is based on cash flow calculations for the life of the defined reserve.

TABLE 1.1 Minimum reporting requirements for feasibility studies

	Conceptual Study	Prefeasibility Study	Feasibility Study
SUMMARY AND RECOMMENDATIONS			
Principal parameters including:	Mostly assumed and factored	Some engineering basis	Sound engineering basis
Ore reserves	Yes	Yes	Yes
Mining and processing rates	Yes	Yes	Yes
Environmental issues and permitting requirements	Yes	Yes	Yes
Development period and mine life	Yes	Yes	Yes
Metal recoveries	Yes, assumed	Yes	Yes
Capital cost estimate	Yes	Yes	Yes
Operating cost estimate	Yes	Yes	Yes
NPV, IRR, and ROI	No	Yes	Yes
INTRODUCTION			
Location, topography, and climate			
Site location map	Yes	Yes	Yes
Detailed topography map	Yes	Yes	Yes
Ownership and royalties			
Claims list	No	No	Yes
Claim map	No	Yes	Yes
Current status and history			
Historical chronology	No	Yes	Yes
Past production, if any	No	Yes	Yes
GEOLOGY AND RESOURCES			
Geologic description			
Geologic map	Yes	Yes	Yes
Geologic cross-sections	No	Yes	Yes
Drilling, sampling, and assaying			
Parameters	Yes	Yes	Yes
Drill hole location map	Yes	Yes	Yes
Sampling/assaying flow diagram	No	Yes	Yes
Assay check graph	No	Yes	Yes
Mineral resource estimate			
Geologic model physical limits	Yes	Yes	Yes
Lithology/tonnage factors/code	No	Yes	Yes
Basic statistics	No	Yes	Yes
Cumulative frequency of samples versus grade	No	Yes	Yes
Variograms	No	Yes	Yes
Resource estimate	Internationally accepted standards*	Internationally accepted standards*	Internationally accepted standards*
MINING			
Ore reserve estimate			
Reserve calculation parameters	Assumed values	Test-based values	Test-based values
Cutoff grade equations	No	Yes	Yes
Reserve estimate	Internationally accepted standards*	Internationally accepted standards*	Internationally accepted standards*
Mining method and plans			
Mining parameters	Yes, minimal engineering basis	Yes, some geotechnical data	Yes

continues next page

TABLE 1.1 Minimum reporting requirements for feasibility studies (continued)

	Conceptual Study	Prefeasibility Study	Feasibility Study
Mining method and plans (continued)			
Hydrology/geotechnical parameters	No	Yes	Yes
Equipment list	No	Yes	Yes
Consumables list	No	Yes	Yes
Personnel list	No	Yes	Yes
Surface mining:			
Final pit and dump outlines	Yes, simple outline	Yes	Yes
Annual pit and dump outlines	No	Yes	Yes
Underground mining:			
General mine development	Yes, simple outline	Yes	Yes
Stoping system	No	Yes	Yes
Production schedule			
Annual ore and waste tonnage and grade	Yes, simple division of total	Yes	Yes
Mining capital and operating cost estimates			
Capital and operating cost estimates	Yes, factored	Yes, from estimating manuals, ~±30%	Yes, from vendor quotes and take-offs, ~±20%
PROCESSING			
Ore sampling and test work			
Test-work data	No, assumed values	Yes, preliminary data	Yes, detailed data
Processing method and plans			
Processing parameters	Yes, minimal engineering basis	Yes	Yes
Equipment list	No	Yes	Yes
Consumables list	No	No	Yes
Flow diagram	Yes, simple block diagram	Yes	Yes
Personnel list	No	Yes	Yes
Material balance	No	No	Yes
Site plan	No	Yes	Yes
General arrangement drawings	No	No	Yes
Processing capital and operations cost estimates			
Capital and operating cost estimates	Yes, factored	Yes, from estimating manuals, ~±30%	Yes, from vendor quotes and take-offs, ~±20%
Freight, smelting, and refining (FSR) costs	Yes	Yes	Yes
INFRASTRUCTURE AND ADMINISTRATION			
Infrastructure facilities			
Facilities list	Yes, with minimal detail	Yes	Yes
Power and water parameters	Yes, preliminary	Yes	Yes
Full site plan	No	Yes	Yes
Infrastructure, capital, and administration operating cost estimates			
Personnel list	No	Yes	Yes
Capital and operating cost estimates (including working capital and owner's preproduction expenses)	Yes, factored	Yes, from estimating manuals, ~±30%	Yes, from vendor quotes and take-offs, ~±20%

continues next page

TABLE 1.1 Minimum reporting requirements for feasibility studies (continued)

	Conceptual Study	Prefeasibility Study	Feasibility Study
ENVIRONMENTAL/PERMITTING STATUS			
Environmental management system			
Permit/regulatory framework	Yes, preliminary	Yes	Yes
Environmental Impact Analysis (EIA)	Yes, preliminary	Yes	Yes, well-defined
Impact mitigation plans	Yes, preliminary	Yes, preliminary	Yes
Mine waste management plan	No	Yes, preliminary	Yes
Solid and hazardous materials handling	No	Yes, preliminary	Yes
Spill prevention and emergency response plan	No	No	Yes
Environmental cost estimates			
Capital and operating cost estimates	Yes, conceptual	Yes, ~±30%	Yes, ~±20%
Closure costs and accounting method	Yes, minimal detail	Yes, ~±30%	Yes, ~±20%
DEVELOPMENT SCHEDULE			
Schedule chart	No	Yes	Yes
Schedule calendar	No	No	Yes
ECONOMICS			
Principal economic parameters	Yes, preliminary assessment	Yes	Yes
Royalties and taxes	No	Yes	Yes
Cash flows	Yes, preliminary assessment	Yes	Yes
Sensitivities	No	Yes	Yes

*Internationally accepted standards include:
1. Canadian National Instrument 43-101 and 43-101 CP.
2. Australasian Code for Reporting of Mineral Resources and Ore Reserves — Prepared by the Joint Ore Reserve Committee (JORC).
3. U.S. Securities & Exchange Commission Industry Guide 7.
4. SME Guide for Reporting Exploration Information, Mineral Resources and Mineral Reserves.

Source: Pincock Allen & Holt 2000 *(reprinted with permission).*

DUE DILIGENCE STUDIES

Due diligence studies are often required when projects or operations are being evaluated for financing, loans, participation by others, or acquisition of a property or company by another firm. The steps involved are similar to those taken in project feasibility studies. An abbreviated checklist of due diligence procedures, adapted with permission from Behre Dolbear and Company, Inc. (1994), follows. Because every mining project has unique characteristics, some activities listed might be eliminated and others added.

DUE DILIGENCE CHECKLIST

Reserves
- Drill hole surveys—spot-check for accuracy.
- Sampling procedures—check that channel, drill cuttings, or core were obtained by normal industry standards.
- Splitting—assess who conducted the splitting, the procedures they used, and any possible biasing of samples.

- Assaying—check lab reputation, procedures used, repeatability, and location of rejects and pulps.
- Logging of drill holes—check procedures and spot-check logs against cores or cuttings for accuracy.
- Plotting of assays/compositing—spot-check for accuracy.
- Cutting/capping of grades—check for consistency and adherence to standard practices.
- Bulk samples—assess how sample grade agrees with included and adjacent drill hole assays.
- Geologic cross-sections—check for inclusion of structures, lithologies, and mineralization. Density factor—check on how the factor was obtained and its adequacy.
- Drill hole spacing—determine if spacing is close enough to confirm continuity of mineralization.
- Cutoff grades—include, at operating mines, all cash costs, general and administrative (G&A), excise taxes, metallurgical recovery factor, and new capital. At new mines, include all capital.
- Dilution—use an adequate dilution factor.
- Geostatistical programs—check search radius and determine if method used is properly done and to industry standards.
- Independent calculation—redo (independently) 15%–20% of reserve blocks geostatistically and check against manually calculated reserves for the same blocks.
- Reserve categories—use local government standards and definitions for proven (measured) and probable (indicated).

Mine Engineering and Planning
- Mining method—check if appropriate mining method is described for geologic, geotechnical, climatic, and labor conditions.
- Mine plan—check production scheduling, grade and tonnages, dilution and tonnage factors, stripping schedule, and labor and equipment capacities.
    - Surface mining—check pit slope justification, sufficiency of slopes and benches, and how final pit limits were determined.
    - Underground mining—check adequacy of stope spacing for ground control, ventilation requirements, escapeways, hoisting and haulage capacities, productivities, development, and economics.
- Equipment—check major pieces for size and quantity, adequacy of design, availability, and utilization. Ensure that support equipment is sufficient; if in feasibility stage, check costs for delivery, setup, and spares.
- Manning—check supervision and labor requirements, appropriate wage rates, and benefits detailed for life of mine.
- Maintenance—check for sufficient facilities, spare parts, procedures, and staff.
- Operating costs—check labor costs, fuel consumption, electricity rate and consumption, consumables, water, maintenance, engineering staff, and grade control.
- Capital costs—check costs for equipment, maintenance shops, electrical support, water acquisition and distribution, fencing, change facilities, engineering facilities, working capital, and crushing and transport (if not in plant costs).
- Contract mining—review description of mining method, costs per ton, mileage over/under charges and mobilization/demobilization charges, grade control staff, and capital and operating costs.
- Mine permits—review permits including Mine Safety and Health Administration identification and training program, state and local permits, federal ATF, and training programs for employee certification.

Metallurgical/Processing
- Exploration phase—review mineral characterization studies including mineralogy, estimation of liberation size, design of sampling to gain representation for ore grade, metallurgical recovery, concentrate grade, and delineation of "trouble minerals."
- Ore reserve estimation/basic engineering phase
  - Laboratory testing—check screen analysis, crushability, grindability, bench-scale extraction, mid-scale extraction and demonstration tests, ore variability analysis, and concentrate and tailings dewatering tests. Develop reagent recommendations and disposal evaluation.
  - Pilot-scale testing—check power, crusher lining grinding power, grinding ball/rod mill, reagent consumption, and water requirements.
  - Operating cost analysis—analyze approximate labor needs, electric power, consumables, maintenance materials, water, high- and low-pressure air, and operating costs.
  - Capital cost analysis—check costs for crushing, grinding/classification; extraction; thickening; filtration; drying; packaging; loading concentrate; tailings disposal; administrative, office, and engineering support; warehousing; shops; and water systems.
- Detailed engineering phase—review definitive estimate of operating costs, final flowsheets and material balances, definitive estimate of construction and capital costs, electrical contracts, water supply contracts, access roads, town site and administrative design, and detailed financial analysis.
- Operating plant
  - Operations—review scheduling, process availability, maintenance costs, staffing, safety programs, inventory controls, warehouse controls, and electric power management. Conduct process audit on metal recovery, concentrate grade, retention times, grinding efficiency, classification efficiency, and reagent consumption.
  - Maintenance—review scheduling, equipment availability, wear rates, staffing, lubrication and inspection programs, preventive maintenance scheduling and control, and safety programs.
  - Engineering—analyze metallurgical, quality control, process improvement, cost reduction, material research, and staffing/control.
  - General and administrative—review labor relations, absenteeism, workers' compensation, work schedules, management systems/access, and employee involvement.

Economics
- Economic analysis input
  - Capital investments—verify that these costs are consistent with capital investments estimated by mining and processing engineers and tabulated elsewhere. The costs should include not only initial purchases but replacement or expansion capital as well. Consider the costs for mobile mining equipment; fixed structures; fixed, nonstructural facilities; process facilities; lightweight vehicles and development, exploration, acquisition, and working capital.
  - Operating costs—ensure that these costs are consistent with operating costs estimated by mining and processing engineers and tabulated elsewhere. The operating costs can be classified by unit operations or as follows: wages, salaries and benefits, materials and supplies, electrical power, diesel fuel and gasoline, and other types of overhead.
  - Royalties—verify how the royalty was determined, the variance of royalty from area to area, and if a minimum royalty is to be paid.
  - Commodity prices—evaluate the method by which the commodity was priced.
  - Tax structure—analyze the following taxes: national income; national minimum; local, state, or provincial income; severance; property; depreciation, amortization, and depletion; and any special taxes.

- Financing—evaluate the amount to be borrowed, the drawdown and payback schedule, the interest rate, and any loan origination fees charged.
- Economic analysis outputs
    - Cash flow spreadsheet—develop a cash flow spreadsheet that shows annual cash flows over the preproduction, production, and postproduction periods. Include production; revenue; operating costs; royalties; depreciation; amortization; income before depletion and taxes; depletion, special, state, and local taxes; national income tax; income after tax; net cash flow from operations; capital investments; and project cash flow. Show a life-of-mine total for all pertinent costs and quantities.
    - Detailed spreadsheets—create additional spreadsheets to detail the determination of capital investment, depreciation by asset or asset class, amortization, and depletion by year.
    - Net present value—determine the net present value of the project's cash flow for various discount rates such as 0%, 5%, 10%, 15%, 20%, and 25%. State the method of discounting.
    - Internal rate of return (IRR)—use this valuation method where there are major capital investments made before production and significant cash flows thereafter. State the method by which the IRR was determined.
    - Sensitivity analysis—run analyses that reflect various commodity prices, along with operating and capital costs, to determine the effect of such variations.
    - Documentation—explain sources of capital and operating input in tables; source, tabulation, or explanation of royalty; source of commodity price; source and explanation of tax data; source or explanation of financing data for base case, for other cases, and for sensitivity analyses
- Environmental
- Project request/identification
- Project team formation
- Background information review
- Project plan development
- Field inspection
- Media-specific investigation procedures
    - Resource Conservation and Recovery Act (RCRA)
    - Clean Water Act (CWA)
    - Clean Air Act (CAA)
    - Safe Drinking Water Act (SDWA)
    - Toxic Substances Control Act (TSCA)
    - Federal Insecticide, Fungicide and Rodenticide Act (FIFRA)
    - Emergency Planning and Community Right-to-Know Act (EPCRA)
    - Comprehensive Emergency Response Compensation and Liability Act (CERCLA)
- Laboratory and quality audits.

See chapter 24 for more information on environmental due diligence.

REFERENCES

Behre Dolbear & Company, Inc. 1994. *Quality Assurance and Checklist/Procedures for Due Diligence Studies*. Denver, CO: Behre Dolbear & Company.

Pincock Allen & Holt. 2000. *Feasibility Studies Minimum Reporting Requirements*. Technical Bulletin 2000-1. Lakewood, CO: Pincock Allen & Holt.

CHAPTER 2

Material Properties

Jack W. Burgess, P.E.

Engineering properties of natural earth-related materials are largely compiled based on their important uses to society. Materials considered here are soils, rocks, minerals, and coal; various properties related to each material are also presented. Although the discussion below is not all-inclusive, materials most commonly encountered in mining are covered.

In many cases, several physical properties allow a particular material to be readily identified. For example, color, particle size, crystal system, hardness, and chemical or metal content often assist in identification.

Table 2.1 on pages 12–13 shows bank and loose densities, angles of repose, and swell factors of common mining-related materials.

Table 2.2 on page 14 presents general swell and void percentages and related load factors.

SOILS

Soils include gravels, sands, silts, clays, organic soils, and permafrost. They are classified on the basis of index properties, such as particle-size distribution and plasticity characteristics. From an engineering standpoint, other important physical properties include natural water content, density, permeability, shear strength, and compressibility (Sherman 1973).

Particle size of a soil is expressed in boulders, cobbles, gravel, sand, silt, and clay; Table 2.3 on page 14 lists customary sizes.

The Unified Soil Classification System, commonly used to classify soils for engineering purposes, is shown in Table 2.4 on pages 15–16.

Table 2.5 on page 17 shows weight (saturated and dry), friction angle, and cohesion for typical soils and rocks. Strength characteristics of soils are in Table 2.6 on page 18, and Table 2.7 on page 18 gives typical soil modulus values.

Table 2.8 on pages 19–20 lists important engineering properties and uses for various soils.

Permafrost, found in northern locations, is frozen ground, no matter what its other soil or rock attributes may be. Associated physical properties can be completely different from unfrozen similar materials and may pose serious problems, such as solifluction or mass creep, in a zone where thawing occurs.

ROCKS

Rock is a naturally occurring solid material consisting of one or more minerals. See Table 2-9 on pages 21–31 for properties of rocks.

See Table 13.1 in Chapter 13, which covers ground control and support, for more information on strength properties.

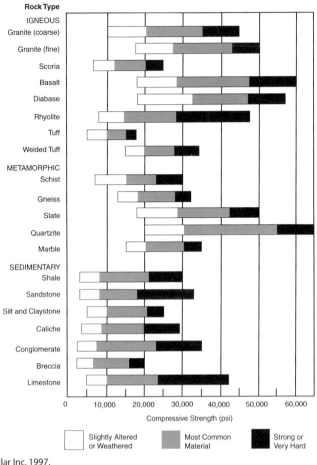

Source: Caterpillar Inc. 1997.

FIGURE 2.1 Compressive strength of common rocks

MINERALS

Minerals are solid, naturally occurring chemical elements or compounds that are homogenous, with a definite chemical composition and a very regular arrangement of atoms. More than 3,000 mineral species are known, most of which are characterized by definite chemical composition, crystalline structure, and physical properties. They are classified primarily by chemical composition, crystal class, hardness, and appearance (color, luster, and opacity). Mineral species are generally limited to solid substances, the only liquids being metallic mercury and water. Metalliferous minerals of economic value, which are mined for their metals, are known as ores. See Table 2.10 on pages 32–49 for properties of minerals. See Table 2.11 on page 50 for properties of major mineral fillers.

COAL

Coal is a generic designation for many solid organic minerals with different compositions and properties. All are rich in carbon and have a dark color. A genetic relationship exists between peat, brown coal, lignite, bituminous coal, and anthracite. The process of coal formation, or coalification, is a continuous transformation of plant material, with each phase characterized by the degree of coalification (Hower and Parekh 1991).

See Table 2.12 on page 51 for classification of coals by rank.

See Table 2.13 on page 51 for typical proximate and ultimates analyses.

Specific gravity of coal ranges from 1.23 to 1.72 depending on rank, moisture, and ash content; and it tends higher in the range as each increases (Hower and Parekh 1991).

See Table 2.14 on pages 52–53 for petrographic and physical properties.

See Table 2.15 on pages 53–54 for sulfur content and forms.

See Table 2.16 on page 54 for ash content and fusion temperature.

See Table 13.1 in Chapter 13, which covers ground control and support, for information on strength properties.

REFERENCES

Anon. 1972. *Steam—Its Generation and Use*. New York: Babcock and Wilcox.

ASTM. 1998. *Classification of Coals by Rank*. D 388, Standard Classification of Coals by Rank. West Conshohocken, PA: ASTM. 176.

Beasley, C.A., M.H. Erten, O.A. Gallegos, V. Joyce, D.E. Beasley, and D.A. Shuman. 1991. Coal characteristics and preparation requirements. In *Coal Preparation*. 5th ed. Edited by J.W. Leonard, III. Littleton, CO: Society for Mining, Metallurgy, and Exploration, Inc. (SME). 145–186.

Bolles, J.L., and E.J. McCullough. 1985. Minerals and their properties. In *SME Mineral Processing Handbook*. Vol. 1. Edited by N.L.Weiss. New York: Society of Mining Engineers of the American Institute of Mining, Metallurgical, and Petroleum Engineers, Inc. 2-4–2-17.

Carmichael, R.S. 1982. *Handbook of Physical Properties of Rocks*. Vol. II. Boca Raton, FL: CRC Press.

Caterpillar, Inc. 1997. *Caterpillar Performance Handbook*. Edition 28. Peoria, IL: Caterpillar, Inc.

Cummins, A.B. 1960. Mineral fillers. In *Industrial Minerals and Rocks*. 3rd ed. Edited by J.L. Gillson. New York: American Institute of Mining, Metallurgical, and Petroleum Engineers, Inc. 567–584.

Hartley, J.D., and J.M. Ducan. 1987. E′ and its variation with depth. *Journal of Transportation*. September.

Hartman, H.L., ed. 1992. Appendix Table E: Material properties and characteristics. In *SME Mining Engineering Handbook*. 2nd ed., Vol. 2. Littleton, CO: SME. A-32–A-33.

Houk, E., and J. Bray. 1977. *Rock Slope Engineering*. Rev. 2nd ed. London: The Institution of Mining & Metallurgy.

Hower, J.C., and B.K. Parekh. 1991. Chemical/physical properties and marketing. In *Coal Preparation*. 5th ed. J.W. Leonard, III. Littleton, CO: SME. 3–94.

Levy, A., R.E. Barrett, R.D. Giammar, and H.R. Hazard. 1981. Coal combustion. In *Coal Handbook*. Edited by R.A. Meyers. New York: Marcel Dekker. 362.

Lindeburg, M.R. 1992. *Civil Engineering Reference Manual*. 6th ed. Belmont, CA: Professional Publications.

Sherman, W.C. 1973. Elements of soil and rock mechanics—soil mechanics. In *SME Mining Engineering Handbook*. Vol. 1. Edited by A.B. Cummins and I.A. Given. New York: Society of Mining Engineers of The American Institute of Mining, Metallurgical, and Petroleum Engineers, Inc. 6-2–6-13.

Trivedi, N.C., and R.W. Hagemeyer. 1994. Fillers and coatings. In *Industrial Minerals and Rocks*. 6th ed. Edited by D.D. Carr. Littleton, CO: SME. 483–495.

U.S. Army Corps of Engineers. 1953. *The Unified Soil Classification System*. Technical Memo. 3-357. Vicksburg, MS: Office, Chief of U.S. Army Corps of Engineers.

Wagner, A.A. 1957. The use of the unified soil classification system for the Bureau of Reclamation. *Proceedings, 4th International Conference of Soil Mechanical Engineering*. 125–134. London.

TABLE 2.1 Properties of common mining-related materials

Material	Bank Density (lb/ft³)	Loose Density (lb/ft³)	Angle of Repose (degrees)	Swell Factor
Alumina	—	60	22	—
Ammonium nitrate	—	45	—	—
Asbestos ore	—	81	30–44	—
Ashes, dry	—	35–40	40	—
Ashes, wet	—	45–50	50	—
Bauxite, crushed, 3 × 0 in.	—	75–85	30–44	—
Bauxite, ground, dry	—	68	20–31	—
Bauxite, run of mine	100–160	75–120	31	0.75
Clay, compact, natural bed	109	82	—	0.75
Clay, dense, tough or wet	111	83	—	0.75
Clay, dry	85	68	—	0.80
Clay, dry excavated	69	—	—	—
Clay, dry in lump, loose	—	60–70	35	—
Clay, fines	—	100–120	35	—
Clay, light (kaolin)	104	80	—	0.77
Clay and gravel, dry	100	71	—	0.71
Clay and gravel, wet	114	81	—	0.71
Chrome ore	—	125–140	30–44	—
Cinders, coal	—	40–45	35	—
Coal, anthracite	81–85	60–63	27	0.74
Coal, anthracite, sized	—	55–60	27	—
Coal, bituminous	80	50–52	45–55	0.62–0.65
Coal, bituminous, mined, run-of-mine	—	45–55	38	—
Coal, bituminous, mined, sized	—	45–55	35	—
Coal, bituminous, mined, slack, ½ in. and under	—	43–50	40	—
Coal, bituminous, strip, not cleaned	—	50–60	—	—
Coal, lignite	—	40–45	38	—
Coke	—	24–31	—	—
Coke, breeze, ¼ in. and under	—	25–34	30–45	—
Coke, loose	—	23–35	30–44	—
Coke, petroleum	—	35–40	—	—
Copper ore	141	100	—	0.71
Earth, dry	104	57–83	35	0.55–0.80
Earth, dry, loam	78	57–68	—	0.73–0.87
Earth, moist	100	75–85	—	0.75–0.85
Earth, rock	93–119	71–91	—	0.76
Earth, sand, gravel	115	98	—	0.85
Earth, wet	125	100–104	—	0.80–0.83
Earth, wet, containing clay	—	100–110	45	—
Feldspar, ½-in. screenings	—	70–85	38	—
Feldspar, 1½ to 3 in.	—	90–100	34	—
Feldspar, 200 mesh	—	100	30–44	—
Gneiss	168	96	—	0.57
Granite	167	90–111	—	0.54–0.66

continues next page

MATERIAL PROPERTIES | 13

TABLE 2.1 Properties of common mining-related materials (continued)

Material	Bank Density (lb/ft³)	Loose Density (lb/ft³)	Angle of Repose (degrees)	Swell Factor
Granite and porphyry	170	97	—	0.57
Graphite ore	—	65–75	30–44	—
Gravel, dry	91–120	46–107	—	0.51–0.89
Gravel, dry, screened	—	90–100	40	—
Gravel, run-of-bank	—	90–100	38	—
Gravel, wet	144	131	—	0.91
Gypsum	163–167	100–111	—	0.61–0.66
Gypsum, ½-in. screenings	—	70–80	40	—
Gypsum, 1½ to 3 in.	—	70–80	30	—
Iron ore	—	100–200	35	—
Iron ore, hematite	241–322	144–145	—	0.60–0.45
Iron ore pellets	—	116–130	30–44	—
Iron ore, taconite	150–200	107–143	—	0.71–0.72
Kaolin	104	80	—	0.77
Lead ore	—	200–270	30	—
Lime, pebble	—	53–56	30	—
Limestone	163	99	—	0.61
Limestone, blasted	156	89–93	—	0.57–0.60
Limestone, crushed	—	85–90	38	—
Limestone, marble	170	97–101	—	0.57–0.59
Manganese ore	—	125–140	39	—
Mud, dry	80–110	66–91	—	0.82–0.83
Mud, wet	110–130	91–108	—	0.83
Nickel–cobalt sulfate ore	—	80–150	30–44	—
Rock, crushed	—	125–145	20–29	—
Rock, soft, excavated with shovel	—	100–110	30–44	—
Rock, stone, crushed	120–145	89–107	—	0.74
Rock, well-blasted	148	99	—	0.67
Sand, bank, damp	—	105–130	45	—
Sand, bank, dry	—	90–110	35	—
Sand, dry	81–126	70–115	—	0.86–0.91
Sand, moist	126	110	—	0.87
Sand and gravel, dry	123	108	—	0.88
Sand and gravel, wet	144	125	—	0.87
Sandstone	144–153	96–110	—	0.67–0.72
Sandstone, broken	—	85–90	30–44	—
Shale, broken	—	90–100	20–29	—
Shale, crushed	—	85–90	39	—
Shale, riprap	104	78	—	0.75
Slag	136	110	—	0.81
Slate	170–180	131–139	—	0.77
Stone, crushed	120–145	89–107	—	0.74
Sulfur ore	—	87	—	—
Trap rock	185	122–124	—	0.66–0.67
Zinc ore, crushed	—	160	38	—

Source: Adapted from Hartman 1992.

TABLE 2.2   Load factors from swell and void percentages

Swell (%)	Voids (%)	Load Factor
5	4.8	0.952
10	9.1	0.909
15	13.0	0.870
20	16.7	0.833
25	20.0	0.800
30	23.1	0.769
35	25.9	0.741
40	28.6	0.714
45	31.0	0.690
50	33.3	0.667
55	35.5	0.645
60	37.5	0.625
65	39.4	0.606
70	41.2	0.588
75	42.9	0.571
80	44.4	0.556
85	45.9	0.541
90	47.4	0.526
95	48.7	0.513
100	50.0	0.500

TABLE 2.3   Particle sizes of soils

Types of Material		Sizes (mm)
Boulders		Over 200
Cobbles		60–200
Gravel	Coarse	20–60
	Medium	6–20
	Fine	2–6
Sand	Coarse	0.6–2
	Medium	0.2–0.6
	Fine	0.06–0.2
Silt	Coarse	0.02–0.06
	Medium	0.006–0.02
	Fine	0.002–0.006
Clay		Less than 0.002

Source: Wagner 1957.

TABLE 2.4 Unified soil classification system

1	2	3	4	5	6	7
Major Divisions		**Group Symbols**	**Typical Names**	**Field Identification Procedures** (Excluding particles larger than 3 in. and basing fractions on estimated weights)	**Information Required for Describing Soils**	**Laboratory Classification Criteria**
Coarse-Grained Soils — More than half of material is *larger* than No. 200 sieve size. The No. 200 sieve size is about the smallest particle visible to the naked eye.	**Gravels** — More than half of coarse fraction is larger than No. 4 sieve size. (For visual classification, the ¼-in. size may be used as equivalent to the No. 4 sieve size.)				Give typical name; indicate approximate percentages of sand and gravel, maximum size; angularity, surface condition, and hardness of the coarse grains; local or geologic name and other pertinent descriptive information; and symbol in parentheses. For undisturbed soils add information on stratification, degree of compactness, cementation, moisture conditions, and drainage characteristics. Example: *Silty sand, gravelly;* about 20% hard, angular gravel particles, ½-in. maximum size; rounded and subangular sand grains, coarse to fine; about 15% nonplastic fines with low dry strength; well compacted and moist in place; alluvial sand (SM).	Use grain-size curve in identifying the fractions as given under field identification. Determine percentages of gravel and sand from grain-size curve. Depending on percentage of fines (fraction smaller than No. 200 sieve size) coarse-grained soils are classified as follows: Less than 5% = GW, GP, SW, SP More than 12% = GM, GC, SM, SC 5% to 12% = *Borderline* cases requiring use of dual symbols
	Clean Gravels (Little or no fines)	GW	Well-graded gravels, gravel-sand mixtures, little or no fines	Wide range in grain sizes and substantial amounts of all intermediate particle sizes		$C_u = \dfrac{D_{60}}{D_{10}}$ Greater than 4 $C_c = \dfrac{(D_{30})^2}{D_{10} \times D_{60}}$ Between 1 and 3
		GP	Poorly graded gravels or gravel-sand mixtures, little or no fines	Predominantly one size or a range of sizes with some intermediate sizes missing		Not meeting all gradation requirements for GW
	Gravels with Fines (Appreciable amount of fines)	GM	Silty gravels, gravel-and-silt mixtures	Nonplastic fines or fines with low plasticity (for identification procedures see ML below)		Atterberg limits below "A" line or PI less than 4 / Atterberg limits above "A" line with PI greater than 7 — *borderline* cases requiring use of dual symbols.
		GC	Clayey gravels, gravel-sand-clay mixtures	Plastic fines (for identification procedures see CL below)		
	Sands — More than half of coarse fraction is *smaller* than No. 4 sieve size.					
	Clean Sands (Little or no fines)	SW	Well-graded sands, gravelly sands, little or no fines	Wide range in grain size and substantial amounts of all intermediate particle sizes		$C_u = \dfrac{D_{60}}{D_{10}}$ Greater than 4 $C_c = \dfrac{(D_{30})^2}{D_{10} \times D_{60}}$ Between 1 and 3
		SP	Poorly graded sands or gravelly sands, little or no fines	Predominantly one size or a range of sizes with some intermediate sizes missing		Not meeting all gradation requirements for SW
	Sands with Fines (Appreciable amount of fines)	SM	Silty sands, sand-silt mixtures	Nonplastic fines or fines with low plasticity (for identification procedures see ML below)		Atterberg limits below "A" line or PI less than 4 / Atterberg limits above "A" line with PI greater than 7 — *borderline* cases requiring use of dual symbols.
		SC	Clayey sands, sand-clay mixtures	Plastic fines (for identification procedures see CL below)		

continues next page

TABLE 2.4 Unified soil classification system (continued)

Major Divisions		Group Symbols	Typical Names	Identification Procedures on Fractions Smaller than No. 40 Sieve Size			Information Required for Describing Soils	Laboratory Classification Criteria
1	2	3	4	Dry Strength (Crushing characteristics)	Dilatancy (Reaction to shaking)	Toughness (Consistency near PL)	6	7
				5				
Fine-Grained Soils (More than half of material is smaller than No. 200 sieve size. The No. 200 sieve size is about the smallest particle visible to the naked eye.)	Silts and Clays Liquid limit is less than 50.	ML	Inorganic silts and very fine sands, rock flour, silty or clayey fine sands or clayey silts with slight plasticity	None to slight	Quick to slow	None	For undisturbed soils add information on structure, stratification, consistency in undisturbed and remolded states, moisture and drainage conditions. Give typical name; indicate degree and character of plasticity; amount and maximum size of coarse grains; color in wet condition; odor, if any; local or geologic name and other pertinent descriptive information; and symbol in parentheses. Example: Clayey silt, brown; slightly plastic; small percentage of fine sand; numerous vertical root holes; firm and dry in place; loess (ML).	Use grain-size curve in identifying the fractions as given under field identification.
		CL	Inorganic clays of low to medium plasticity, gravelly clays, sandy clays, silty clays, lean clays	Medium to high	None to very slow	Medium		
		OL	Organic silts and organic silty clays of low plasticity	Slight to medium	Slow	Slight		
	Silts and Clays Liquid limit is greater than 50.	MH	Inorganic silts, micaceous or diatomaceous fine sandy or silty soils, elastic silts	Slight to medium	Slow to none	Slight to medium		
		CH	Inorganic clays of high plasticity, fat clays	High to very high	None	High		
		OH	Organic clays of medium to high plasticity, organic silts	Medium to high	None to very slow	Slight to medium		
Highly Organic Soils		Pt	Peat and other highly organic soils	Readily identified by color, odor, spongy feel, and frequently by fibrous texture				

Plasticity Chart — For laboratory classification of fine-grained soils. A-Line separates CL, CH (above) from ML, OL, MH, OH (below). CL-ML zone at PI 4–7, LL ~20.

Notes: (1) Boundary classifications: Soils possessing characteristics of two groups are designated by combinations of group symbols. For example, GW-GC, well-graded gravel-sand mixture with clay binder. (2) All sieve sizes on this chart are U.S. standard.

Source: U.S. Army Corps of Engineers 1953.

MATERIAL PROPERTIES | 17

TABLE 2.5 Unit weight, friction angle, and cohesion for soils and rocks

Type	Material	Unit Weight (Saturated/dry) (lb/ft³)	(kN/m³)	Friction Angle (degrees)	Cohesion (lb/ft²)	(kPa)
Cohesionless						
Sand	Loose sand, uniform grain size	118/90	19/14	28–34*		
	Dense sand, uniform grain size	130/109	21/17	32–40*		
	Loose sand, mixed grain size	124/99	20/16	34–40*		
	Dense sand, mixed grain size	135/116	21/18	38–46*		
Gravel	Gravel, uniform grain size	140/130	22/20	34–37*		
	Sand and gravel, mixed grain size	120/110	19/17	48–45*		
Blasted/broken rock	Basalt	140/110	22/17	40–50*		
	Chalk	80/62	13/10	30–40*		
	Granite	125/110	20/17	45–50*		
	Limestone	120/100	19/16	35–40*		
	Sandstone	110/80	17/13	35–45*		
	Shale	125/100	20/16	30–35*		
Cohesive						
Clay	Soft bentonite	80/30	13/6	7–13	200–400	10–20
	Very soft organic clay	90/40	14/6	12–16	200–600	10–30
	Soft, slightly organic clay	100/60	16/10	22–27	400–1,000	20–50
	Soft glacial clay	110/76	17/12	27–32	600–1,500	30–70
	Stiff glacial clay	130/105	20/17	30–32	1,500–3,000	70–150
	Glacial till, mixed grain size	145/130	23/20	32–35	3,000–5,000	150–250
Rock	Hard igneous rocks—granite, basalt, porphyry	160 to 190†	25 to 30	35–45	720,000–1,150,000	35,000–55,000
	Metamorphic rocks—quartzite, gneiss, slate	160 to 180	25 to 28	30–40	400,000–800,000	20,000–40,000
	Hard sedimentary rocks—limestone, dolomite, sandstone	150 to 180	23 to 28	35–45	200,000–600,000	10,000–30,000
	Soft sedimentary rock—sandstone, coal, chalk, shale	110 to 150	17 to 23	25–35	20,000–400,000	1,000–20,000

* Higher friction angles in cohesionless materials occur at low confining or normal stresses.
† For intact rock, the density of the material does not vary significantly between saturated and dry states with the exception of some materials such as porous sandstones.

Source: Houk and Bray 1977 *(reprinted with permission from the Institution of Mining & Metallurgy).*

TABLE 2.6 Typical strength characteristics of soils

Group Symbol	Cohesion (as Compacted) psf c	Cohesion (Saturated) psf c_{sat}	Effective Stress Envelope Degrees ϕ
GW	0	0	>38
GP	0	0	>37
GM	—	—	>34
GC	—	—	>31
SW	0	0	38
SP	0	0	37
SM	1,050	420	34
SM–SC	1,050	300	33
SC	1,550	230	31
ML	1,400	190	32
ML–CL	1,350	460	32
CL	1,800	270	28
OL	—	—	—
MH	1,500	420	25
CH	2,150	230	19
OH	—	—	—

Source: Lindeburg 1992.

TABLE 2.7 Typical soil modulus values

Type of Soil	Depth of Cover (ft)	Standard AASHTO* Relative Compaction			
		85%	90%	95%	100%
Fine-grained soils with less than 25% sand content (CL, ML, CL–ML)	0–5	500	700	1,000	1,500
	5–10	600	1,000	1,400	2,000
	10–15	700	1,200	1,600	2,300
	15–20	800	1,300	1,800	2,600
Coarse-grained soils with fines (SM, SC)	0–5	600	1,000	1,200	1,900
	5–10	900	1,400	1,800	2,700
	10–15	1,000	1,500	2,100	3,200
	15–20	1,100	1,600	2,400	3,700
Coarse-grained soils with little or no fines (SP, SW, GP, GW)	0–5	700	1,000	1,600	2,500
	5–10	1,000	1,500	2,200	3,300
	10–15	1,050	1,600	2,400	3,600
	15–20	1,100	1,700	2,500	3,800

* AASHTO = American Association of State Highway and Transportation Officials.
Notes: Values of modulus of soil reaction, E' (psi) based on depth of cover, type of soil, and relative compaction. Soil type symbols are from the Unified Classification System.
Source: Hartley and Ducan 1987.

MATERIAL PROPERTIES | 19

TABLE 2.8 Soil engineering properties and uses

		Important Properties				Relative Desirability for Various Users (Graded From 1 [Highest] to 14 [Lowest])									
						Rolled Earth Dams			Canal Sections			Foundations		Fills Roadways		
Typical Names of Soil Groups	Group Symbols	Permeability When Compacted	Shearing Strength When Compacted and Saturated	Compressibility When Compacted and Saturated	Workability as a Construction Material	Homogeneous Embankment	Core	Shell	Erosion Resistance	Compacted Earth Lining	Seepage Important	Seepage Not Important	Frost Heave Not Possible	Frost Heave Possible	Surfacing
Well-graded gravels, gravel-sand mixtures, little or no fines	GW	Pervious	Excellent	Negligible	Excellent	—	—	1	1	—	—	1	1	1	1
Poorly graded gravels, gravel-sand mixtures, little or no fines	GP	Very pervious	Good	Negligible	Good	—	—	2	2	—	—	3	3	3	—
Silty gravels, poorly graded gravel-sand-silt mixtures	GM	Semipervious to impervious	Good	Negligible	Good	2	4	—	4	4	1	4	4	9	5
Clayey gravels, poorly graded gravel-sand-clay mixtures	GC	Impervious	Good to fair	Very low	Good	1	1	—	3	1	2	6	5	5	1
Well-graded sands, gravelly sands, little or no fines	SW	Pervious	Excellent	Negligible	Excellent	—	—	3 if gravelly	6	—	—	2	2	2	4
Poorly graded sands, gravelly sands, little or no fines	SP	Pervious	Good	Very low	Fair	—	—	4 if gravelly	7 if gravelly	—	—	5	6	4	—
Silty sands, poorly graded sand-silt mixtures	SM	Semipervious to impervious	Good	Low	Fair	4	5	—	8 if gravelly	5 erosion critical	3	7	8	10	6
Clayey sands, poorly graded sand-clay mixtures	SC	Impervious	Good to fair	Low	Good	3	2	—	5	2	4	8	7	6	2

continues next page

TABLE 2.8 Soil engineering properties and uses (continued)

Typical Names of Soil Groups	Group Symbols	Important Properties					Relative Desirability for Various Users (Graded From 1 [Highest] to 14 [Lowest])										
							Rolled Earth Dams			Canal Sections			Foundations		Roadways		
															Fills ||
		Permeability When Compacted	Shearing Strength When Compacted and Saturated	Compressibility When Compacted and Saturated	Workability as a Construction Material	Homogeneous Embankment	Core	Shell	Erosion Resistance	Compacted Earth Lining	Seepage Important	Seepage Not Important	Frost Heave Not Possible	Frost Heave Possible	Surfacing
Inorganic silts and very fine sands, rock flour, silty or clayey fine sands with slight plasticity	ML	Semipervious to impervious	Fair	Medium	Fair	6	6	—	—	6 erosion critical	6	9	10	11	—
Inorganic clays of low to medium plasticity, gravelly clays, sandy clays, silty clays, lean clays	CL	Impervious	Fair	Medium	Good to fair	5	3	—	9	3	5	10	9	7	7
Organic silts and organic silt-clays of low plasticity	OL	Semipervious to impervious	Poor	Medium	Fair	8	8	—	—	7 erosion critical	7	11	11	12	—
Inorganic silts, micaceous or diatomaceous fine sandy or silty soils, elastic silts	MH	Semipervious to impervious	Fair to poor	High	Poor	9	9	—	—	—	8	12	12	13	—
Inorganic clays of high plasticity, fat clays	CH	Impervious	Poor	High	Poor	7	7	—	—	8 volume change	9	13	13	8	—
Organic clays of medium to high plasticity	OH	Impervious	Poor	High	Poor	10	10	—	10	—	10	14	14	14	—
Peat and other highly organic soils	Pt	—	—	—	—	—	—	—	—	—	—	—	—	—	—

Source: Wagner 1957.

MATERIAL PROPERTIES | 21

TABLE 2.9 Properties of rocks[*]

Rock Type: Geologic Unit	Location	γ (kg/m³)	C_o (Pa)	Ref.	H_a	V_p (km/s)	Ref.	u	G (Pa)	Ref.	E_r (Pa)	Ref.	E_s (Pa)	Ref.	Ref.
Igneous															
Amphibolite	McLeese Lake, British Columbia	2920	2.65	7				0.24			1.75	10			7
Amphibolite, fine	Oorgaum, Mysore State, India	3070	4.23	8	92	5.79	3		4.58	10	1.04	11			20
Andesite, hypersthene	Palisades Dam, Idaho	2570	1.29–1.32	9				0.18					5.45	10	1
Anorthosite, Labradorite, C.	Ukrainian Shield, Union of Soviet Socialist Republics	2770	2.27	8				0.36	3.41	10	9.28	10			2
Basalt, Lower Granite	Pullman, Washington	2727	2.27–3.55	8	57:412:116	5.27	3				5.02	10			6
Basalt, Olivine, dense	Nevada Test Site, Nevada	2720									2.47	10			14
Basalt, Olivine, sl. vesicular		2660									2.86	10			14
Basalt, Olivine, Western Cascade	Medford, Oregon	2730	1.69–2.20	8				0.25					4.21	10	1
Basalt	Painesdale, Michigan	2850	2.30	8	69	4.63	3		2.68	10	6.15	10			20
	Ahmeek, Michigan	2940	2.58–3.59	8	79	5.15	3		3.17	10	7.79	10			20
Basalt, subaqueous	Eniwetok, PTT	2860	1.94	8	71			0.18			6.93	10			4
Basalt, vesicular	Bergstrom, Texas	2550	7.44	7		4.65	3	0.13			3.74	10			19
		2580	8.34	7		5.04	3	0.19			4.05	10			19
Basalt, dense		2593	1.13	8				0.20			5.21	10			19
		2761	1.32	8		5.56	3	0.17			7.65	10			19
		2752	1.25	8		4.70	3				5.79	10			19
Charnokite (hypersthene granite)	Ukrainian Shield, Union of Soviet Socialist Republics	2730	2.47	8				0.22	2.75	10	6.73	10			2
Diabase, Medford	Cambridge, Massachusetts	2882	1.77	8	44.2,13,60										8
Diabase, Palisades	W. Nyack, New York	2932	2.41	8	59						8.19	10			8
Diabase; French Creek	St. Peters, Pennsylvania	3060	3.01	8	58						9.94	10			8
Diabase, altered	Clinton County, New York	2940	3.21	8	92	5.70	3		3.73	10	9.58	10			20
Diorite; Kennsington	Washington, District of Columbia	2820	8.09(7)–2.76[8]		50.8;7:150										8
Diorite, gneissic	Mineville, New York	3030	1.86	8	90	4.27	3		2.78	10	5.53	10			20
Diorite, augite, fresh	Keetley, Utah	2740	3.33	8	82	5.55	3	0.25	3.37	10	8.41	10			21
Diorite, augite, sl. altered		2720	2.79	8	83	5.43	3	0.26	3.18	10	8.00	10			21
Diorite, augite, altered		2720	2.15	8	71	4.94	3	0.30	2.56	10	6.64	10			21

continues next page

TABLE 2.9 Properties of rocks* (continued)

Rock Type: Geologic Unit	Location	γ (kg/m³)	C_o (Pa)	Ref.	H_a	V_p (km/s)	Ref.	u	G (Pa)	Ref.	E_r (Pa)	Ref.	E_s (Pa)	Ref.	Ref.
Diorite, biotite, porph., sl. altered		2690	2.28	8	77	4.97	3	0.27	2.83	10	6.68	10			21
Diorite, biotite, porph., s. altered		2660	1.80	8	67	4.75	3	0.22	2.45	10	6.01	10			21
Diorite, hornblende	Ishpeming, Michigan	3010	2.74	8	84	6.00	3	0.29	4.22	10	1.07	11			4
Gabbro; Salem	Beverly, Massachusetts	3060	1.33–1.49	8	52:6,47:129						8.76	10			8
Gabbro, altered	Clinton County, New York	2930	2.77	8	82	5.36	3		3.36	10	8.48	10			20
Gabbro/diabase	Ukrainian Shield, Union of Soviet Socialist Republics	3000	3.09	8				0.33	4.41	10	1.19	11			2
Gabbro/diabase	Karelian SSR, Union of Soviet Socialist Republics	3190	3.14	8							1.17	11			2
Granite, f.	Grand Coulee, Washington	2571	1.94	8	53:10,5:172	4.64	3				5.48	10			5
Granite, c.		2627	1.61	8	52:9,5:161	4.08	3				5.24	10			5
Granite, Pikes Peak	Colorado Springs, Colorado	2675	1.57	8	58						7.06	10			8
Granite, Barre	Barre, Vermont	2643	1.94	8	53						6.15	10			15
Granite, Pre-Cambrian	Loveland, Colorado	2630	7.21	7				0.14					2.69	10	1
Granite	Woodstock, Maryland	2650	2.51	8	98	4.51	3	–0.19	2.54	10	5.46	10			20
	Tem Piute District, Nevada	2630	2.72	8	100	4.42	3	–.023	2.25	10	5.13	10			20
	Mount Airy, North Carolina	2600	2.10	8	90	2.44	3	0.02	1.02	10	1.57	10			20
Granite, biotite, m.	Karelian SSR, Union of Soviet Socialist Republics	2700	2.39	8				0.25	2.41	10	6.93	10			2
Granite, gneissic; Lithonia	Lithonia, Georgia	2640	1.93	8	85	2.71	3	–0.28	1.18	10	1.91	10			3
		2640	2.13	8	85	2.50	3	–0.19	1.09	10	1.64	10			3
		2660	2.09	8	89	2.62	3	0.00	8.96	9	1.86	10			3
		2620	2.05	8	85	1.08	3	0.12	7.10	9	1.04	10			3
Granite, f.-m; unaweep	Grand Junction, Colorado	2670	1.74	8	59	3.17	3	–0.13	1.68	10	2.72	10			4
		2710	1.59	8	53	3.75	3	0.29	1.91	10	3.82	10			4
Granite, par. to foliation; unaweep		2730	1.61	8	44	3.93	3		1.90	10	4.23	10			4
Granite; unaweep		2660	1.74	8	37	3.17	3		1.55	10	2.72	10			4
Granite, pink	Bergstrom, Texas	2710				6.47	3		6.84	10	8.57	10			19
		2650				5.83	10		4.72	10	5.75	10			19
Granite, weathered		2620				5.33	10		4.65	10	5.36	10			19

continues next page

MATERIAL PROPERTIES

TABLE 2.9 Properties of rocks* (continued)

Rock Type: Geologic Unit	Location	γ (kg/m³)	C_o (Pa)	Ref.	H_a	V_p (km/s)	Ref.	u	G (Pa)	Ref.	E_r (Pa)	Ref.	E_s (Pa)	Ref.	Ref.
Granodiorite	Bergstrom, Texas	2689	4.07	7		5.72	3	0.70			5.84	10			19
		2703	1.39	8		5.89	3	0.22			7.99	10			19
		2699	8.51	7		5.80	3	0.17			6.87	10			19
		2702	1.29	8		5.72	3	0.19			7.3	10			19
		2700	1.15	8		5.93	3	0.19			7.10	10			19
Magnetite, ore	Mineville, New York	4230	1.41	8	72	2.72	3		1.86	10	3.14	10			20
Monzonite, porphyritic; Colville	Grand Coulee, Washington	2575	1.49	8				0.18					4.14	10	1
		2575	1.71	8				0.15					4.21	10	1
Pegmatite	Star Lake, New York	2590	2.14	8	87	4.88	3		2.28	10	6.16	10			20
Pyroxenite	Clinton County, New York	3450	1.70	8	70	1.98	3		1.03	10	1.31	10			20
Pyroxenite, fresh	Star Lake, New York	3430	1.82	8	60	6.03	3		5.03	10	1.24	11			20
Pyroxenite, heavily altered		2530	5.86	7	28	2.96	3		7.58	9	2.20	10			20
Quartz, diorite	Mountain Home, Idaho		8.74	7				0.05					2.14	10	1
Quartz, monzonite	Bergstrom, Texas	2669	1.48	8				0.70			6.74	10			19
		2680	1.55	8				0.22			7.24	10			19
		2670	1.30	8				0.17			6.68	10			19
		2673	1.29	8				0.19			7.65	10			19
		2667	1.39	8				0.19			7.72	10			19
Rapakivi (granite)	Ukrainian Shield, Union of Soviet Socialist Republics	2640	2.72	8				0.20	2.43	10	5.81	10			2
Shonkinite (dark syenite)	Clinton County, New York	3350	1.85	8	78	3.23	3		1.94	10	3.54	10			20
Syenite	Kirkland Lake, Ontario	2820	3.03	8		5.12	3		2.83	10	7.38	10			20
Syenite, porphyritic		2700	4.34	8		5.12	3		3.03	10	7.10	10			20
Metamorphic															
Argillite, Cambridge	Dorchester, Massachusetts	2810	1.36	8							8.41	10			8
	Cambridge, Massachusetts	2642	6.61	7											8
		2510	3.15	7	150:45:10										8
		2759	1.55	8	26:1.2						4.83	10			8
		2715	1.55	8							3.86	10			8

continues next page

TABLE 2.9 Properties of rocks* (continued)

Rock Type: Geologic Unit	Location	γ (kg/m³)	C_o (Pa)	Ref.	H_a	V_p (km/s)	Ref.	u	G (Pa)	Ref.	E_r (Pa)	Ref.	E_s (Pa)	Ref.	Ref.
Gneiss, quartz diorite	Bethesda, Maryland	2775	9.60	7	64.5;17:139						7.24	10			8
Gneiss, schistose; Wissahickon	Washington, District of Columbia	2980	7.01	7	46:2.90:79										8
Gneiss, Dworssak	Orofino, Idaho	2804	1.62	8	48:						5.36	10			8
Gneiss, diorite: Idaho Springs	Montezuma Quad., Colorado	2865	8.41	7				0.06			6.41	10			18
Gneiss, granite	Mineville, New York	2750	2.12	8	99	3.63	3		1.96	10	3.85	10			20
Gneiss, granite, pegmatitic	Star Lake, New York	3040	1.53	8	75	4.66	3		2.88	10	6.67	10			20
Gneiss, pegmatitic		2650	1.96	8	81	4.11	3		2.12	10	4.46	10			20
Gneiss, augite	Hackettstown, New Jersey	3360	2.19	8	74	5.55	3	0.27	4.07	10	1.03	11			21
Gneiss, biotite		2910	1.61	8	74	4.79	3	0.24	2.71	10	6.72	10			21
Gneiss	Bergstrom, Texas	2710				4.58	3	0.15	2.59	10	5.38	10			19
		2810				6.28	3	0.29	6.74	10	8.32	10			19
Greenstone	Mount Weather, Virginia	3020	2.69	8	81	5.85	3		4.21	10	1.05	11			20
		2960	3.05	8	80	5.21	3		3.86	10	8.07	10			20
Greenstone, amygdaloidal	Catoctin, Pennsylvania	3040	2.01	8	64	3.99	3		3.07	10	4.90	10			3
Hematite, ore	Soudan, Minnesota	5070	6.07	8	74	6.28	3		7.79	10	2.00	11			20
Hematite, ore; par. bedding	Bessemer, Alabama	3780	1.19	8	51	4.30	3		2.69	10	6.69	10			20
		3670	1.39	8	50	4.30	3		2.70	10	6.73	10			20
Hornfels	Tem Piute District, Nevada	3190	5.33	8		5.49	3		4.09	10	9.58	10			20
Marble, Cherokee	Tate, Georgia	2707	6.69	7	36						5.59	10			8
Marble, taconic	Rutland, Vermont	2707	6.21	7	31						4.79	10			8
Marble, perp. bedding	Cockeysville, Maryland	2870	2.12	8	56	4.18	3		2.61	10	4.93	10			20
Marble, par. bedding		2870	2.23	8	27				2.83	10	6.74	10			20
Marble, paleozoic	Ural Mountains, Union of Soviet Socialist Republics	2710	1.49	8							7.67	10			2
Marble, dolomitic, f.	Karelian SSR, Union of Soviet Socialist Republics	2820	2.74	8				0.26	3.00	10	8.94	10			2
Marble, Oro Grande	Oro Grande, California	2720	1.65	8	56	5.40	3	0.30	3.03	10	7.86	10			3
		2680	5.52	7	42	4.90	3	0.16	2.80	10	6.52	10			3
Metarhyolite	Soudan, Minnesota	2840	1.25	8	47	5.06	3		3.16	10	7.86	10			20

continues next page

MATERIAL PROPERTIES | 25

TABLE 2.9 Properties of rocks* (continued)

Rock Type: Geologic Unit	Location	γ (kg/m³)	C₀ (Pa)	Ref.	H_a	V_p (km/s)	Ref.	u	G (Pa)	Ref.	E_r (Pa)	Ref.	E_s (Pa)	Ref.	Ref.
Phyllite, sericite	El Dorado County, California	2340	9.79	6									1.79	10	1
Phyllite, quartzose		2180	9.38	6									7.58	9	1
Phyllite, graphitic	El Dorado County, California	2350	6.69	6									9.65	9	1
Phyllite, green	Ishpeming, Michigan	3240	1.26	8	40	4.85	3		3.28	10	7.65	10			20
Quartzite, Wissahickon	Washington, District of Columbia	2804	4.71	7	38:2.78:63										8
Quartzite, phyllite lenses	Raven, Yugoslavia	2590				8.22	2				1.27	9			13
Quartzite, altered		2590				2.5	3				1.21	10			13
Quartzite, ferruginous	Kursk, Union of Soviet Socialist Republics	3510	3.43	8							1.71	11			2
Quartzite, Biwabik	Babbitt, Minnesota	2750	6.29	8		5.55	3	0.10	3.86	10	8.48	10			3
Quartzite, hematitic	Ishpeming, Michigan	4070	2.93	8	71	5.21	3	0.20	4.06	10	9.79	10			4
Schist, chlorite	Bethesda, Maryland	2813	2.53	7	37:1.89:51						3.10	10			8
Schist, biotite, Idaho Springs	Montezuma Quad, Colorado	2720	2.09	7									2.48	10	1
Schist, sericite	Superior, Arizona	2700	1.62	8	82	4.72	3		2.62	10	6.00	10			20
Skam, garnet-pyroxene	Star Lake, New York	3280	1.30	8	61	5.12	3		3.48	10	8.62	10			20
Slate, par. bedding, calcareous	Bangor, Pennsylvania	2740	1.83	8	56						8.88	10			20
Tactite, epidote	Ophir, Utah	2870	2.66	8	65	4.60	3	0.11	2.77	10	6.14	10			21
Sedimentary															
Borax, ore: Ricardo	Boron, California	2140	4.41	7	22								4.21	9	4
Chert, chalcedonic; Boone	Picker, Oklahoma	2560	3.60	8	96			0.09			5.34	10			3
Chert, dolomitic; Fort Payne	Smithville, Tennessee	2630	2.10	8	74	3.35	3	0.00	1.65	10	3.54	10			4
		2670	2.02	8	67	4.48	3	0.14	2.37	10	5.62	10			4
Conglomerate; Roxbury	Boston, Massachusetts	2679	8.28	7	41:6.37:102										8
Conglomerate	Kirkland Lake, Ontario	2670	1.65	8		5.40	3		3.24	10	7.79	10			20
Dolomite, Lockport	Rochester, New York	2765	2.12	8	50:3.24:86						4.48	10			8
	Niagara Falls, New York	2579	9.10	7	44:						5.10	10			8
Dolomite, Bonne Terre	Bonne Terre, Missouri	2673	1.52	8	49:						6.63	10			8
Dolomite	Jefferson City, Tennessee	2760	3.59	8	69:	5.30	3		3.17	10	7.79	10			20

continues next page

TABLE 2.9 Properties of rocks* (continued)

Rock Type: Geologic Unit	Location	γ (kg/m³)	C_o (Pa)	Ref.	H_a	V_p (km/s)	Ref.	u	G (Pa)	Ref.	E_r (Pa)	Ref.	E_s (Pa)	Ref.	Ref.
Dolomite, Beekmantown	Wood County, West Virginia	2833						0.22	2.95	10	7.23	10			17
		3004						0.22	3.05	10	7.50	10			17
		2783						0.26	3.75	10	9.50	10			17
		2832						0.19	3.64	10	8.65	10			17
Dolomite, Maple Mill	Omaha, Nebraska		3.47	7				0.36			4.79	10			18
		2827	4.32	7				0.05			2.74	10			18
		2818	1.13	8				0.12			6.13	10			18
		2528	4.45	7				0.51			4.39	10			18
		2507	7.08	7				0.09			2.14	10			18
		2531	6.09	7				0.40			8.62	10			18
Dolomite, jointed; Jurassic	Gojak, Yugoslavia	2800				2.51	3				1.27	10			13
Dolomite	Mascot, Tennessee	2840	3.22	8	74	5.46	3		3.52	10	8.48	10			20
Graywacke, m.; Chico	Monticello Dam, California	2440	4.88	7				0.03					1.24	10	1
		2490	5.07	7				0.02					9.65	9	1
Gypsum	Buffalo, New York	2262	1.25	7	18										8
Jaspillite, ferrugtinous, siliceous sandstone	Ishpeming, Michigan	3390	3.42	8	85	5.55	3		4.83	10	1.03	11			20
Limestone	Bedford, Indiana	2206	5.10	7	330;43;20	3.91	3				2.85	10			6
Limestone, Solenhofen	Bavaria, FRG	2621	2.45	8	54;1.75;72	5.78	3				6.38	10			6
Limestone, Ozark tavern	Carthage, Missouri	2659	9.79	7	49						5.59	10			8
Limestone, porous; redwall	Lee's Ferry, Arizona	2440	1.33	8				0.18					1.65	10	1
Limestone, reef	Eniwetok, PTT	2300	3.42	7				0.16					3.79	10	1
Limestone, fossiliferous	Bedford, Indiana	2370	7.52	7	27	3.78	3		1.42	10	3.34	10			20
Limestone, fossiliferous, par. bed.	Bedford, Indiana	2370	6.85	7	27				1.56	10	3.91	10			20
Limestone, limonitic	Bessemer, Alabama	2920	1.72	8	61	4.75	3		2.82	10	4.54	10			20
Limestone, marly	Rifle, Colorado	2250	1.10	8	56	2.38	3		6.90	9	1.25	10			20
Limestone, marly; par. bed.	Rifle, Colorado	2180				3.11	3		6.76	9	2.14	10			20
Limestone, Martinsburg	Martinsburg, West Virginia	2680	1.59	8	61	5.00	3	0.21	2.73	10	6.59	10			21
Limestone, Black River	Trenton, West Virginia	2688						0.16	2.45	10	5.70	10			17

continues next page

MATERIAL PROPERTIES | 27

TABLE 2.9 Properties of rocks* (continued)

Rock Type: Geologic Unit	Location	γ (kg/m³)	C₀ (Pa)	Ref.	H_a	V_p (km/s)	Ref.	u	G (Pa)	Ref.	E_r (Pa)	Ref.	E_s (Pa)	Ref.	Ref.
Limestone, dolomitic; Mesozoic	Turkmenian SSR, Union of Soviet Socialist Republics	2700	2.10	8							7.62	10			2
Limestone, detrital	Moscow Syncline, Union of Soviet Socialist Republics	2160	5.20	7							2.90	10			2
Limestone, chalky; Smokey Hill	Pickstown, South Dakota	1410	8.27	6	10	1.34	3	0.30	1.59	9	2.90	9			3
		1710	1.65	7	13	1.74	3	−0.13	2.55	9	4.48	9			3
Limestone, dolomitic; Bonne Terre	Bonne Terre, Missouri	2660	1.75	8	51	5.09	3	0.22	2.85	10	6.96	10			3
		2780	1.98	8	59	5.88	3	0.29	3.76	10	9.72	10			3
		2710	1.96	8	49			0.05			1.99	10			3
		2690	1.96	8	33	5.36	3	0.22	3.13	10	7.65	10			3
		2670	1.46	8	48	3.78	3	−0.07	2.10	10	3.87	10			3
Limestone, fossiliferous; St. Louis	St. Genevieve, Missouri	2670	1.64	8	48	5.00	3	0.24	2.68	10	6.67	10			4
Limestone; Wyandotte	Omaha, Nebraska	2546	1.15	7				0.24			2.11	9			18
		2605	4.90	7				0.64			1.61	10			18
Limestone, silurian	Omaha, Nebraska	2352	9.60	7				0.19			3.07	10			18
Limestone, Chickamauga	Smithville, Tennessee	2740	1.73	8	53	4.39	3	0.14	2.33	10	5.30	10			4
		2730	1.73	8	52	3.08	3	0.22	1.17	10	2.72	10			4
Limestone, dolomitic, well-cemented	Pondera County, Montana	2710	1.68	8				0.31					7.65	10	16
Limestone, jointed; Jurassic	Gojak, Yugoslavia	2700				1.92	3				9.16	9			13
Marlstone, mahagony	Rifle, Colorado	2220	8.14	7	49	3.20	3	0.17	1.02	10	2.41	10			3
Marlstone, par. bed; mahagony	Rifle, Colorado	2360	1.72	8	61	4.18	3	0.33	1.53	10	4.10	10			3
Marlstone, Maxville	E. Fultonham, Ohio	2190	5.59	7	23	3.38	3	0.13	1.10	10	2.50	10			4
Oil Shale, Parachute Creek	Rio Blanco, Colorado	2044	8.28	7				0.33					6.24	9	12
		2220	1.10	8				0.37					1.12	10	12
		2190	1.81	8				0.30					1.08	10	12
		2124	9.35	7				0.24					7.03	9	12
Quartzite, Baraboo	Baraboo, Wisconsin	2627	3.21	8	59						8.84	10			8

continues next page

TABLE 2.9 Properties of rocks* (continued)

Rock Type: Geologic Unit	Location	γ (kg/m³)	C_o (Pa)	Ref.	Hᵃ	V_p (km/s)	Ref.	u	G (Pa)	Ref.	E_r (Pa)	Ref.	E_s (Pa)	Ref.
Quarzite	Bergstrom, Texas	2610	6.45	7							2.76	6		19
		2570	1.26	8							3.56	10		19
		2610	1.75	8							5.91	10		19
		264	2.23	8							6.36	10		19
		2570	1.64	8							5.44	10		19
Salt; diamond crystal	Jefferson Island, Louisiana	2163	2.14	7	23						4.90	9		8
Salt	Bergstrom, Texas	2167	1.81	7		3.76	3				6.14	9		19
		2168	1.89	7		3.37	3	0.06			3.45	9		19
		2167	2.85	7		4.08	3				3.45	10		19
		2298	2.20	7		4.07	3	0.189			2.05	10		19
		2317	3.07	7				0.03			3.28	10		19
Sandstone, Navajo	Page, Arizona	2015	4.35	7	30:0.04:6	2.52	3						1.53	10	6
Sandstone, Cambridge	Cambridge, Massachusetts	2647	4.93	7	27:0.44:18										8
Sandstone, Crab orchard	Crossville, Tennessee	2531	2.14	8	47						3.92	10	1.31	10	6
Sandstone, f.; Tensleep	Casper, Wyoming	2325	7.25	7				0.06							1
Sandstone, c.	Amherst, Ohio	2170	4.21	7	20	1.20	3		4.00	9	7.10	9		20
Sandstone, c., par. bed.	Amherst, Ohio	2170	3.55	7	20				4.65	9	1.09	10		20
Sandstone, ferruginous	Bessemer, Alabama	2930	2.35	8	65	4.05	3		2.42	10	4.96	10		20
	Monogalia County, West Virginia	2600	1.32	8	53	3.42	3		1.51	10	3.83	10		21
Sandstone	Huntington, Utah	2200	1.07	8		2.44	3	0.22	7.03	10	1.31	10		21
		2170	7.93	7		2.56	3	-0.10	7.03	9	1.45	10		21
		2140	9.79	7		2.19	3	0.04	4.83	9	1.01	10		21
		2350	2.23	8		2.96	3	-0.11	1.17	10	2.07	10		21
		2330	1.91	8		2.87	3	-0.07	1.02	10	1.86	10		21
Sandstone; carboniferous	Donets Basin, Union of Soviet Socialist Republics	2650	2.56	8	62			0.14	2.43	10	5.55	10		21
Sandstone, Thorold	Niagara Falls, Ontario	2460						-0.12			2.13	10		11
		2510						-0.18			3.31	9		11
Sandstone, calcareous, nonesuch	White Pine, Montana	2600	1.58	8		4.63	3	0.16	2.39	10	5.53	10		3

continues next page

MATERIAL PROPERTIES | 29

TABLE 2.9 Properties of rocks* (continued)

Rock Type:Geologic Unit	Location	γ (kg/m³)	C₀ (Pa)	C₀ Ref.	Hᵃ	Vp (km/s)	Vp Ref.	u	G (Pa)	G Ref.	Er (Pa)	Er Ref.	Es (Pa)	Es Ref.
Sandstone, cemented; Navajo	Huntington, Utah	2880	1.24	8	50	2.77	3	−0.07	9.45	9	1.75	10			3
Sandstone, cemented; obl. bed.; Navajo	Huntington, Utah	2370	3.38	7	54	3.38	3	0.05	1.41	10	2.71	10			3
Sandstone, uncemented; obl. bed.; Navajo	Huntington, Utah	2130	5.59	7	32	2.29	3	−0.05	5.86	9	1.12	10			3
Sandstone, uncemented, par. bed.; Navajo	Huntington, Utah	2130	3.31	7	36	2.10	3	−0.04	4.96	9	9.58	9			3
Sandstone, Graywacke, Kanawha	DeHue, West Virginia	2600	1.41	8	55	2.93	3	−0.17	1.34	10	2.23	10			3
Sandstone, f.; Morrison/Bushy Basin	Long Park, Colorado	2540	3.73	7		2.62	3	−0.04	9.10	9	1.76	10			4
Sandstone, Shaly; St. Peter	Omaha, Nebraska	2344	3.46	7				0.05			7.19	9			18
		2450		7				0.06			1.25	10			18
Sandstone, silty; Seminole	Tulsa, Oklahoma	2500	7.45	7	31	2.87	3		1.08	10	2.19	9			4
Sandstone; Homewood	Franklin, Pennsylvania	2200	8.69	7	43	1.92	3	−0.11	4.69	9	8.27	9			4
Sandstone, Berea	Amherst, Ohio	2182	7.38	7	42:0.47:29	2.64	3				1.93	10			6
Shale, Rochester	Rochester, New York	2738	1.22	8	45:0.73:39						3.79	10			8
Shale, Brunswick	Highland Park, New Jersey	2631	8.29	7	38:0.70:31						1.38	10			8
Shale, Bertie	Buffalo, New York	2712	1.97	8	42:1.92:59						5.03	10			8
Shale, siderite, banded; Kanawha	DeHue, West Virginia	2760	1.12	8	38	2.16	3	−0.43	1.17	10	1.33	10			3
Shale, calcareous; Wyandotte	Omaha, Nebraska	2177	1.19	7				0.32			1.97	9			
Shale, sl. weathered; Cherokee	Omaha, Nebraska	2496	8.34	6				0.15			1.67	9			18
Shale, calcareous; Sheffeild	Omaha, Nebraska	2602	5.98	6				0.14			3.09	10			18
Shale, Maqueketa	Omaha, Nebraska	2618	4.25	7				0.01			7.32	9			18
Shale, carbonaceous; Chattanooga	Smithville, Tennessee	2300	1.12	8	50	2.38	3	00.00	6.55	9	1.39	10			4
		2300	1.10	8	48	2.38	3	−0.02	7.10	9	1.34	10			4
Siltstone, Hackensack	Hackensack, New Jersey	2595	1.23	8	47:1.54:58	3.99	3	0.13			2.63	10			6
Siltstone, par. bedding; Maxville	E. Fultonham, Ohio	2660	3.65	7	20			0.13					4.81	10	4
		2680	3.45	7	19			0.26					8.68	10	4
Siltstone, poorly cemented; Bandera	Omaha, Nebraska	2304	3.54	6				0.35			1.25	8			18

continues next page

TABLE 2.9 Properties of rocks* (continued)

Legend

Nature: Field (F), Laboratory (L), Derived (D)

Engineering Property	Symbol	Nature	Method of Measurement	Units of Measure	Use in Engineered Construction	Limitations
Unit weight	γ	L	Volumetric displacement in water; weight per unit volume; usually weighed as oven dried but may be specified on several other bases	kg/m³	Weight, per volume, of entire rock; primary term in many computations; useful in computing in situ stress	Expected statistical variances in the more porous rocks
Compressive strength	C_o	L	Uniaxial or triaxial conditions, in universal test machine; strain gauges used for moduli determinations	Pa	Index classification test; load-bearing capacity; other properties according to Mohr failure concept; slope stability; mine pillar stress; subsidence; excavation, blasting, drilling and mole-boring performance	Avoid unrepresentative anisotropic fabric elements of discontinuities; select representative sample; consider data scatter; L/D ratio is quite important (standard is 2.1); peak strength is obtained
Hardness*	H_s	L	Small laboratory test holding devices for impact; height or rebound of small diamond-tipped device	Dimensionless	Indicator or relative hardness; useful in tunnel boring rate estimates	
Schmidt rebound		L		Dimensionless	As above	Softer rock breaks on impact; must use Type L device of minimal energy
Taber	H_a	L		Dimensionless	As above	
Total	H_t	L		Dimensionless		
Seismic velocity Compressional Shear	V_p V_s	F	Mechanical or explosive energy wave arrival sensed by geophone, timer, and recorder; measured on ground surface or in borehole configurations	km/s	As above, indicator of overall maturity of rock due to averaging effect on wave travel paths, depending on geophone/energy source array	Degree of saturation important; test does not introduce nonlinear, time-dependent strain variation in derived properties; most valid in homogeneous and isotropic rock
Poisson's ratio	u	D	Calculated from sonic/seismic velocity tests or by use of electrical resistance strain gauges on compression tests	Dimensionless; in range 0 to 0.5	Computational input for calculation of stress distribution patterns and of predicted strain in elastic media; required for finite element modeling	Difficult to extroplate laboratory measurement to field conditions; often estimated without testing; best approximation is from triaxial compression test at confinement equivalent to in situ conditions
Modulus of rigidity	G	D	As above	Pa	Indicator of seismic design stiffness	Strain-related
Tangent modulus of elasticity (Young's modulus, or modulus of deformation)	E_t	L	Triaxial compression in universal test machine; electrical resistance strain gauges	Pa	The fundamental stress-strain relationship; input for static displacement computations and for dynamic, seismic analyses	Requires accommodation of any anisotropy of rock fabric and model in situ conditions
Secant modulus of elasticity	E_s	L	As above	Pa	Alternate expression of the fundamental stress-strain relationship	As above

* Hardness values are in order: Schmidt rebound (H_s); Tabor (H_a); Total (H_t).

continues next page

TABLE 2.9 Properties of rocks* (continued)

References

1. Balmer, G.C. 1953. *Physical Properties of Some Typical Foundation Rocks, Concrete Lab.* Report No. SP-39. Denver, CO: U.S. Bureau of Reclamation.
2. Belikow, B.P. 1962. Elastic properties of rock. *Studies in Geophysics and Geology*, 6:75.
3. Blair, B.E. 1955. *Physical Properties of Mine Rock, Part 3.* Invest. Report No. 5130. Washington, DC: U.S. Bureau of Mines.
4. Blair, B.E. 1956. *Physical Properties of Mine Rock, Part 4.* Invest. Report No. 5244. Washington, DC: U.S. Bureau of Mines.
5. Coulson, J.H. 1971. Shear strength of flat surfaces in rock. *Proceedings 13th Symposium Rock Mechanics.* New York: American Society of Civil Engineers. 77 pp.
6. Deere, D.U., and R.P. Miller. 1966. *Engineering Classification and Index Properties for Intact Rock.* Report No. AFWL-TR-65-116. Albuquerque, NM: U.S. Air Force Weapons Laboratory, Kirtland Air Force Base. 324 pp.
7. Gyenge, M., and G. Herget. 1977. Mechanical properties (rock). In *Pit and Slope Manual*, Rep. No. 7Z-12. Ottawa, Canada: Canada Centre for Mineral and Energy Technology.
8. Brierley, G.S., and B.E. Beverly, eds. 1980. *ROTEDA Computer File of Rock Properties.* Cambridge, MA: Haley & Aldrich, Inc.
9. Hatheway, A.W. 1971. *Lava tubes and collapse depressions.* Ph.D. thesis, University of Arizona.
10. Hatheway, A.W., and W.C. Paris Jr. 1979. Geologic conditions and considerations for underground construction in rock, Boston, Massachusetts. In *Engineering Geology in New England.* Preprint 3602. Edited by A.W. Hatheway. New York: American Society of Civil Engineers.
11. Hogg, A.D. 1959. Some engineering studies of rock movement in the Niagara area (Canada). In *Engineering Geology Case Histories.* No. 3. Boulder, CO: Geological Society of America.
12. Horino, F.G., and V.E. Hooker. 1978. *Mechanical Properties of Cores Obtained from the Unleached Saline Zone, Piceance Creek Basin, Rio Blanco County, Colorado.* Invest. Report No. 8297. Washington, DC: U.S. Bureau of Mines.
13. Kunundzic, B., and B. Colic. 1961. Determination of the elasticity modulus of rock and the depth of the loose zone in hydraulic tunnels by seismic refraction method, *Proceedings Water Resources Engineering Institute.* OTS 60-21614. Sarajevo, Yugoslavia.
14. Lutton, R.J., F.E. Girucky, and R.W. Hunt. 1967. *Project Pre-Schooner; Geologic and Engineering Properties Investigations.* Report No. PNE-505F. Vicksburg, MS: Waterways Experiment Station, U.S. Army Corps of Engineers.
15. Obert, L., S.L. Windes, and W.I. Duvall. 1946. *Standardized Tests for Determining the Physical Properties of Mine Rock.* Invest. Report No. 3891. Washington, DC: U.S. Bureau of Mines.
16. Ortel, W.J. 1965. *Laboratory Investigations for Foundation Rock, Swift Damsite-Pondera County Canal and Reservoir Company, MT.* Report No. C-1153. Denver, CO: U.S. Bureau of Reclamation, Concrete and Structural Branch.
17. Robertson, E.C. 1959. *Physical Properties of Limestone and Dolomite Cores from the Sandhill Well, Wood County, W. Va.* Invest. Report No. 18. Charleston, WV: West Virginia Geological Survey.
18. U.S. Army Engineer District. 1961. *Subsurface Investigation Report, Headquarters, SAC Combat Operations Center, Offutt AFB.* Omaha, NE: US Army Engineer District.
19. U.S. Army. 1969. Report of Data, Rock Property Test/Program, Bergstrom area (near Austin, TX). Waterways Experiment Station, Concrete Division, Vicksburg, MS. Letters of July 30 and August 11, 1969. (Note: Actual locations may vary; not strictly identified.)
20. Windes, S.L. 1949. *Physical Properties of Mine Rock, Part 1.* Invest. Report No. 4459. Washington, DC: U.S. Bureau of Mines.
21. Windes, S.L. 1950. *Physical Properties of Mine Rock, Part 2.* Invest. Report No. 4727. Washington, DC: U.S. Bureau of Mines.

Source: Carmichael 1982 (reprinted with permission of CRC Press).

TABLE 2.10 Properties of minerals

Name	Chemical Formula	Metal (Percentage)	Crystal Structure	Color	Luster	Streak	Degree of Transparency	Tenacity	Cleavage	Fracture	Mohs Hardness	Specific Gravity	Occurrence	Common Names or Synonyms
Albite feldspar	$NaSi_3AlO_8$		Triclinic	Colorless or white; sometimes yellow, pink, green, or black	Vitreous to pearly	White	Transparent to subtranslucent	Brittle	Perfect	Uneven to conchoidal	6.3	2.61	In igneous rocks	Soda feldspar, cryptoclase
Alunite	$KAl_3(SO_4)_2(OH)_6$	K_2O 11.37 Al_2O_3 36.92	Hexagonal ditrigonal pyramidal	White, grayish, yellow, or reddish brown	Vitreous	White	Transparent to subtranslucent	Brittle	Distinct	Flat conchoidal uneven	3.7	2.67	Secondary mineral in acid volcanic rocks which have been altered	Alum stone
Amblygonite	$LiAl(PO_4)F$	Li_2O 10.10 Al_2O_3 34.46 P_2O_5 48.00 F 12.85	Triclinic pinacoidal	White, yellowish, beige, salmon pink, greenish, bluish, gray	Vitreous to greasy	White	Subtransparent to translucent	Brittle	Perfect	Uneven to subconchoidal	6.0	3.05	In granite pegmatite veins	Hebronite
Andalusite	$Al_2O(SiO_4)$	Al_2O_3 60.16–62.70	Orthorhombic	Pink, white, or rose red	Vitreous	White	Transparent to opaque	Brittle	Distinct	Uneven to subconchoidal	7.5	3.18	Contact mineral in clay, slate, and argillaceous schists	Chiastolite, viridine
Anglesite	$PbSO_4$	PbO 73.6	Orthorhombic dipyramidal	Colorless, white, often tinged gray	Adamantine to resinous	Colorless	Transparent to opaque	Very brittle	Distinct but interrupted	Conchoidal	2.8	6.35	Secondary mineral in the oxidation zone of lead veins. Alteration product of galena	
Anhydrite	$CaSO_4$	CaO 41.19	Orthorhombic dipyramidal	Colorless to violet. Also white, mauve, rose, brownish	Pearly to vitreous	White or grayish white	Translucent to opaque	Brittle	Very perfect	Uneven, sometimes splintery	3.3	2.95	As evaporite usually associated with gypsum	Cube spar, tripe stone
Anorthite feldspar	$CaAl_2Si_2O_8$		Triclinic	Colorless or white; sometimes yellow, pink, green, or black	Vitreous to pearly	White	Transparent to translucent	Brittle	Perfect	Conchoidal to uneven	6.3	2.75	In basic igneous rocks	Lime feldspar, calciclase

continues next page

MATERIAL PROPERTIES | 33

TABLE 2.10 Properties of minerals (continued)

Name	Chemical Formula	Metal (Percentage)	Crystal Structure	Color	Luster	Streak	Degree of Transparency	Tenacity	Cleavage	Fracture	Mohs Hardness	Specific Gravity	Occurrence	Common Names or Synonyms
Antlerite	$Cu_3(SO_4)(OH)_4$	Cu 54.0	Orthorhombic dipyramidal	Green to blackish green	Vitreous	Pale green			Perfect		3.0	3.90	Copper deposits, an alteration of brochantite	Stelznerite, arnimite
Apatite	$Ca_5(PO_4)_3(F,Cl,OH)$	CaO 55.38 P_2O_5 42.06 F 1.25 Cl 2.33	Hexagonal hexagonal-dipyramidal	Usually seagreen	Vitreous, to subresinous	White	Transparent to opaque	Brittle	Imperfect	Conchoidal and uneven	5.0	3.20	Most common in metamorphic crystalline rocks, often associated with beds of iron ore	Asparagus stone, collophane
Aragonite	$CaCO_3$	CaO 56.03	Orthorhombic dipyramidal	Colorless to white; also gray, yellowish, blue, green, rose red	Vitreous, resinous on fracture	Uncolored	Transparent to translucent	Brittle	Distinct	Subconchoidal	3.7	2.94	Hot springs deposit, precipitate from saline solution with gypsum in cavities in lavas	Flowers of iron, oserskit
Argentite	Ag_2S	Ag 87.06	Isometric hexoctahedral	Blackish lead gray	Metallic	Blackish lead gray	Opaque	Sectile	Traces	Conchoidal	2.5	7.30	Usually with galena and other sulfide ores	Silver glance, argyrite
Arsenic	As	100	Hexagonal scalenohedral	Tin white tarnishing to dark gray	Nearly metallic on fresh surface	Same as color	Opaque	Brittle	Perfect	Granular	3.5	5.70	Metallic veins with silver, cobalt, nickel ores	
Arsenopyrite	FeAsS	Fe 34.30 As 46.01	Monoclinic prismatic	Silver white to steel gray	Metallic	Dark grayish black	Opaque	Brittle	Distinct	Uneven	6.0	6.10	Usually veins with other sulfides	Mispickel, arsenical pyrites
Atacamite	$Cu_2(OH)_3Cl$	Cu 14.88	Orthorhombic dipyramidal	Green to blackish green	Adamantine to vitreous	Apple green	Transparent to translucent	Brittle	Highly perfect	Conchoidal	3.3	3.77	Secondary mineral derived from malachite and cuprite	Remolinite, halochalzit
Autunite	$Ca(UO_2)_2(PO_4)_2 \cdot 10-12H_2O$	CaO 5.69 UO_3 58.00 P_2O_5 14.39	Tetragonal ditetragonal-dipyramidal	Lemon yellow to sulfur yellow, sometimes greenish	Vitreous, pearly	Yellowish	Transparent to translucent	Brittle	Eminent		2.3	3.10	Secondary mineral usually associated with uraninite and other uranium minerals	Lime uranite

continues next page

TABLE 2.10 Properties of minerals (continued)

Name	Chemical Formula	Metal (Percentage)	Crystal Structure	Color	Luster	Streak	Degree of Transparency	Tenacity	Cleavage	Fracture	Mohs Hardness	Specific Gravity	Occurrence	Common Names or Synonyms
Azurite	$Cu_3(OH)_2(CO_3)_2$	Cu 55.3	Monoclinic prismatic	Azure blue, very dark blue	Vitreous, almost adamantine	Blue, lighter than color	Transparent to subtranslucent	Brittle	Perfect, but interrupted	Conchoidal	3.8	3.83	With malachite as secondary mineral in the oxidized zone of copper deposits	Chessylite, blue spar
Barite	$BaSO_4$	BaO 65.70	Orthorhombic dipyramidal	Colorless, white, yellow, brown, red	Vitreous, to resinous	White	Transparent to opaque	Brittle	Perfect	Uneven	3.0	4.45	Veins or beds, gangue mineral in veins, cement in sandstones	Barytes, heavy spar, desert roses
Beryl	$Be_3Al_2(Si_6O_{18})$	BeO 10.54–13.76, Al_2O_3 17.10–19.00	Hexagonal dihexagonal-dipyramidal	White, bluish green, greenish yellow, yellow	Vitreous	White	Transparent to subtranslucent	Brittle	Imperfect	Conchoidal to uneven	7.6	2.70	In granitic rocks and pegmatites	Aquamarine, emerald, goshenite
Biotite mica	$K_2(Mg, Fe^{+2})_{6-4}(Fe^{+3}, Al, Ti)_{0-2}Al_{2-3}O_{20}O_{0-2}(OH,F)_{4-2}$		Monoclinic	Colorless, shades of pink, purple	Vitreous to pearly		Transparent to opaque	Elastic	Basal, highly perfect		2.7	2.9	Important constituent of many igneous rocks, and as an alteration product	Black mica
Bismuth	Bi	100	Hexagonal scalenohedral	Silver white, with reddish hue. Iridescent tarnish	Metallic	Same as color	Opaque	Sectile	Perfect	Uneven	2.5	9.80	Veins in granite, gneiss, with ores of cobalt, nickel, silver, lead	
Bismuthinite	Bi_2S_3	Bi 81.3	Orthorhombic dipyramidal	Lead gray to tin white	Metallic	Lead gray	Opaque	Sectile	Perfect		2.0	6.40	With igneous rocks, magnetite, garnet, pyrite, tin, and tungsten	Bismuthine, wismuthglanz
Borax	$Na_2B_4O_7 \cdot 10H_2O$	Na_2O 16.26, B_2O_3 36.51	Monoclinic prismatic	Colorless, white, also grayish, bluish or greenish	Vitreous to resinous	White	Translucent to opaque	Rather brittle	Perfect	Conchoidal	2.3	1.70	In the waters of saline lakes and in the beds resulting from the evaporation of these lakes	Tincal
Bornite	Cu_5FeS_4	Cu 63.33	Isometric hexoctahedral	Brownish bronze	Metallic	Pale grayish black	Opaque		Traces	Uneven	3.0	5.20	Usually primary with other copper minerals	Purple copper ore, peacock ores

continues next page

MATERIAL PROPERTIES | 35

TABLE 2.10 Properties of minerals (continued)

Name	Chemical Formula	Metal (Percentage)	Crystal Structure	Color	Luster	Streak	Degree of Transparency	Tenacity	Cleavage	Fracture	Mohs Hardness	Specific Gravity	Occurrence	Common Names or Synonyms
Bournonite	$PbCuSbS_3$	Pb 42.40 Cu 13.01 Sb 24.91	Orthorhombic dipyramidal	Steel gray to iron black	Metallic, often brilliant	Same as color	Opaque	Brittle	Imperfect	Uneven	3.0	5.80	Veins with other sulfide minerals	Cogwheel ore, endellione
Brucite	$Mg(OH)_2$	41.7	Hexagonal scalenohedral	White to pale green, gray or blue	Pearly on cleavages, elsewhere, waxy	White	Transparent to translucent	Fibrous, folia flexible sectile	Basal eminent	Fibrous	2.5	2.40	Secondary origin in serpentine and chlorites	Nemalite, texalite
Calaverite	$AuTe_2$	Au 43.59	Monoclinic prismatic	Brass yellow to silver white	Metallic	Yellowish to greenish gray					2.5	9.00	Veins with gold, pyrite, and quartz	None
Calcite	$CaCO_3$	CaO 56.03	Hexagonal hexagonal-scalenohedral	Colorless or white when pure	Vitreous; sometimes pearly or iridescent	White to grayish	Transparent to opaque	Brittle	Highly perfect	Conchoidal	3.0	2.71	Minor secondary constituent of igneous rocks; widespread constituent of sedimentary rocks; generally deposited by limebearing fluids	Iceland spar, limestone
Carnallite	$KMgCl_3 \cdot 6H_2O$	K 14.07 Mg 8.75	Orthorhombic dipyramidal	Colorless to milk white; often reddish	Greasy, dull to shining		Transparent to translucent	Brittle	Not distinct	Conchoidal	2.5	1.60	A deposit from low-temperature solutions	None
Carnotite	$K_2(UO_2)_2(VO_4)_2 \cdot 3H_2O$	K_2O 10.44 U 52.7 V 28.5	Orthorhombic or monoclinic	Bright yellow, lemon yellow, greenish yellow	Earthy				Basal				Occurs as a yellow crystalline powder or in loosely cohering masses mixed with quartzose material	Kalio-carnotite, potassio-carnotite
Cassiterite	SnO_2	Sn 78.6	Tetragonal ditetragonal dipyramidal	Yellowish or reddish brown to black	Adamantine to metallic	White, grayish, brownish	Transparent to opaque	Brittle	Imperfect	Uneven	6.5	6.90	Veins of quartz near granite or pegmatite, often in gravel sands	Tin stone, stannolite

continues next page

TABLE 2.10 Properties of minerals (continued)

Name	Chemical Formula	Metal (Percentage)	Crystal Structure	Color	Luster	Streak	Degree of Transparency	Tenacity	Cleavage	Fracture	Mohs Hardness	Specific Gravity	Occurrence	Common Names or Synonyms
Celestite	$SrSO_4$	SrO 56.42	Orthorhombic dipyramidal	Colorless to pale blue	Vitreous, pearly on cleavages	White	Transparent to subtranslucent	Brittle	Perfect	Uneven	3.3	3.96	Usually in limestone or sandstone with gypsum, rock salt, etc.	Celestine, coelestine
Cerussite	$PbCO_3$	Pb 76.5	Orthorhombic dipyramidal	Colorless to white and gray or smoky	Adamantine to vitreous	Colorless to white	Transparent to subtranslucent	Very brittle	Distinct	Conchoidal	3.3	6.52	Oxidized zones of lead veins	White lead, lead spar
Chalcanthite	$CuSO_4 \cdot 5H_2O$	CuO 31.87	Triclinic pinacoidal	Sky blue	Vitreous	Colorless	Subtransparent to translucent	Brittle	Imperfect	Conchoidal	2.5	2.21	Formed by the oxidation of chalcopyrite and other copper sulfides	Blue vitriol, blue stone, cyanose
Chalcocite	Cu_2S	Cu 79.86	Orthorhombic dipyramidal	Blackish lead gray	Metallic	Blackish lead gray	Opaque	Sectile	Indistinct	Conchoidal	2.5	5.70	Secondary, usually with pyrite, chalcopyrite, etc.	Copper glance, vitreous copper
Chalcopyrite	$CuFeS_2$	Cu 34.64 Fe 30.42	Tetragonal scalenohedral	Brass yellow, often tarnished irridescent	Metallic	Greenish black	Opaque	Brittle	Distinct	Uneven	3.5	4.20	Primary, veins or disseminated often with pyrite, quartz	Copper pyrites, cupropyrite
Chlorite	$(Mg, Al, Fe)_{12}[(Si, Al)_8 O_{20}](OH)_{16}$		Monoclinic	Green, white, yellow, pink, red, brown	Vitreous to pearly	White, pale green	Transparent to subtranslucent	Flexible	Perfect		2.3	2.72	In chlorite schist and other crystalline schists	
Chromite	$FeCr_2O_4$	Cr 46.4	Isometric hexoctahedral	Black	Metallic	Brown	Translucent to opaque	Brittle		Uneven	5.5	4.50	Occurs in veins in peridotites or serpentines derived from them	Eisenchrom, chromoferrite
Chrysoberyl	$BeAl_2O_4$	BeO 19.71 Al_2O_3 80.29	Orthorhombic dipyramidal	Green, greenish white, yellowish green, yellow	Vitreous	Colorless	Transparent to translucent	Brittle	Quite distinct	Uneven to conchoidal	8.5	3.67	In granite rocks and pegmatites and in sands and gravels	Alexandrite, cat's-eye, cymophane

continues next page

MATERIAL PROPERTIES | 37

TABLE 2.10 Properties of minerals (continued)

Name	Chemical Formula	Metal (Percentage)	Crystal Structure	Color	Luster	Streak	Degree of Transparency	Tenacity	Cleavage	Fracture	Mohs Hardness	Specific Gravity	Occurrence	Common Names or Synonyms
Chrysocolla	$CuSiO_3 \cdot nH_2O$	CuO 32.4–42.2	Amorphous	Green to greenish blue	Vitreous		Translucent to opaque	Sectile or brittle		Conchoidal	2.4	2.12	Secondary mineral found in the upper portions of copper veins	Kieselmalachit, chalcostaktite
Cinnabar	HgS	Hg 86.2	Hexagonal trigonal-trapezohedral	Cochineal red, often brownish red	Adamantine to metallic when dark colored to dull	Scarlet	Transparent to opaque	Sectile	Perfect	Uneven	2.5	8.10	Veins in sediments often with pyrite and marcasite	Cinnabarite, zinnober, hepatic-cinnabar
Cobaltite	CoAsS	Co 35.53 As 45.15	Isometric tetragonal	Silver white, to red. Also steel gray, with violet tinge	Metallic	Grayish, black	Opaque	Brittle	Cubic perfect	Uneven	5.5	6.20	Contact deposits, in gneiss, schists and diopside, with cobalt and nickel, copper and silver	Cobaltine, sehta, bright white cobalt
Copper	Cu	100	Isometric hexoctahedral	Fresh copper red. Tarnishes to brown, red, black, green	Metallic	Copper red metallic and shining	Opaque	Malleable and ductile	None	Hackly	2.7	8.80	Secondary, associated with copper minerals frequently near igneous rocks	
Cordierite	$Al_3(Mg, Fe^{2+})_2(Si_5AlO_{18})$		Orthorhombic dipyramidal	Grayish blue, lilac blue, dark blue	Vitreous	White	Transparent to translucent	Brittle	Distinct	Subconchoidal	7.3	2.63	In contact metamorphic zones	Iolite, dichroite, water sapphire
Corundum	Al_2O_3	Al 52.91	Hexagonal scalenohedral	Blue to colorless, yellow to golden, pink to deep red	Adamantine to vitreous, sometimes pearly	Uncolored	Transparent	Brittle	Interrupted	Uneven	9.0	4.00	Usually in limestone, dolomite, or gneiss, with minerals of chlorite group	Ruby, sapphire, emery, oriental amethyst
Covellite	CuS	Cu 66.48	Hexagonal dihexagonal dipyramidal	Indigo blue. Often highly iridescent	Crystals—submetallic inclining to resinous	Lead gray to black, shining	Opaque	Flexible	Perfect	Uneven	2.0	4.60	Secondary with other copper sulfides	Blue copper, indigo copper, covelline
Cristobalite	SiO_2		Tetragonal, trapezohedral	White, colorless	Vitreous	White	Transparent to opaque	Brittle to tough	None	Conchoidal	7.0	2.27	In acidic volcanic rocks and in meteorites	Lussatite

continues next page

TABLE 2.10 Properties of minerals (continued)

Name	Chemical Formula	Metal (Percentage)	Crystal Structure	Color	Luster	Streak	Degree of Transparency	Tenacity	Cleavage	Fracture	Mohs Hardness	Specific Gravity	Occurrence	Common Names or Synonyms
Crocoite	$Pb(CrO)_4$	PbO 69.06 CrO 30.94	Monoclinic prismatic	Hyacinth red, deep orange red, orange, yellow	Adamantine to vitreous	Orange yellow	Translucent	Sectile	Rather distinct	Small conchoidal to uneven	2.7	6.00	Secondary mineral from hot solutions	Callochrome, crocoise, red lead ore
Cryolite	Na_3AlF_6	Na 32.86 Al 12.85	Monoclinic prismatic	Colorless to white; also brownish; reddish or brick red	Vitreous to greasy	White	Transparent to translucent	Brittle		Uneven	2.5	2.97	In granite veins	Eisstein, ice-stone
Cuprite	Cu_2O	Cu 88.82	Isometric gyriodal	Cochineal red; sometimes almost black	Adamantine or submetallic to earthy	Several shades of brownish red	Translucent	Brittle	Interrupted	Conchoidal	3.5	6.00	Secondary, often with malachite, azurite, limonite	Ruby copper, ruberite, red copper ore
Diamond	C	100	Isometric hexetrahedral	Pale yellow to deep yellow, pale to deep brown, white to blue white	Adamantine to greasy		Transparent	Brittle	Perfect	Conchoidal	10.0	3.50	Alluvial deposits of sand and clay. Volcanic pipes	Bort, carbonado
Diaspore	$HAlO_2$	Al_2O_3 84.98	Orthorhombic dipyramidal	White, grayish white, colorless	Brilliant; pearly on cleavage		Transparent to subtranslucent	Very brittle	Eminent	Conchoidal	6.7	3.40	Alteration product of corrundum. Also in limestones	
Dolomite	$CaMg(CO_3)_2$	CaO 30.41 MgO 21.86	Hexagonal rhombohedral	Colorless or white, sometimes gray or greenish	Vitreous to pearly	White	Transparent to translucent	Brittle	Perfect	Subconchoidal	3.7	2.85	Vein mineral or altered limestone	Pearl spar, rhomb spar, bitter spar
Enargite	Cu_3AsS_4	Cu 48.42 As 19.02	Orthorhombic pyramidal	Grayish black to iron black	Metallic, tarnishing dull	Grayish black	Opaque	Brittle	Distinct	Uneven	3.0	4.40	Primary, usually with other copper minerals	Garbyite, clarite, guayacanite
Epidote	$Ca_2Fe^{+3}Al_2O$ $OH(Si_2O_7)$ (SiO_4)	CaO 22, 18–24.15 Fe_2O_3 11.07–23.42 Al_2O_3 13.10–24.36	Monoclinic prismatic	Green, yellow, gray	Vitreous	White or grayish white	Transparent to opaque	Brittle	Perfect	Uneven	6.5	3.42	Formed by the metamorphism of impure calcareous sedimentary rocks	Pistacite

continues next page

MATERIAL PROPERTIES | 39

TABLE 2.10 Properties of minerals (continued)

Name	Chemical Formula	Metal (Percentage)	Crystal Structure	Color	Luster	Streak	Degree of Transparency	Tenacity	Cleavage	Fracture	Mohs Hardness	Specific Gravity	Occurrence	Common Names or Synonyms
Epsomite	$MgSO_4 \cdot 7H_2O$	MgO 16.36	Orthorhombic disphenoidal	Colorless white, pink, or greenish	Vitreous	White	Transparent to translucent		Very perfect	Conchoidal	2.3	1.75	In mineral waters and on cave and mine walls	Epsom salt, bitter salt, gletschersalz
Fluorite	CaF_2	Ca 51.33	Isometric hexoctahedral	Yellow, green, greenish blue, violet blue; also white, gray, yellow	Vitreous; glimmering to dull in massive varieties		Transparent to subtranslucent	Brittle	Perfect	Flat-conchoidal or splintery	4.0	3.13	In veins and sedimentary rocks	Fluorspar, fluor, chlorephane
Fosterite	Mg_2SiO_4	SiO_2 41.72 FeO 1.11 MgO 57.83	Orthorhombic	Green, lemon yellow to greenish yellow, yellow amber	Vitreous	White or gray	Transparent to translucent	Brittle	Rather distinct	Conchoidal	6.7	3.32	In igneous rocks from low silica melts and metamorphic rocks formed from impure dolomites	
Franklinite	$ZnFe_2O_4$	Zn 5.4–18.7	Isometric hexoctahedral	Black to brownish black	Metallic to semimetallic	Reddish brown	Opaque	Brittle	Pseudo-cleavage (parting) octahedral	Conchoidal to uneven	6.0	5.14	Crystallized from igneous melts	Zinkoferrite, isophane, francklinite
Galena	PbS	Pb 86.60	Isometric hexoctahedral	Lead gray	Metallic	Lead gray	Opaque	Brittle	Cubic	Even	2.5	7.50	Veins, often with pyrite, sphalerite, chalcopyrite, intrusive replacement	Gelenite, lead glance, plumbago
Garnierite	(Ni, Mg) $SiO_3 \cdot nH_2O$	NiO 15.56	Amorphous-monoclinic	Apple green to white	Earthy and dull		Opaque	Soft and friable				2.52	A variation of serpentine	Noumeite, nickel gymnite, genthite, nepouite
Gibbsite	$Al(OH)_3$		Monoclinic prismatic	White; grayish, greenish, or reddish white	Pearly on cleavage, vitreous other surfaces		Translucent	Tough	Eminent		3.0	2.30	Usually with bauxite	
Glauberite	$Na_2Ca(SO_4)_2$	Na_2O 22.29 CaO 20.16	Monoclinic prismatic	Gray or yellowish	Vitreous, pearly on cleavage	White		Brittle	Perfect	Conchoidal	2.7	2.77	In salt deposits	Brongniartine

continues next page

TABLE 2.10 Properties of minerals (continued)

Name	Chemical Formula	Metal (Percentage)	Crystal Structure	Color	Luster	Streak	Degree of Transparency	Tenacity	Cleavage	Fracture	Mohs Hardness	Specific Gravity	Occurrence	Common Names or Synonyms
Goethite	$HFeO_2$	Fe 62.9	Orthorhombic dipyramidal	Crystals—blackish brown. Massive—yellowish or reddish brown. Earthy—brownish yellow, ocher yellow	Crystals—imperfect adamantine metallic, sometimes dull; fibrous variety often silky	Brownish yellow, orange yellow, ocher yellow	Opaque	Brittle	Very perfect	Uneven	5.3	4.28	Found with limonite as alteration product of a sulfide, usually pyrite	Bog iron ore, yellow ocher
Gold	Au	100	Isometric hexoctahedral	Gold yellow when pure. Silver white to orange red when impure	Metallic	Same as color	Opaque	Malleable and ductile	None	Hackly	2.7	19.3	In significant amounts in hydrothermal veins and related rocks, in consolidated placer deposits and unconsolidated placer deposits	Moss gold, wire or sponge gold
Graphite	C	100	Hexagonal dihexagonal-dipyramidal	Black to steel gray	Metallic, sometimes dull, earthy	Black	Opaque	Flexible	Perfect		1.5	2.1	Veins in granite, gneiss, quartzite, and limestone	Plumbago, black lead, graphitite
Greenockite	CdS	Cd 77.81	Hexagonal dihexagonal-pyramidal	Yellow, orange	Adamantine to resinous	Orange yellow	Transparent	Brittle	Distinct	Conchoidal	3.5	5.0	Usually coating on sphalerite	Cadmium-blended, cadmium ocher
Gypsum	$CaSO_4 \cdot 5H_2O$	CaO 32.57	Monoclinic prismatic	Colorless; also white, gray, yellowish or brownish	Subvitreous	White	Transparent to opaque	Flexible to brittle	Eminent	Conchoidal	1.7	2.32	Forms extensive sedimentary beds	Satin spar, alabaster, selenite
Halite	NaCl	Na 39.34	Isometric hexoctahedral	Colorless; also white, red, yellow, blue, purple	Vitreous	Colorless	Transparent to translucent	Rather brittle	Cubic, perfect	Conchoidal	2.5	2.3	An evaporite	Rock salt, muriate of soda

continues next page

TABLE 2.10 Properties of minerals (continued)

Name	Chemical Formula	Metal (Percentage)	Crystal Structure	Color	Luster	Streak	Degree of Transparency	Tenacity	Cleavage	Fracture	Mohs Hardness	Specific Gravity	Occurrence	Common Names or Synonyms
Hematite	Fe_2O_3	Fe 69.94	Hexagonal scalenohedral	Steel gray (crystals) dull red to bright red for earthy material	Metallic to submetallic to dull	Cherry red or reddish brown	Opaque	Brittle	Parts due to lamellar structure	Uneven	6.0	5.1	Often in granites, syenites, andesites. Altered limonite	Martite, red ocher, specularite
Ilmenite	$FeTiO_3$	Fe 36.8 Ti 31.6	Hexagonal rhombohedral	Iron black	Metallic to submetallic	Black	Opaque	Brittle		Conchoidal	5.5	4.7	Veins near igneous rocks	Titanic iron ore, menaccanite
Jamesonite	$Pb_4FeSb_6S_{14}$	Pb 40.16 Fe 2.71 Sb 35.39	Monoclinic prismatic	Gray black, tarnishes iridescent	Metallic	Gray black	Opaque	Brittle	Perfect	Uneven	2.5	5.8	Veins with galena, sphalerite, quartz	Brittle feather ore
Kaolinite	$Al_4(Si_4O_{10})(OH)_8$		Triclinic	White with reddish, brownish, or bluish tints	Dull earthy	White	Transparent to translucent	Flexible	Perfect		2.3	2.6	A result of decomposition of aluminous minerals	Kaolin, China clay, smelite
Kernite	$Na_2B_4O_7 \cdot 4H_2O$	Na_2O 22.66 B_2O_3 51.02	Monoclinic prismatic	Colorless; white	Vitreous	White	Transparent	Brittle	Perfect		3.0	1.95	In salt marshes as an evaporite	Rasorite
Kyanite	$Al_2O(SiO_4)$	Al_2O_3 60.43–62.74	Triclinic	Blue to white	Vitreous to pearly	White	Translucent to transparent	Brittle	Very perfect		6.2	3.6	In gneiss and mica schist	Diathene, cyanite
Lazulite	$(Mg, Fe)Al_2(PO_4)_2(OH)_2$	MgO:Fe = 1:0 MgO 13.34 Al_2O_3 33.73 P_2O_5 46.97	Monoclinic prismatic	Azure blue, bluish white, or bluish green	Vitreous	White	Subtranslucent to opaque	Brittle	Prismatic indistinct	Uneven	5.5	3.1	In quartz or pegmatite veins	blue spar, blue feldspar
Lazurite	$(Na, Ca)_8(Al_6Si_6O_{24})(SO_4, S, Cl)_2$		Isometric	Deep azure blue, greenish blue	Vitreous	White	Translucent	Brittle	Dodecahedral	Uneven	5.3	2.4	Contact metamorphism limestone	
Lepidolite mica	$K_2(Li, Al)_{5-6}(Si_{6-7}Al_{2-1}O_{20})(OH, F)_4$		Monoclinic	Colorless, shades of pink, purple	Vitreous to pearly	White	Translucent	Elastic	Basal, highly eminent		3.2	3.0	In granite pegmatites	Lithia mica
Leucite	$K(AlSi_2O_6)$		Tetragonal (pseudocubic)	White or gray	Vitreous to dull	White	Translucent to opaque	Brittle	Very imperfect	Conchoidal	5.7	2.50	In recent lavas	

continues next page

TABLE 2.10 Properties of minerals (continued)

Name	Chemical Formula	Metal (Percentage)	Crystal Structure	Color	Luster	Streak	Degree of Transparency	Tenacity	Cleavage	Fracture	Mohs Hardness	Specific Gravity	Occurrence	Common Names or Synonyms
Limonite	Largely $HFeO_2 \cdot nH_2O$ also $Fe_2O_3 \cdot nH_2O$ and other hydrous iron oxides		Amorphous or cryptocrystalline	Shades of brown, commonly dark brown to brownish black. When earthy, dull brown, yellow, ocher	Vitreous to dull	Yellowish brown to reddish	Opaque	Brittle	None	Uneven	5.3	3.8	Secondary iron mineral	Brown ocher, bog iron ore
Magnesite	$MgCO_3$	MgO 47.81	Hexagonal hexagonal-scalenohedral	Colorless, white, grayish white, yellowish to brown	Vitreous	Nearly white	Transparent to opaque	Brittle	Perfect	Flat conchoidal	4.0	3.06	Alteration product of magnesium-rich rocks by carbonic fluids	Baudisserite, magnesianite
Magnetite	$FeFe_2O_4$	Fe 72.4	Isometric hexoctahedral	Black to brownish black	Metallic to semimetallic	Black	Opaque	Brittle	Not distinct	Subconchoidal to uneven	6.0	5.17	Common constituent of crystalline rocks	Lode stone, siderite
Malachite	$Cu_2(OH)_2(CO_3)$	Cu 57.4	Monoclinic	Bright green, blackish green	Adamantine to vitreous	Pale green	Translucent to opaque	Brittle	Perfect	Subconchoidal, uneven	3.7	3.96	Oxidation zone of copper deposits. Alteration product of other copper minerals	Mountain green, molochite
Manganite	MnO(OH)	Mn 62.4	Monoclinic prismatic	Dark steel gray to iron black	Submetallic	Reddish brown to black	Opaque	Brittle	Perfect	Uneven	4.0	4.3	With other manganese oxides, barite, calcite	Sphenomanganite, newkirkite
Marcasite	FeS_2	Fe 46.55	Orthorhombic dipyramidal	Pale bronze yellow	Metallic	Grayish or brownish black	Opaque	Brittle	Poor	Uneven	6.5	4.9	Formed near surface, with galena, sphalerite, calcite, dolomite	White iron pyrites, cockscomb
Microcline feldspar	$K(AlSi_3O_8)$		Triclinic	Colorless or white; sometimes pink, yellow, red, or green	Vitreous	White	Transparent to translucent	Brittle	Perfect	Uneven	6.3	2.55	In igneous rocks	Amazon stone, moonstone

continues next page

MATERIAL PROPERTIES | 43

TABLE 2.10 Properties of minerals (continued)

Name	Chemical Formula	Metal (Percentage)	Crystal Structure	Color	Luster	Streak	Degree of Transparency	Tenacity	Cleavage	Fracture	Mohs Hardness	Specific Gravity	Occurrence	Common Names or Synonyms
Millerite	NiS	Ni 64.67	Hexagonal scalenohedral	Pale brass yellow	Metallic	Greenish black	Opaque	Elastic	Perfect	Uneven	3.5	5.4	Capillary crystals among other sulfides	Harkise, capillose
Molybdenite	MoS_2	Mo 59.94	Hexagonal dihexagonal dipyramidal	Lead gray	Metallic	Greenish on porcelain; bluish gray on paper	Opaque	Flexible, sectile	Perfect		1.5	4.7	Veins often with quartz, and copper sulfides	Moly, molybdaena
Muscovite mica	K_2Al_4 $(Si_6Al_2O_{20})$ $(OH,F)_4$		Monoclinic prismatic	Colorless; light shades of green, red, or brown	Vitreous to silky or pearly	White	Transparent to translucent	Elastic	Basal eminent		2.3	2.88	Original constituent of granite permatites and other potash and alumina-rich rocks	White mica, adamsite, didymite isinglass
Niccolite	NiAs	Ni 43.92	Hexagonal dihexagonal dipyramidal	Pale copper red	Metallic	Pale brownish black	Opaque	Brittle	None	Uneven	5.0	7.5	With sulfides and silver-arsenic minerals	Copper nickel, nickeline
Niter	KNO_3	K_2O 46.5	Orthorhombic dipyramidal	Colorless to white, gray	Vitreous	Colorless to white	Transparent to nearly opaque	Brittle	Perfect		2.0	2.1	Occurs on the surface of the earth	Saltpeter, nitrokalite
Opal	$SiO_2 \cdot nH_2O$		Submicrocrystalline aggregate	Milky white or bluish white; also yellow to brown, orange, green, and blue	Vitreous, often somewhat resinous	White	Transparent to nearly opaque	Brittle	None	Conchoidal	6.0	2.1	In seams and fissures of igneous rocks; deposited at low temperature by silica-bearing waters	Girasol, hydrophane, tabasheer, geyserite
Orpiment	As_2S_3	As 60.91	Monoclinic prismatic	Lemon yellow; golden yellow; brownish yellow	Pearly on cleavage surfaces, elsewhere resinous	Pale lemon yellow	Translucent	Sectile, flexible	Perfect		2.0	3.5	Vein with realgar	Yellow arsenic, arsenblende

continues next page

TABLE 2.10 Properties of minerals (continued)

Name	Chemical Formula	Metal (Percentage)	Crystal Structure	Color	Luster	Streak	Degree of Transparency	Tenacity	Cleavage	Fracture	Mohs Hardness	Specific Gravity	Occurrence	Common Names or Synonyms
Orthoclase feldspar	$K(AlSi_3O_8)$		Monoclinic	Colorless or white; sometimes pink, yellow, red, or green	Vitreous	White	Transparent to translucent	Brittle	Perfect	Conchoidal to uneven	6.0	2.57	In igneous rocks	Common feldspar, moonstone
Pentlandite	$(Fe,Ni)_9S_8$	Fe 32.55 Ni 34.22	Isometric hexoctahedral	Light bronze yellow	Metallic	Light bronze brown	Opaque	Brittle	Octahedral	Uneven	4.0	5.0	Intergrown with pyrrhotite	Micropyrite, folgerite
Phenacite	$Be_2(SiO_4)$	BeO 45.55	Hexagonal rhombohedral	Colorless, white, yellow, rose red, brown	Vitreous		Transparent to subtranslucent	Brittle	Distinct	Conchoidal	7.8	2.98	Granite pegmatites	Phenakite
Phlogopite mica	$K_2(Mg,Fe^{+2})_2(Si_6Al_2O_{20})(OH,F)_4$		Monoclinic prismatic	Colorless, yellowish, brown, green, reddish brown, dark brown	Pearly to vitreous	White to gray	Transparent to opaque	Elastic	Basal highly eminent		2.8	2.8	In crystalline limestone or dolomite and also in serpentine	Amber mica, flogopite
Platinum	Pt	100	Isometric hexoctahedral	Whitish steel gray to dark	Metallic	Whitish steel gray	Opaque	Malleable and ductile	None	Hackly	4.3	19.0	Placer deposits, with gold, chromite	
Polyhalite	$K_2Ca_2Mg(SO_4)_4 \cdot 2H_2O$	K_2O 15.62 CaO 18.60 MgO 6.69	Triclinic pinacoidal	White or gray; often salmon pink to brick red	Vitreous to resinous				Distinct		2.8	2.78		Mamanite, ischelite
Proustite	Ag_3AsS_3	Ag 65.42 As 15.14	Hexagonal ditrigonal pyramidal	Scarlet vermilion	Adamantine	Vermilion	Translucent	Brittle	Distinct	Uneven	2.5	5.6	Usually with pyrargerite	Light ruby silver
Psilomelane	$BaMn_2Mn_8^4O_{16}(OH)_4$	Ba 14.35 Mn 51.75	Amorphous	Black	Submetallic	Brownish black	Opaque		None		6.0	4.0	Usually with pyrolucite	Black hematite, psilomelanite
Pyrargyrite	Ag_3SbS_3	Ag 59.76 Sb 22.48	Hexagonal ditrigonal pyramidal	Deep red	Adamantine	Purplish red	Opaque	Brittle	Distinct	Uneven	2.5	5.8	Veins with silver, galena, sphalerite	Dark ruby silver
Pyrite	FeS_2	Fe 46.55	Isometric diploidal	Pale brass yellow	Metallic, splendent to glistening	Greenish or brownish black	Opaque	Brittle	Indistinct	Uneven	6.0	5.0	Primary, veins or disseminated, usually crystalline	Fool's gold, iron pyrites, mundic

continues next page

MATERIAL PROPERTIES | 45

TABLE 2.10 Properties of minerals (continued)

Name	Chemical Formula	Metal (Percentage)	Crystal Structure	Color	Luster	Streak	Degree of Transparency	Tenacity	Cleavage	Fracture	Mohs Hardness	Specific Gravity	Occurrence	Common Names or Synonyms
Pyrolusite	MnO_2	Mn 63.19	Tetragonal ditetragonal dipyramidal	Light steel gray	Metallic	Black, bluish black	Opaque	Soft, soils fingers	None		2.0	4.8	Usually secondary, often in clays	Polianite, varvicite
Pyrophyllite	$Al_4(Si_8O_{20})(OH)_2$		Monoclinic	White, yellow, pale blue, grayish or brownish green	Pearly	White	Subtransparent to opaque	Flexible	Eminent		1.5	2.8	In schistose rocks	Pyrauxite
Pyrrhotite	$Fe_{1-x}S$ (x between 0 and 2)	Fe 63.53	Hexagonal dihexagonal dipyramidal	Bronze yellow to pinchbeck brown	Metallic	Dark grayish black	Opaque	Brittle	Distinct	Uneven	4.0	4.6	In basic igneous, with sulfides and magnetite	Magnetic pyrites, pyrrhotine
Quartz	SiO_2	Si 46.71	Hexagonal trigonal trapezohedral	Varies widely	Vitreous; sometimes greasy	White	Transparent to translucent	Brittle	None	Conchoidal	7.0	2.65	In igneous rocks, sediments, and metamorphics	Rock crystal, chalcedony, agate, flint, chert, jasper
Realgar	AsS	As 70.0	Monoclinic	Aurora red to orange yellow	Resinous to greasy	Orange red to aurora red	Transparent to translucent	Sectile	Fair	Conchoidal	2.0	3.6	Vein with orpiment, stibnite lead, silver, and gold	Red orpiment, red arsenic, ruby sulfur
Rhodochrosite	$MnCO_3$	Mn 61.71	Hexagonal hexagonal-scalenohedral	Pink, rose, red; fawn colored; brown	Vitreous, inclining to pearly	White	Translucent to subtranslucent	Brittle	Perfect	Uneven	4.0	3.52	Gangue mineral of primary origin in sediments and mete-sediments	Manganese spar, dialogite
Rutile	TiO_2	Ti 60	Tetragonal ditetragonal dipyramidal	Reddish brown, passing into red	Metallic to adamantine	Pale brown to yellowish	Transparent to opaque	Brittle	Distinct	Uneven	6.5	4.2	Frequently secondary in micas or igneous rocks. Black sands	Edisonite, titanite
Scheelite	$CaWO_4$	WO_3 80.53	Tetragonal dipyramidal	Yellowish white, pale yellow, or brownish	Vitreous	White	Transparent to translucent	Brittle	Distinct	Uneven	4.8	6.0	Pegmatite veins or in veins associated with granite or gneiss	Tungstein, schellspath

continues next page

TABLE 2.10 Properties of minerals (continued)

Name	Chemical Formula	Metal (Percentage)	Crystal Structure	Color	Luster	Streak	Degree of Transparency	Tenacity	Cleavage	Fracture	Mohs Hardness	Specific Gravity	Occurrence	Common Names or Synonyms
Serpentine	$Mg_3(Si_2O_5)(OH)_2$	MgO 43.0 Si_2O 44.1 H_2O 12.9	Monoclinic prismatic	Green, green blue, white, gray, yellow	Waxy, greasy, or silky	White	Translucent to opaque		Distinct	Conchoidal or splintery	4.0	2.58	Secondary mineral formed by alteration of nonaluminous silicates containing magnesia	Verd antique, bowenite, ophite
Siderite	$FeCO_3$	Fe 48.2	Hexagonal hexagonal-scalenohedral	Yellowish brown to reddish brown	Vitreous, inclining to pearly or silky	White	Translucent to subtranslucent	Brittle	Perfect	Uneven or subconchoidal	3.7	3.85	Sometimes in sedimentary deposits and in veins	Spathic iron, clay ironstone
Silver	Ag	100	Isometric hexoctahedral	Silver white, gray to black due to tarnish	Metallic	Silver white	Opaque	Malleable and ductile	None	Hackly	2.8	10.5	Usually secondary in upper part of silver bearing veins	White gold
Smithsonite	$ZnCO_3$	Zn 52.3	Hexagonal hexagonal-scalenohedral	Grayish white to dark gray, greenish or brownish white	Vitreous	White	Subtransparent to translucent	Brittle	Perfect	Uneven to imperfectly conchoidal	5.5	4.38	Both in veins and beds with galena and sphalerite in calcareous rocks	Dry bone ore, turkey-fat ore, zinc spar
Sodalite	$Na_8(Al_6Si_6O_{24})Cl_2$		Isometric	Pale pink, gray, yellow, blue, green	Vitreous	White	Transparent to translucent	Brittle	Dodecahedral distinct	Conchoidal to uneven	5.8	2.17	Igneous rocks of the nepheline-syenite groups	
Soda niter	$NaNO_3$	Na_2O 36.5	Hexagonal scalenohedral	Colorless; also white	Vitreous	White	Transparent	Rather sectile	Perfect	Conchoidal	1.8	2.27	In deserts as an evaporite	Chile saltpeter, nitratine
Sphalerite	ZnS	Zn 67.10	Isometric hextetrahedral	Commonly brown, black, yellow; also red, green to white to nearly colorless	Resinous to adamantine	Brownish to light yellow and white	Translucent	Brittle	Perfect	Conchoidal	3.5	4.0	Often in limestone with other sulfides	Zinc blende, black jack, ruby zinc
Spinel	$MgAl_2O_4$	MgO 27.49–13.65	Isometric hexoctahedral	Variable; red to blue, green, brown to nearly colorless	Vitreous, splendent to nearly dull	White	Transparent to nearly opaque	Brittle	Imperfect	Conchoidal	8.0	3.8	In sands and gravels, accessory mineral in basic rocks	Balas ruby, picotite, rubicelle

continues next page

MATERIAL PROPERTIES | 47

TABLE 2.10 Properties of minerals (continued)

Name	Chemical Formula	Metal (Percentage)	Crystal Structure	Color	Luster	Streak	Degree of Transparency	Tenacity	Cleavage	Fracture	Mohs Hardness	Specific Gravity	Occurrence	Common Names or Synonyms
Stannite	Cu_2FeSnS_4	Cu 29.58 Fe 12.99 Sn 27.61	Tetragonal scalenohedral	Steel gray to iron black	Metallic	Blackish	Opaque	Brittle	Indistinct	Uneven	3.5	4.4	Veins with cassiterite, chalcopyrite, pyrite	Bell metal ore, zinnkies
Staurolite	$(Fe^{+2}, Mg)_2 (Al, Fe^{+3})_9 O_6(SiO_4)_4 (O,OH)_2$		Monoclinic prismatic	Dark brown, reddish brown, yellow brown	Vitreous to resinous	Gray	Translucent to nearly opaque	Brittle	Distinct but interrupted	Subconchoidal	7.3	3.70	In crystalline schists or phyllites as a result of regional metamorphosis	Cross stones, fairy stone crosses
Stephanite	Ag_5SbS_4	Ag 68.33 Sb 15.42	Orthorhombic pyramidal	Iron black	Metallic	Iron black	Opaque	Brittle	Imperfect	Uneven	2.5	6.2	Veins with silver, galena, sphalerite	Brittle silver, malanglanz
Stibnite	Sb_2S_3	Sb 71.69	Orthorhombic dipyramidal	Lead gray to steel gray	Metallic, splendent on cleavage	Lead gray	Opaque	Sectile	Perfect	Subconchoidal	2.0	4.6	Veins with quartz often in granite	Antimonite, antimony glance
Strontianite	$SrCO_3$	SrO 70.19	Orthorhombic dipyramidal	Colorless to gray, yellowish or greenish	Vitreous, resinous on fracture	White	Transparent to translucent	Brittle	Nearly perfect	Uneven	3.8	3.70	In veins in limestones and marls	Strontian
Sulfur	S	100	Orthorhombic dipyramidal	Yellow, brown, green, red, gray	Resinous to greasy	White	Translucent	Brittle	Imperfect	Uneven	2.0	2.07	Volcanic activity, usually with gypsum, limestone	
Sylvanite	$(Au, Ag)Te_2$	Au 24.19 Ag 13.22	Monoclinic prismatic	Steel gray to silver white	Metallic brilliant	Same as color	Opaque	Brittle	Perfect	Uneven	2.0	8.1	Veins with gold, pyrite, and quartz	Aurotellurite
Sylvite	KCl	K 52.44	Isometric hexoctahedral	Colorless or white; also grayish, bluish, yellowish red, or red	Vitreous	White	Transparent to translucent	Brittle	Cubic perfect	Uneven	2.0	1.98	An evaporite	Muriate of potash, hoevelite
Talc	$Mg_6(Si_8O_{20})(OH)_4$	MgO 29.13–31.76 SiO_2 60.06–62.67	Monoclinic	Colorless, white, green, brown	Pearly to greasy	White	Subtransparent to translucent	Sectile	Perfect		1.3	2.75	Secondary mineral formed by alteration of nonaluminous magnesian silicates	Steatite, soapstone

continues next page

TABLE 2.10 Properties of minerals (continued)

Name	Chemical Formula	Metal (Percentage)	Crystal Structure	Color	Luster	Streak	Degree of Transparency	Tenacity	Cleavage	Fracture	Mohs Hardness	Specific Gravity	Occurrence	Common Names or Synonyms
Tennantite	$(Cu,Fe)_{12}As_4S_{13}$ $Cu_{13}As_4S_{13}$	Cu 51.57 As 20.26	Isometric hextetrahedral	Flint gray to iron black to dull black	Metallic, often splendent	Black to brown	Opaque	Brittle			3.5	4.4	Veins with other copper minerals	Arsenicalfahlerz
Tetrahedrite	$(Cu,Fe)_{12}Sb_4S_{13}$ $Cu_{12}Sb_4S_{13}$	Cu 45.77 Sb 29.22	Isometric hextetrahedral	Flint gray to iron black to dull black	Metallic, often splendent	Black to brown	Opaque	Brittle	None	Uneven	3.5	4.7	Veins with copper, silver, pyrite, galena, sphalerite, quartz	Gray copper, fahlore, panabase
Topaz	$Al_2(SiO_4)(OH,F)_4$	Al_2O_3 55.67–56.76 F 13.23–20.7	Orthorhombic	Colorless, white, yellow, light shades of gray, green, red, or blue	Vitreous	White	Transparent to subtranslucent	Brittle	Highly perfect	Subconchoidal to uneven	8.0	3.5	In veins and cavities in igneous rocks	Brazilian ruby, chrysolithos
Tourmaline	$NaMg_3Al_6B_3Si_6O_{27}(OH,F)_4$		Hexagonal ditrigonal-pyramidal	Black to brown	Vitreous to resinous	White	Transparent to opaque	Brittle	Difficult	Subconchoidal to uneven	7.8	3.10	In granites or gneisses or in pegmatite veins	Brazilian emerald, peridot, robellite, siberite, schorl
Turquoise	$CuAl_6(PO_4)_4(OH)_8 \cdot 4H_2O$	CuO 9.78 Al_2O_3 37.60	Triclinic pinacoidal	Sky blue, bluish green, to apple green, greenish gray	Waxy to vitreous	White or greenish	Subtranslucent to opaque	Rather brittle	Two directions in crystals, none in massive material	Small conchoidal	5.5	2.7	Secondary mineral occurring in veins in highly altered rocks	Agaphite, calaite
Uraninite	UO_2 to U_3O_8	UO_2 varies from 70.09 to 23.07 while UO_3 varies 22.69 to 40.60	Isometric hexoctahedral	Steely to velvety black and brownish black, grayish, greenish	Submetallic to pitchlike or greasy and dull	Brownish black, grayish olive green, a little shining	Opaque	Brittle		Uneven	5.5	9.4	Granitic pegmatites, or with ores of silver, lead, copper	Pitchblende, ulrichite
Vanadinite	$Pb_5(VO_4)_3Cl$	PbO 78.80 V_2O_5 19.26	Hexagonal hexagonal-dipyramidal	Orange red, ruby red, brownish red	Subresinous	White or yellowish	Subtranslucent to opaque			Uneven, flat or conchoidal	2.9	6.88	In altered lead deposits	Vanadate of lead
Wavellite	$Al_3(OH)_3(PO_4)_2 \cdot 5H_2O$	Al_2O_3 37.11 P_2O_5 34.4	Orthorhombic	Greenish white, green, yellow	Vitreous to resinous	White	Translucent	Brittle	Rather perfect	Uneven to subconchoidal	3.6	2.32	Secondary mineral associated with many rock types	Devonite, hydrargillite

continues next page

MATERIAL PROPERTIES | 49

TABLE 2.10 Properties of minerals (continued)

Name	Chemical Formula	Metal (Percentage)	Crystal Structure	Color	Luster	Streak	Degree of Transparency	Tenacity	Cleavage	Fracture	Mohs Hardness	Specific Gravity	Occurrence	Common Names or Synonyms
Willemite	$Zn_2(SiO_4)$	ZnO 73.0	Hexagonal rhombohedral	White or greenish yellow	Vitreous to resinous	White or colorless	Transparent to opaque	Brittle	Easy	Conchoidal to uneven	5.5	4.04	In zinc ore deposits	Villemite, hebertine
Witherite	$BaCO_3$	BaO 77.70	Orthorhombic dipyramidal	Colorless to milky white or grayish	Vitreous, resinous on fracture	White	Subtransparent to translucent	Brittle	Distinct	Uneven	3.4	4.31	In veins with galena direct crystallization from barium carbonate rich fluids	Barolite
Wolframite	$(Fe,Mn)WO_4$	WO 74.78–76.58	Monoclinic prismatic	Grayish or brownish black	Submetallic	Reddish brown to black	Opaque	Brittle	Very perfect	Uneven	5.3	7.3	In granite and pegmatite veins	Walfram, mock-lead
Wulfenite	$PbMoO_4$	PbO 60.79 MoO_3 39.21	Tetragonal pyramidal	Orange yellow to yellowish gray, grayish white	Resinous	White	Subtransparent to subtranslucent	Brittle	Very smooth	Subconchoidal	2.8	6.85	Oxidation zone of lead and zinc deposits	Yellow lead ore, melinose
Zincite	ZnO	Zn 80.34	Hexagonal dihexagonal pyramidal	Orange yellow to deep red	Subadamantine	Orange yellow	Translucent	Brittle	Perfect	Subconchoidal	4.5	5.5	Usually with franklinite and willemite. Sometimes in calcite	Spartalite
Zircon	$Zr(SiO)_4$	ZrO_2 67.2	Tetragonal	Reddish brown, yellow, gray, green, or colorless	Adamantine	Uncolored	Transparent to subtranslucent or opaque	Brittle	Imperfect	Conchoidal	7.5	4.69	Accessory mineral in igneous rocks	Hyacinth, azorite

Source: Bolles and McCullough 1985.

TABLE 2.11   Properties of major mineral fillers

	Theoretical Chemical Composition	Specific Gravity	Bulk Density kg/m³*	Hardness Mohs Scale	Refractive Index	Reaction, pH	Oil Absorption Cc per 100 g	Particle Characteristics
Asbestos (chrysotile)	$3MgO \cdot 2SiO_2 \cdot 2H_2O$	2.5–2.6	160–640	2.5–4.0	1.51–1.55	8.5–10.3	40–90	Fibers fine, easily separable; fibrils hexagonally close packed 100–300Å
Barite	$BaSO_4$	4.3–4.6	1280–2400	2.5–3.5	1.64	7	6–10	Generally equi-dimensional
Bentonite	$(Mg,Ca)O \cdot Al_2O_3 5SiO_2 \cdot xH_2O$	2.3–2.8	800–960	1.5+	1.55–1.56	6.2–9.0	20–30	Porous microaggregates, irregular shapes; ultimate plate structure
Diatomite	$SiO_2 \cdot xH_2O$	2.0–2.35	96–320	4.5–6.0	1.42–1.49	6–8.5	100–300	Unique diatom structure; micro and ultramicro porosity
Fuller's earth	$(Mg,Ca)O \cdot Al_2O_3 5SiO_2 \cdot xH_2O$	2.2–2.6	432–608	4	1.50	7.5–8.2	30	Apparently equi-dimensional; electron-microscopically fibrous, lath-like
Gypsum	$CaSO_4 \cdot 2H_2O$	2.3	400–640	1.5–2.0	1.52	6.5–7	17–25	Irregular, roughly equi-dimensional
Kaolin	$Al_2O_3 \cdot 2SiO_2 \cdot 2H_2O$	2.6	320–640	2.0–2.5	1.56–1.58	4.5–7	25–50	Thin, flat hexagonal plates, 0.05–2μ size and stacks of same
Limestone	$CaCO_3$	2.7	640–1600	3	1.63–1.66	7.8–8.5	6–30	Variable size particles; ultimate rhombs
Mica (muscovite)	$H_2KAl_3(SiO_4)_3$	2.7–3.0	192–320	2.0–3.0	1.59±	7.4–9.4	25–50	Platelike particles
Nepheline syenite	$K_2O \cdot 7d_2O \cdot 4.5Al_2O_3 \cdot 2OSiO_2$	2.61	800–1280	5.5–6.0	1.53	9.9	21–29	Nodular and irregular
Perlite	Like rhyolite	2.5–2.6	64–320	5.0	1.48–1.49	9	50–275	Expanded "glass" bubbles and fragments
Portland cement	Essentially carbon silicates and aluminates	2.9–3.15	1440–1600	5.6	17.2±	11.0–12.6	20	Variable, smooth, rounded, angular, and flake particles
Pumicite	A silicate, like rhyolite	2.2–2.63	640–800	5–6	1.49–1.50	7–9	30–40	Vesicular
Pyrophyllite	$Al_2O_3 \cdot 4SiO_2 \cdot H_2O$	2.8–2.9	400–480	1–2	1.57–1.59	6–8	40–70	Minute foliated plates or scales and extra-long particles
Rock dusts	Variable	2.6–3.3	800–1600	4–6.5	Variable	Usually above 7	20–40	Variable
Silicas, crystalline and microcrystalline	SiO_2	2.60–2.65	800–1280	6.5–7.0	1.53–1.54	6–7	20–50	Variable sized, angular and equi-dimensional particles, or minute particles to porous masses
Slate	Mixture of mineral silicates	2.7–2.8	640–1280	4–6	—	6.8	20–25	Flat or wedge-shaped, or spherical grains
Talc	$H_2Mg_3(SiO_3)_4$	2.6–3.0	416–960	1–1.5	1.57–1.59	8.1–9.0	20–50	Lamellar, foliated, or microfibrous
Vermiculite	$(Mg,Fe)_3(Si,Al)_4O_{10}(OH)_2 \cdot 4H_2O$	2.2–2.7	96–160; fines-320	1.5	1.56	Pract. neutral	—	Platelets or lamellar structure
Wollastonite	$CaSiO_3$	2.8–3.0	320–640	4.5–5.0	1.63	9.9	25–30	Brilliant white powder with acicular nature

* Metric equivalent: 1 lb/ft³ × 16.01846 = Kg/m³.
Source: Trivedi and Hagemeyer 1994.

MATERIAL PROPERTIES | 51

TABLE 2.12 Classification of coals by rank*

Class/Group	Fixed Carbon Limits (dry, mineral-matter-free basis, %)		Volatile Matter Limits (dry, mineral-matter-free basis, %)		Gross Calorific Value Limits (moist† mineral-matter-free basis)				Agglomerating Character
					Btu/lb		Mj/kg‡		
	Equal or Greater Than	Less Than	Greater Than	Equal or Less Than	Equal or Greater Than	Less Than	Equal or Greater Than	Less Than	
Anthracitic									
Meta-anthracite	98	—	—	2	—	—	—	—	
Anthracite	92	98	2	8	—	—	—	—	Nonagglomerating
Semianthracite§	86	92	8	14	—	—	—	—	
Bituminous									
Low volatile bituminous coal	78	86	14	22	—	—	—	—	
Medium volatile bituminous coal	69	78	22	31	—	—	—	—	
High volatile A bituminous coal	—	69	31	—	14,000††	—	32.6	—	Commonly agglomerating**
High volatile B bituminous coal	—	—	—	—	13,000††	14,000	30.2	32.6	
High volatile C bituminous coal	—	—	—	—	11,500	13,000	26.7	30.2	
					10,500	11,500	24.4	26.7	Agglomerating
Subbituminous									
Subbituminous A coal	—	—	—	—	10,500	11,500	24.4	26.7	
Subbituminous B coal	—	—	—	—	9,500	10,500	22.1	24.4	
Subbituminous C coal	—	—	—	—	8,300	9,500	19.3	22.1	Nonagglomerating
Lignitic									
Lignite A	—	—	—	—	6,300	8,300	14.7	19.3	
Lignite B	—	—	—	—	—	6,300	—	14.7	

* This classification does not apply to certain coals.
† Moist refers to coal that contains its natural inherent moisture but not including visible water on the coal's surface.
‡ Megajoules per kilogram; to convert British thermal units per pound to megajoules per kilogram, multiply by 0.002326.
§ If agglomerating, classify in low volatile group of the bituminous class.
** It is recognized that there may be nonagglomerating varieties in these groups of the bituminous class, and that there are notable exceptions in the high volatile C bituminous group.
†† Coals having 69% or more fixed carbon on the dry, mineral-matter-free basis are classified according to fixed carbon, regardless of gross calorific value.
Source: ASTM 1998 *(reprinted with permission).*

TABLE 2.13 Typical proximate and ultimate analyses and heating values of coals in the United States

Coal Type	Ultimate Analysis (Dry) (wt %)						Proximate Analysis (As Received) (wt %)				Heating Value (As Received) (Btu/lb)
	C	H	S	N	Ash	O	Volatile Matter	Fixed Carbon	Moisture	Ash	
Lignite	65.7	4.5	1.0	1.2	9.2	18.4	31.4	25.9	35.5	7.2	7,100
Western subbituminous	62.6	4.0	1.0	1.0	13.6	17.8	36.6	42.8	8.1	12.5	9,400
Illinois No. 6	70.0	4.9	3.8	1.4	9.2	10.7	41.1	39.6	11.2	8.1	11,300
Eastern bituminous (high-sulfur)	70.4	4.6	4.6	1.4	10.5	8.5	37.0	46.4	6.9	9.7	11,700
Eastern bituminous (low-sulfur)	79.9	5.5	1.3	1.5	5.4	6.4	36.9	53.9	4.0	5.2	14,000

Source: Levy et al. 1981 *(reprinted with permission of Marcel Dekker).*

TABLE 2.14 Petrographic and physical properties for various U.S. coal seams

Coal Seam	Location of Band Within Seam (From Top to Bottom)	MM*	VM*	FSI*	Btu*	BSG*	T*	HGI*
Kittanning Coal, Preston County, West Virginia	10 in. of coal at top of seam	16.3	34.9	9	13020	1.42	116	78
	7 in. of dark gray shale†	—	—	—	—	—	—	—
	2¾ in. of coal and shale lenses	—	—	—	—	—	—	—
	5½ in. of dark brown shale	—	—	—	—	—	—	—
	7½ in. friable dull coal	16.1	37.7	9	13192	1.38	121	85
	1 in. of shale	—	—	—	—	—	—	—
	9 in. of coal	26.5	34.3	5	11525	1.48	102	73
	3¼ in. of coal with shale lenses	—	—	—	—	—	—	—
	8 in. of coal grading into shale	—	—	—	—	—	—	—
	24 in. of cross-bedded shale	—	—	—	—	—	—	—
	18 in. of blocky coal	6.5	32.2	9	14655	1.30	107	89
	1 in. of black shale	—	—	—	—	—	—	—
	9 in. of blocky coal	11.1	33.7	9	14016	1.35	104	86
Pond Creek Coal, Stone County, Kentucky	4 in. of dull blocky coal†	—	—	—	—	—	—	—
	8 in. of coal with lenses of clay	—	—	—	—	—	—	—
	12½ in. of coal with nodules and bonds of shale	—	—	—	—	—	—	—
	19½ in. of blocky detrital coal	18.2	33.3	1	12331	1.43	7	58
	6 in. of blocky coal	15.4	34.6	1	12936	1.39	32	52
	2½ in. of dull blocky coal	13.1	31.0	1	13354	1.41	41	52
	5½ in. of finely banded coal	6.9	35.8	1	14157	1.30	51	58
Jawbone Coal, Dickenson County, Virginia	10½ in. of blocky coal	3.6	27.9	8.5	15366	1.30	89	93
	11 in. of bright banded coal	11.6	27.3	8	13957	1.38	84	88
	16 in. of blocky coal	9.6	28.5	8.5	14359	1.32	93	84
Sewickley Coal, Fayette County, Pennsylvania	Bulk sample of coal	12.7	38.9	7.5	13485	1.37	84	72
Pittsburgh No. 8 Coal, Athens County, Ohio	Bulk sample of coal 48 in. thick	15.5	48.7	3	12950	—	—	—
Herrin No. 6 Coal, Jefferson County, Illinois	Bulk sample of coal 6 ft thick	10.5	42.1	—	12890	—	—	—
Crowburg Coal, Randolph County, Missouri	4 in. bright coal	—	—	—	—	—	—	—
	1 in. pyritized coal	—	—	—	—	—	—	—
	1 in. bright coal	—	—	—	—	—	—	—
	2 in. pyritized coal	—	—	—	—	—	—	—
	11 in. banded coal analysis of bulk sample	16.7	48.3	4	11708	—	—	—
Blue Creek Coal, Jefferson County, Alabama	Bulk sample of coal	16.8	28.9	8.5	14008	—	—	—

continues next page

MATERIAL PROPERTIES | 53

TABLE 2.14 Petrographic and physical properties for various U.S. coal seams (continued)

Coal Seam	Location of Band Within Seam (From Top to Bottom)	MM*	VM*	FSI*	Btu*	BSG*	T*	HGI*
Anderson-Canyon (WYODAK) Coal, Campbell County, Wyoming	1 ft coal; vitrain bands + attritus	—	—	—	—			
	21 ft coal; vitrain bands + bright attritus	—	—	—	—			
	13 ft banded to nonbanded coal; vitrain, attritus, wood grain	—	—	—	—			
	25 ft coal; vitrain bands, attritus, fusain	—	—	—	—			
	0.1 ft carbonaceous claystone	—						
	6.5 ft banded coal; vitrain, attritus	—	—	—	—			
	7.8 ft banded to nonbanded coal; vitrain, attritus, wood grain	—	—	—	—			
	Bulk sample coal	10.1	48.8	—	11500			

* MM = Mineral matter = 1.1 (ash) + 0.1 (sulfur), VM = dry mineral matter-free volatile matter, FSI = free-swelling index, Btu = moisture-free heating value, BSG = bulk specific gravity, T = Gieseler maximum plastic temperature range, HGI = hardgrove grindability index.
† Bands that are reported but not analyzed have specific gravities above 1.60.
Source: Beasley et al. 1991.

TABLE 2.15 Sulfur content and sulfur forms for various coals

		Percentage, Moisture-Free Basis*			
Mine Location	Coal Seam	Total Sulfur	Pyritic Sulfur	Organic Sulfur	Organic Sulfur as Percentage of Total Sulfur
Washington County, Pennsylvania†	Pittsburgh	1.13	0.35	0.78	69.0
Clearfield County, Pennsylvania	Upper Freeport	3.56	2.82	0.74	20.8
Allegheny County, Pennsylvania	Thick Freeport	0.92	0.46	0.45	48.9
Somerset County, Pennsylvania	B	0.78	0.19	0.57	73.1
Somerset County, Pennsylvania	C prime	2.00	1.43	0.54	27.0
Clearfield County, Pennsylvania	B	1.90	1.12	0.75	39.5
Cambria County, Pennsylvania	Miller	1.25	0.56	0.65	52.0
Franklin County, Illinois	No. 6	2.52	1.50	1.02	40.5
Franklin County, Illinois	No. 6	1.50	0.81	0.69	46.0
Montgomery County, Illinois	No. 6	4.97	2.53	2.40	48.3
Williamson County, Illinois	No. 6	4.01	2.17	1.80	44.9
Union County, Kentucky	No. 9	3.28	1.05	2.23	68.0
Union County, Kentucky	No. 9	3.46	1.65	1.81	52.3
Webster County, Kentucky	No. 12	1.48	0.70	0.78	52.7
Pike County, Kentucky	Freeburn	0.46	0.13	0.33	71.7
Letcher County, Kentucky	Elkhorn	0.68	0.13	0.51	75.0
McDowell County, West Virginia	Pocahontas No. 3	0.55	0.08	0.46	83.6
Boone County, West Virginia	Eagle	2.48	1.47	1.01	40.7
Walker County, Alabama	Pratt	1.62	0.81	0.81	50.0
Jefferson County, Alabama	Pratt	1.72	0.97	0.72	41.9
Jefferson County, Alabama	Mary Lee	1.05	0.33	0.69	65.7
Clay County, Indiana	No. 3	3.92	2.13	1.79	45.7
Cumnock, North Carolina	Deep River	2.32	1.52	0.80	34.5
Cumnock, North Carolina	Deep River	2.08	1.53	0.55	26.4
Allegany County, Maryland	Big Vein	0.86	0.18	0.67	77.9
Meigs County, Ohio	8-A	2.51	1.61	0.86	34.3
Natal, South Africa		1.51	0.47	0.97	64.2

continues next page

TABLE 2.15 Sulfur content and sulfur forms for various coals (continued)

		Percentage, Moisture-Free Basis*			
Mine Location	Coal Seam	Total Sulfur	Pyritic Sulfur	Organic Sulfur	Organic Sulfur as Percentage of Total Sulfur
Transvaal, South Africa		1.39	0.59	0.70	50.4
Transvaal, South Africa		0.44	0.06	0.37	84.1
Brazil, South America		2.39	1.78	0.50	20.9
Istria, Italy‡		9.01	1.09	7.90	87.7
Germany, bituminous		1.78	0.92	0.76	42.7
Germany, brown		3.15	0.02	3.06	97.1
Germany		4.77	0.15	4.57	95.8
Czechoslovakia, Bohemia, brown		0.76	0.27	0.46	60.5
Great Britain, Tamworth		4.30	2.11	1.87	43.5
Great Britain, Derbyshire		2.61	1.55	0.87	33.3
Great Britain, Parkgate		3.15	2.71	0.36	11.4
Great Britain, Anthracite		1.06	0.75	0.23	21.7

* Sulfate sulfur values are not recorded in this table. Where the sum of pyritic and organic sulfur is not equal to total sulfur, the difference is sulfate sulfur. In other cases, sulfate sulfur is included with the pyritic sulfur. Organic sulfur by difference.
† Average of two mines.
‡ Total iron in ash calculated to pyrite.
Source: Hower and Parekh 1991.

TABLE 2.16 Ash content and fusion temperature of various coals

Rank	Low Volatile Bituminous		High Volatile Bituminous			Subbituminous	Lignite
Seam	Pocahontas No. 3	No. 9	Pittsburgh	No. 6			
Location	West Virginia	Ohio	West Virginia	Illinois	Utah	Wyoming	Texas
Ash, dry basis, %	12.3	14.10	10.87	17.36	6.6	6.6	12.8
Sulfur, dry basis, %	0.7	3.30	3.53	4.17	0.5	1.0	1.1
Analysis of ash, % by wt							
SiO_2	60.0	47.27	37.64	47.52	48.0	24.0	41.8
Al_2O_3	30.0	22.96	20.11	17.87	11.5	20.0	13.6
TiO_2	1.6	1.00	0.81	0.78	0.6	0.7	1.5
Fe_2O_3	4.0	22.81	29.28	20.13	7.0	11.0	6.6
CaO	0.6	1.30	4.25	5.75	25.0	26.0	17.6
MgO	0.6	0.85	1.25	1.02	4.0	4.0	2.5
Na_2O	0.5	0.28	0.80	0.36	1.2	0.2	0.6
K_2O	1.5	1.97	1.60	1.77	0.2	0.5	0.1
Total	98.8	98.44	95.74	95.20	97.5	86.4	84.3
Ash fusibility							
Initial deformation							
Temperature, °F							
Reducing	2900+	2030	2030	2000	2060	1990	1975
Oxidizing	2900+	2420	2265	2300	2120	2190	2070
Softening temperature, °F							
Reducing		2450	2175	2160		2180	2130
Oxidizing		2605	2385	2430		2220	2190
Hemispherical temperature, °F							
Reducing		2480	2225	2180	2140	2250	2150
Oxidizing		2620	2450	2450	2220	2240	2210
Fluid temperature, °F							
Reducing		2620	2370	2320	2250	2290	2240
Oxidizing		2670	2540	2610	2460	2300	2290

Source: Anon. 1972.

CHAPTER 3

Exploration and Geology

William K. Smith, P.E.

GEOLOGIC TIME

Eon	Era	Period, Subperiod		Epoch	Age estimate of boundaries (Ma)
Phanerozoic	Cenozoic (Cz)	Quaternary (Q)		Holocene	0.010
				Pleistocene	1.6
		Tertiary (T)	Neogene (N)	Pliocene	5
				Miocene	24
			Paleogene (PE)	Oligocene	38
				Eocene	55
				Paleocene	66
	Mesozoic (Mz)	Cretaceous (K)		Late	96
				Early	138
		Jurassic (J)		Late	
				Middle	
				Early	205
		Triassic (₸)		Late	
				Middle	
				Early	~240
	Paleozoic (Pz)	Permian (P)		Late	
				Early	290
		Carboniferous* (C)	Pennsylvanian (℗)	Late	
				Middle	
				Early	~330
			Mississippian (M)	Late	
				Early	360
		Devonian (D)		Late	
				Middle	
				Early	410
		Silurian (S)		Late	
				Middle	
				Early	435
		Ordovician (O)		Late	
				Middle	
				Early	500
		Cambrian (€)		Late	
				Middle	
				Early	~570[†]
Proterozoic (P)		Late Proterozoic (Z)		None defined	900
		Middle Proterozoic (Y)		None defined	1600
		Early Proterozoic (X)		None defined	2500
Archean (A)		Late Archean (W)		None defined	3000
		Middle Archean (V)		None defined	3400
		Early Archean (U)		None defined	3800?
		pre-Archean (pA)[‡]			

* Carboniferous includes both the Mississippian and Pennsylvanian periods. The term is not widely used in the United States. In European usage, the boundary between Early and Late Carboniferous does not correspond to the boundary between Mississippian and Pennsylvanian periods in the United States.
[†] Rocks older than 570 Ma are also called Precambrian (p€), a time term without specific rank.
[‡] pA is an informal time term without specific rank.

Source: Adapted from Hansen 1991.

FIGURE 3.1 Geologic time scale

GEOLOGIC MAP SYMBOLS

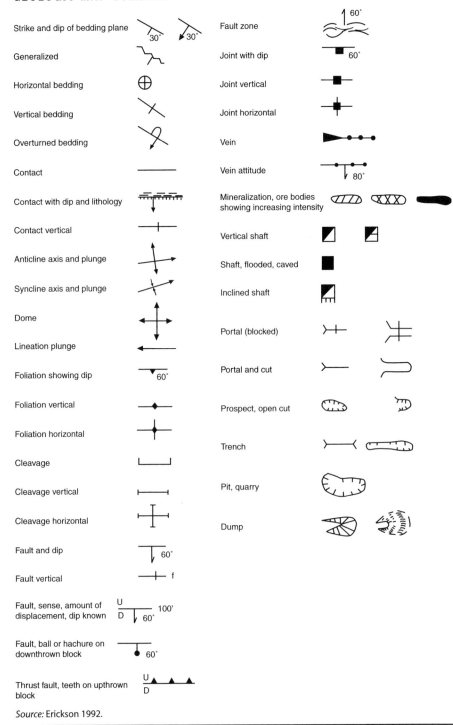

Source: Erickson 1992.

FIGURE 3.2 Geologic map symbols

ROCK CLASSIFICATION

Igneous Rocks

Color		Light				Dark					
						Plagioclase > 2/3 Total Feldspar				Chiefly Pyroxene and/or Olivine	Special Types
Essential Minerals	Feldspar	K-Feldspar > 2/3 Total Feldspar		K-Feldspar 1/3 – 2/3 Total Feldspar		K-Feldspar > 10% Total Feldspar	Sodic Plagioclase		Calcic Plagioclase		
	Quartz	>10%	<10%	>10%	<10%	>10%		<10%			
Chief Accessory Minerals		Hornblende Biotite Muscovite		Hornblende Biotite Pyroxene		Hornblende Biotite Pyroxene			Pyroxene Olivine	Serpentine Iron Ore	Pegmatite Very coarse-grained, normally silicic dike rock, crystals 1 cm to 1 m in size
Phaneritic >0.1 mm	Equigranular	Granite	Syenite	Quartz Monzonite	Monzonite	Granodiorite	Quartz Diorite	Diorite	Gabbro	Peridotite	
Porphyritic	Phaneritic Groundmass	Granite Porphyry	Syenite Porphyry	Quartz Monzonite Porphyry	Monzonite Porphyry	Granodiorite Porphyry	Quartz Diorite Porphyry	Diorite Porphyry	Gabbro Porphyry	Peridotite Porphyry	Aplite Fine-grained, sugary texture
	Aphanitic Groundmass	Rhyolite Porphyry	Trachyte Porphyry	Quartz Latite Porphyry	Latite Porphyry	Dacite Porphyry		Andesite Porphyry	Basalt Porphyry	(Rare)	
Aphanitic <0.1 mm	Microcrystalline	Rhyolite	Trachyte	Quartz Latite	Latite	Dacite		Andesite	Basalt		
	Glassy	Obsidian — dark colored Pitchstone — resinous Vitrophyre — porphyritic Perlite — concentric fractures Pumice — light colored, finely vesicular, floats in water Scoria — dark colored, coarsely vesicular									

Source: Adapted from Travis 1955 *and* U.S. Bureau of Reclamation 1998 *(shown here with permission of Colorado School of Mines).*

FIGURE 3.3 Classification of igneous rocks

Sedimentary Rocks

Grain Size	<4 μm	4 μm–2 mm						Volcanic Ejecta	>2 mm
	Crystalline, Clastic or Amorphous	Clastic — Size Grades: 4–62.5 μm: Silt; 62.5–125 μm: Very fine sand; 125–250 μm: Fine sand; 0.25–0.5 mm: Medium sand; 0.5–1.0 mm: Coarse sand; 1–2 mm: Very coarse sand							Clastic — 2–4 mm: Granules; 4–64 mm: Pebbles; 64–256 mm: Cobbles; >256mm: Boulders
Texture	Crystalline, Clastic, Bioclastic, Oolitic, Etc.	Chiefly Calcite or Dolomite	Chiefly Quartz			Quartz with >25% Feldspar	Quartz, Feldspar, and Rock Fragments		
			>90% Quartz	10–25% Feldspar	>10% Rock Fragments				
Composition of Major Fraction	Composition as indicated in left column for minor fraction								Sedimentary rocks with predominant grain size greater than 2 mm are generally classified as Conglomerate, if the fragments are rounded, or Breccia if the fragments are angular. Attach appropriate modifiers, for example, Quartz Cobble Conglomerate or Limestone Pebble Breccia. Special types are Fanglomerate, indurated alluvial fan deposits, and Tillite, indurated glacial till.
Composition of Minor Fraction									
<10% Minor Fraction — Clay Minerals or clay-size materials (<4 μm)	Clay Minerals or clay-size materials (<4 μm)	Limestone Dolomite	Quartz Sandstone	Feldspathic Sandstone	Lithic Sandstone	Arkose	Graywacke	Ash — unconsolidated fragments < 4 mm	
Clay Minerals or clay-size materials (<4 μm)	Claystone, Siltstone, Mudstone — non-fissile Shale — may include much silt Bentonite — sodium montmorillonite, swells and disaggregates in water	Argillaceous Limestone	Argillaceous Quartz Sandstone	Argillaceous Feldspathic Sandstone	Argillaceous Lithic Sandstone	Argillaceous Arkose	Argillaceous Graywacke	Tuff — consolidated ash Volcanic Breccia — angular fragments > 4 mm	
Silica Opal Chalcedony Quartz Chert	Siliceous Shale Siliceous Claystone Chert Diatomite Radiolarite	Siliceous Limestone Cherty Limestone	Orthoquartzite (Siliceous Quartz Sandstone)	Feldspathic Orthoquartzite	Siliceous Lithic Sandstone	Siliceous Arkose	Siliceous Graywacke	Agglomerate — >25% volcanic bombs	
Calcite or Dolomite	Calcareous Shale Marlstone	Limestone Dolomite	Calcareous Quartz Sandstone	Calcareous Feldspathic Sandstone	Calcareous Lithic Sandstone	Calcareous Arkose	Calcareous Graywacke		
Carbon	Coal Bituminous Anthracite	Carbonaceous Shale Oil Shale	Add appropriate modifier to rock names in vertical columns above, for example, Carbonaceous Limestone or Bituminous Quartz Sandstone. Humus yields carbonaceous derivatives, sapropel yields bituminous derivatives.						
Misc. Phosphate Evaporites Halite Anhydrite Gypsum	Phosphorite Rock Salt Anhydrite Gypsum	Phosphatic Shale, etc.	Add appropriate modifier to rock names in vertical columns above, for example, Phosphatic Limestone.						

Source: Adapted from Travis 1955 and U.S. Bureau of Reclamation 1998 (shown with permission of Colorado School of Mines).

FIGURE 3.4 Classification of sedimentary rocks

Metamorphic Rocks

Color	Chief Minerals	Accessory Minerals	Nondirectional Structure		Directional Structure (Lineate or Foliate)					
			Contact Metamorphism		Mechanical Metamorphism	Regional Metamorphism				Plutonic Metamorphism
						Highly Foliated		Less Foliated		
			Fine	Fine to Coarse	Cataclastic	Slaty	Phyllitic	Schistose	Gneissose	Migmatitic
						Aphanitic	Fine	Fine to Coarse		
Lighter	Feldspar	Actinolite Albite Andalusite Anthophyllite Biotite	Hornfels	Metaquartzite	Formed by crushing and shearing with only minor recrystallization Cataclasite — non-directional, fine-grained	Slate — Most slates are dark colored	Phyllite — Intermediate between slate and schist. Recrystallization of micaceous minerals gives a sheen to the rock but grains are too small for megascopic determination.	Schist — Finely foliated because of parallel orientation of phaneritic flaky or lamellar minerals such as mica.	Gneiss — Coarsely banded, alternating schistose and granulose layers Augen Gneiss — lensoid grains in finer grained groundmass	Migmatite — These rocks have a gneissose, streaked, or irregular structure produced by intimate mixing of metamorphic and magmatic materials. They may originate by injection or by differential fusion. Many migmatites probably originate by partical granitization or by metamorphic differentiation. Migmatites are named by prefixing the rock name of the granitic material to the appropriate root, for example, "granite migmatite," "monzonite injection migmatite," etc.
Darker	Quartz	Chiastolite Chlorite Cordierite Diopside Enstatite Epidote Garnet		Amphibolite	Crush Breccia — non-directional, coarse-grained Mylonite — finely ground, foliate Flaser Granite, Flaser Diorite, Flaser Sandstone, etc. —Flaser structure: lenses or layers of original or relatively unaltered granular minerals surrounded by matrix of highly sheared and crushed material Augen Gneiss — Augen structure					
	Mica									
	Hornblende									
	Chlorite									
Lighter	Actinolite	Gluacophane Graphite Kyanite Muscovite Olivine								
	Tremolite									
	Talc	Phlogopite Pyrophyllite Scapolite Sericite Serpentine Sillimanite Staurolite Tourmaline Tremolite Wollastonite		Soapstone						
	Calcite or Dolomite			Marble						
	Calcsilicates			Skarn						
Darker	Serpentine			Serpentinite				Serpentinite		

Notes: Naming a metamorphic rock consists mainly of prefixing the appropriate structural term with mineral names or an appropriate name. The rock name indicates either the original rock, if recognizable, or the new mineral composition. The prefix "meta," as in "metagabbro," "metasandstone," "metatuff," is applied to rocks that have undergone considerable recrystallization but have largely retained their original fabric. Most of the minerals listed as accessories are genetically important and if present should be included in the rock name regardless of their quantity.

Source: *Adapted from Travis 1955 and U.S. Bureau of Reclamation 1998 (shown here with permission of Colorado School of Mines).*

FIGURE 3.5 Classification of metamorphic rocks

Mohs Hardness Scale

Minerals
1. Talc
2. Gypsum
3. Calcite
4. Fluorite
5. Apatite
6. Orthoclase
7. Quartz
8. Topaz
9. Corundum
10. Diamond.

Common Materials
- Fingernail–2.5
- U.S. bronze cent (pre-1982)–3.0
- Window glass–5.5
- Knife blade–6.0
- Hardened steel file–6.5.

STRUCTURAL GEOLOGY

Fault Classification and Terminology
- Dip—inclination of a plane, measured from the horizontal (dip direction is always perpendicular to the strike)
- Dip slip—component of fault motion in direction of dip, measured in the plane of the fault (cb in Figure 3.6 C)
- Footwall—lower block of an inclined fault
- Hade—complement of the dip angle (obsolete term)
- Hanging wall—upper block of an inclined fault
- Heave—horizontal component of the dip slip, measured in vertical section perpendicular to the strike
- Net slip—total displacement on a fault, measured in the plane of the fault (ab in Figure 3.6 C and E)
- Normal fault—hanging wall moves down relative to footwall; also called gravity fault (indicates extension of the earth's crust; see Figure 3.6 A)
- Rake—angle between a line in a plane and a horizontal line in that plane; sometimes called pitch (angle abc in Figure 3.6 E is the rake of the net slip)
- Reverse fault—hanging wall moves up relative to the footwall; also (usually) called thrust fault, especially if dip is less than 45 degrees (indicates shortening of the earth's crust; see Figure 3.6 D)
- Right lateral, left lateral—descriptions applied to strike-slip component of fault movement; right and left refer to the apparent movement of the block opposite the observer (Figure 3.6 B is left lateral)
- Strike—bearing of a horizontal line in a plane
- Strike slip—horizontal component of fault movement (ac in Figure 3.6 C)
- Strike-slip fault—blocks move horizontally with respect to each other with little or no vertical component of movement (see Figure 3.6 B)
- Throw—vertical component of the net slip and dip slip, measured in vertical section perpendicular to the strike.

EXPLORATION AND GEOLOGY

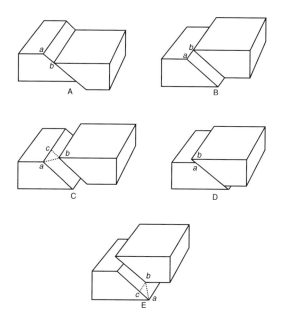

Notes: Net slip, Dip slip, and Strike slip. A. *ab* = net slip = dip slip; strike slip is zero.
B. *ab* = net slip = strike slip; dip slip is zero. C. *ab* = net slip; *cb* = dip slip; *ac* = strike slip.
D. *ab* = net slip = dip slip; strike slip is zero. E. *ab* = net slip; *bc* = strike slip; *ac* = dip slip.

Source: Billings 1954 (reprinted with permission of Pearson Education, Inc., Upper Saddle River, New Jersey).

FIGURE 3.6   Fault terminology

Apparent Dip

In a vertical cross-section drawn at other than a right angle to the strike of a planar feature such as a stratigraphic bed, fault, or vein, the apparent dip in the line of section will be less than the true dip of the plane. The apparent dip is given by

$$\tan \alpha = \tan \delta \sin \theta \qquad (EQ\ 3.1)$$

where:
- α = apparent dip
- δ = true dip
- θ = angle between strike of plane and line of section

NOTE: Equation 3.1 does not apply to projection of linear features such as fault intersections or boreholes, in which the apparent inclination is greater than the true inclination.

Determining the Strike and Dip of a Plane (Three-Point Problem)

To determine the strike and dip of a plane given the location and elevation of three noncolinear points on the plane, proceed as follows (Figure 3.7):

1. Select a suitable scale and plot a plan view of the three points (*A*, *B*, *C*). Label the points with their elevations.
2. Connect the high and the low points with a straight line (*AC*).
3. Compute the proportion $\Delta h_{AB}/\Delta h_{AC}$ or $\Delta h_{BC}/\Delta h_{AC}$, where Δh is the elevation difference between the points indicated by the subscripts.
4. Lay off the proportionate distance of *AC* calculated in step 3 from the appropriate end (*A* or *C*) to locate point *D*. Line *BD* is the strike of the plane.

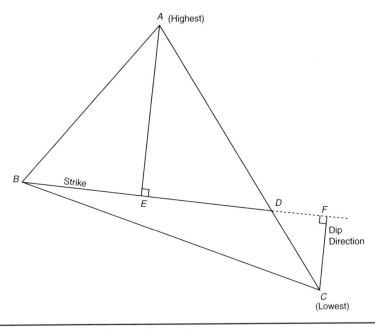

FIGURE 3.7 Determining strike and dip of a plane

5. To determine the dip, draw a line AE or CF perpendicular to BD. Scale the horizontal distance. Using the elevation difference Δh_{AB} or Δh_{BC}, calculate the dip from

$$\tan \delta = \Delta h_{AB}/AE = \Delta h_{BC}/CF \quad \text{(EQ 3.2)}$$

where:
  δ is the angle of dip
The direction of dip can be determined by inspection.

Determining the Intersection of Two Planes

To determine the bearing and plunge of the intersection of two planes given their strikes and dips, proceed as follows (Figure 3.8):

1. Plot the strikes of the two planes so that they intersect at a convenient point (A).
2. Select a suitable scale and a suitable elevation drop Δh (100 ft or 100 m, for example) for constructing structure contours on each plane.
3. For each plane calculate the horizontal distance x corresponding to the selected elevation drop Δh from

$$x = \Delta h/\tan \delta \quad \text{(EQ 3.3)}$$

where:
  δ is the dip of the respective plane
4. Plot lines DE and FG parallel to AB and AC, respectively, at the respective horizontal distances x in the down-dip directions. Label the intersection of DE and FG as point H. Line AH gives the bearing of the line of intersection of the two planes.
5. Scale the horizontal distance AH. Calculate the plunge ρ of the intersection from

$$\tan \rho = \Delta h/AH \quad \text{(EQ 3.4)}$$

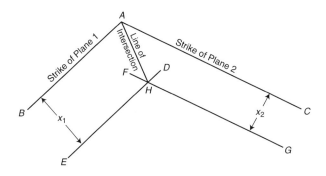

FIGURE 3.8 Determining the intersection of two planes, graphical method

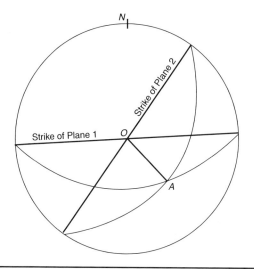

FIGURE 3.9 Determining the intersection of two planes, stereonet method

If a stereonet is available, the bearing and plunge can be determined with a quicker method. Equal-angle (Wulff) nets are preferred to equal-area nets for this purpose. Proceed as follows (Figure 3.9):

1. Plot the strikes and lower hemisphere traces of the two planes on tracing paper laid over the stereonet. The vector OA from the center of the net to the intersection of the lower hemisphere traces represents the intersection of the two planes.
2. Rotate the paper to align the north indexes and read the bearing on the outer circle of the net.
3. Rotate the paper so that OA lies on the east-west line of the stereonet. Determine the plunge by counting the meridian lines inward from the outer circle of the net to point A.

Determination of Strike and Dip from Two Apparent Dips

To determine the strike and dip of a plane given apparent dips in two directions, proceed as follows (Figure 3.10):

1. Plot the apparent dip directions intersecting at point O.
2. Select a suitable scale and elevation drop Δh, such as 100 ft or 100 m.

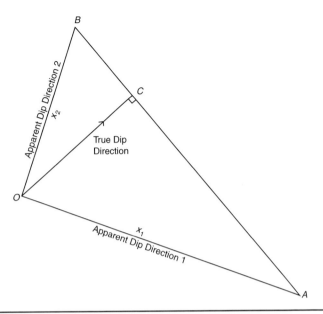

FIGURE 3.10 Strike and dip from two apparent dips, graphical method

3. Calculate and plot horizontal distances OA and OB using

$$x = \Delta h/\tan \alpha \qquad \text{(EQ 3.5)}$$

where:
  α is the respective apparent dip
4. Line AB is the strike of the plane.
5. Plot line OC perpendicular to AB and scale horizontal distance OC. Calculate the dip from

$$\tan \delta = \Delta h/OC \qquad \text{(EQ 3.6)}$$

where:
  δ is the dip of the plane
  The directed line OC is the dip direction.

The resolution of strike and dip from two apparent dips can also be solved using a stereonet, as follows (Figure 3.11):

1. Plot the apparent dip vectors OA and OB on the tracing paper from the center of the net.
2. Rotate the tracing paper about O until points A and B lie on the same meridian line on the net.
3. Plot the strike line of the plane through the north and south poles of the net and trace the meridian line N-A-B-S.
4. Measure the dip of the plane by counting from the outer circle of the net to the trace of the plane.
5. Rotate the tracing paper about O to align the north index on the paper with the north pole of the net.
6. Read the strike of the plane on the outer circle of the net.

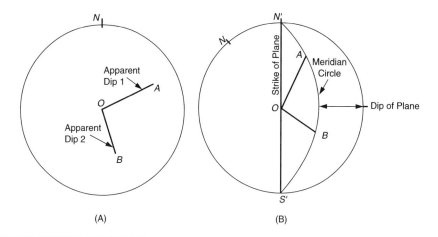

FIGURE 3.11 Strike and dip from two apparent dips, stereonet method

GEOPHYSICS

Plate Tectonics
See Figure 3.12.

Interpretation of Seismic Refraction Surveys
On a plot of arrival times versus distance from the shot point (Figure 3.13), the seismic velocity of successive layers is the inverse slope of the time-distance plot (slope = $1/V$). For the two-horizontal-layer case, the depth to the interface is given by

$$z = \frac{x_c}{2}\sqrt{\frac{V_1 - V_0}{V_1 + V_0}} x_c \qquad \text{(EQ 3.7; Dobrin 1960)}$$

where:
- z = depth to interface
- V_0 = seismic velocity of upper layer
- V_1 = seismic velocity of lower layer
- x_c = critical distance (the distance to the break in slope)

NOTE: A low-velocity layer underlying a higher velocity layer will not be detected by the seismic refraction method and, if present, will introduce errors into depth calculations of deeper layers.

AERIAL PHOTOGRAPHY

The scale of a vertical photograph from the air is given by

$$S = f/H \qquad \text{(EQ 3.8)}$$

where:
- S = the scale of the air photograph at a given elevation
- f = the focal length of the aerial camera
- H = the flying height above the given elevation

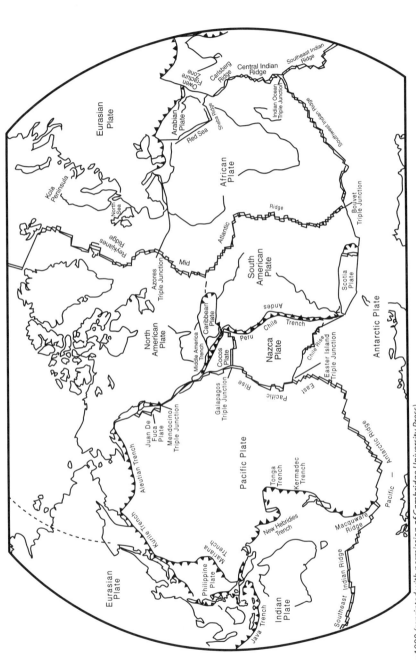

FIGURE 3.12 Plate tectonic map of the world

Source: Fowler 1990 (reprinted with permission of Cambridge University Press).

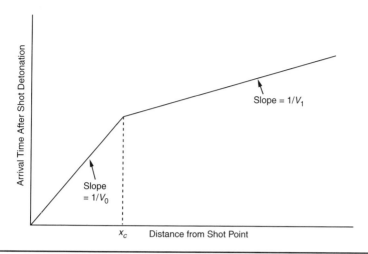

FIGURE 3.13   Idealized time-distance plot from seismic refraction survey

DRILLING

See Chapter 7, which covers sampling and analysis, for information on drilling, including commonly used wireline core drilling sizes and the spacing of drill holes in coalfields.

REFERENCES

Billings, M.P. 1954. *Structural Geology*. 2nd ed. New York: Prentice-Hall.
Dobrin, M.B. 1960. *Introduction to Geophysical Prospecting*. 2nd ed. New York: McGraw-Hill.
Erickson, A.J. Jr. 1992. Geologic data collection and recording. In *SME Mining Engineering Handbook*. 2nd ed., Vol. 1. Edited by H.L. Hartman. Littleton, CO: Society for Mining, Metallurgy, and Exploration, Inc. 288–313.
Fowler, C.M.R. 1990. *The Solid Earth*. Cambridge: Cambridge University Press.
Hansen, W.R. 1991. *Suggestions to Authors of the Reports of the United States Geological Survey*. 7th ed. Washington DC: U.S. Government Printing Office.
Travis, R.B. 1955. Classification of rocks. *Quarterly of the Colorado School of Mines*. 50:1. Golden, CO: Colorado School of Mines.
U.S. Bureau of Reclamation. 1998. *Engineering Geology Field Manual*. 2nd ed., Vol. 1. Washington, DC: U.S. Government Printing Office.

CHAPTER 4
Physical Science and Engineering

R. Karl Zipf, Jr., P.E.

CHEMISTRY

TABLE 4.1 Chemical elements

Name	Symbol	Atomic Number	Atomic Weight
Actinium	Ac	89	(227)
Aluminum	Al	13	26.981538
Americium	Am	95	(243)
Antimony	Sb	51	121.760
Argon	Ar	18	39.948
Arsenic	As	33	74.92160
Astatine	At	85	(210)
Barium	Ba	56	137.327
Berkelium	Bk	97	(247)
Beryllium	Be	4	9.012182
Bismuth	Bi	83	208.98038
Boron	B	5	10.811
Bromine	Br	35	79.904
Cadmium	Cd	48	112.411
Calcium	Ca	20	40.078
Californium	Cf	98	(251)
Carbon	C	6	12.0107
Cerium	Ce	58	140.116
Cesium	Cs	55	132.90545
Chlorine	Cl	17	35.4527
Chromium	Cr	24	51.9961
Cobalt	Co	27	58.933200
Copper	Cu	29	63.546
Curium	Cm	96	(247)
Dysprosium	Dy	66	162.50
Einsteinium	Es	99	(252)
Erbium	Er	68	167.26
Europium	Eu	63	151.964
Fermium	Fm	100	(257)

continues next page

TABLE 4.1 Chemical elements (continued)

Name	Symbol	Atomic Number	Atomic Weight
Fluorine	F	9	18.9984032
Francium	Fr	87	(223)
Gadolinium	Gd	64	157.25
Gallium	Ga	31	69.723
Germanium	Ge	32	72.61
Gold	Au	79	196.96655
Hafnium	Hf	72	178.49
Helium	He	2	4.002602
Holmium	Ho	67	164.93032
Hydrogen	H	1	1.00794
Indium	In	49	114.818
Iodine	I	53	126.90447
Iridium	Ir	77	192.217
Iron	Fe	26	55.845
Krypton	Kr	36	83.80
Lanthanum	La	57	138.9055
Lawrencium	Lr	103	(262)
Lead	Pb	82	207.2
Lithium	Li	3	6.941
Lutetium	Lu	71	174.967
Magnesium	Mg	12	24.3050
Manganese	Mn	25	54.938049
Mendelevium	Md	101	(258)
Mercury	Hg	80	200.59
Molybdenum	Mo	42	95.94
Neodymium	Nd	60	144.24
Neon	Ne	10	20.1797
Neptunium	Np	93	(237)
Nickel	Ni	28	58.6934
Niobium	Nb	41	92.90638
Nitrogen	N	7	14.00674
Nobelium	No	102	(259)
Osmium	Os	76	190.23
Oxygen	O	8	15.9994
Palladium	Pd	46	106.42
Phosphorus	P	15	20.973761
Platinum	Pt	78	195.078
Plutonium	Pu	94	(244)
Polonium	Po	84	(209)
Potassium	K	19	39.0983
Praseodymium	Pr	59	140.90765
Promethium	Pm	61	(145)
Protactinium	Pa	91	231.03588
Radium	Ra	88	(226)
Radon	Rn	86	(222)
Rhenium	Re	75	186.207
Rhodium	Rh	45	102.90550
Rubidium	Rb	37	85.4678

continues next page

TABLE 4.1 Chemical elements (continued)

Name	Symbol	Atomic Number	Atomic Weight
Ruthenium	Ru	44	101.07
Samarium	Sm	62	150.36
Scandium	Sc	21	44.955910
Selenium	Se	34	78.96
Silicon	Si	14	28.0855
Silver	Ag	47	107.8682
Sodium	Na	11	22.989770
Strontium	Sr	38	87.62
Sulfur	S	16	32.066
Tantalum	Ta	73	180.9479
Technetium	Tc	43	(98)
Tellurium	Te	52	127.60
Terbium	Tb	65	158.92534
Thallium	Tl	81	204.3833
Thorium	Th	90	232.0381
Thullium	Tm	69	168.93421
Tin	Sn	50	118.710
Titanium	Ti	22	47.867
Tungsten	W	74	183.84
Uranium	U	92	238.0289
Vanadium	V	23	50.9415
Xenon	Xe	54	131.29
Ytterbium	Yb	70	173.04
Yttrium	Y	39	88.90585
Zinc	Zn	30	65.39
Zirconium	Zr	40	91.224

Source: Lide 1997.

STATICS

Basic Principles and Definitions

Force Force is a vector quantity characterized by a point of application, a magnitude, and a direction. The magnitude and direction of the force resulting from two or more forces may be determined graphically using the parallelogram law or trigonometrically using the law of cosines and the law of sines.

Components Any force acting on a particle or rigid body can be resolved into two or more components that have the same effect on the body. Denoting by θ_x, θ_y, and θ_z, the angles that F forms with the x, y, and z coordinate axes, the rectangular components of F are

$$F_x = F\cos\theta_x \qquad F_y = F\cos\theta_y \qquad F_z = F\cos\theta_z$$

When the rectangular components F_x, F_y, and F_z of a force F are given, the magnitude F of the resultant force is found as

$$F = \sqrt{F_x^2 + F_y^2 + F_z^2}$$

The direction cosines are

$$\cos\theta_x = \frac{F_x}{F} \qquad \cos\theta_y = \frac{F_y}{F} \qquad \cos\theta_z = \frac{F_z}{F}$$

Moments (Couples) A couple is a system of two forces that are equal in magnitude, opposite in direction, parallel to each other, and separated by a perpendicular distance. Moment is a vector defined as the cross product of vector r with components r_x, r_y, and r_z, and force F with components F_x, F_y, and F_z.

$$M = r \times F$$
$$M_x = r_y F_z - r_z F_y$$
$$M_y = r_z F_x - r_x F_z$$
$$M_z = r_x F_y - r_y F_x$$

Equilibrium By resolving each force and each moment into its rectangular components, the necessary and sufficient conditions for equilibrium of a particle or rigid body are the six scalar equations that follow:

$$\sum F_x = 0 \qquad \sum F_y = 0 \qquad \sum F_z = 0$$
$$\sum M_x = 0 \qquad \sum M_y = 0 \qquad \sum M_z = 0$$

Use these equations to determine unknown forces applied to a rigid body or unknown reactions at the supports.

For a two-dimensional body, the above six equations reduce to three as follows:

$$\sum F_x = 0 \qquad \sum F_y = 0 \qquad \sum M_A = 0$$

where the moment is taken about an arbitrary point A in the plane of the structure. These equations can be solved for a maximum of three unknowns. The three equilibrium equations above cannot be augmented by additional equations, but any of them can be replaced by another equation. Alternative equilibrium equations are

$$\sum F_x = 0 \qquad \sum M_A = 0 \qquad \sum M_B = 0$$

where A and B are in a line different from the y direction, and

$$\sum M_A = 0 \qquad \sum M_B = 0 \qquad \sum M_C = 0$$

where A, B, and C are not in a straight line.

Centroids of Lines, Areas, Volumes, and Masses

The centroid of a line is

$$x_{lc} = \left(\sum x_n l_n\right)/L$$
$$y_{lc} = \left(\sum y_n l_n\right)/L$$
$$z_{lc} = \left(\sum z_n l_n\right)/L$$

where:

$L = \sum l_n$
l_n = length of line segment
x_n, y_n, z_n = distance from x, y, and z axes, respectively

The centroid of an area is

$$x_{ac} = \left(\sum x_n a_n\right)/A$$

$$y_{ac} = \left(\sum y_n a_n\right)/A$$

$$z_{ac} = \left(\sum z_n a_n\right)/A$$

where:

$A = \sum a_n$
a_n = elemental area
x_n, y_n, z_n = distance from x, y, and z axes, respectively

The centroid of a volume is

$$x_{vc} = \left(\sum x_n v_n\right)/V$$

$$y_{vc} = \left(\sum y_n v_n\right)/V$$

$$z_{vc} = \left(\sum z_n v_n\right)/V$$

where:

$V = \sum v_n$
v_n = elemental volume
x_n, y_n, z_n = distance from x, y, and z axes, respectively

The moment of area (M_a) is defined as

$$M_{ay} = \sum x_n a_n, \text{ with respect to the } y \text{ axis}$$

$$M_{ax} = \sum y_n a_n, \text{ with respect to the } x \text{ axis}$$

The centroid of a mass is

$$r_c = \sum m_n r_n / \sum m_n$$

where:

r_c = radius vector from reference point to center of mass
m_n = mass of each particle in system
r_n = radius vector from reference point to particle

Moment of Inertia

The moment of inertia or second moment of area is

$$I_x = \int y^2 dA$$
$$I_y = \int x^2 dA$$

The polar moment of inertia of an area A with respect to a pole at O is

$$J_O = \int r^2 dA$$

where:
   r = the distance from O to the element of area dA

Since $r^2 = x^2 + y^2$, then $J_O = I_x + I_y$

The radius of gyration r_O, r_x, r_y is the distance from a reference axis to a point where all the area is imagined to be concentrated to produce the moment of inertia.

$$r_O = \sqrt{J_O/A} \qquad r_x = \sqrt{I_x/A} \qquad I_y = \sqrt{I_y/A}$$

The moment of inertia of an area about any axis parallel to a centroidal axis is

$$I'_x = I_{xc} + d^2 A, \text{ and}$$
$$I'_y = I_{yc} + d^2 A$$

where:
   d = the distance from the centroidal axis to the other axis
   I_{xc}, I_{yc} = the moments of inertia about the centroidal axes

The product of inertia with respect to a particular coordinate system is defined as

$$I_{xy} = \int xy\, dA$$
$$I_{xz} = \int xz\, dA$$
$$I_{yz} = \int yz\, dA$$

Area, Centroid, Moment of Inertia, Section Modulus, and Radius of Gyration for Selected Shapes

Rectangular Section

Area

$$A = bh$$

Centroid

$$x_c = b/2$$
$$y_c = h/2$$

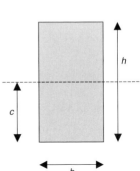

Moment of inertia

$$I = \frac{bh^3}{12}$$

Section modulus

$$\frac{I}{c} = \frac{bh^2}{6}$$

Radius of gyration

$$r_O = \frac{h}{\sqrt{12}}$$

Circular Section
Area

$$A = \frac{\pi}{4}d^2 = \pi r^2$$

Centroid

$$x_c = r$$
$$y_c = r$$

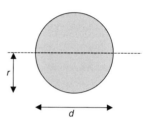

Moment of inertia

$$I = \frac{\pi d^4}{64} = \frac{\pi r^4}{4}$$

Section modulus

$$\frac{I}{c} = \frac{\pi d^3}{32} = \frac{\pi r^3}{4}$$

Radius of gyration

$$r_O = \frac{r}{2}$$

Pipe Section
Area

$$A = \frac{\pi}{4}(D^2 - d^2) = \pi(R^2 - r^2)$$

Centroid

$$x_c = r$$
$$y_c = r$$

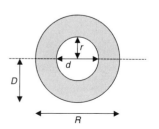

Moment of inertia

$$I = \frac{\pi(D^4 - d^4)}{64} = \frac{\pi(R^4 - r^4)}{4}$$

Section modulus

$$\frac{I}{c} = \frac{\pi(D^4 - d^4)}{32D} = \frac{\pi(R^4 - r^4)}{4R}$$

Radius of gyration

$$r_O = \frac{\sqrt{D^2 + d^2}}{4} = \frac{\sqrt{R^2 + r^2}}{2}$$

"H" Section 1
Area

$$A = HB + hb$$

Centroid

$$x_c = \frac{B + b}{2}$$
$$y_c = \frac{H}{2}$$

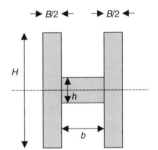

Moment of inertia

$$I = \frac{BH^3 + bh^3}{12}$$

Section modulus

$$\frac{I}{c} = \frac{BH^3 + bh^3}{6H}$$

Radius of gyration

$$r_O = \sqrt{\frac{BH^3 + bh^3}{12(BH + bh)}}$$

"H" Section 2
Area

$$A = HB - hb$$

Centroid

$$x_c = \frac{B}{2}$$
$$y_c = \frac{H}{2}$$

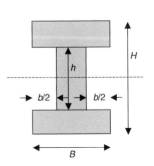

Moment of inertia

$$I = \frac{BH^3 - bh^3}{12}$$

Section modulus

$$\frac{I}{c} = \frac{BH^3 - bh^3}{6H}$$

Radius of gyration

$$r_O = \sqrt{\frac{BH^3 - bh^3}{12(BH - bh)}}$$

Channel Section 1
Area

$$A = HB - b(H - d)$$

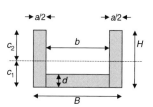

Centroid

$$x_c = \frac{B}{2}$$

$$y_c = \frac{aH^2 + (B-a)d^2}{HB - b(H-d)}$$

Moment of inertia

$$I = \frac{1}{3}(Bc_1^3 - bh^3 + ac_2^3)$$

Section modulus

$$\frac{I}{c} = \frac{(Bc_1^3 - bh^3 + ac_2^3)}{3c}$$

$$c = c_1 \text{ or } c_2$$

$$c_1 = \frac{1}{2}\frac{aH^2 + bd^2}{aH + bd}$$

$$c_2 = H - c_1$$

Radius of gyration

$$r_O = \sqrt{\frac{I}{[Bd + a(H-d)]}}$$

Channel Section 2
Area

$$A = HB - hb$$

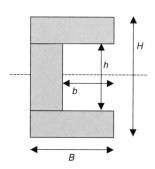

Centroid

$$x_c = \frac{B^2(H-h) + (B-b)^2 h}{2(HB - hb)}$$

$$y_c = \frac{H}{2}$$

Moment of inertia

$$I = \frac{BH^3 - bh^3}{12}$$

Section modulus

$$\frac{I}{c} = \frac{BH^3 - bh^3}{6H}$$

Radius of gyration

$$r_0 = \sqrt{\frac{BH^3 - bh^3}{12(BH - bh)}}$$

Friction

Plane Friction The largest friction force possible on a surface before it starts to move is called the limiting friction and is given by

$$F_s = \mu_s N$$

where:
F_s = static friction force
μ_s = coefficient of static friction
N = normal force between surfaces in contact

Once motion has begun, the magnitude of F may decrease to a lower value given as

$$F_k = \mu_k N$$

where:
F_k = kinetic friction force
μ_k = coefficient of kinetic friction
N = normal force between surfaces in contact

Belt Friction

$$F_1 = F_2 e^{\mu \theta}$$

where:
F_1 = force applied in direction of impending motion
F_2 = force applied to resist impending motion
μ = coefficient of static friction
θ = angle of contact between surfaces in contact in radians

Analysis of Structures

Analysis of Statically Determinant Trusses   A truss is a rigid framework that satisfies the following conditions:
- All members lie in a plane.
- All members are connected at the ends with frictionless pins.
- All applied loads lie in the plane of the truss.
- Reactions and all member forces are determined using equilibrium equations for statically determinant forces.
- Trusses that cannot be analyzed using equilibrium equations are statically indeterminant.

Method of Joints   This method for analysis of trusses begins by solving for all support reactions. Next, the two equilibrium equations are solved at each joint in the truss beginning at the support joints.

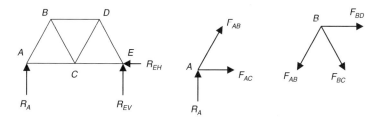

Method of Sections   This method begins by solving for all support reactions. Place an imaginary cut through any portion of the truss so that the unknown forces in particular truss members are exposed as external forces. Next, compute the unknown forces using the equilibrium equations.

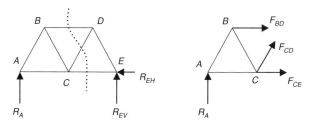

Analysis of Beams

To determine the shear V and bending moment M at a particular cross section of a beam, first determine the support reactions. Make an imaginary cut at the cross section and solve for the shear V and bending moment M.

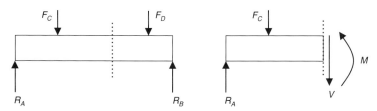

DYNAMICS

Basic Kinematics

Let $r(t)$ be the position of a particle as a function of time. The instantaneous velocity is

$$v = \frac{dr}{dt}$$

The instantaneous acceleration is

$$a = \frac{dv}{dt} = \frac{d^2r}{dt^2} \qquad a = v\frac{dv}{dr}$$

Straight-Line Motion If the acceleration of a body is a constant a_0, then

$$s = s_0 + v_0 t + (a_0 t^2)/2$$
$$v = v_0 + a_0 t$$
$$v^2 = v_0^2 + 2a_0(s - s_0)$$

where:
- s = displacement along straight line
- s_0 = initial position
- v_0 = initial velocity at time $t = 0$
- t = time
- a_0 = constant acceleration
- v = velocity at time t

For a free-falling body, $a_0 = g$, which is the acceleration caused by gravity (32.17 ft/s² or 9.8 m/s²).

For variable acceleration, $a(t)$

$$v = v_0 + \int_0^t a(t)dt$$

For variable velocity, $v(t)$

$$s = s_0 + \int_0^t v(t)dt$$

Plane Circular Motion For rotation of a body about the origin with constant radius r, the angular velocity is

$$\omega = \dot{\theta} = v_t/r$$

Angular acceleration is

$$\alpha = \dot{\omega} = \ddot{\theta} = a_t/r$$
$$s = r\theta$$
$$v_t = r\omega$$

Tangential acceleration is

$$a_t = r\alpha = dv_t/dt$$

Normal acceleration is

$$a_n = v_t^2/r = r\omega^2$$

Newton's Second Law of Motion

$$\sum F = \frac{d(mv)}{dt}$$

where:

$\sum F$ = sum of all applied forces acting on a body

mv = momentum of a body

For a fixed mass:

$$\sum F = m\frac{dv}{dt} = ma$$

Impulse and Momentum

Assuming mass is constant, the equation of motion in the x direction is

$$m\, dv_x/dt = F_x$$
$$m\, dv_x = F_x dt$$
$$m[v_x(t) - v_x(0)] = \int_0^t F_x(t')dt'$$

The left side of this equation is the change in linear momentum, and the right side is the impulse of the force that acts from time 0 to t.

Impact Momentum is conserved while energy may or may not be conserved. For impact with no external forces or dissipation of energy:

$$m_1 v_1 + m_2 v_2 = m_1 v_1' + m_2 v_2'$$

where:

m_1, m_2 = masses of two bodies

v_1, v_2 = velocities before impact

v_1', v_2' = velocities after impact

The relative velocities before and after impact when energy dissipation occurs is

$$v_{1n}' - v_{2n}' = -e(v_{1n} - v_{2n})$$

where:

e = coefficient of restitution for the materials

n = components normal to the plane of impact ($0 \le e \le 1$; $e = 1$ is perfectly elastic; $e = 0$ is perfectly plastic with no rebound)

Knowing e, the velocities after rebound are

$$v_1' = \frac{m_2 v_2(1+e) + (m_1 - em_2)v_1}{m_1 + m_2}$$

$$v_2' = \frac{m_1 v_1(1+e) - (em_1 - m_2)v_2}{m_1 + m_2}$$

Work and Energy

Work (W) is a scalar quantity defined as the integral of the scalar product of the force vector and the force's displacement vector (dr), or

$$W = \int F \cdot dr$$

The kinetic energy (KE) of a particle is the work done accelerating the particle from rest to velocity v:

$$KE = \frac{1}{2}mv^2$$

The change in kinetic energy in going from velocity v_1 to v_2 is

$$KE_2 - KE_1 = \frac{1}{2}mv_2^2 - \frac{1}{2}mv_1^2$$

Potential energy (PE) is the work done by a force acting within a conservative field. In a gravitational field, the potential energy is

$$PE = mgh$$

where:
h = elevation above an arbitrary datum

Elastic potential energy is the recoverable strain energy stored in an elastic body. For a linear elastic spring with modulus k, force F as a function of deformation x is

$$F_s = kx$$

The elastic potential energy stored in the spring is

$$PE = \frac{1}{2}kx^2$$

The change in potential energy in deforming the spring from position x_1 to position x_2 is

$$PE_2 - PE_1 = \frac{1}{2}kx_2^2 - \frac{1}{2}kx_1^2$$

Conservation of Work and Energy

If PE_1 and KE_1 are potential and kinetic energy at state i, conservation of energy for a conservative system, meaning one without energy dissipation, is

$$PE_1 + KE_1 = PE_2 + KE_2$$

If friction is present, the work done by friction W_f is accounted for in the conservation equation as

$$PE_1 + KE_1 + W_f = PE_2 + KE_2$$

MECHANICS OF MATERIALS

Stress

If a force vector ΔP acts on an area ΔA, stress is defined as the limiting value of $\Delta P/\Delta A$ as ΔA goes to zero. In three dimensions, the nine components of stress at a point are

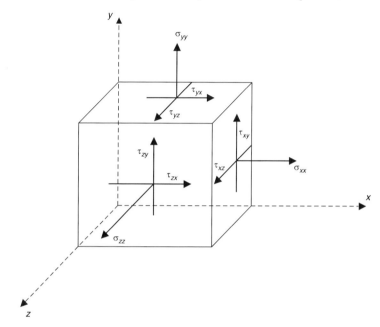

The nine components of stress at a point are represented by the symmetric 3×3 matrix:

$$\begin{bmatrix} \sigma_{xx} & \tau_{xy} & \tau_{xz} \\ \tau_{xy} & \sigma_{yy} & \tau_{yz} \\ \tau_{xz} & \tau_{yz} & \sigma_{zz} \end{bmatrix}$$

Strain

Engineering strain is defined as the change in distance between two points divided by the distance between those points:

$$\varepsilon = \frac{\Delta L}{L_0}$$

Similar to stress, the nine components of strain at a point are written as the following symmetric 3×3 matrix:

$$\begin{bmatrix} \varepsilon_{xx} & \varepsilon_{xy} & \varepsilon_{xz} \\ \varepsilon_{xy} & \varepsilon_{yy} & \varepsilon_{yz} \\ \varepsilon_{xz} & \varepsilon_{yz} & \varepsilon_{zz} \end{bmatrix}$$

Stress–Strain Relations

Typical stress–strain relations for brittle and ductile materials are shown below:

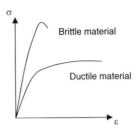

The slope of the initial linear portion of the stress–strain curve is the modulus of elasticity, also known as Young's Modulus. Ductile materials deform elastically at first and then yield. Beyond the yield point, deformation proceeds with little or no increase in applied stress. A large amount of plastic deformation can occur before final rupture. Brittle materials deform elastically at first and then rupture suddenly with relatively little plastic deformation. Most metals, including steel and aluminum as well as many plastics, are examples of ductile materials. Rock, concrete, glass, and ceramics are examples of brittle materials.

For a three-dimensional linear isotropic solid, strain is related to applied stress as follows:

$$\varepsilon_x = \frac{\sigma_x}{E} - \nu\frac{\sigma_y}{E} - \nu\frac{\sigma_z}{E}$$
$$\varepsilon_y = -\nu\frac{\sigma_x}{E} + \frac{\sigma_y}{E} - \nu\frac{\sigma_z}{E}$$
$$\varepsilon_z = -\nu\frac{\sigma_x}{E} - \nu\frac{\sigma_y}{E} + \frac{\sigma_z}{E}$$

$$\gamma_{xy} = \tau_{xy}/G$$
$$\gamma_{yz} = \tau_{yz}/G$$
$$\gamma_{zx} = \tau_{zx}/G$$

In the above equations, E is the modulus of elasticity, elastic modulus, or Young's modulus. E is the proportionality constant relating normal stress to linear strain determined under uniaxial stress conditions.

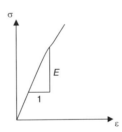

Bodies subject to uniaxial stress also deform in the lateral direction; that is, perpendicular to the applied stress. Poisson's ratio, ν, is defined as the ratio between lateral strain and axial strain under uniaxial stress conditions:

$$\nu = -\frac{\varepsilon_y}{\varepsilon_x} = -\frac{\varepsilon_z}{\varepsilon_x} = \frac{\text{lateral strain}}{\text{axial strain}}$$

The shear modulus G relates resultant shear strain to applied shear stress. For linear elastic isotropic solids, G is found from Young's modulus and Poisson's ratio as

$$G = \frac{E}{2(1+v)}$$

The bulk modulus K relates hydrostatic compressive stress to decrease in volume. For linear elastic isotropic solids, K is found from Young's modulus and Poisson's ratio as

$$K = \frac{E}{3(1-2v)}$$

Uniaxial Loading and Deformation

Given a body with length L and cross-section area A with an applied uniaxial load of P, the axial stress is

$$\sigma_a = \frac{P}{A}$$

In the uniaxial case, all stress components except σ_x are zero. The general three-dimensional stress–strain equations above simplify to an axial strain of

$$\varepsilon_a = \frac{\sigma_a}{E}$$

Lateral strain is

$$\varepsilon_l = -v\frac{\sigma_a}{E}$$

Axial deformation is

$$\delta_a = \varepsilon_a L = \frac{\sigma_a}{E}L = \frac{PL}{AE}$$

$$\text{or } P = \frac{AE}{L}\delta_a = k\delta_a$$

where:

$k = \dfrac{AE}{L}$ is known as stiffness

Plane Stress

In this special case, all stresses in the z direction are zero, such as in a thin plate ($\sigma_z = \tau_{xz} = \tau_{yz} = 0$). The general three-dimensional stress–strain equations become

$$\varepsilon_x = \frac{\sigma_x}{E} - v\frac{\sigma_y}{E}$$

$$\varepsilon_y = -v\frac{\sigma_x}{E} + \frac{\sigma_y}{E}$$

$$\varepsilon_z = -v\frac{\sigma_x}{E} - v\frac{\sigma_y}{E}$$

$$\gamma_{xy} = \tau_{xy}/G$$

Plane Strain

In this special case, strain and shear stress components in the z direction are zero ($\varepsilon_z = \tau_{xz} = \tau_{yz} = 0$). The general three-dimensional stress–strain equations become

$$\sigma_{xx} = \frac{E(1-\nu)}{(1+\nu)(1-2\nu)}\left(\varepsilon_{xx} + \left(\frac{\nu}{1-\nu}\right)\varepsilon_{yy}\right)$$

$$\sigma_{yy} = \frac{E(1-\nu)}{(1+\nu)(1-2\nu)}\left(\varepsilon_{yy} + \left(\frac{\nu}{1-\nu}\right)\varepsilon_{xx}\right)$$

$$\sigma_{zz} = \frac{E(1-\nu)}{(1+\nu)(1-2\nu)}\left(\frac{\nu}{1-\nu}\right)(\varepsilon_{xx} + \varepsilon_{yy})$$

$$\gamma_{xy} = \tau_{xy}/G$$

Torsional Stresses

$$\tau_{max} = \frac{Tc}{J}$$

where:
- T = applied torque or moment
- c = shaft radius
- J = polar moment of inertia
- $J = \pi d^4/32$ for solid shaft
- $J \cong \pi c^3 t$ for thin-walled tube (c = tube diameter and t = wall thickness)

Transformation of Stresses in Two Dimensions

Mohr's circle is a convenient way to compute principal stresses and to transform stresses from one coordinate system to another. Figure 4.1 illustrates construction of Mohr's circle:

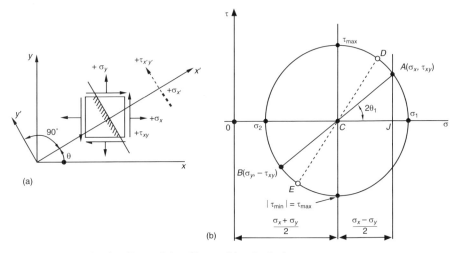

Source: Popov 1968 (reprinted by permission of Pearson Education, Inc.).

FIGURE 4.1 Mohr's circle of stress

Failure Criteria

Maximum Normal Stress Criterion This criterion states that failure occurs when one of the three principal stresses equals the strength of the material. If $\sigma_1 > \sigma_2 > \sigma_3$, this criterion predicts failure when $\sigma_1 \geq S_t$ (the tensile strength of the material), or $\sigma_3 \leq -S_c$ (the compressive strength of the material). This criterion is typically applied to metals and other manufactured materials but not rock.

Maximum Shear Stress Criterion The Mohr-Coulomb failure criterion is frequently applied to rock (and soils), and it assumes that shear failure occurs when the applied shear stress exceeds the strength:

$$\tau = c + \sigma \tan \phi$$

where:

   c = cohesion
   σ = confining stress or normal stress on failure plane
   ϕ = friction angle

The Mohr-Coulomb failure criterion applies both to intact material and discontinuities such as fault surfaces or joint planes. When applied to intact rock, c is the cohesion of the material or inherent shear strength and ϕ is the internal friction angle of the material. When applied to a discontinuity, c is the cohesion of the discontinuity surface and ϕ is the friction angle of that surface.

Analysis of Beams

Bending moment is positive if it deforms the beam concave upward and causes downward deflection. The top fibers are in compression, and the bottom fibers are in tension.

Shearing force is positive if the right portion of a beam tends to move downward with respect to the left portion.

Bending Stresses in Beams Consider a section of an elastic beam in pure flexure.

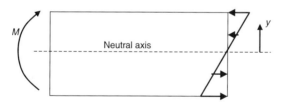

Normal bending stresses in the beam section are

$$\sigma_x = -\frac{M y}{I}$$

where:

   M = moment at the beam section
   y = distance above (+) or below (−) the beam centroid or neutral axis
   I = moment of inertia

Maximum bending stress occurs where

$$|y| = y_{max} = c$$

$$\sigma_{max} = \frac{Mc}{I}$$

Shear Stresses in Beams Maximum shear occurs along the neutral axis of a beam and is

$$\tau_{max} = \frac{3V}{2A}$$

where:
V = total shear force at the section
A = total area of the section

Deflection of Beams The differential equation of the beam deflection curve is

$$EI\frac{d^2y}{dx^2} = M$$

$$EI\frac{d^3y}{dx^3} = \frac{dM(x)}{dx} = V$$

$$EI\frac{d^4y}{dx^4} = \frac{dV(x)}{dx} = -w$$

Double integration of the first relation and application of the appropriate boundary conditions gives the deflection curve. Deflection curves for common beam-loading conditions and end constraints are given in Figure 4.2.

Buckling of Columns
The critical, or Euler, buckling load for a column pinned at both ends is

$$P_{cr} = \frac{\pi^2 EI}{L_{eff}^2}$$

The effective length (L_{eff}) for various column end conditions is shown in Figure 4.3.

Elastic Strain Energy
If strain is within the elastic limit, the work done by loading and deforming a member is transformed into elastic strain energy and can be recovered. If the final load and deflection of the member are P and δ, respectively, the elastic strain energy (SE) is

$$W_{SE} = \frac{P\delta}{2}$$

The strain energy per unit volume (strain energy density) is

$$u = \frac{U}{AL} = \frac{\sigma^2}{2E}$$

Strength Properties of Common Materials
See Chapter 2, which covers material properties, for strength properties of soils and rocks.

Symbols used:

L = length of beam
I = second moment of area
w = load per unit length
W = total load = wL for distributed loads
E = Young's modulus

Maximum bending moment $M_m = k_1 WL$
Maximum slope $i_m = k_2 WL^2/EI$
Maximum deflection $y_m = k_3 WL^3/EI$

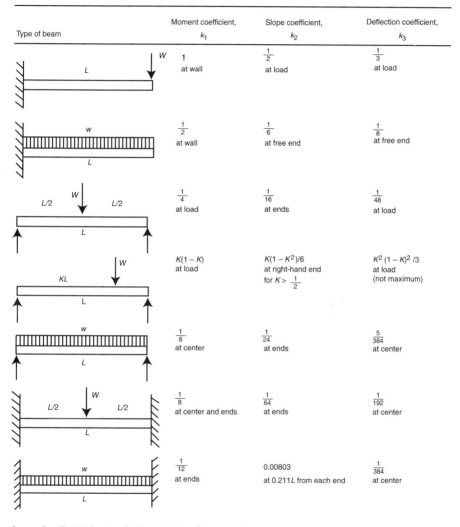

Type of beam	Moment coefficient, k_1	Slope coefficient, k_2	Deflection coefficient, k_3
Cantilever, point load at end	1 at wall	$\frac{1}{2}$ at load	$\frac{1}{3}$ at load
Cantilever, UDL	$\frac{1}{2}$ at wall	$\frac{1}{6}$ at free end	$\frac{1}{8}$ at free end
Simply supported, central point load	$\frac{1}{4}$ at load	$\frac{1}{16}$ at ends	$\frac{1}{48}$ at load
Simply supported, off-center point load	$K(1-K)$ at load	$K(1-K^2)/6$ at right-hand end for $K > \frac{1}{2}$	$K^2(1-K)^2/3$ at load (not maximum)
Simply supported, UDL	$\frac{1}{8}$ at center	$\frac{1}{24}$ at ends	$\frac{5}{384}$ at center
Fixed ends, central point load	$\frac{1}{8}$ at center and ends	$\frac{1}{64}$ at ends	$\frac{1}{192}$ at center
Fixed ends, UDL	$\frac{1}{12}$ at ends	0.00803 at 0.211L from each end	$\frac{1}{384}$ at center

Source: Carvill 1993 (reprinted with permission of Butterworth-Heinemann).

FIGURE 4.2 Maximum moment, slope, and deflection of uniform beams

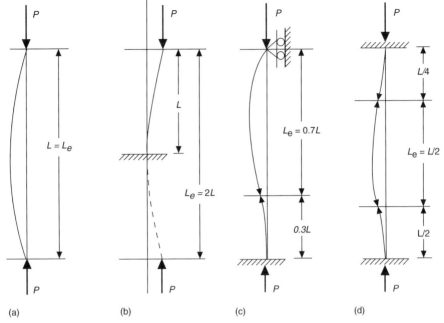

Source: Popov 1968 (reprinted by permission of Pearson Education, Inc.).

FIGURE 4.3 Effective lengths of columns with different restraints

TABLE 4.2 Strength properties of common materials

Material	E (10^6 psi)	E (10^9 Pa)	G (10^6 psi)	G (10^9 Pa)	v
Steel	30.0	207	11.5	83	0.30
Cast iron	14.5	100	6	41.4	0.21
Aluminum	10.0	69	4	28	0.33
Wood	1.6	11	0.6	4.1	0.33
Concrete	3	20.7	—	—	—

Note: E = modulus of elasticity, G = modulus of rigidity, v = Poisson's ratio
Source: Adapted from Popov 1968.

FLUID MECHANICS

For more information about fluids see Chapter 14 on ventilation, Chapter 15 on pumping, and Chapter 18 on site structures and hydrology.

Basic Concepts

$$\text{Mass density} = \rho = \text{mass / volume}$$

$$\text{Specific volume} = \upsilon = 1/\rho$$

$$\text{Specific weight} = \gamma = \rho g$$

$$\text{Specific gravity} = SG = \rho/\rho_{H_2O}$$

Viscosity relates the rate of shearing strain $\frac{du}{dy}$ in a fluid to shear stress (τ) as

$$\tau = \mu \frac{du}{dy}$$

where:
 μ = absolute viscosity, dynamic viscosity, or viscosity of the fluid ($N\text{-}s/m^2$)

The kinematic viscosity is defined as

$$\nu = \frac{\mu}{\rho} \text{ and has units of } m^2/s$$

The Reynolds number is a dimensionless combination of variables defined as

$$Re = \frac{\rho VD}{\mu}$$

where:
 ρ = mass density
 V = fluid velocity
 D = pipe diameter
 μ = absolute viscosity

The Reynolds number relates inertial forces to viscous forces and is used to distinguish laminar from turbulent flow. A Reynolds number above 2,000 usually indicates a fully turbulent flow.

The Froude number is another dimensionless group, and is defined as

$$Fr = \frac{V}{\sqrt{gl}}$$

where:
 V = fluid velocity
 g = acceleration caused by gravity
 l = characteristic length such as fluid depth

The Froude number relates inertial forces to gravitational forces and is used to characterize open channel flow.

Pressure

Pressure variation for an incompressible fluid at rest is

$$\frac{dP}{dh} = \gamma$$

$$P = P_0 + \gamma h$$

where:
 P = pressure
 P_0 = reference pressure
 γ = specific weight
 h = change in elevation

The difference in pressure between two different points is

$$P_2 - P_1 = -\gamma(z_2 - z_1) = -\gamma h$$

where:
z = elevation

Pressure variation for a compressible fluid at rest is

$$P_2 = P_1 \exp\left[-\frac{g(z_2 - z_1)}{RT_o}\right]$$

where:
- g = acceleration caused by gravity
- z = elevation
- R = the gas constant (286.7 J/kg-°K for air)
- T = the temperature (°K)

Hydrostatic force on a plane surface is

$$F_R = \gamma h_c A$$

Pressure Prism

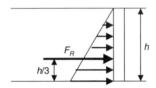

The resultant force acting on a vertical surface extending depth h into a fluid is

$$F_R = P_{ave}A = \gamma\frac{h}{2}A \text{ acting } \frac{h}{3} \text{ above the base}$$

Buoyant force is

$$F_B = \gamma V_B$$

where V_B is the volume of the submerged body

Continuity Equation

The continuity equation for incompressible flow is

$$Q_1 = A_1 V_1 = Q_2 = A_2 V_2$$

where:
- Q = flow quantity (volume/time)
- A = cross-sectional area of flow
- V = flow velocity at points 1 and 2 along the flow path

Fluid Momentum

Force of a flowing fluid is found as

$$F = \rho Q V$$

The resultant force equals the rate of change of fluid momentum or

$$\sum F = \rho_1 Q_1 V_1 - \rho_2 Q_2 V_2$$

where:
$\sum F$ = resultant external force acting on control volume
$\rho_1 Q_1 V_1$ = fluid momentum entering control volume
$\rho_2 Q_2 V_2$ = fluid momentum exiting control volume

Bernoulli Equation

The Bernoulli equation is

$$P + \frac{1}{2}\rho V^2 + \gamma z = \text{constant along streamline}$$

where:
P = pressure
ρ = mass density
V = flow velocity
γ = specific weight (= ρg)
z = elevation above some arbitrary datum point

Another form of the Bernoulli equation is

$$\frac{P}{\gamma} + \frac{V^2}{2g} + z = \text{constant}$$

where:
$\frac{P}{\gamma}$ = pressure head
$\frac{V^2}{2g}$ = velocity head
z = elevation head

The Bernoulli equation between two points along a flow path with no friction losses is

$$\frac{P_1}{\gamma} + \frac{V_1^2}{2g} + z_1 = \frac{P_2}{\gamma} + \frac{V_2^2}{2g} + z_2$$

Friction Losses

The Bernoulli equation between two points along a flow path with friction losses is

$$\frac{P_1}{\gamma} + \frac{V_1^2}{2g} + z_1 = \frac{P_2}{\gamma} + \frac{V_2^2}{2g} + z_2 + h_l$$

In this equation, friction loss is determined with the Darcy-Weisbach equation as

$$h_l = f\frac{l}{D}\frac{V^2}{2g} \text{ or } \Delta P = \rho g h_l = f\frac{l}{D}\frac{\rho V^2}{2}$$

where friction factor f is found from a Moody diagram and depends on the Reynolds number and relative roughness (ε/D) of the pipe. ε is the equivalent roughness of the pipe and D is pipe diameter. See Moody diagram (Figure 4.4) for equivalent roughness for new pipes.

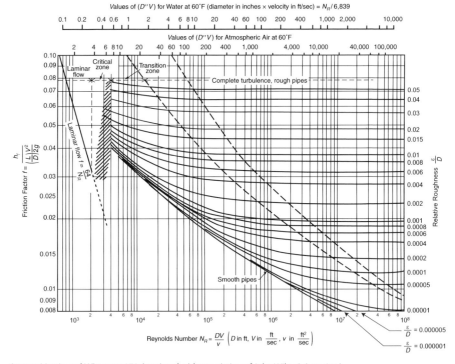

Source: Morris and Wiggert 1972 (reprinted with permission of John Wiley & Sons, Inc.).

FIGURE 4.4 Moody diagram (friction factor for pipes)

For turbulent flow (Re > 2,000), the Colebrooke formula can be used to calculate friction factor f. This formula is the basis for the entire turbulent flow portion of the Moody diagram.

$$\frac{1}{\sqrt{f}} = -2.0\log_{10}\left(\frac{\varepsilon/D}{3.7} + \frac{2.51}{\text{Re}\sqrt{f}}\right)$$

For laminar flow (Re < 2,000), the Poiseville equation applies:

$$Q = \frac{\pi D^4 \Delta P}{128\mu l}$$

where:
- Q = flow quantity
- D = pipe diameter
- ΔP = pressure drop
- μ = absolute viscosity
- l = pipe length

Hydraulic Diameter or Hydraulic Radius

Flow through conduits with noncircular cross sections uses hydraulic diameter or hydraulic radius, defined as

$$R_H = \frac{\text{cross-sectional area}}{\text{wetted perimeter}} = \frac{D_H}{4}$$

Airway Resistance
See Chapter 14, which covers ventilation, for more information.
For airflow the R factor is used to compute pressure losses as

$$\Delta P = RQ^2 = R\frac{V^2}{A^2}$$

where:
- ΔP = pressure drop
- R = airway resistance factor
- Q = flow quantity
- V = flow velocity
- A = flow area

Airway resistance factor R is computed as

$$R = \frac{fl\rho}{2DA^2} = \frac{4\bar{f}l\rho}{2DA^2}$$

where:
- l = pipe length
- ρ = mass density
- D = pipe diameter
- A = flow area

The friction factors f and $\bar{f}$ are related as $f = 4\bar{f}$.

Flow Networks
Pipes and Airways in Series   If three pipes are in series, the flow quantity Q through each is $Q_1 = Q_2 = Q_3$ and the total head loss is $h_{L\text{total}} = h_{L1} + h_{L2} + h_{L3}$.

Pipes and Airways in Parallel   If three pipes (or airways) are in parallel, the total flow quantity is $Q_{\text{total}} = Q_1 + Q_2 + Q_3$ and the head loss in each is $h_{L1} = h_{L2} = h_{L3}$.

Network Flow Analysis   At each junction, $\sum Q = 0$, which is analogous to Kirchoff's 1st law.

Around each loop or flow path, $\sum h_L = 0$, which is analogous to Kirchoff's 2nd law.

Open Channel Flow
The Manning equation describes velocity and flow in open channels:

$$V = \frac{K}{n}R_h^{2/3}S_0^{1/2}$$

$$Q = A\frac{K}{n}R_h^{2/3}S_0^{1/2}$$

where:
- K = 1 in the SI system and 1.49 in English units
- n = Manning coefficient
- R_h = hydraulic radius = A/P
- S_0 = slope of channel
- A = flow area
- P = wetted perimeter

A consistent unit system is required when using the Manning equation (see Table 4.3).

TABLE 4.3 Manning coefficients of channel roughness

Constructed Channel Condition	Values of n		
	Minimum	Maximum	Average
Earth channels, straight and uniform	0.017	0.025	0.0225
Dredged earth channels	0.025	0.033	0.0275
Rock channels, straight and uniform	0.025	0.035	0.033
Rock channels, jagged and irregular	0.035	0.045	0.045
Concrete lined, regular finish	0.012	0.018	0.014
Concrete lined, smooth finish	0.010	0.013	—
Grouted rubble paving	0.017	0.030	—
Corrugated metal	0.023	0.025	0.024
Natural Channel Condition	**Value of n**		
Smoothest natural earth channels, free from growth with straight alignment.	0.017		
Smooth natural earth channels, free from growth, little curvature.	0.020		
Average, well-constructed, moderate-sized earth channels in good condition.	0.0225		
Small earth channels in good condition, or large earth channels with some growth on banks or scattered cobbles in bed.	0.025		
Earth channels with considerable growth, natural streams with good alignment and fairly constant section or large, well-maintained floodway channels.	0.030		
Earth channels considerably covered with small growth, or cleared but not continuously maintained floodways.	0.035		
Mountain streams in clean loose cobbles, rivers with variable cross section and some vegetation growing on banks, or earth channels with thick aquatic growths.	0.050		
Rivers with fairly straight alignment and cross section, badly obstructed by small trees, very little underbrush or aquatic growth.	0.075		
Rivers with irregular alignment and cross section, moderately obstructed by small trees and underbrush.	0.100		
Rivers with fairly regular alignment and cross section, heavily obstructed by small trees and underbrush.	0.100		
Rivers with irregular alignment and cross section, covered with growth of virgin timber and occasional dense patches of bushes and small trees, some logs and dead fallen trees.	0.125		
Rivers with very irregular alignment and cross section, many roots, trees, large logs, and other drift on bottom, trees continually falling into channel due to bank caving.	0.200		

Source: Office of Surface Mining 1982.

THERMODYNAMICS AND HEAT TRANSFER

For more information, see Chapter 14 on ventilation.

Nomenclature

P = absolute pressure (lbf/in.^{2} or Pa)
T = absolute temperature (°R or °K)
v = specific volume (ft^3/lbm or m^3/kg)
u = internal energy (Btu/lbm or kJ/kg)
h = $u + Pv$ = enthalpy (Btu/lbm or kJ/kg)
s = entropy (Btu/(lbm-°R) or kJ/(kg-°K))
$c_P = \left(\dfrac{\partial h}{\partial T}\right)_P$ = heat capacity at constant pressure
$c_v = \left(\dfrac{\partial u}{\partial T}\right)_v$ = heat capacity at constant volume

Ideal Gas Law
The ideal gas law is

$$P = \rho RT$$

where:
- P = absolute pressure
- ρ = density
- R = gas constant
- T = absolute temperature

The behavior of gases undergoing compression or expansion depends on the nature of the process.

R is specific to each gas and is found as

$$R = \frac{\bar{R}}{(\text{molecular weight of gas})}$$

where $\bar{R}$ = universal gas constant = 1,545 ft-lbf/(lbmol-°R) = 8,314 J/(kmol-°K)

For ideal gases,

$$c_p - c_v = R$$

$$\left(\frac{\partial h}{\partial P}\right)_T = 0$$

$$\left(\frac{\partial u}{\partial P}\right)_T = 0$$

At constant temperature (Boyle's law):

$$\frac{P_1}{\rho_1} = \frac{P_2}{\rho_2}$$

At constant pressure (Charles' law):

$$\frac{T_1}{T_2} = \frac{\rho_2}{\rho_1}$$

At constant volume (ρ = constant):

$$\frac{P_1}{P_2} = \frac{T_1}{T_2}$$

For an ideal frictionless compression or expansion with no heat exchange with the surroundings (an isentropic process):

$$\frac{P}{\rho^k} = \text{constant or } \frac{T_2}{T_1} = \left(\frac{P_2}{P_1}\right)^{\frac{k-1}{k}}$$

where k is defined as c_p/c_v or the ratio of specific heat at constant pressure to specific heat at constant volume. ($k \approx 1.4$ for air.)

First Law of Thermodynamics
The First Law of Thermodynamics or conservation of energy states that the net energy crossing a system boundary equals the change in energy inside the system.

For a closed thermodynamic system:

$$Q - W = \Delta U$$

where:
- Q = heat energy transferred because of temperature difference (positive if inward or added to system)
- W = work done by the system (positive if outward)
- ΔU = change in internal energy of the system (positive if an increase)

For an open thermodynamic system in which mass enters and exits the system, the flow work is given by

$$w_{rev} = \int v \, dP$$

The energy balance for an open thermodynamic system is

$$\sum \dot{m}_i [h_i + V_i^2/(2\alpha) + gZ_i] - \sum \dot{m}_e [h_e + V_e^2/(2\alpha) + gZ_e] + \dot{Q}_{in} - \dot{W}_{net} = \frac{d(m_s u_s)}{dt}$$

where:
- $\dot{m}$ = mass flow rate
- α = 1 for turbulent flow and ½ for laminar flow
- $\dot{Q}_{in}$ = rate of heat transfer
- $\dot{W}_{net}$ = rate of work transfer
- m_s = mass within system
- u_s = specific internal energy within system

TABLE 4.4 Summary of ideal gas processes (reversible with constant specific heats)

Process	Isothermal T = constant	Constant Pressure P = constant	Constant Volume v = constant	Isentropic S = constant
PvT relations	Pv = constant	$\dfrac{v}{T}$ = constant	$\dfrac{P}{T}$ = constant	Pv^k = constant Tv^{k-1} = constant $\dfrac{P^{\frac{k-1}{k}}}{T}$ = constant
Nonflow work $-\int P \, dv =$	$Pv \ln \dfrac{v_f}{v_i}$	$P\Delta v$	0	$nc_v \Delta T$
Steady-flow work $-\int v \, dP =$	$Pv \ln \dfrac{P_i}{P_f}$	0	$v\Delta P$	$nc_p \Delta T$
Heat $\int T \, dS =$	$Pv \ln \dfrac{P_i}{P_f}$	$nc_p \Delta T$	$nc_v \Delta T$	0
$\Delta U =$	0	$nc_v \Delta T$	$nc_v \Delta T$	$nc_v \Delta T$
$\Delta H =$	0	$nc_p \Delta T$	$nc_p \Delta T$	$nc_p \Delta T$
$\Delta S =$	$nR \ln \dfrac{P_i}{P_f}$	$nc_p \ln \dfrac{T_f}{T_i}$	$nc_v \ln \dfrac{T_f}{T_i}$	0

Source: Developed from Reynolds and Perkins 1970.

Conduction
Fourier's Law of Conduction is

$$\dot{Q} = -kA\left(\frac{dT}{dx}\right)$$

where:
$\dot{Q}$ = rate of heat transfer

For conduction through a plane wall:

$$\dot{Q} = -kA(T_2 - T_1)/L$$

where:
k = thermal conductivity of wall
A = surface area of wall
L = wall thickness
T_1 = temperature on near side of wall
T_2 = temperature on far side of wall

Thermal resistance of a wall is

$$R = L/(kA)$$

For composite walls, thermal resistances in series are added as

$$R_{total} = R_1 + R_2$$

Convection
Convective heat transfer is determined as

$$\dot{Q} = hA(T_w - T_\infty)$$

where:
h = convective heat transfer coefficient
A = heat transfer area
T_w = wall temperature
T_∞ = bulk fluid temperature

Radiation
Radiant heat transfer is given by

$$\dot{Q} = \varepsilon \sigma A T^4$$

where:
ε = emissivity of the body
σ = 5.67×10^{-8} W/(m²-°K⁴) or 0.173×10^{-8} Btu/(h-ft²-°R⁴)
A = surface area of the body
T = absolute temperature (°R or °K)

ELECTRICITY AND MAGNETISM
See Chapter 16 on power for more information.

Basic Definitions

Electromotive force (E) in volts is

$$E = \frac{W}{Q}$$

where:
- W = energy or work (joules)
- Q = charge quantity (coulombs)

Current (I) in amperes is

$$I = \frac{Q}{t} \text{ or charge quantity per unit time (seconds)}$$

Resistance (R) in ohms is

$$R = \frac{\rho l}{A}$$

where:
- ρ = resistivity in ohm-m, which is a material property
- l = length of resistor along direction of current flow
- A = cross-sectional area perpendicular to current flow

Voltage drop (V) through a resistor is

$$V = IR \text{ (Ohm's law)}$$

Power (P) or energy per unit time in watts is

$$P = VI$$

Power dissipation in a resistor is

$$P = I^2 R = \frac{V^2}{R} \text{ (Joule's law)}$$

Resistors, Capacitors, and Inductors in Series and in Parallel

Resistors in series:

$$R_T = \sum_{i=1}^{n} R_i$$

Resistors in parallel:

$$\frac{1}{R_T} = \sum_{i=1}^{n} R_i$$

Capacitance in farads is the energy stored in an electric field and is defined as

$$C = \frac{Q}{V}$$

The energy stored in a capacitor is

$$W = \frac{CV^2}{2}$$

Capacitors in series:

$$\frac{1}{C_T} = \sum_{i=1}^{n} C_i$$

Capacitors in parallel:

$$C_T = \sum_{i=1}^{n} C_i$$

In an alternating current circuit, the capacitive reactance (X_C) or reactance is

$$X_C = \frac{1}{2\pi f C}$$

where f is the frequency in hertz

The time required to charge a capacitor connected to a direct current source (called the time constant) is

$$t = RC$$

Inductance L in henrys is the energy stored in a magnetic field. In an alternating current circuit, the inductive reactance (X_L) or impedance is

$$X_L = 2\pi f L$$

Inductors in series:

$$L_T = \sum_{i=1}^{n} L_i$$

Inductors in parallel:

$$\frac{1}{L_T} = \sum_{i=1}^{n} L_i$$

The energy stored in an inductor is

$$W = \frac{LI^2}{2}$$

The time constant for an RL circuit is

$$t = \frac{L}{R}$$

Kirchoff's Laws

The algebraic sum of all currents entering a junction is zero.

$$\sum I_{in} = \sum I_{out}$$

The algebraic sum of the potential drops around any closed loop in a conductor network is zero.

$$\sum V_{rises} = \sum V_{drops}$$

REFERENCES

Carvill, J. 1993. *Mechanical Engineer's Data Handbook.* Boca Raton, FL: CRC Press.

Lide, D.R., ed. 1997. *Handbook of Chemistry and Physics.* 78th ed. New York: CRC Press.

Morris, H.M., and J.M. Wiggert. 1972. *Applied Hydraulics in Engineering.* New York: Ronald Press Company.

Office of Surface Mining. 1982. *Surface Mining Water Diversion Design Manual.* OSM/TR-82/2. Prepared under contract J5101050 by Simons, Li & Associates, Inc. Washington, DC: Office of Surface Mining. 4.11.

Popov, E.P. 1968. *Introduction to Mechanics of Solids.* Englewood Cliffs, NJ: Prentice Hall.

Reynolds, W.C., and H.P. Perkins. 1970. *Engineering Thermodynamics.* New York: McGraw-Hill.

CHAPTER 5

Mathematics, Statistics, and Probability

R. Karl Zipf, Jr., P.E.

ELEMENTARY ANALYSIS

Basic Coordinate Relations

Distance Between Two Points

$$P_1(x_1, y_1) \text{ and } P_2(x_2, y_2)$$

$$d = \sqrt{(x_2 - x_1)^2 + (y_2 - y_1)^2}$$

Slope m of Line Joining Two Points

$$P_1(x_1, y_1) \text{ and } P_2(x_2, y_2)$$

$$m = \frac{y_2 - y_1}{x_2 - x_1} = \tan\theta$$

Equation of Line Joining Two Points

$$P_1(x_1, y_1) \text{ and } P_2(x_2, y_2)$$

$$y = mx + b$$

$$m = \frac{y_2 - y_1}{x_2 - x_1}$$

$$b = y_1 - mx_1$$

Area of Triangle With Vertices at

$$P_1(x_1, y_1), P_2(x_2, y_2) \text{ and } P_3(x_3, y_3)$$

$$\text{Area} = \pm\frac{1}{2}(x_1 y_2 - x_1 y_3 + x_2 y_3 - x_2 y_1 + x_3 y_1 - x_3 y_2)$$

Distance Between Points

$$P_1(x_1, y_1, z_1) \text{ and } P_2(x_2, y_2, z_2)$$

$$d = \sqrt{(x_2 - x_1)^2 + (y_2 - y_1)^2 + (z_2 - z_1)^2}$$

General Equation of a Plane

$$Ax + By + Cz + D = 0$$

Equation of Plane Passing Through Points

$$\begin{vmatrix} x-x_1 & y-y_1 & z-z_1 \\ x_2-x_1 & y_2-y_1 & z_2-z_1 \\ x_3-x_1 & y_3-y_1 & z_3-z_1 \end{vmatrix} = 0$$

Vectors

Components of a Vector

$$A = A_1 i + A_2 j + A_3 k$$
$$B = B_1 i + B_2 j + B_3 k$$

Addition and Subtraction

$$A + B = (A_1 + B_1)i + (A_2 + B_2)j + (A_3 + B_3)k$$
$$A - B = (A_1 - B_1)i + (A_2 - B_2)j + (A_3 - B_3)k$$

Dot or Scalar Product

$$A \cdot B = AB \cos\theta = A_1 B_1 + A_2 B_2 + A_3 B_3$$

where θ = angle between A and B

Cross or Vector Product

$$A \times B = \begin{vmatrix} i & j & k \\ A_1 & A_2 & A_3 \\ B_1 & B_2 & B_3 \end{vmatrix} = (A_2 B_3 - A_3 B_2)i + (A_3 B_1 - A_1 B_3)j + (A_1 B_2 - A_2 B_1)k$$

Triple Product

$$A \cdot (B \times C) = \begin{vmatrix} A_1 & A_2 & A_3 \\ B_1 & B_2 & B_3 \\ C_1 & C_2 & C_3 \end{vmatrix} = A_1 B_2 C_3 + A_2 B_3 C_1 + A_3 B_1 C_2 - A_3 B_2 C_1 - A_2 B_1 C_3 - A_1 B_3 C_2$$

Regression Analysis

The best-fit linear relationship through n (x,y) pairs is

$$Y = mX + b$$
$$m = \frac{SPXY}{SSX} \qquad b = \bar{Y} - m\bar{X}$$

where:

$$\bar{X} = \frac{\sum x}{n} \qquad \bar{Y} = \frac{\sum y}{n}$$

$$SPXY = \sum (x - \bar{X})(y - \bar{Y}) = \sum xy - n\bar{X}\bar{Y}$$

$$SSX = \sum(x-\bar{X})^2 = \sum x^2 - n\bar{X}^2$$

$$SSY = \sum(y-\bar{Y})^2 = \sum y^2 - n\bar{Y}^2$$

The correlation coefficient is

$$r_{xy} = \frac{SPXY}{[(SSX)(SSY)]^{1/2}}$$

ALGEBRA

Basic Laws

Commutative $a + b = b + a$; $ab = ba$
Associative $a + (b + c) = (a + b) + c$; $a(bc) = (ab)c$
Distributive $c(a + b) = ca + cb$

Special Products and Factors

$$(x + y)^2 = x^2 + 2xy + y^2 \,;\, (x - y)^2 = x^2 - 2xy + y^2$$

$$(x + y)^3 = x^3 + 3x^2y + 3xy^2 + y^3 \,;\, (x - y)^3 = x^3 - 3x^2y + 3xy^2 - y^3$$

Binomial Formula for Positive Integer n

$$(x + y)^n = x^n + nx^{n-1}y + \frac{n(n-1)}{2!}x^{n-2}y^2 + \frac{n(n-1)(n-2)}{3!}x^{n-3}y^3 + \ldots + y^n$$

Factorial n

$$n! = 1 \cdot 2 \cdot 3 \ldots n$$

Proportion

If $\frac{a}{b} = \frac{c}{d}$, then

$$\frac{a+b}{b} = \frac{c+d}{d},$$

$$\frac{a-b}{b} = \frac{c-d}{d},$$

$$\frac{a-b}{a+b} = \frac{c-d}{c+d}$$

Basic Equations

Equation of Straight Line   The general form of the equation is $Ax + By + C = 0$; the standard form of the equation is $y = mx + b$.

Quadratic Equation

$$ax^2 + bx + c = 0$$

$$\text{roots} = \frac{-b \pm \sqrt{b^2 - 4ac}}{2a}$$

if: $b^2 - 4ac > 0$, two real roots,
$b^2 - 4ac = 0$, two equal roots,
$b^2 - 4ac < 0$, two complex conjugate roots

Equation of Circle

$$(x-h)^2 + (y-k)^2 = r^2$$

where r = radius of circle with center at (h,k)

Radius of Circle

$$r = \sqrt{(x-h)^2 + (y-k)^2}$$

Matrices

A matrix is a rectangular array of numbers with m rows and n columns. Element a_{ij} is in row i and column j.

Multiplication   If matrix $A = (a_{ik})$ is an $m \times n$ matrix and $B = (b_{kj})$ is an $n \times s$ matrix; the matrix product AB is an $m \times s$ matrix:

$$C = (c_{ij}) = \left(\sum_{l=1}^{n} a_{il} b_{lj} \right)$$

where n is the common integer representing the number of columns of A and the number of rows of B (l and $k = 1, 2, ..., n$).

Addition   If $A = (a_{ij})$ and $B = (b_{ij})$ are two matrices of the same size $m \times n$, the sum $A + B$ is an $m \times n$ matrix $C = (c_{ij})$ where $c_{ij} = a_{ij} + b_{ij}$.

Identity   The matrix $I = (a_{ij})$ is a square $n \times n$ identity matrix when $a_{ii} = 1$ for $i = 1, 2, ..., n$ and $a_{ij} = 0$ for $i \neq j$.

Transpose   The matrix B is the transpose of the matrix A if each entry b_{ji} in B is the same as the entry a_{ij} in A. $B = A^T$.

Inverse   The inverse B of a square $n \times n$ matrix A is $B = A^{-1}$ such that $AB = I$.

Determinants   A determinant of order n consists of n^2 numbers arranged in n rows and n columns and enclosed by two vertical lines.

For a second-order determinant:

$$\begin{vmatrix} a_1 & a_2 \\ b_1 & b_2 \end{vmatrix} = a_1 b_2 - a_2 b_1$$

For a third-order determinant:

$$\begin{vmatrix} a_1 & a_2 & a_3 \\ b_1 & b_2 & b_3 \\ c_1 & c_2 & c_3 \end{vmatrix} = a_1 b_2 c_3 + a_2 b_3 c_1 + a_3 b_1 c_2 - a_3 b_2 c_1 - a_2 b_1 c_3 - a_1 b_3 c_2$$

TRIGONOMETRY

Definitions
Triangle ABC is a right triangle with $C = 90°$ and sides of length a, b, c. The trigonometric functions of angle A are defined as

sine of $A = \sin A = \dfrac{a}{c} = \dfrac{\text{opposite}}{\text{hypotenuse}}$

cosine of $A = \cos A = \dfrac{b}{c} = \dfrac{\text{adjacent}}{\text{hypotenuse}}$

tangent of $A = \tan A = \dfrac{a}{b} = \dfrac{\text{opposite}}{\text{adjacent}}$

cotangent of $A = \cot A = \dfrac{b}{a} = \dfrac{\text{adjacent}}{\text{opposite}}$

secant of $A = \sec A = \dfrac{c}{b} = \dfrac{\text{hypotenuse}}{\text{adjacent}}$

cosecant of $A = \csc A = \dfrac{c}{a} = \dfrac{\text{hypotenuse}}{\text{opposite}}$

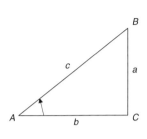

$\sin z = \dfrac{e^{iz} - e^{-iz}}{2i} \quad (z = x + iy)$

$\cos z = \dfrac{e^{iz} + e^{-iz}}{2}$

$\tan z = \dfrac{\sin z}{\cos z}$

$\csc z = \dfrac{1}{\sin z}$

$\sec z = \dfrac{1}{\cos z}$

$\cot z = \dfrac{1}{\tan z}$

Periodic Properties

$\sin(z + 2k\pi) = \sin z \quad (k = \text{any integer})$
$\cos(z + 2k\pi) = \cos z$
$\tan(z + 2k\pi) = \tan z$
$\sin^2 z - \cos^2 z = 1$
$\sec^2 z - \tan^2 z = 1$
$\csc^2 z - \cot^2 z = 1$

Negative Angle Formulas

$\sin(-z) = -\sin z$
$\cos(-z) = \cos z$
$\tan(-z) = -\tan z$

Addition Formulas

$$\sin(z_1 + z_2) = \sin z_1 \cos z_2 + \cos z_1 \sin z_2$$
$$\cos(z_1 + z_2) = \cos z_1 \cos z_2 - \sin z_1 \sin z_2$$
$$\tan(z_1 + z_2) = \frac{\tan z_1 + \tan z_2}{1 - \tan z_1 \tan z_2}$$
$$\cot(z_1 + z_2) = \frac{\cot z_1 \cot z_2 - 1}{\cot z_2 + \cot z_1}$$

Half-Angle Formulas

$$\sin\frac{z}{2} = \pm\left(\frac{1 - \cos z}{2}\right)^{1/2}$$
$$\cos\frac{z}{2} = \pm\left(\frac{1 + \cos z}{2}\right)^{1/2}$$
$$\tan\frac{z}{2} = \pm\left(\frac{1 - \cos z}{1 + \cos z}\right)^{1/2} = \frac{1 - \cos z}{\sin z} = \frac{\sin z}{1 + \cos z}$$

Multiple-Angle Formulas

$$\sin 2z = 2\sin z \cos z = \frac{2\tan z}{1 + \tan^2 z}$$
$$\cos 2z = 2\cos^2 z - 1 = 1 - 2\sin^2 z$$
$$\cos 2z = \cos^2 z - \sin^2 z = \frac{1 - \tan^2 z}{1 + \tan^2 z}$$

Products of Sines and Cosines

$$2\sin z_1 \sin z_2 = \cos(z_1 - z_2) - \cos(z_1 + z_2)$$
$$2\cos z_1 \cos z_2 = \cos(z_1 - z_2) + \cos(z_1 + z_2)$$
$$2\sin z_1 \cos z_2 = \sin(z_1 - z_2) + \sin(z_1 + z_2)$$

Addition and Subtraction of Two Circular Functions

$$\sin z_1 + \sin z_2 = 2\sin\left(\frac{z_1 + z_2}{2}\right)\cos\left(\frac{z_1 - z_2}{2}\right)$$
$$\sin z_1 - \sin z_2 = 2\cos\left(\frac{z_1 + z_2}{2}\right)\sin\left(\frac{z_1 - z_2}{2}\right)$$
$$\cos z_1 + \cos z_2 = 2\cos\left(\frac{z_1 + z_2}{2}\right)\cos\left(\frac{z_1 - z_2}{2}\right)$$
$$\cos z_1 - \cos z_2 = -2\sin\left(\frac{z_1 + z_2}{2}\right)\sin\left(\frac{z_1 - z_2}{2}\right)$$
$$\tan z_1 \pm \tan z_2 = \frac{\sin(z_1 \pm z_2)}{\cos z_1 \cos z_2}$$
$$\cot z_1 \pm \cot z_2 = \frac{\sin(z_1 \pm z_2)}{\sin z_1 \sin z_2}$$

GEOMETRY

Right Triangle

$A + B = C = 90°$
$c^2 = a^2 + b^2$ (Pythagorean theorem)
Area $= \dfrac{1}{2}ab$

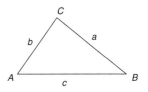

General Triangle

$A + B + C = 180°$
$c^2 = a^2 + b^2 - 2ab\cos C$ (law of cosines)
$\dfrac{\sin A}{a} = \dfrac{\sin B}{b} = \dfrac{\sin C}{c}$ (law of sines)
Area $= \dfrac{1}{2}ab\sin C$

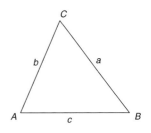

Rectangle

$A = B = C = D = 90°$
Area $= a \cdot b$
Diagonal $= p = \sqrt{a^2 + b^2}$

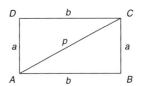

Trapezoid

Area $= \dfrac{1}{2}(a + b)h$

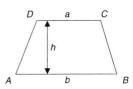

Parallelogram

$A = C,\; B = D,\; A + B = 180°$
Area $= bh = ab\sin A = ab\sin B$
$h = a\sin A = a\sin B$
$p = \sqrt{a^2 + b^2 - 2ab\cos A}$
$q = \sqrt{a^2 + b^2 - 2ab\cos B}$

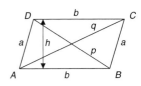

General Quadrilateral

Area $= \dfrac{1}{2}pq\sin\theta$

Area $= \sqrt{(s-a)(s-b)(s-c)(s-d) - abcd\cos^2\left(\dfrac{A+B}{2}\right)}$

where $s = \dfrac{1}{2}(a + b + c + d)$

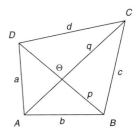

Circle of Radius r

Area = πr^2

Perimeter = $2\pi r$

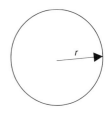

Ellipse

Area = $\pi(OA)(OC) = \dfrac{\pi}{4}(AB)(CD)$

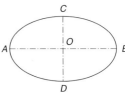

Sphere

Volume = $\dfrac{4}{3}\pi r^3 = \dfrac{1}{6}\pi d^3$

Area = $4\pi r^2 = \pi d^2$

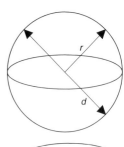

Cylinder

Volume = $\pi r^2 L$

Area = $\pi dL + 2\left(\dfrac{\pi}{4}d^2\right)$

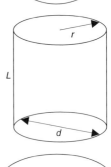

Cone

Volume = $\dfrac{1}{3}\pi r^2 h$

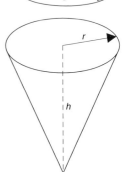

Frustum of Cone

Volume = $\frac{1}{3}\pi h(B^2 + b^2 + Bb)$

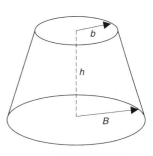

Cube

Volume = a^3
Diagonal = $d = a\sqrt{3}$
Area = $6a^2$

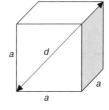

Rectangular Prism

Volume = abc
Diagonal = $d = \sqrt{a^2 + b^2 + c^2}$
Area = $2(ab + bc + ca)$

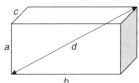

Pyramid

Volume = $\frac{1}{3}abh$

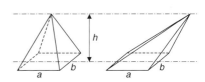

Frustum of Pyramid

Volume = $\frac{1}{3}h\left(ab + cd + \sqrt{(ab)(cd)}\right)$

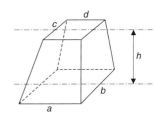

Planar Areas by Approximation

Divide the planar area into n strips by equidistant parallel chords of length $y_0, y_1, y_2, ..., y_n$ (where y_0 and y_n may be zero), and let h denote the common distance between chords.

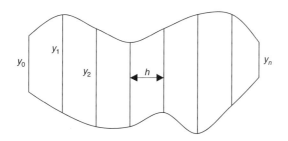

Trapezoidal rule:

$$\text{Area} = h\left(\frac{1}{2}y_0 + y_1 + y_2 + \ldots + y_{n-1} + \frac{1}{2}y_n\right)$$

Simpson's rule (n even):

$$\text{Area} = \frac{1}{3}h(y_0 + 4y_1 + 2y_2 + 4y_3 + 2y_4 + \ldots + 4y_{n-2} + 2y_{n-1} + y_n)$$

DERIVATIVES

Basic Definitions
If $y = f(x)$, the derivative of y or $f(x)$ with respect to x is defined as

$$\frac{dy}{dx} = \lim_{h \to 0}\frac{f(x+h) - f(x)}{h} = \lim_{\Delta x \to 0}\frac{f(x + \Delta x) - f(x)}{\Delta x}$$

where $h = \Delta x$. The derivative is also denoted by y', df/dx, or $f'(x)$. The process of taking a derivative is called differentiation.

General rules of differentiation: in the following equations u, v, and w are functions of x; a, b, c, and n are constants (restricted as indicated). All angles are in radians.

$$\frac{d}{dx}(c) = 0$$

$$\frac{d}{dx}(cx) = c$$

$$\frac{d}{dx}(cx^n) = ncx^{n-1}$$

$$\frac{d}{dx}(u \pm v \pm w \pm \ldots) = \frac{du}{dx} \pm \frac{dv}{dx} \pm \frac{dw}{dx} \pm \ldots$$

$$\frac{d}{dx}(cu) = c\frac{du}{dx}$$

$$\frac{d}{dx}(uv) = u\frac{dv}{dx} + v\frac{du}{dx}$$

$$\frac{d}{dx}(uvw) = uv\frac{dw}{dx} + uw\frac{dv}{dx} + vw\frac{du}{dx}$$

$$\frac{d}{dx}\left(\frac{u}{v}\right) = \frac{v(du/dx) - u(dv/dx)}{v^2}$$

$$\frac{d}{dx}(u^n) = nu^{n-1}\frac{du}{dx}$$

$$\frac{dy}{dx} = \frac{dy}{du}\frac{du}{dx} \quad \text{(Chain rule)}$$

$$\frac{du}{dx} = \frac{1}{dx/du}$$

$$\frac{dy}{dx} = \frac{dy/du}{dx/du}$$

$$\text{Second derivative} = \frac{d}{dx}\left(\frac{dy}{dx}\right) = \frac{d^2y}{dx^2} = f''(x) = y''$$

$$\text{Third derivative} = \frac{d}{dx}\left(\frac{d^2y}{dx^2}\right) = \frac{d^3y}{dx^3} = f'''(x) = y'''$$

$$n\text{th derivative} = \frac{d}{dx}\left(\frac{d^{n-1}y}{dx^{n-1}}\right) = \frac{d^ny}{dx^n} = f^{(n)}(x) = y^{(n)}$$

Maxima, Minima, and Inflection Points of Functions

$y = f(x)$ is a maximum for $x = a$, if $f'(a) = 0$ and $f''(a) < 0$

$y = f(x)$ is a minimum for $x = a$, if $f'(a) = 0$ and $f''(a) > 0$

$y = f(x)$ has an inflection point at $x = a$, if $f''(a) = 0$ and $f''(x)$ changes sign as x increases through $x = a$

Derivatives of Common Functions

$$\frac{d(\log_a u)}{dx} = (\log_a e)\frac{1}{u}\frac{du}{dx}$$

$$\frac{d(\ln u)}{dx} = \frac{1}{u}\frac{du}{dx}$$

$$\frac{d(a^u)}{dx} = (\ln a)a^u\frac{du}{dx}$$

$$\frac{d(e^u)}{dx} = e^u\frac{du}{dx}$$

$$\frac{d(u^v)}{dx} = vu^{v-1}\frac{du}{dx} + (\ln u)u^v\frac{dv}{dx}$$

$$\frac{d(\sin u)}{dx} = \cos u\frac{du}{dx}$$

$$\frac{d(\cos u)}{dx} = -\sin u\frac{du}{dx}$$

$$\frac{d(\tan u)}{dx} = \sec^2 u\frac{du}{dx}$$

$$\frac{d(\cot u)}{dx} = -\csc^2 u\frac{du}{dx}$$

$$\frac{d(\sec u)}{dx} = \sec u\tan u\frac{du}{dx}$$

$$\frac{d(\csc u)}{dx} = -\csc u\cot u\frac{du}{dx}$$

$$\frac{d(\sin^{-1} u)}{dx} = \frac{1}{\sqrt{1-u^2}}\frac{du}{dx} \quad (-\pi/2 \le \sin^{-1} u \le \pi/2)$$

$$\frac{d(\cos^{-1} u)}{dx} = \frac{1}{\sqrt{1-u^2}}\frac{du}{dx} \quad (0 \le \cos^{-1} u \le \pi)$$

$$\frac{d(\tan^{-1} u)}{dx} = \frac{1}{1+u^2}\frac{du}{dx} \quad (-\pi/2 < \tan^{-1} u < \pi/2)$$

$$\frac{d(\cot^{-1} u)}{dx} = \frac{1}{1+u^2}\frac{du}{dx} \quad (0 < \cot^{-1} u < \pi)$$

INTEGRALS

The fundamental theorem of integral calculus is

$$\lim_{n \to \infty} \sum_{i=1}^{n} f(x_i)\Delta x_i = \int_a^b f(x)dx$$

$\Delta x_i \to 0$ for all i

$$\int df(x) = f(x)$$

$$\int dx = x$$

$$\int af(x)dx = a\int f(x)dx$$

$$\int [u(x) \pm v(x)]dx = \int u(x)dx \pm \int v(x)dx$$

$$\int x^m dx = \frac{x^{m+1}}{m+1} \quad (m \neq -1)$$

$$\int u(x)dv(x) = u(x)v(x) - \int v(x)du(x)$$

$$\int \frac{dx}{ax+b} = \frac{1}{a}\ln|ax+b|$$

$$\int \frac{dx}{\sqrt{x}} = 2\sqrt{x}$$

$$\int a^x dx = \frac{a^x}{\ln a}$$

$$\int \sin x\, dx = -\cos x$$

$$\int \cos x\, dx = \sin x$$

$$\int \sin^2 x\, dx = \frac{x}{2} - \frac{\sin 2x}{4}$$

$$\int \frac{dx}{ax^2+c} = \frac{1}{\sqrt{ac}}\tan^{-1}\left(x\sqrt{\frac{a}{c}}\right) \quad (a>0, c>0)$$

$$\int \frac{dx}{ax^2+bx+c} = \frac{2}{\sqrt{4ac-b^2}}\tan^{-1}\left(\frac{2ax+b}{\sqrt{4ac-b^2}}\right) \quad (4ac-b^2>0)$$

$$\int \frac{dx}{ax^2+bx+c} = \frac{1}{\sqrt{b^2-4ac}}\ln\left|\frac{2ax+b-\sqrt{b^2-4ac}}{2ax+b+\sqrt{b^2-4ac}}\right| \quad (b^2-4ac>0)$$

$$\int \frac{dx}{ax^2+bx+c} = -\frac{2}{2ax+b} \quad (b^2-4ac=0)$$

$$\int \cos^2 x\, dx = \frac{x}{2} + \frac{\sin 2x}{4}$$

$$\int x \sin x\, dx = \sin x - x\cos x$$

$$\int x \cos x\, dx = \cos x + x\sin x$$

$$\int \sin x \cos x\, dx = (\sin^2 x)/2$$

$$\int \tan x\, dx = -\ln|\cos x| = \ln|\sec x|$$

$$\int \cot x\, dx = -\ln|\csc x| = \ln|\sin x|$$

$$\int \tan^2 x\, dx = \tan x - x$$

$$\int \cot^2 x\, dx = -\cot x - x$$

$$\int e^{ax} dx = (1/a)e^{ax}$$

$$\int xe^{ax} dx = \frac{e^{ax}}{a^2}(ax-1)$$

$$\int \ln x\, dx = x[\ln(x)-1] \quad (x>0)$$

$$\int \frac{dx}{a^2+x^2} = \frac{1}{a}\tan^{-1}\frac{x}{a} \quad (a \neq 0)$$

PROBABILITY AND STATISTICS

Permutations and Combinations

A permutation is a particular ordered sequence from a given set of objects.

A combination is the set itself without reference to order.

The number of different permutations of n objects taken r at a time is

$$P(n, r) = \frac{n!}{(n-r)!}$$

The number of different combinations of n objects taken r at a time is

$$C(n, r) = \frac{P(n, r)}{r!} = \frac{n!}{r!(n-r)!}$$

Laws of Probability

Probability of an Event $P(E)$ is a real number in the range 0 to 1. The probability of an impossible event is 0 and that of a certain event is 1.

Law of Total Probability

$$P(A + B) = P(A) + P(B) - P(A,B)$$

where:

$P(A+B)$ = probability that either A or B occur alone or that both occur together
$P(A)$ = probability that A occurs
$P(B)$ = probability that B occurs
$P(A,B)$ = probability that both A and B occur simultaneously

Law of Compound or Joint Probability

If neither $P(A)$ nor $P(B)$ is zero, then $P(A,B) = P(A)P(B|A) = P(B)P(A|B)$

where:

$P(B|A)$ = probability that B occurs given that A has already occurred
$P(A|B)$ = probability that A occurs given that B has already occurred

If either $P(A)$ or $P(B)$ is zero, then $P(A,B) = 0$

Means

If $a_1, a_2, ..., a_n$ represent the values of n items or observations from a population, the means of these items or observations are

Arithmetic mean:

$$\bar{A} = \frac{a_1 + a_2 + ... + a_n}{n} = \frac{1}{n}\sum_{i=1}^{n} a_i$$

$\bar{A} \to \mu$, μ = population mean for sufficiently large n

Weighted arithmetic mean:

$$\bar{A}_w = \frac{\sum w_i a_i}{\sum w_i}$$

where w_i is the weight applied to the a_i value.

Geometric mean:

$$\bar{G} = (a_1 a_2 ... a_n)^{\frac{1}{n}} \quad (a_k > 0, k = 1, 2, ..., n)$$

Harmonic mean:

$$\frac{1}{\bar{H}} = \frac{1}{n}\left(\frac{1}{a_1} + \frac{1}{a_2} + \ldots + \frac{1}{a_n}\right) \qquad (a_k > 0, k = 1, 2, \ldots, n)$$

Standard Deviation

The variance of the observations is the arithmetic mean of the squared deviations from the population mean μ.

Population variance:

$$\sigma^2 = \frac{1}{n}[(a_1 - \mu)^2 + (a_2 - \mu)^2 + \ldots + (a_n - \mu)^2] = \frac{1}{n}\sum_{i=1}^{n}(a_i - \mu)^2$$

Standard deviation of population:

$$\sigma = \sqrt{\frac{1}{n}\sum_{i=1}^{n}(a_i - \mu)^2}$$

Sample variance:

$$s^2 = \frac{1}{n-1}\sum_{i=1}^{n}(a_i - \bar{A})^2$$

Sample standard deviation:

$$s = \sqrt{\frac{1}{n-1}\sum_{i=1}^{n}(a_i - \bar{A})^2}$$

Confidence Intervals

Confidence intervals for mean μ of normal distribution Z:

Standard Deviation σ Known

$$\bar{A} - Z_{\alpha/2}\frac{\sigma}{\sqrt{n}} \leq \mu \leq \bar{A} + Z_{\alpha/2}\frac{\sigma}{\sqrt{n}}$$

Standard Deviation σ Not Known

$$\bar{A} - t_{\alpha/2}\frac{s}{\sqrt{n}} \leq \mu \leq \bar{A} + t_{\alpha/2}\frac{s}{\sqrt{n}}$$

where $t_{\alpha/2}$ corresponds to $n - 1$ degrees of freedom.

Confidence intervals for difference in means μ_1 and μ_2 of normal distribution Z:

Standard Deviation σ_1 and σ_2 Known

$$\bar{A}_1 - \bar{A}_2 - Z_{\alpha/2}\sqrt{\frac{\sigma_1^2}{n_1} + \frac{\sigma_2^2}{n_2}} \leq \mu_1 - \mu_2 \leq \bar{A}_1 - \bar{A}_2 + Z_{\alpha/2}\sqrt{\frac{\sigma_1^2}{n_1} + \frac{\sigma_2^2}{n_2}}$$

MATHEMATICS, STATISTICS, AND PROBABILITY

Standard Deviation σ_1 and σ_2 Not Known

$$\bar{A}_1 - \bar{A}_2 - t_{\alpha/2}\sqrt{\frac{\left(\frac{1}{n_1} + \frac{1}{n_2}\right)[(n_1-1)S_1^2 + (n_2-1)S_2^2]}{n_1+n_2-2}} \leq \mu_1 - \mu_2$$

$$\leq \bar{A}_1 - \bar{A}_2 + t_{\alpha/2}\sqrt{\frac{\left(\frac{1}{n_1} + \frac{1}{n_2}\right)[(n_1-1)S_1^2 + (n_2-1)S_2^2]}{n_1+n_2-2}}$$

where $t_{\alpha/2}$ corresponds to $n_1 + n_2 - 2$ degrees of freedom.

Normal or Gaussian Distribution

For a unit normal distribution, $\mu = 0$ and $\sigma = 1$.

$$Z(x) = \frac{1}{\sqrt{2\pi}} e^{-x^2/2}$$

Area under curve from $-\infty$ to x

$$P(x) = \int_{-\infty}^{x} Z(t)\,dt$$

Area under curve from x to ∞

$$Q(x) = \int_{x}^{\infty} Z(t)\,dt$$

Area under curve between $-x$ and x

$$A(x) = \int_{-x}^{x} Z(t)\,dt$$

$$P(x) + Q(x) = 1$$
$$P(-x) = Q(x)$$
$$A(x) = 2P(x) - 1$$

Chi-Square Probability Function

Let $X_1, X_2, \ldots, X_\nu$ be independent and identically distributed random variables, each following a normal distribution with mean zero and unit variance. Then

$$X^2 = \sum_{i=1}^{\nu} X_i^2$$

is said to follow the chi-square distribution with ν degrees of freedom and the probability that $X^2 \leq \chi^2$ is given by $P(\chi^2|\nu)$.

$$P(\chi^2|\nu) = \left[2^{\nu/2}\Gamma\left(\frac{\nu}{2}\right)\right]^{-1} \int_{0}^{\chi^2} (t)^{\frac{\nu}{2}-1} e^{-\frac{t}{2}}\,dt \qquad (0 \leq \chi^2 < \infty)$$

$$Q(\chi^2|\nu) = 1 - P(\chi^2|\nu) = \left[2^{\nu/2}\Gamma\left(\frac{\nu}{2}\right)\right]^{-1} \int_{\chi^2}^{\infty} (t)^{\frac{\nu}{2}-1} e^{-\frac{t}{2}}\,dt \qquad (0 \leq \chi^2 < \infty)$$

TABLE 5.1 Normal probability distribution

x	P(x)	x	P(x)	x	P(x)	x	P(x)
0.00	.50000	0.90	.81594	1.80	.96407	2.70	.99653
0.02	.50798	0.92	.82121	1.82	.96562	2.72	.99674
0.04	.51595	0.94	.82639	1.84	.96712	2.74	.99693
0.06	.52392	0.96	.83147	1.86	.96856	2.76	.99711
0.08	.53188	0.98	.83646	1.88	.96995	2.78	.99728
0.10	.53983	1.00	.84134	1.90	.97128	2.80	.99744
0.12	.54776	1.02	.84614	1.92	.97257	2.82	.99760
0.14	.55567	1.04	.85083	1.94	.97381	2.84	.99774
0.16	.56356	1.06	.85543	1.96	.97500	2.86	.99788
0.18	.57142	1.08	.85993	1.98	.97615	2.88	.99801
0.20	.57926	1.10	.86433	2.00	.97725	2.90	.99813
0.22	.58706	1.12	.86864	2.02	.97831	2.92	.99825
0.24	.59483	1.14	.87286	2.04	.97932	2.94	.99836
0.26	.60257	1.16	.87698	2.06	.98030	2.96	.99846
0.28	.60126	1.18	.88100	2.08	.98124	2.98	.99856
0.30	.61791	1.20	.88493	2.10	.98214	3.00	.99865
0.32	.62552	1.22	.88877	2.12	.98300	3.05	.99886
0.34	.63307	1.24	.89251	2.14	.98382	3.10	.99903
0.36	.64058	1.26	.89617	2.16	.98461	3.15	.99918
0.38	.64803	1.28	.89973	2.18	.98537	3.20	.99931
0.40	.65542	1.30	.90320	2.20	.98610	3.25	.99942
0.42	.66276	1.32	.90658	2.22	.98679	3.30	.99952
0.44	.67003	1.34	.90988	2.24	.98745	3.35	.99960
0.46	.67724	1.36	.91309	2.26	.98809	3.40	.99966
0.48	.68439	1.38	.91621	2.28	.98870	3.45	.99972
0.50	.69146	1.40	.91924	2.30	.98928	3.50	.99977
0.52	.69847	1.42	.92220	2.32	.98983	3.55	.99981
0.54	.70540	1.44	.92507	2.34	.99036	3.60	.99984
0.56	.71226	1.46	.92785	2.36	.99086	3.65	.99987
0.58	.71904	1.48	.93056	2.38	.99134	3.70	.99989
0.60	.72575	1.50	.93319	2.40	.99180	3.75	.99991
0.62	.73237	1.52	.93574	2.42	.99224	3.80	.99993
0.64	.73891	1.54	.93822	2.44	.99266	3.85	.99994
0.66	.74537	1.56	.94062	2.46	.99305	3.90	.99995
0.68	.75175	1.58	.94295	2.48	.99343	3.95	.99996
0.70	.75804	1.60	.94520	2.50	.99379	4.00	.99997
0.72	.76424	1.62	.94738	2.52	.99413		
0.74	.77035	1.64	.94950	2.54	.99446		
0.76	.77637	1.66	.95154	2.56	.99477		
0.78	.78230	1.68	.95352	2.58	.99506		
0.80	.78814	1.70	.95543	2.60	.99534		
0.82	.79389	1.72	.95728	2.62	.99560		
0.84	.79955	1.74	.95907	2.64	.99585		
0.86	.80511	1.76	.96080	2.66	.99609		
0.88	.81057	1.78	.96246	2.68	.99632		

Source: Pearson and Hartley 1954 *(reprinted with the permission of Biometrica Trustees).*

F-(Variance Ratio) Distribution Function

If X_1^2 and X_2^2 are independent random variables following the chi-square distribution with v_1 and v_2 degrees of freedom, respectively, the distribution of $F = \dfrac{X_1^2/v_1}{X_2^2/v_2}$ is said to follow the variance ratio or F-distribution with v_1 and v_2 degrees of freedom. The distribution function is

$$P(F|v_1, v_2) = \frac{v_1^{1/2v_1} v_2^{1/2v_2}}{B\left(\frac{1}{2}v_1, \frac{1}{2}v_2\right)} \int_0^F t^{1/2(v_1-2)} (v_2 + v_1 t)^{-1/2(v_1+v_2)} dt \quad (F \geq 0)$$

$$Q(F|v_1, v_2) = 1 - P(F|v_1, v_2) = I\left(\frac{v_2}{2}, \frac{v_1}{2}\right)$$

where $x = \dfrac{v_2}{v_2 + v_1 F}$

Student's t-Distribution

If X is a random variable following a normal distribution with mean zero and variance unity, and χ^2 is a random variable following an independent chi-square distribution with v degrees of freedom, the distribution of the ratio $\dfrac{X}{\sqrt{\chi^2/v}}$ is called Student's t-distribution with v degrees of freedom. The probability that $\dfrac{X}{\sqrt{\chi^2/v}}$ will be less in absolute value than a fixed constant t is

$$A(t|v) = P_T\left\{\left|\frac{X}{\sqrt{\chi^2/v}}\right| \leq t\right\} = \left[\sqrt{v}B\left(\frac{1}{2}, \frac{v}{2}\right)\right]^{-1} \int_{-t}^{t} (1 + \chi^2/v)^{-\frac{v+1}{2}} dx = 1 - I_x\left(\frac{v}{2}, \frac{1}{2}\right) \quad (0 \leq t < \infty)$$

where $x = \dfrac{v}{v + i^2}$

MISCELLANEOUS

Critical Path Method

Critical path method (CPM) is a deterministic method and requires only an estimate of the activity duration. Using CPM requires
1. Listing and sequencing project tasks
2. Estimating task duration
3. Evaluating the project.

TABLE 5.2 Percentage points of the χ^2-distribution (values of χ^2 in terms of Q and v)

v\Q	0.995	0.990	0.975	0.950	0.900	0.100	0.050	0.025	0.010	0.005
1	0.00004	0.00016	0.00098	0.00393	0.01579	2.706	3.841	5.024	6.635	7.879
2	0.0100	0.0201	0.0506	0.103	0.211	4.605	5.991	7.378	9.210	10.60
3	0.0717	0.115	0.216	0.352	0.584	6.251	7.815	9.348	11.35	12.84
4	0.207	0.297	0.484	0.711	1.064	7.779	9.488	11.14	13.28	14.86
5	0.412	0.554	0.831	1.145	1.610	9.236	11.07	12.83	15.09	16.75
6	0.676	0.872	1.237	1.635	2.204	10.64	12.59	14.45	16.81	18.55
7	0.989	1.239	1.690	2.167	2.833	12.02	14.07	16.01	18.48	20.28
8	1.344	1.646	2.180	2.733	3.490	13.36	15.51	17.54	20.09	21.96
9	1.735	2.088	2.700	3.325	4.168	14.68	16.92	19.02	21.67	23.59
10	2.156	2.558	3.247	3.940	4.865	15.99	18.31	20.48	23.21	25.19
11	2.603	3.053	3.816	4.575	5.578	17.28	19.68	21.92	24.73	26.76
12	3.074	3.571	4.404	5.226	6.304	18.55	21.03	23.34	26.22	28.30
13	3.565	4.107	5.009	5.892	7.042	19.81	22.36	24.74	27.69	29.82
14	4.075	4.660	5.629	6.571	7.790	21.06	23.69	26.12	29.14	31.32
15	4.601	5.229	6.262	7.261	8.547	22.31	25.00	27.49	30.58	32.80
16	5.142	5.812	6.908	7.962	9.312	23.54	26.30	28.85	32.00	34.27
17	5.697	6.408	7.564	8.672	10.09	24.77	27.59	30.19	33.41	35.72
18	6.265	7.015	8.231	9.390	10.86	25.99	28.87	31.53	34.81	37.16
19	6.844	7.633	8.907	10.12	11.65	27.20	30.14	32.85	36.19	38.58
20	7.434	8.260	9.591	10.85	12.44	28.41	31.41	34.17	37.57	40.00
21	8.034	8.897	10.28	11.59	13.24	29.62	32.67	35.48	38.93	41.40
22	8.643	9.542	10.98	12.34	14.04	30.81	33.92	36.78	40.29	42.80
23	9.260	10.20	11.69	13.09	14.85	32.01	35.17	38.08	41.64	44.18
24	9.886	10.86	12.40	13.85	15.66	33.20	36.42	39.36	42.98	45.56
25	10.52	11.52	13.12	14.61	16.47	34.38	37.65	40.65	44.31	46.93
26	11.16	12.20	13.84	15.38	17.29	35.56	38.89	41.92	45.64	48.29
27	11.81	12.88	14.57	16.15	18.11	36.74	40.11	43.19	46.96	49.65
28	12.46	13.57	15.31	16.93	18.94	37.92	41.34	44.46	48.28	50.99
29	13.12	14.26	16.05	17.71	19.77	39.08	42.56	45.72	49.59	52.34
30	13.79	14.95	16.79	18.49	20.60	40.26	43.77	46.98	50.89	53.67
40	20.71	22.16	24.43	26.51	29.05	51.81	55.76	59.34	63.69	66.77
50	27.99	29.71	32.36	34.76	37.69	63.17	67.50	71.42	76.15	79.49
60	35.53	37.48	40.48	43.19	46.46	74.40	79.08	83.30	88.38	91.95
70	43.28	45.44	48.76	51.74	55.33	85.53	90.53	95.02	100.4	104.2
80	51.17	53.54	57.15	60.39	64.28	96.58	101.9	106.6	112.3	116.3
90	59.20	61.75	65.65	69.13	73.29	107.6	113.1	118.1	124.1	128.3
100	67.33	70.06	74.22	77.93	82.36	118.5	124.3	129.6	135.8	140.1

Source: Pearson and Hartley 1954 *(reprinted with the permission of Biometrica Trustees).*

MATHEMATICS, STATISTICS, AND PROBABILITY | 121

TABLE 5.3 Percentage points of Student's *t*-distribution (values of *t* in terms of *A* and *v*)

v\A	0.200	0.500	0.800	0.900	0.950	0.980	0.990	0.995	0.998	0.999
1	.325	1.000	3.078	6.314	12.70	31.82	63.66	127.3	318.3	636.6
2	.289	.816	1.886	2.920	4.303	6.965	9.925	14.09	22.33	31.60
3	.277	.765	1.638	2.353	3.182	4.541	5.841	7.453	10.21	12.92
4	.271	.741	1.533	2.132	2.776	3.747	4.604	5.598	7.173	8.610
5	.267	.727	1.476	2.015	2.571	3.365	4.032	4.773	5.893	6.869
6	.265	.718	1.440	1.943	2.447	3.143	3.707	4.317	5.208	5.959
7	.263	.711	1.415	1.895	2.365	2.998	3.499	4.029	4.785	5.408
8	.262	.706	1.397	1.860	2.306	2.896	3.355	3.833	4.501	5.041
9	.261	.703	1.383	1.833	2.262	2.821	3.250	3.690	4.297	4.781
10	.260	.700	1.372	1.812	2.228	2.764	3.169	3.581	4.144	4.587
11	.260	.697	1.363	1.796	2.201	2.718	3.106	3.497	4.025	4.437
12	.259	.695	1.356	1.782	2.179	2.681	3.055	3.428	3.930	4.318
13	.259	.694	1.350	1.771	2.160	2.650	3.012	3.372	3.852	4.221
14	.258	.692	1.345	1.761	2.145	2.624	2.977	3.326	3.787	4.140
15	.258	.691	1.341	1.753	2.131	2.602	2.947	3.286	3.733	4.073
16	.258	.690	1.337	1.746	2.120	2.583	2.921	3.252	3.686	4.015
17	.257	.689	1.333	1.740	2.110	2.567	2.898	3.223	3.646	3.965
18	.257	.688	1.330	1.734	2.101	2.552	2.878	3.197	3.610	3.922
19	.257	.688	1.328	1.729	2.093	2.539	2.861	3.174	3.579	3.883
20	.257	.687	1.325	1.725	2.086	2.528	2.845	3.153	3.552	3.850
21	.257	.686	1.323	1.721	2.080	2.518	2.831	3.135	3.527	3.819
22	.256	.686	1.321	1.717	2.074	2.508	2.819	3.119	3.505	3.792
23	.256	.685	1.319	1.714	2.069	2.500	2.807	3.104	3.485	3.768
24	.256	.685	1.318	1.711	2.064	2.492	2.797	3.090	3.467	3.745
25	.256	.684	1.316	1.708	2.060	2.485	2.787	3.078	3.450	3.725
26	.256	.684	1.315	1.706	2.056	2.479	2.779	3.067	3.435	3.707
27	.256	.684	1.314	1.703	2.052	2.473	2.771	3.057	3.421	3.690
28	.256	.683	1.313	1.701	2.048	2.467	2.763	3.047	3.408	3.674
29	.256	.683	1.311	1.699	2.045	2.462	2.756	3.038	3.396	3.659
30	.256	.683	1.310	1.697	2.042	2.457	2.750	3.030	3.385	3.646
40	.255	.681	1.303	1.684	2.021	2.423	2.704	2.971	3.307	3.551
60	.254	.679	1.296	1.671	2.000	2.390	2.660	2.915	3.232	3.460
120	.254	.677	1.289	1.658	1.980	2.358	2.617	2.860	3.160	3.373
∞	.253	.674	1.282	1.645	1.960	2.326	2.576	2.807	3.090	3.291

Source: Pearson and Hartley 1954 *(reprinted with the permission of Biometrica Trustees).*

TABLE 5.4 Percentage points of the F-distribution

[values of F in terms of Q, v_1, v_2
$Q(F|v_1, v_2) = 0.05$ (95% Confidence Level)]

$v_2\backslash v_1$	1	2	3	4	5	6	8	12	15	20	30	60	∞
1	161.4	199.5	215.7	224.6	230.2	234.0	238.9	243.9	245.9	248.0	250.1	252.2	254.3
2	18.51	19.00	19.16	19.25	19.30	19.33	19.37	19.41	19.43	19.45	19.46	19.48	19.50
3	10.13	9.55	9.28	9.12	9.01	8.94	8.85	8.74	8.70	8.66	8.62	8.57	8.53
4	7.71	6.94	6.59	6.39	6.26	6.16	6.04	5.91	5.86	5.80	5.75	5.69	5.63
5	6.61	5.79	5.41	5.19	5.05	4.95	4.82	4.68	4.62	4.56	4.50	4.43	4.36
6	5.99	5.14	4.76	4.53	4.39	4.28	4.15	4.00	3.94	3.87	3.81	3.74	3.67
7	5.59	4.74	4.35	4.12	3.97	3.87	3.73	3.57	3.51	3.44	3.38	3.30	3.23
8	5.32	4.46	4.07	3.84	3.69	3.58	3.44	3.28	3.22	3.15	3.08	3.01	2.93
9	5.12	4.26	3.86	3.63	3.48	3.37	3.23	3.07	3.01	2.94	2.86	2.79	2.71
10	4.96	4.10	3.71	3.48	3.33	3.22	3.07	2.91	2.85	2.77	2.70	2.62	2.54
11	4.84	3.98	3.59	3.36	3.20	3.09	2.95	2.79	2.72	2.65	2.57	2.49	2.40
12	4.75	3.89	3.49	3.26	3.11	3.00	2.85	2.69	2.62	2.54	2.47	2.38	2.30
13	4.67	3.81	3.41	3.18	3.03	2.92	2.77	2.60	2.53	2.46	2.38	2.30	2.21
14	4.60	3.74	3.34	3.11	2.96	2.85	2.70	2.53	2.46	2.39	2.31	2.22	2.13
15	4.54	3.68	3.29	3.06	2.90	2.79	2.64	2.48	2.40	2.33	2.25	2.16	2.07
16	4.49	3.63	3.24	3.01	2.85	2.74	2.59	2.42	2.35	2.28	2.19	2.11	2.01
17	4.45	3.59	3.20	2.96	2.81	2.70	2.55	2.38	2.31	2.23	2.15	2.06	1.96
18	4.41	3.55	3.16	2.93	2.77	2.66	2.51	2.34	2.27	2.19	2.11	2.02	1.92
19	4.38	3.52	3.13	2.90	2.74	2.63	2.48	2.31	2.23	2.16	2.07	1.98	1.88
20	4.35	3.49	3.10	2.87	2.71	2.60	2.45	2.28	2.20	2.12	2.04	1.95	1.84
21	4.32	3.47	3.07	2.84	2.68	2.57	2.42	2.25	2.18	2.10	2.01	1.92	1.81
22	4.30	3.44	3.05	2.82	2.66	2.55	2.40	2.23	2.15	2.07	1.98	1.89	1.78
23	4.28	3.42	3.03	2.80	2.64	2.53	2.37	2.20	2.13	2.05	1.96	1.86	1.76
24	4.26	3.40	3.01	2.78	2.62	2.51	2.36	2.18	2.11	2.03	1.94	1.84	1.73
25	4.24	3.39	2.99	2.76	2.60	2.49	2.34	2.16	2.09	2.01	1.92	1.82	1.71
26	4.23	3.37	2.98	2.74	2.59	2.47	2.32	2.15	2.07	1.99	1.90	1.80	1.69
27	4.21	3.35	2.96	2.73	2.57	2.46	2.31	2.13	2.06	1.97	1.88	1.79	1.67
28	4.20	3.34	2.95	2.71	2.56	2.45	2.29	2.12	2.04	1.96	1.87	1.77	1.65
29	4.18	3.33	2.93	2.70	2.55	2.43	2.28	2.10	2.03	1.94	1.85	1.75	1.64
30	4.17	3.32	2.92	2.69	2.53	2.42	2.27	2.09	2.01	1.93	1.84	1.74	1.62
40	4.08	3.23	2.84	2.61	2.45	2.34	2.18	2.00	1.92	1.84	1.74	1.64	1.51
60	4.00	3.15	2.76	2.53	2.37	2.25	2.10	1.92	1.84	1.75	1.65	1.53	1.39
120	3.92	3.07	2.68	2.45	2.29	2.17	2.02	1.83	1.75	1.66	1.55	1.43	1.25
∞	3.84	3.00	2.60	2.37	2.21	2.10	1.94	1.75	1.67	1.57	1.46	1.32	1.00

Source: Pearson and Hartley 1954 *(reprinted with the permission of Biometrica Trustees).*

Activity Graphs

Two classes of activity graphs are shown in Figures 5.1 and 5.2.

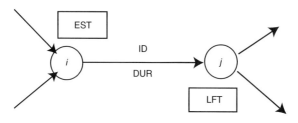

FIGURE 5.1 Activity-oriented graph

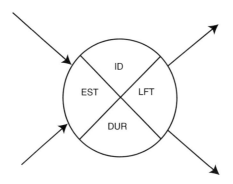

FIGURE 5.2 Event-oriented graph

Where:
- ID = identity of project task
- DUR = time duration of project task
- EST = estimated start time
- EFT = estimated finish time = EST + duration
- LFT = latest finish time
- LST = latest start time = LFT − duration
- TF = total float = LFT − EFT = LST − EST
- FF = free float = EST − EFT

Activities with TF = 0 are on a *critical path*.

Identification of the critical path with CPM provides an estimate of minimum project duration.

Activity Network

Activity networks are shown in Figures 5.3 and 5.4.

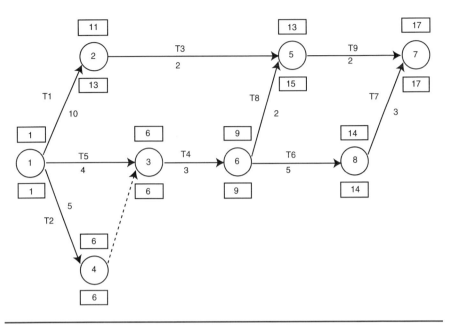

FIGURE 5.3 Activity-oriented network

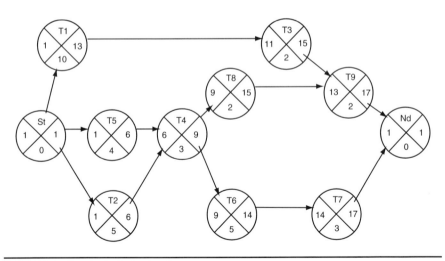

FIGURE 5.4 Event-oriented network

TABLE 5.5 Activity times and floats

ID	Duration	EST	EFT	LST	LFT	TF	FF
T1	10	1	11	3	13	2	0
T2*	5	1	6	1	6	0	0
T3	2	11	13	13	15	2	0
T4*	3	6	9	6	9	0	0
T5	4	1	5	2	6	1	1
T6*	5	9	14	9	14	0	0
T7*	3	14	17	14	17	0	0
T8	2	9	11	13	15	4	2
T9	2	13	15	15	17	2	2

* Activities on critical path.

PERT

The program evaluation and review technique (PERT) requires three activity duration estimates: a most likely time (m), a minimum time (a), and a maximum time (b). Expected completion time for the activity is approximately

$$t_e = \frac{1}{3}\left[2m + \frac{1}{2}(a+b)\right]$$

The variance in completion time for an activity is

$$\sigma^2 = \left[\frac{1}{6}(b-a)\right]^2$$

REFERENCES

Pearson, E.S., and H.O. Hartley, eds. 1954. *Biometrica Tables for Statisticians*, Vol. I. Cambridge, England: Cambridge University Press.

CHAPTER 6

Weights, Measures, Conversions, Constants, and Symbols

Raymond L. Lowrie, P.E.

INTERNATIONAL SYSTEM OF UNITS

The International System of Units, abbreviated "SI," is the modern metric system. Table 6.1 lists SI conversion factors. Although many engineers prefer SI units, English units remain in widespread use in the United States.

TABLE 6.1 SI conversion factors (Factors in **boldface** are exact)

To Convert From	To	Multiply By	
Acceleration			
acceleration of free fall, standard (g_n)	meter per second squared (m/s^2)	**9.806 65**	**E+00**
foot per second squared (ft/s^2)	meter per second squared (m/s^2)	**3.048**	**E−01**
gal (Gal)	meter per second squared (m/s^2)	**1.0**	**E−02**
inch per second squared (in/s^2)	meter per second squared (m/s^2)	**2.54**	**E−02**
Angle			
degree (°)	radian (rad)	1.745 329	E−02
gon (also called grade) (gon)	radian (rad)	1.570 796	E−02
gon (also called grade) (gon)	degree (°)	**9.0**	**E−01**
mil	radian (rad)	9.817 477	E−04
mil	degree (°)	**5.625**	**E−02**
minute (′)	radian (rad)	2.908 882	E−04
revolution (r)	radian (rad)	6.283 185	E+00
second (″)	radian (rad)	4.848 137	E−06
Area and second moment of area			
acre (based on U.S. survey foot)[1]	square meter (m^2)	4.046 873	E+03
are (a)	square meter (m^2)	**1.0**	**E+02**
barn (b)	square meter (m^2)	**1.0**	**E−28**
circular mil	square meter (m^2)	5.067 075	E−10
circular mil	square millimeter (mm^2)	5.067 075	E−04
foot to the fourth power (ft^4)[2]	meter to the fourth power (m^4)	8.630 975	E−03
hectare (ha)	square meter (m^2)	**1.0**	**E+04**
inch to the fourth power (in^4)[2]	meter to the fourth power (m^4)	4.162 314	E−07

continues next page

TABLE 6.1 SI conversion factors (Factors in **boldface** are exact) **(continued)**

To Convert From	To	Multiply By	
square foot (ft^2)	square meter (m^2)	**9.290 304**	**E–02**
square inch (in^2)	square meter (m^2)	**6.4516**	**E–04**
square inch (in^2)	square centimeter (cm^2)	**6.4516**	**E+00**
square mile (mi^2)	square meter (m^2)	2.589 988	E+06
square mile (mi^2)	square kilometer (km^2)	2.589 988	E+00
square mile (based on U.S. survey foot) (mi^2)[1]	square meter (m^2)	2.589 998	E+06
square mile (based on U.S. survey foot) (mi^2)[1]	square kilometer (km^2)	2.589 998	E+00
square yard (yd^2)	square meter (m^2)	8.361 274	E–01
Capacity (see Volume)			
Density (i.e., mass density—see Mass divided by volume)			
Electricity and magnetism			
abampere	ampere (A)	**1.0**	**E+01**
abcoulomb	coulomb (C)	**1.0**	**E+01**
abfarad	farad (F)	**1.0**	**E+09**
abhenry	henry (H)	**1.0**	**E–09**
abmho	siemens (S)	**1.0**	**E+09**
abohm	ohm (Ω)	**1.0**	**E–09**
abvolt	volt (V)	**1.0**	**E–08**
ampere hour (A · h)	coulomb (C)	**3.6**	**E+03**
biot (Bi)	ampere (A)	**1.0**	**E+01**
EMU of capacitance (abfarad)	farad (F)	**1.0**	**E+09**
EMU of current (abampere)	ampere (A)	**1.0**	**E+01**
EMU of electric potential (abvolt)	volt (V)	**1.0**	**E–08**
EMU of inductance (abhenry)	henry (H)	**1.0**	**E–09**
EMU of resistance (abohm)	ohm (Ω)	**1.0**	**E–09**
ESU of capacitance (statfarad)	farad (F)	1.112 650	E–12
ESU of current (statampere)	ampere (A)	3.335 641	E–10
ESU of electric potential (statvolt)	volt (V)	2.997 925	E+02
ESU of inductance (stathenry)	henry (H)	8.987 552	E+11
ESU of resistance (statohm)	ohm (Ω)	8.987 552	E+11
faraday (based on carbon 12)	coulomb (C)	9.648 531	E+04
franklin (Fr)	coulomb (C)	3.335 641	E–10
gamma (γ)	tesla (T)	**1.0**	**E–09**
gauss (Gs, G)	tesla (T)	**1.0**	**E–04**
gilbert (Gi)	ampere (A)	7.957 747	E–01
maxwell (Mx)	weber (Wb)	**1.0**	**E–08**
mho	siemens (S)	**1.0**	**E+00**
oersted (Oe)	ampere per meter (A/m)	7.957 747	E+01
ohm centimeter (Ω · m)	ohm meter (Ω · m)	**1.0**	**E–02**
ohm circular-mil per foot	ohm meter (Ω · m)	1.662 426	E–09
ohm circular-mil per foot	ohm square millimeter per meter (Ω · mm^2/m)	1.662 426	E–03
statampere	ampere (A)	3.335 641	E–10
statcoulomb	coulomb (C)	3.335 641	E–10
statfarad	farad (F)	1.112 650	E–12
stathenry	henry (H)	8.987 552	E+11
statmho	siemens (S)	1.112 650	E–12

continues next page

TABLE 6.1 SI conversion factors (Factors in **boldface** are exact) **(continued)**

To Convert From	To	Multiply By	
statohm	ohm (Ω)	8.987 552	E+11
statvolt	volt (V)	2.997 925	E+02
unit pole	weber (Wb)	1.256 637	E−07
Energy (includes Work)			
British thermal unit$_{IT}$ (Btu$_{IT}$)[3]	joule (J)	1.055 056	E+03
British thermal unit$_{th}$ (Btu$_{th}$)[3]	joule (J)	1.054 350	E+03
British thermal unit (mean) (Btu)	joule (J)	1.055 87	E+03
British thermal unit (39°F) (Btu)	joule (J)	1.059 67	E+03
British thermal unit (59°F) (Btu)	joule (J)	1.054 80	E+03
British thermal unit (60°F) (Btu)	joule (J)	1.054 68	E+03
calorie$_{IT}$ (cal$_{IT}$)[3]	joule (J)	**4.1868**	**E+00**
calorie$_{th}$ (cal$_{th}$)[3]	joule (J)	**4.184**	**E+00**
calorie (mean) (cal)	joule (J)	4.190 02	E+00
calorie (15°C) (cal$_{15}$)	joule (J)	4.185 80	E+00
calorie (20°C) (cal$_{20}$)	joule (J)	4.181 90	E+00
calorie$_{IT}$, kilogram (nutrition)[4]	joule (J)	**4.1868**	**E+03**
calorie$_{th}$, kilogram (nutrition)[4]	joule (J)	**4.184**	**E+03**
calorie (mean), kilogram (nutrition)[4]	joule (J)	4.190 02	E+03
electronvolt (eV)	joule (J)	1.602 177	E−19
erg (erg)	joule (J)	**1.0**	**E−07**
foot poundal	joule (J)	4.214 011	E−02
foot pound-force (ft · lbf)	joule (J)	1.355 818	E+00
kilocalorie$_{IT}$ (kcal$_{IT}$)	joule (J)	**4.1868**	**E+03**
kilocalorie$_{th}$ (kcal$_{th}$)	joule (J)	**4.184**	**E+03**
kilocalorie (mean) (kcal)	joule (J)	4.190 02	E+03
kilowatt hour (kW · h)	joule (J)	**3.6**	**E+06**
kilowatt hour (kW · h)	megajoule (MJ)	**3.6**	**E+00**
quad (10^{15} Btu$_{IT}$)[3]	joule (J)	1.055 056	E+18
therm (EC)[5]	joule (J)	**1.055 06**	**E+08**
therm (U.S.)[5]	joule (J)	**1.054 804**	**E+08**
ton of TNT (energy equivalent)[6]	joule (J)	**4.184**	**E+09**
watt hour (W · h)	joule (J)	**3.6**	**E+03**
watt second (W · s)	joule (J)	**1.0**	**E+00**
Energy divided by area time			
erg per square centimeter second [erg/(cm² · s)]	watt per square meter (W/m²)	1.0	E−03
watt per square centimeter (W/cm²)	watt per square meter (W/m²)	1.0	E+04
watt per square inch (W/in²)	watt per square meter (W/m²)	1.550 003	E+03
Flow (see Mass divided by time or Volume divided by time)			
Force			
dyne (dyn)	newton (N)	**1.0**	**E−05**
kilogram-force (kgf)	newton (N)	**9.806 65**	**E+00**
kilopond (kilogram-force) (kp)	newton (N)	**9.806 65**	**E+00**
kip (1 kip = 1,000 lbf)	newton (N)	4.448 222	E+03
kip (1 kip = 1,000 lbf)	kilonewton (kN)	4.448 222	E+00
ounce (avoirdupois)-force (ozf)	newton (N)	2.780 139	E−01
poundal	newton (N)	1.382 550	E−01
pound-force (lbf)[7]	newton (N)	4.448 222	E+00

continues next page

TABLE 6.1 SI conversion factors (Factors in **boldface** are exact) **(continued)**

To Convert From	To	Multiply By	
pound-force per pound (lbf/lb) (thrust to mass ratio)	newton per kilogram (N/kg)	**9.806 65**	**E+00**
ton-force (2,000 lbf)	newton (N)	8.896 443	E+03
ton-force (2,000 lbf)	kilonewton (kN)	8.896 443	E+00
Force divided by area (see Pressure)			
Force divided by length			
pound-force per foot (lbf/ft)	newton per meter (N/m)	1.459 390	E+01
pound-force per inch (lbf/in)	newton per meter (N/m)	1.751 268	E+02
Heat			
Available energy			
British thermal unit$_{IT}$ per cubic foot (Btu$_{IT}$/ft^3)	joule per cubic meter (J/m^3)	3.725 895	E+04
British thermal unit$_{th}$ per cubic foot (Btu$_{th}$/ft^3)	joule per cubic meter (J/m^3)	3.723 403	E+04
British thermal unit$_{IT}$ per pound (Btu$_{IT}$/lb)	joule per kilogram (J/kg)	**2.326**	**E+03**
British thermal unit$_{th}$ per cubic foot (Btu$_{th}$/lb)	joule per kilogram (J/kg)	2.324 444	E+03
calorie$_{IT}$ per gram (cal$_{IT}$/g)	joule per kilogram (J/kg)	**4.1868**	**E+03**
calorie$_{th}$ per gram (cal$_{th}$/g)	joule per kilogram (J/kg)	**4.184**	**E+03**
Coefficient of heat transfer			
British thermal unit$_{IT}$ per hour square foot degree Fahrenheit [Btu$_{IT}$/(h · ft^2 · °F)]	watt per square meter kelvin [W/(m^2 · K)]	5.678 263	E+00
British thermal unit$_{th}$ per hour square foot degree Fahrenheit [Btu$_{th}$/(h · ft^2 · °F)]	watt per square meter kelvin [W/(m^2 · K)]	5.674 466	E+00
British thermal unit$_{IT}$ per second square foot degree Fahrenheit [Btu$_{IT}$/(s · ft^2 · °F)]	watt per square meter kelvin [W/(m^2 · K)]	2.044 175	E+04
British thermal unit$_{th}$ per second square foot degree Fahrenheit [Btu$_{th}$/(s · ft^2 · °F)]	watt per square meter kelvin [W/(m^2 · K)]	2.042 808	E+04
Density of heat			
British thermal unit$_{IT}$ per square foot (Btu$_{IT}$/ft^2)	joule per square meter (J/m^2)	1.135 653	E+04
British thermal unit$_{th}$ per square foot (Btu$_{th}$/ft^2)	joule per square meter (J/m^2)	1.134 893	E+04
calorie$_{th}$ per square centimeter (cal$_{th}$/cm^2)	joule per square meter (J/m^2)	**4.184**	**E+04**
langley (cal$_{th}$/cm^2)	joule per square meter (J/m^2)	**4.184**	**E+04**
Density of heat flow rate			
British thermal unit$_{IT}$ per square foot hour [Btu$_{IT}$/(ft^2 · h)]	watt per square meter (W/m^2)	3.154 591	E+00
British thermal unit$_{th}$ per square foot hour [Btu$_{th}$/(ft^2 · h)]	watt per square meter (W/m^2)	3.152 481	E+00
British thermal unit$_{th}$ per square foot minute [Btu$_{th}$/(ft^2 · min)]	watt per square meter (W/m^2)	1.891 489	E+02
British thermal unit$_{IT}$ per square foot second [Btu$_{IT}$/(ft^2 · s)]	watt per square meter (W/m^2)	1.135 653	E+04
British thermal unit$_{th}$ per square foot second [Btu$_{th}$/(ft^2 · s)]	watt per square meter (W/m^2)	1.134 893	E+04
British thermal unit$_{th}$ per square inch second [Btu$_{th}$/(in^2 · s)]	watt per square meter (W/m^2)	1.634 246	E+06
calorie$_{th}$ per square centimeter minute [cal$_{th}$/(cm^2 · min)]	watt per square meter (W/m^2)	6.973 333	E+02
calorie$_{th}$ per square centimeter second [cal$_{th}$/(cm^2 · s)]	watt per square meter (W/m^2)	**4.184**	**E+04**

continues next page

WEIGHTS, MEASURES, CONVERSIONS, CONSTANTS, AND SYMBOLS | **131**

TABLE 6.1 SI conversion factors (Factors in **boldface** are exact) **(continued)**

To Convert From	To	Multiply By	
Fuel consumption			
gallon (U.S.) per horsepower hour [gal/(hp · h)]	cubic meter per joule (m³/J)	1.410 089	E–09
gallon (U.S.) per horsepower hour [gal/(hp · h)]	liter per joule (L/J)	1.410 089	E–06
mile per gallon (U.S.) (mpg) (mi/gal)	meter per cubic meter (m/m³)	4.251 437	E+05
mile per gallon (U.S.) (mpg) (mi/gal)	kilometer per liter (km/L)	4.251 437	E–01
mile per gallon (U.S.) (mpg) (mi/gal)	liter per 100 kilometer (L/100 km)	divide 235.215 by number of miles per gallon	
pound per horsepower hour [lb/(hp · h)]	kilogram per joule (kg/J)	1.689 659	E–07
Heat capacity and entrophy			
British thermal unit$_{IT}$ per degree Fahrenheit (Btu$_{IT}$/°F)	joule per kelvin (J/k)	1.899 101	E+03
British thermal unit$_{th}$ per degree Fahrenheit (Btu$_{th}$/°F)	joule per kelvin (J/k)	1.897 830	E+03
British thermal unit$_{IT}$ per degree Rankine (Btu$_{IT}$/°R)	joule per kelvin (J/k)	1.899 101	E+03
British thermal unit$_{th}$ per degree Rankine (Btu$_{th}$/°R)	joule per kelvin (J/k)	1.897 830	E+03
Heat flow rate			
British thermal unit$_{IT}$ per hour (Btu$_{IT}$/h)	watt (W)	2.930 711	E–01
British thermal unit$_{th}$ per hour (Btu$_{th}$/h)	watt (W)	2.928 751	E–01
British thermal unit$_{th}$ per minute (Btu$_{th}$/min)	watt (W)	1.757 250	E+01
British thermal unit$_{IT}$ per second (Btu$_{IT}$/s)	watt (W)	1.055 056	E+03
British thermal unit$_{th}$ per second (Btu$_{th}$/s)	watt (W)	1.054 350	E+03
calorie$_{th}$ per minute (cal$_{th}$/min)	watt (W)	6.973 333	E–02
calorie$_{th}$ per second (cal$_{th}$/s)	watt (W)	**4.184**	**E+00**
kilocalorie$_{th}$ per minute (kcal$_{th}$/min)	watt (W)	6.973 333	E+01
kilocalorie$_{th}$ per second (kcal$_{th}$/s)	watt (W)	**4.184**	**E+03**
ton of refrigeration (12,000 Btu$_{IT}$/h)	watt (W)	3.516 853	E+03
Specific heat capacity and specific entropy			
British thermal unit$_{IT}$ per pound degree Fahrenheit [Btu$_{IT}$/(lb · °F)]	joule per kilogram kelvin [J/(kg · K)]	**4.1868**	**E+03**
British thermal unit$_{th}$ per pound degree Fahrenheit [Btu$_{th}$/(lb · °F)]	joule per kilogram kelvin [J/(kg · K)]	**4.184**	**E+03**
British thermal unit$_{IT}$ per pound degree Rankine [Btu$_{IT}$/(lb · °R)]	joule per kilogram kelvin [J/(kg · K)]	**4.1868**	**E+03**
British thermal unit$_{th}$ per pound degree Rankine [Btu$_{th}$/(lb · °R)]	joule per kilogram kelvin [J/(kg · K)]	**4.184**	**E+03**
calorie$_{IT}$ per gram degree Celsius [cal$_{IT}$/(g · °C)]	joule per kilogram kelvin [J/(kg · K)]	**4.1868**	**E+03**
calorie$_{th}$ per gram degree Celsius [cal$_{th}$/(g · °C)]	joule per kilogram kelvin [J/(kg · K)]	**4.184**	**E+03**
calorie$_{IT}$ per gram kelvin [cal$_{IT}$/(g · K)]	joule per kilogram kelvin [J/(kg · K)]	**4.1868**	**E+03**
calorie$_{th}$ per gram kelvin [cal$_{th}$/(g · K)]	joule per kilogram kelvin [J/(kg · K)]	**4.184**	**E+03**
Thermal conductivity			
British thermal unit$_{IT}$ foot per hour square foot degree Fahrenheit [Btu$_{IT}$ · ft/(h · ft² · °F)]	watt per meter kelvin [W/(m · K)]	1.730 735	E+00
British thermal unit$_{th}$ foot per hour square foot degree Fahrenheit [Btu$_{th}$ · ft/(h · ft² · °F)]	watt per meter kelvin [W/(m · K)]	1.729 577	E+00
British thermal unit$_{IT}$ inch per hour square foot degree Fahrenheit [Btu$_{IT}$ · in/(h · ft² · °F)]	watt per meter kelvin [W/(m · K)]	1.442 279	E–01

continues next page

TABLE 6.1 SI conversion factors (Factors in **boldface** are exact) **(continued)**

To Convert From	To	Multiply By	
British thermal unit$_{th}$ inch per hour square foot degree Fahrenheit [Btu$_{th}$ · in/(h · ft^2 · °F)]	watt per meter kelvin [W/(m · K)]	1.441 314	E−01
British thermal unit$_{IT}$ inch per second square foot degree Fahrenheit [Btu$_{IT}$ · in/(s · ft^2 · °F)]	watt per meter kelvin [W/(m · K)]	5.192 204	E+02
British thermal unit$_{th}$ inch per second square foot degree Fahrenheit [Btu$_{th}$ · in/(s · ft^2 · °F)]	watt per meter kelvin [W/(m · K)]	5.188 732	E+02
calorie$_{th}$ per centimeter second degree Celsius [cal$_{th}$/(cm · s · °C)]	watt per meter kelvin [W/(m · K)]	**4.184**	**E+02**
Thermal diffusivity			
square foot per hour (ft^2/h)	square meter per second (m^2/s)	**2.580 64**	**E−05**
Thermal insulance			
clo	square meter kelvin per watt (m^2 · K/W)	1.55	E−01
degree Fahrenheit hour square foot per British thermal unit$_{IT}$ [°F · h · ft^2/Btu$_{IT}$)]	square meter kelvin per watt (m^2 · K/W)	1.761 102	E−01
degree Fahrenheit hour square foot per British thermal unit$_{th}$ [°F · h · ft^2/Btu$_{th}$)]	square meter kelvin per watt (m^2 · K/W)	1.762 280	E−01
Thermal resistance			
degree Fahrenheit hour per British thermal unit$_{IT}$ (°F · h/Btu$_{IT}$)	kelvin per watt (K/W)	1.895 634	E+00
degree Fahrenheit hour per British thermal unit$_{th}$ (°F · h/Btu$_{th}$)	kelvin per watt (K/W)	1.896 903	E+00
degree Fahrenheit second per British thermal unit$_{IT}$ (°F · s/Btu$_{IT}$)	kelvin per watt (K/W)	5.265 651	E−04
degree Fahrenheit second per British thermal unit$_{th}$ (°F · s/Btu$_{th}$)	kelvin per watt (K/W)	5.269 175	E−04
Thermal resistivity			
degree Fahrenheit hour square foot per British thermal unit$_{IT}$ inch [°F · h · ft^2/(Btu$_{IT}$ · in)]	meter kelvin per watt (m · K/W)	6.933 472	E+00
degree Fahrenheit hour square foot per British thermal unit$_{th}$ inch [(°F · h · ft^2/(Btu$_{th}$ · in)]	meter kelvin per watt (m · K/W)	6.938 112	E+04
Length			
ångström (Å)	meter (m)	**1.0**	**E−10**
ångström (Å)	nanometer (nm)	**1.0**	**E−01**
astronomical unit (AU)	meter (m)	1.495 979	E+11
chain (based on U.S. survey foot) (ch)[1]	meter (m)	2.011 684	E+01
fathom (based on U.S. survey foot)[1]	meter (m)	1.828 804	E+00
fermi	meter (m)	**1.0**	**E−15**
fermi	femtometer (fm)	**1.0**	**E+00**
foot (ft)	meter (m)	**3.048**	**E−01**
foot (U.S. survey) (ft)[1]	meter (m)	3.048 006	E−01
inch (in)	meter (m)	**2.54**	**E−02**
inch (in)	centimeter (cm)	**2.54**	**E+00**
kayser (K)	reciprocal meter (m^{-1})	1	E+02
light year (l.y.)[8]	meter (m)	9.460 73	E+15
microinch	meter (m)	**2.54**	**E−08**
microinch	micrometer (μm)	**2.54**	**E−02**
micron (μ)	meter (m)	**1.0**	**E−06**
micron (μ)	micrometer (μm)	**1.0**	**E+00**

continues next page

TABLE 6.1 SI conversion factors (Factors in **boldface** are exact) **(continued)**

To Convert From	To	Multiply By	
mil (0.001 in)	meter (m)	**2.54**	**E–05**
mil (0.001 in)	millimeter (mm)	**2.54**	**E–02**
mile (mi)	meter (m)	**1.609 344**	**E+03**
mile (mi)	kilometer (km)	**1.609 344**	**E+00**
mile (based on U.S. survey foot) (mi)[1]	meter (m)	1.609 347	E+03
mile (based on U.S. survey foot) (mi)[1]	kilometer (km)	1.609 347	E+00
mile, nautical[9]	meter (m)	**1.852**	**E+03**
parsec (pc)	meter (m)	3.085 678	E+16
pica (computer) (1/6 in)	meter (m)	4.233 333	E–03
pica (computer) (1/6 in)	millimeter (mm)	4.233 333	E+00
pica (printer's)	meter (m)	4.217 518	E–03
pica (printer's)	millimeter (mm)	4.217 518	E+00
point (computer) (1/72 in)	meter (m)	3.527 778	E–04
point (computer) (1/72 in)	millimeter (mm)	3.527 778	E–01
point (printer's)	meter (m)	3.514 598	E–04
point (printer's)	millimeter (mm)	3.514 598	E–01
rod (based on U.S. survey foot) (rd)[1]	meter (m)	5.029 210	E+00
yard (yd)	meter (m)	**9.144**	**E–01**
Light			
candela per square inch (cd/in^2)	candela per square meter (cd/m^2)	1.550 003	E+03
footcandle	lux (lx)	1.076 391	E+01
footlambert	candela per square meter (cd/m^2)	3.426 259	E+00
lambert[10]	candela per square meter (cd/m^2)	3.183 099	E+03
lumen per square foot (lm/ft^2)	lux (lx)	1.076 391	E+01
phot (ph)	lux (lx)	**1.0**	**E+04**
stilb (sb)	candela per square meter (cd/m^2)	**1.0**	**E+04**
Mass and moment of inertia			
carat, metric	kilogram (kg)	**2.0**	**E–04**
carat, metric	gram (g)	**2.0**	**E–01**
grain (gr)	kilogram (kg)	**6.479 891**	**E–05**
grain (gr)	milligram (mg)	**6.479 891**	**E+01**
hundredweight (long, 112 lb)	kilogram (kg)	5.080 235	E+01
hundredweight (short, 100 lb)	kilogram (kg)	4.535 924	E+01
kilogram-force second squared per meter ($kgf \cdot s^2/m$)	kilogram (kg)	**9.806 65**	**E+00**
ounce (avoirdupois) (oz)	kilogram (kg)	2.834 952	E–02
ounce (avoirdupois) (oz)	gram (g)	2.834 952	E+01
ounce (troy or apothecary) (oz)	kilogram (kg)	3.110 348	E–02
ounce (troy or apothecary) (oz)	gram (g)	3.110 348	E+01
pennyweight (dwt)	kilogram (kg)	1.555 174	E–03
pennyweight (dwt)	gram (g)	1.555 174	E+00
pound (avoirdupois) (lb)[11]	kilogram (kg)	4.535 924	E–01
pound (troy or apothecary) (lb)	kilogram (kg)	3.732 417	E–01
pound foot squared ($lb \cdot ft^2$)	kilogram meter squared ($kg \cdot m^2$)	4.214 011	E–02
pound inch squared ($lb \cdot in^2$)	kilogram meter squared ($kg \cdot m^2$)	2.926 397	E–04
slug (slug)	kilogram (kg)	1.459 390	E+01
ton, assay (AT)	kilogram (kg)	2.916 667	E–02

continues next page

TABLE 6.1 SI conversion factors (Factors in **boldface** are exact) **(continued)**

To Convert From	To	Multiply By	
ton, assay (AT)	gram (g)	2.916 667	E+01
ton, long (2,240 lb)	kilogram (kg)	1.016 047	E+03
ton, metric (t)	kilogram (kg)	**1.0**	**E+03**
tonne (called "metric ton" in U.S.) (t)	kilogram (kg)	**1.0**	**E+03**
ton, short (2,000 lb)	kilogram (kg)	9.071 847	E+02
Mass density (see Mass divided by volume)			
Mass divided by area			
ounce (avoirdupois) per square foot (oz/ft^2)	kilogram per square meter (kg/m^2)	3.051 517	E−01
ounce (avoirdupois) per square inch (oz/in^2)	kilogram per square meter (kg/m^2)	4.394 185	E+01
ounce (avoirdupois) per square yard (oz/yd^2)	kilogram per square meter (kg/m^2)	3.390 575	E−02
pound per square foot (lb/ft^2)	kilogram per square meter (kg/m^2)	4.882 428	E+00
pound per square inch (*not* pound force) (lb/in^2)	kilogram per square meter (kg/m^2)	7.030 696	E+02
Mass divided by capacity (see Mass divided by volume)			
Mass divided by length			
denier	kilogram per meter (kg/m)	1.111 111	E−07
denier	gram per meter (g/m)	1.111 111	E−04
pound per foot (lb/ft)	kilogram per meter (kg/m)	1.488 164	E+00
pound per inch (lb/in)	kilogram per meter (kg/m)	1.785 797	E+01
pound per yard (lb/yd)	kilogram per meter (kg/m)	4.960 546	E−01
tex	kilogram per meter (kg/m)	**1.0**	**E−06**
Mass divided by time (includes flow)			
pound per hour (lb/h)	kilogram per second (kg/s)	1.259 979	E−04
pound per minute (lb/min)	kilogram per second (kg/s)	7.559 873	E−03
pound per second (lb/s)	kilogram per second (kg/s)	4.535 924	E−01
ton, short, per hour	kilogram per second (kg/s)	2.519 958	E−01
Mass divided by volume (includes mass density and mass concentration)			
grain per gallon (U.S.) (gr/gal)	kilogram per cubic meter (kg/m^3)	1.711 806	E−02
grain per gallon (U.S.) (gr/gal)	milligram per liter (mg/L)	1.711 806	E+01
gram per cubic centimeter (g/cm^3)	kilogram per cubic meter (kg/m^3)	**1.0**	**E+03**
ounce (avoirdupois) per cubic inch (oz/in^3)	kilogram per cubic meter (kg/m^3)	1.729 994	E+03
ounce (avoirdupois) per gallon [Canadian and U.K. (Imperial)] (oz/gal)	kilogram per cubic meter (kg/m^3)	6.236 023	E+00
ounce (avoirdupois) per gallon [Canadian and U.K. (Imperial)] (oz/gal)	gram per liter (g/L)	6.236 023	E+00
ounce (avoirdupois) per gallon (U.S.) (oz/gal)	kilogram per cubic meter (kg/m^3)	7.489 152	E+00
ounce (avoirdupois) per gallon (U.S.) (oz/gal)	gram per liter (g/L)	7.489 152	E+00
pound per cubic foot (lb/ft^3)	kilogram per cubic meter (kg/m^3)	1.601 846	E+01
pound per cubic inch (lb/in^3)	kilogram per cubic meter (kg/m^3)	2.767 990	E+04
pound per cubic yard (lb/yd^3)	kilogram per cubic meter (kg/m^3)	5.932 764	E−01
pound per gallon [Canadian and U.K. (Imperial)] (lb/gal)	kilogram per cubic meter (kg/m^3)	9.977 637	E+01
pound per gallon [Canadian and U.K. (Imperial)] (lb/gal)	kilogram per liter (kg/L)	9.977 637	E−02
pound per gallon (U.S.) (lb/gal)	kilogram per cubic meter (kg/m^3)	1.198 264	E+02
pound per gallon (U.S.) (lb/gal)	kilogram per liter (kg/L)	1.198 264	E−01
slug per cubic foot (slug/ft^3)	kilogram per cubic meter (kg/m^3)	5.153 788	E+02
ton, long, per cubic yard	kilogram per cubic meter (kg/m^3)	1.328 939	E+03

continues next page

TABLE 6.1 SI conversion factors (Factors in **boldface** are exact) **(continued)**

To Convert From	To	Multiply By	
ton, short, per cubic yard	kilogram per cubic meter (kg/m³)	1.186 553	E+03
Moment of force or torque			
dyne centimeter (dyn · cm)	newton meter (N · m)	**1.0**	**E−07**
kilogram-force meter (kgf · m)	newton meter (N · m)	**9.806 65**	**E+00**
ounce (avoirdupois)-force inch (ozf · in)	newton meter (N · m)	7.061 552	E−03
ounce (avoirdupois)-force inch (ozf · in)	millinewton meter (mN · m)	7.061 552	E+00
pound-force foot (lbf · ft)	newton meter (N · m)	1.355 818	E+00
pound-force in (lbf · in)	newton meter (N · m)	1.129 848	E−01
Moment of force or torque, divided by length			
pound-force foot per inch (lbf · ft/in)	newton meter per meter (N · m/m)	5.337 866	E+01
pound-force inch per inch (lbf · in/in)	newton meter per meter (N · m/m)	4.448 222	E+00
Permeability			
darcy[12]	meter squared (m²)	9.869 233	E−13
perm (0°C)	kilogram per pascal second square meter [kg/(Pa · s · m²)]	5.721 35	E−11
perm (23°C)	kilogram per pascal second square meter [kg/(Pa · s · m²)]	5.745 25	E−11
perm inch (0°C)	kilogram per pascal second meter [kg/(Pa · s · m)]	1.453 22	E−12
perm inch (23°C)	kilogram per pascal second meter [kg/(Pa · s · m)]	1.459 29	E−12
Power			
erg per second (erg/s)	watt (W)	**1.0**	**E−07**
foot pound-force per hour (ft · lbf/h)	watt (W)	3.766 161	E−04
foot pound-force per minute (ft · lbf/min)	watt (W)	2.259 697	E−02
foot pound-force per second (ft · lbf/s)	watt (W)	1.355 818	E+00
horsepower (550 ft · lbf/s)	watt (W)	7.456 999	E+02
horsepower (boiler)	watt (W)	9.809 50	E+03
horsepower (electric)	watt (W)	**7.46**	**E+02**
horsepower (metric)	watt (W)	7.354 988	E+02
horsepower (U.K.)	watt (W)	7.4570	E+02
horsepower (water)	watt (W)	7.460 43	E+02
Pressure or stress (force divided by area)			
atmosphere, standard (atm)	pascal (Pa)	**1.013 25**	**E+05**
atmosphere, standard (atm)	kilopascal (kPa)	**1.013 25**	**E+02**
atmosphere, technical (at)[13]	pascal (Pa)	**9.806 65**	**E+04**
atmosphere, technical (at)[13]	kilopascal (kPa)	**9.806 65**	**E+01**
bar (bar)	pascal (Pa)	**1.0**	**E+05**
bar (bar)	kilopascal (kPa)	**1.0**	**E+02**
centimeter of mercury (0°C)[14]	pascal (Pa)	1.333 22	E+03
centimeter of mercury (0°C)[14]	kilopascal (kPa)	1.333 22	E+00
centimeter of mercury, conventional (cmHg)[14]	pascal (Pa)	1.333 224	E+03
centimeter of mercury, conventional (cmHg)[14]	kilopascal (kPa)	1.333 224	E+00
centimeter of water (4°C)[14]	pascal (Pa)	9.806 38	E+01
centimeter of water, conventional (cmH₂O)[14]	pascal (Pa)	**9.806 65**	**E+01**
dyne per square centimeter (dyn/cm²)	pascal (Pa)	**1.0**	**E−01**
foot of mercury, conventional (ftHg)[14]	pascal (Pa)	4.063 666	E+04

continues next page

TABLE 6.1 SI conversion factors (Factors in **boldface** are exact) **(continued)**

To Convert From	To	Multiply By	
foot of mercury, conventional (ftHg)[14]	kilopascal (kPa)	4.063 666	E+01
foot of water (39.2°F)[14]	pascal (Pa)	2.988 98	E+03
foot of water (39.2°F)[14]	kilopascal (kPa)	2.988 98	E+00
foot of water, conventional (ftH$_2$O)[14]	pascal (Pa)	2.989 067	E+03
foot of water, conventional (ftH$_2$O)[14]	kilopascal (kPa)	2.989 067	E+00
gram-force per square centimeter (gf/cm^2)	pascal (Pa)	**9.806 65**	**E+01**
inch of mercury (32°F)[14]	pascal (Pa)	3.386 38	E+03
inch of mercury (32°F)[14]	kilopascal (kPa)	3.386 38	E+00
inch of mercury (60°F)[14]	pascal (Pa)	3.376 85	E+03
inch of mercury (60°F)[14]	kilopascal (kPa)	3.376 85	E+00
inch of mercury, conventional (inHg)[14]	pascal (Pa)	3.386 389	E+03
inch of mercury, conventional (inHg)[14]	kilopascal (kPa)	3.386 389	E+00
inch of water (39.2°F)[14]	pascal (Pa)	2.490 82	E+02
inch of water (60°F)[14]	pascal (Pa)	2.4884	E+02
inch of water, conventional (inH$_2$O)[14]	pascal (Pa)	2.490 889	E+02
kilogram-force per square centimeter (kgf/cm^2)	pascal (Pa)	**9.806 65**	**E+04**
kilogram-force per square centimeter (kgf/cm^2)	kilopascal (kPa)	**9.806 65**	**E+01**
kilogram-force per square meter (kgf/m^2)	pascal (Pa)	**9.806 65**	**E+00**
kilogram-force per square millimeter (kgf/mm^2)	pascal (Pa)	**9.806 65**	**E+06**
kilogram-force per square millimeter (kgf/mm^2)	megapascal (MPa)	**9.806 65**	**E+00**
kip per square inch (ksi) (kip/in^2)	pascal (Pa)	6.894 757	E+06
kip per square inch (ksi) (kip/in^2)	kilopascal (kPa)	6.894 757	E+03
millibar (mbar)	pascal (Pa)	**1.0**	**E+02**
millibar (mbar)	kilopascal (kPa)	**1.0**	**E−01**
millimeter of mercury, conventional (mmHg)[14]	pascal (Pa)	1.333 224	E+02
millimeter of water, conventional (mmH$_2$O)[14]	pascal (Pa)	**9.806 65**	**E+00**
poundal per square foot	pascal (Pa)	1.488 164	E+00
pound-force per square foot (lbf/ft^2)	pascal (Pa)	4.788 026	E+01
pound-force per square inch (psi) (lbf/in^2)	pascal (Pa)	6.894 757	E+03
pound-force per square inch (psi) (lbf/in^2)	kilopascal (kPa)	6.894 757	E+00
psi (pound-force per square inch) (lbf/in^2)	pascal (Pa)	6.894 757	E+03
psi (pound-force per square inch) (lbf/in^2)	kilopascal (kPa)	6.894 757	E+00
torr (Torr)	pascal (Pa)	1.333 224	E+02
Radiology			
curie (Ci)	becquerel (Bq)	**3.7**	**E+10**
rad (absorbed dose) (rad)	gray (Gy)	**1.0**	**E−02**
rem (rem)	sievert (Sv)	**1.0**	**E−02**
roentgen (R)	coulomb per kilogram (C/kg)	2.58	E−04
Speed (see Velocity)			
Stress (see Pressure)			
Temperature			
degree Celsius (°C)	kelvin (K)	$T/K = t/°C +$ **273.15**	
degree centigrade[15]	degree Celsius (°C)	$t/°C \approx t/\text{deg. cent.}$	
degree Fahrenheit (°F)	degree Celsius (°C)	$t/°C = (t/°F -$ **32**$)/$**1.8**	
degree Fahrenheit (°F)	kelvin (K)	$T/K = (t/°F +$ **459.67**$)/$**1.8**	

continues next page

WEIGHTS, MEASURES, CONVERSIONS, CONSTANTS, AND SYMBOLS

TABLE 6.1 SI conversion factors (Factors in **boldface** are exact) **(continued)**

To Convert From	To	Multiply By	
degree Rankine (°R)	kelvin (K)	$T/K = (T/°R)$/**1.8**	
kelvin (K)	degree Celsius (°C)	$t/°C = T/K -$ **273.15**	
Temperature interval			
degree Celsius (°C)	kelvin (K)	**1.0**	**E+00**
degree centigrade[15]	degree Celsius (°C)	1.0	E+00
degree Fahrenheit (°F)	degree Celsius (°C)	5.555 556	E−01
degree Fahrenheit (°F)	kelvin (K)	5.555 556	E−01
degree Rankine (°R)	kelvin (K)	5.555 556	E−01
Time			
day (d)	second (s)	**8.64**	**E+04**
day (sidereal)	second (s)	8.616 409	E+04
hour (h)	second (s)	**3.6**	**E+03**
hour (sidereal)	second (s)	3.590 170	E+03
minute (min)	second (s)	**6.0**	**E+01**
minute (sidereal)	second (s)	5.983 617	E+01
second (sidereal)	second (s)	9.972 696	E−01
shake	second (s)	**1.0**	**E−08**
shake	nanosecond (ns)	**1.0**	**E+01**
year (365 days)	second (s)	**3.1536**	**E+07**
year (sidereal)	second (s)	3.155 815	E+07
year (tropical)	second (s)	3.155 693	E+07
Torque (see Moment of force)			
Velocity (includes speed)			
foot per hour (ft/h)	meter per second (m/s)	8.466 667	E−05
foot per minute (ft/min)	meter per second (m/s)	**5.08**	**E−03**
foot per second (ft/s)	meter per second (m/s)	**3.048**	**E−01**
inch per second (in/s)	meter per second (m/s)	**2.54**	**E−02**
kilometer per hour (km/h)	meter per second (m/s)	2.777 778	E−01
knot (nautical mile per hour)	meter per second (m/s)	5.144 444	E−01
mile per hour (mi/h)	meter per second (m/s)	**4.4704**	**E−01**
mile per hour (mi/h)	kilometer per hour (km/h)	**1.609 344**	**E+00**
mile per minute (mi/min)	meter per second (m/s)	**2.682 24**	**E+01**
mile per second (mi/s)	meter per second (m/s)	**1.609 344**	**E+03**
revolution per minute (rpm) (r/min)	radian per second (rad/s)	1.047 198	E−01
rpm (revolution per minute) (r/min)	radian per second (rad/s)	1.047 198	E−01
Viscosity, dynamic			
centipoise (cP)	pascal second (Pa · s)	**1.0**	**E−03**
poise (P)	pascal second (Pa · s)	**1.0**	**E−01**
poundal second per square foot	pascal second (Pa · s)	1.488 164	E+00
pound-force second per square foot (lbf · s/ft^2)	pascal second (Pa · s)	4.788 026	E+01
pound-force second per square in (lbf · s/in^2)	pascal second (Pa · s)	6.894 757	E+03
pound per foot hour [lb/(ft · h)]	pascal second (Pa · s)	4.133 789	E−04
pound per foot second [lb/(ft · s)]	pascal second (Pa · s)	1.488 164	E+00
rhe	reciprocal pascal second [(Pa · s)$^{-1}$]	**1.0**	**E+01**
slug per foot second [slug/(ft · s)]	pascal second (Pa · s)	4.788 026	E+01

continues next page

TABLE 6.1   SI conversion factors (Factors in **boldface** are exact) **(continued)**

To Convert From	To	Multiply By	
Viscosity, kinematic			
centistokes (cSt)	meter squared per second (m²/s)	**1.0**	**E−06**
square foot per second (ft²/s)	meter squared per second (m²/s)	9.290 304	E−02
stokes (St)	meter squared per second (m²/s)	**1.0**	**E−04**
Volume (includes capacity)			
acre-foot (based on U.S. survey foot)[1]	cubic meter (m³)	1.233 489	E+03
barrel [for petroleum, 42 gallons (U.S.)] (bbl)	cubic meter (m³)	1.589 873	E−01
barrel [for petroleum, 42 gallons (U.S.)] (bbl)	liter (L)	1.589 873	E+02
bushel (U.S.) (bu)	cubic meter (m³)	3.523 907	E−02
bushel (U.S.) (bu)	liter (L)	3.523 907	E+01
cord (128 ft³)	cubic meter (m³)	3.624 556	E+00
cubic foot (ft³)	cubic meter (m³)	2.831 685	E−02
cubic inch (in³)[16]	cubic meter (m³)	1.638 706	E−05
cubic mile (mi³)	cubic meter (m³)	4.168 182	E+09
cubic yard (yd³)	cubic meter (m³)	7.645 549	E−01
cup (U.S.)	cubic meter (m³)	2.365 882	E−04
cup (U.S.)	liter (L)	2.365 882	E−01
cup (U.S.)	milliliter (mL)	2.365 882	E+02
fluid ounce (U.S.) (fl oz)	cubic meter (m³)	2.957 353	E−05
fluid ounce (U.S.) (fl oz)	milliliter (mL)	2.957 353	E+01
gallon [Canadian and U.K. (Imperial)] (gal)	cubic meter (m³)	**4.546 09**	**E−03**
gallon [Canadian and U.K. (Imperial)] (gal)	liter (L)	**4.546 09**	**E+00**
gallon (U.S.) (gal)	cubic meter (m³)	3.785 412	E−03
gallon (U.S.) (gal)	liter (L)	3.785 412	E+00
gill [Canadian and U.K. (Imperial)] (gi)	cubic meter (m³)	1.420 653	E−04
gill [Canadian and U.K. (Imperial)] (gi)	liter (L)	1.420 653	E−01
gill (U.S.) (gi)	cubic meter (m³)	1.182 941	E−04
gill (U.S.) (gi)	liter (L)	1.182 941	E−01
liter (L)[17]	cubic meter (m³)	**1.0**	**E−03**
ounce [Canadian and U.K. fluid (Imperial)] (fl oz)	cubic meter (m³)	2.841 306	E−05
ounce [Canadian and U.K. fluid (Imperial)] (fl oz)	milliliter (mL)	2.841 306	E+01
ounce (U.S. fluid) (fl oz)	cubic meter (m³)	2.957 353	E−05
ounce (U.S. fluid) (fl oz)	milliliter (mL)	2.957 353	E+01
peck (U.S.) (pk)	cubic meter (m³)	8.809 768	E−03
peck (U.S.) (pk)	liter (L)	8.809 768	E+00
pint (U.S. dry) (dry pt)	cubic meter (m³)	5.506 105	E−04
pint (U.S. dry) (dry pt)	liter (L)	5.506 105	E−01
pint (U.S. liquid) (liq pt)	cubic meter (m³)	4.731 765	E−04
pint (U.S. liquid) (liq pt)	liter (L)	4.731 765	E−01
quart (U.S. dry) (dry qt)	cubic meter (m³)	1.101 221	E−03
quart (U.S. dry) (dry qt)	liter (L)	1.101 221	E+00
quart (U.S. liquid) (liq qt)	cubic meter (m³)	9.463 529	E−04
quart (U.S. liquid) (liq qt)	liter (L)	9.463 529	E−01
stere (st)	cubic meter (m³)	**1.0**	**E+00**
tablespoon	cubic meter (m³)	1.478 676	E−05
tablespoon	milliliter (mL)	1.478 676	E+01
teaspoon	cubic meter (m³)	4.928 922	E−06

continues next page

TABLE 6.1 SI conversion factors (Factors in **boldface** are exact) **(continued)**

To Convert From	To	Multiply By	
teaspoon	milliliter (mL)	4.928 922	E+00
ton, register	cubic meter (m³)	2.831 685	E+00
Volume divided by time (includes flow)			
cubic foot per minute (ft³/min)	cubic meter per second (m³/s)	4.719 474	E–04
cubic foot per minute (ft³/min)	liter per second (L/s)	4.719 474	E–01
cubic foot per second (ft³/s)	cubic meter per second (m³/s)	2.831 685	E–02
cubic inch per minute (in³/min)	cubic meter per second (m³/s)	2.731 177	E–07
cubic yard per minute (yd³/min)	cubic meter per second (m³/s)	1.274 258	E–02
gallon (U.S.) per day (gal/d)	cubic meter per second (m³/s)	4.381 264	E–08
gallon (U.S.) per day (gal/d)	liter per second (L/s)	4.381 264	E–05
gallon (U.S.) per minute (gpm) (gal/min)	cubic meter per second (m³/s)	6.309 020	E–05
gallon (U.S.) per minute (gpm) (gal/min)	liter per second (L/s)	6 309 020	E–02
Work (see Energy)			

[1] The U.S. Metric Law of 1866 gave the relationship 1 m = 39.37 in. (in. is the unit symbol for the inch). From 1893 until 1959, the yard was defined as being exactly equal to (3600/3937) m, and thus the foot was defined as being exactly equal to (1200/3937) m.
In 1959, the definition of the yard was changed to bring the U.S. yard and the yard used in other countries into agreement. Since then the yard has been defined as exactly equal to 0.9144 m, and thus the foot has been defined as exactly equal to 0.3048 m. At the same time it was decided that any data expressed in feet derived from geodetic surveys within the United States would continue to bear the relationship as defined in 1893; that is, 1 ft = (1200/3937) m (ft is the unit symbol for the foot). The name of this foot is "U.S. survey foot"; the name of the new foot defined in 1959 is "international foot." The two are related to each other through the expression 1 international foot = 0.999 998 U.S. survey foot exactly.

[2] This is a unit for the quantity second moment of area, which is sometimes called the "moment of section" or "area moment of inertia" of a plane section about a specified axis.

[3] The Fifth International Conference on the Properties of Steam (London, July 1956) defined the International Table calorie as 4.1868 J. Therefore, the exact conversion factor for the International Table Btu is 1.055 055 852 62 kJ. Note that the notation for International Table used in this listing is subscript "IT." Similarly, the notation for thermochemical is subscript "th." Further, the thermochemical Btu, Btu_{th}, is based on the thermochemical calorie, cal_{th}, where cal_{th} = 4.184 J exactly.

[4] The kilogram calorie or "large calorie" is an obsolete term used for the kilocalorie, which is the calorie used to express the energy content of foods. However, in practice, the prefix "kilo" is usually omitted.

[5] The therm (EC) is legally defined in the Council Directive of 20 December 1979, Council of the European Communities (now the European Union [EU]). The therm (U.S.) is legally defined in the *Federal Register* of July 27, 1968. Although the therm (EC), which is based on the International Table Btu, is frequently used by engineers in the United States, the therm (U.S.) is the legal unit used by the U.S. natural gas industry.

[6] Defined (not measured) value.

[7] If the local value of the acceleration of free fall is taken as g_n = 9.806 65 m/s² (the standard value), the exact conversion factor is 4.448 221 615 260 5 E+00.

[8] This conversion factor is based on 1 d = 86,400 s; and 1 Julian century = 36,525 d. (See U.S. Naval Observatory 1994.)

[9] The value of this unit, 1 nautical mile = 1852 m, was adopted by the First International Extraordinary Hydrographic Conference, Monaco, 1929, under the name "international nautical mile."

[10] The exact conversion factor is $10^4/\pi$.

[11] The exact conversion factor is 4.535 923 7 E–01.

[12] The darcy is a unit for expressing the permeability of porous solids, not area.

[13] One technical atmosphere equals one kilogram-force per square centimeter (1 at = kgf/cm²).

[14] Conversion factors for mercury manometer pressure units are calculated using the standard value for the acceleration of gravity and the density of mercury at the stated temperature. Additional digits are not justified because the definitions of the units do not take into account the compressibility of mercury or the change in density caused by the revised practical temperature scale, ITS-90. Similar comments also apply to water manometer pressure units.

[15] The centigrade temperature scale is obsolete; the degree centigrade is only approximately equal to the degree Celsius.

[16] The exact conversion factor is 1.638 706 4 E–05.

[17] In 1964 the General Conference on Weights and Measures reestablished the name "liter" as a special name for the cubic decimeter. Between 1901 and 1964 the liter was slightly larger (1.000 028 dm³); when using high-accuracy volume data from that time, keep this fact in mind.

Source: Taylor 1995 *(reprinted courtesy of the National Institute of Standards and Technology).*

FUNDAMENTAL PHYSICAL CONSTANTS

TABLE 6.2 Fundamental physical constants

Quantity	Symbol	Value and Unit	Relative Standard Uncertainty (u_r)
Speed of light in a vacuum	c, c_0	299 792 458 ms^{-1}	(exact)
Magnetic constant	μ_0	$4\pi \times 10^{-7}$ NA^{-2} = 12.566 370 614... $\times 10^{-7}$ NA^{-2}	(exact)
Electric constant, $1/\mu_0 c^2$	ϵ_0	8.854 187 817... $\times 10^{-12}$ Fm^{-1}	(exact)
Newtonian constant of gravitation	G	6.673(10) $\times 10^{-11}$ m^{3}kg^{-1}s^{-2}	1.5×10^{-3}
Planck constant	h	6.626 068 76(52) $\times 10^{-34}$ Js	7.8×10^{-8}
$h/2\pi$	$\hbar$	1.054 571 596(82) $\times 10^{-34}$ Js	7.8×10^{-8}
Elementary charge	e	1.602 176 462(63) $\times 10^{-19}$ C	3.9×10^{-8}
Magnetic flux quantum $h/2e$	Φ_0	2.067 833 636(81) $\times 10^{-15}$ Wb	3.9×10^{-8}
Conductance quantum $2e^2/h$	G_0	7.748 091 696(28) $\times 10^{-5}$ S	3.7×10^{-9}
Electron mass	m_e	9.109 381 88(72) $\times 10^{-31}$ kg	7.9×10^{-8}
Proton mass	m_p	1.672 621 58(13) $\times 10^{-27}$ kg	7.9×10^{-8}
Proton-electron mass ratio	m_p/m_e	1 836.152 6675(39)	2.1×10^{-9}
Fine structure constant $e^2/4\pi\epsilon_0 \hbar c$	α	7.297 352 533(27) $\times 10^{-3}$	3.7×10^{-9}
Inverse fine structure constant	α^{-1}	137.035 999 76(50)	3.7×10^{-9}
Rydberg constant $\alpha^2 m_e c/2h$	R_∞	10 973 731.568 549(83) m^{-1}	7.6×10^{-12}
Avogadro constant	N_A, L	6.022 141 99(47) $\times 10^{23}$ mol^{-1}	7.9×10^{-8}
Faraday constant $N_A e$	F	96 485.3415(39) C mol^{-1}	4.0×10^{-8}
Molar gas constant	R	8.314 472(15) J mol^{-1} K^{-1}	1.7×10^{-6}
Boltzmann constant R/N_A	k	1.380 6503(24) $\times 10^{-23}$ J K^{-1}	1.7×10^{-6}
Stefan-Boltzmann constant ($\pi^2/60)k^4/\hbar^3 c^2$	σ	5.670 400(40) $\times 10^{-8}$ W m^{-2} K^{-4}	7.0×10^{-6}
Electron volt: (e/C)J	eV	1.602 176 462(63) $\times 10^{-19}$ J	3.9×10^{-8}
(Unified) Atomic mass unit $1u = m_u = (1/12)m(^{12}C) = 10^{-3}$ kg mol$^{-1}/N_A$	u	1.660 538 73(13) $\times 10^{-27}$ kg	7.9×10^{-8}

Source: Mohr and Taylor 1999 (reprinted with permission of the American Institute of Physics and the National Institute of Standards and Technology [NIST]).

SELECTED CONSTANTS, MEASURES, AND TIME

TABLE 6.3 Selected constants, measures, and time

Item	Quantity	Sources
Acceleration of gravity, standard	32.174 0 ft/s^2 = 9.806 65 m/s^2	Taylor 1995
Density of dry air (sea level, 70°F)	0.075 0 lb/ft^3 = 1.201 kg/m^3	Ramani 1992
Density of water	62.4 lb/ft^3 = 8.345 lb/gal = 1 g/cm^3 = 1 000 kg/m^3	
e (natural logarithm base)	2.718 281 828 459 045	Liepman 1972
pi (π)	3.141 592 653 589 793	Liepman 1972
Speed of sound in dry air at 20°C	343.4 m/s	Sytchev et al. 1987
Speed of sound in water at 10°C	1 447.8 m/s	Lide 1997
Speed of sound in sea water at 10°C (salinity = 3.5%)	1 490.4 m/s	Lide 1997
Time to nearest second (coordinated universal)		Telephone (303) 499-7111; in the United States; Web site is www.boulder.nist.gov

SELECTED UNIT EQUIVALENCIES AND APPROXIMATIONS

TABLE 6.4 Selected unit equivalencies and approximations (Factors in **boldface** type are exact)

1 acre = **43 560** ft^2 (U.S. survey) = **160** rod^2 = 4 046. 873 m^2 = 0.404 687 3 hectare
1 acre-ft = **43 560** ft^3 (U.S. survey) = 325 851 gal (U.S.) = 1 233.489 m^3
1 assay ton = 29.166 667 g = 0.029 166 667 kg
1 atmosphere (standard) = = **1.013 25** bar = **1 013.25** millibar = 760 mm Hg = 33.90 ft H$_2$O = 29.92 in Hg = 14.696 lb/in^2 = 10 332.3 kgf/m^2 = **101 325** pascal (Pa)
1 atmosphere (technical) = **1** kgf/cm^2 = **98 066.5** Pa
1 barrel (U.S. oil) = **42** gal (U.S.) = 0.158 987 3 m^3
1 board foot = 1 ft × 1 ft × 1 in.
1 British thermal unit per pound (Btu/lb) = **2.326** kJ/kg
1 bushel [U.S.] (bu) = **4** peck (pk) = **32** qt (U.S. dry) = 0.035 239 07 m^3
1 bushel [heaped] (cone ≥ 6 in) ≈ 1.25 struck bushel
1 carat (metric) = **200** mg
1 chain (Gunter's) = **4** rod = **66** ft (U.S. survey) = **100** link = **0.012 5** mile (U.S. statute) = 20.116 84 m
1 cord = 4 ft × 4 ft × 8 ft = 128 ft^3 = 3.625 m^3
1 cubic foot (ft^3) = **1 728** in^3 = 0.037 037 04 yd^3 = 7.480 52 gal (U.S.) = 0.028 316 85 m^3
1 cubic foot per second (cusec or ft^3/sec) = 448.831 169 gal/min = 2.222 222 yd^3/min = 1.983 474 acre-ft/day = 0.028 316 85 m^3/s
1 cubic inch (in^3) = **16.387 064** cm^3
1 cubic yard (yd^3) = **27** ft^3 = **46 656** in^3 = 0.764 554 9 m^3
1 cup (U.S.) = **2** gill (U.S.) = **8** fl oz (U.S.) = **16** tablespoon = 236.588 2 mL
1 day = **24** h = **1 440** min = **86 400** s
1 degree (°) = **60** min (arc) = **3 600** sec (arc) = π/180 rad = 0.017 453 29 radian
1 fathom = **6** ft (U.S. survey) = 1.828 804 m
1 fluid ounce [U.S.] (fl oz) = **8** fluid dram (fl dr) = 1.804 69 in^3 = 29.573 53 mL
1 foot [International] (ft) = **12** inch (in) = **0.304 8** m
1 foot [U.S. survey] (ft) = **1 200/3 937** m = 0.304 800 6 m
1 foot pound-force (ft·lbf) = 1.355 818 newton meter (N·m)
1 foot of rock (sp. gr. 2.7) = 1.170 53 lbf/in^2 = 0.082 3 kgf/cm^2
1 foot of water (ft$_{H20}$) = 0.433 528 lb/in^2 = 2 989.067 Pa
1 flask of mercury ≈ 34.5 kg ≈ 76 lb (av) [historical variation]
1 furlong = **10** chain = **660** ft (U.S. survey) = 201.168 m
1 g_n (standard acceleration of free fall) = 32.174 05 ft/s^2 = **9.806 65** m/s^2
1 gallon (Imperial) = **8** pints (Imperial) = 4.546 09 L
1 gallon [U.S.] (gal) = **4** qt (U.S. liquid) = **231** in^3 = 0.133 681 ft^3 = 3.785 412 L
1 gill (U.S.) = **2^{-1}** cup = **4** fl oz (U.S.) = 118.294 1 mL
1 grain (av) = **1** grain (tr) = **1** grain (ap) = **7 000^{-1}** lb (av) = **5 760^{-1}** lb (tr) = **64.798 91** mg
1 hand = 4 in = 0.101 6 m
1 hertz (Hz) = **1** cycle per second (c/s) = **10^{-6}** MHz = **10^{-12}** fresnel
1 hogshead (U.S.) = 63 gal (U.S.) = 1.5 barrel (U.S. oil)
1 horsepower (electric) = **746** watt (W)
1 horsepower (hp) = **550** ft·lbf/s = **33 000** ft·lbf/min = 745.699 9 watt (W)
1 inch (in) = **12^{-1}** ft = **2.54** cm
1 karat (1 part in 24 of gold) = 41.666 667 mg/g
1 kilowatt hour (kW·h) = **3 600 000** joule (J) = 3412.141 16 Btu$_{IT}$
l league = **3** nautical mile = **5 556** m
1 link (surveyor's) = **0.66** ft (U.S. survey) = 0.201 168 m
1 mile [international] (mi) = **5 280** ft (international) = **1 609. 344** m.

continues next page

TABLE 6.4 Selected unit equivalencies and approximations (Factors in **boldface** type are exact) (continued)

1 mile [nautical] (nmi) = 1.150 78 mile (international) = 6 076. 115 ft (international) = **1 852** m
1 mile [U.S. statute based on U.S. survey foot] (mi) = **5,280** ft (U.S. survey) = **880** fathom = **320** rod = **8** furlong = 1 609.347 m
1 mile per hour (mi/h) = **88** ft/min = 1.466 667 ft/s = 1.609 344 km/h = 0.447 04 m/s
1 mile per minute (mi/min) = **5 280** ft/min = **60** mi/h = **88** ft/s = 26.822 4 m/s
1 miner's inch ≈ 1.5 ft^3 water per minute (historical variation with mining district and state)
1 ounce [apothecary] (ap oz) = **8** dram = **24** scruple = **480** grain = 31.103 48 gram (g)
1 ounce [avoirdupois] (av oz) = **16** dram (dr av) = **437.5** grain (gr) = 0.911 458 tr oz = 28.349 52 gram (g)
1 ounce [troy] (tr oz) = **20** pennyweight (dwt) = **480** grain (gr) = 1.097 143 av oz = 31.103 48 gram (g)
1 ounce (troy) per short ton = 34.285 718 gram/tonne
1 part per million (ppm) = **0.0001** percent = **1** g/m^3
1 peck [U.S.] (pk) = **8** qt (U.S. dry) = **4^{-1}** bushel = 537.605 in^3 = 8.809 768 liter (L)
1 pennyweight (dwt) = **24** grain = **20^{-1}** tr oz = 1.555 174 gram (g)
1 perch of masonry = 16.5 ft × 1.5 ft × 1 ft = 24.75 ft^3 ≈ 25 ft^3
1 pint (Imperial) = 0.568 261 25 L
1 pint [U.S. liquid] (liq pt) = **4** gill (U.S.) = **2^{-1}** qt (U.S. liquid) = **2** cup (U.S.) = **16** fl oz (U.S.) = 28.875 0 in^3 = 0.473 176 5 liter (L)
1 pound [avoirdupois] (lb av) = **16** oz av = **256** dram av = **7 000** grain = 1.215 278 lb tr = **453.592 37** g
1 pound [troy] (lb tr) = **12** oz tr = **240** pennyweight (dwt) = **5 760** grain (gr) = 0.822 857 lb av = 373.241 7 g
1 pound (av) per cubic foot (lb/ft^3) = 16.018 463 kg/m^3
1 pound per square inch (lb/in^2) = 2.306 659 ft$_{H2O}$ = 6894.757 Pa
1 quart [U.S. liquid] (liq qt) = **2** pt (U.S. liquid) = **32** fl oz (U.S.) = **57.75** in^3 = 0.946 352 9 liter (L)
1 radian = 57.295 78 degree (°) = 57° 17' 44.8" = **180**/π (°)
1 revolution (rev) = **1** turn = **360** degree (°) = **21 600** minute (') = **1 296 000** second (") = **2π** rad = 6.283 185 307 rad
1 rod (based on U.S. survey foot) = **25** link = **16.5** ft (U.S. survey) = 5.029 210 m
1 slug = 32.174 05 lb = 14.593 90 kg
1 span = 9 in = 0.228 6 m
1 square foot (ft^2) = **144** in^2 = **0.092 903 04** m^2
1 square inch (in^2) = **6.451 6** cm^2
1 square mile [international] (mi^2) = **27 878 400** ft^2 (international) = 2 589 988 m^2
1 square mile [U.S. statute] (mi^2) = **1** section = **640** acre = **36^{-1}** township = **27 878 400** ft^2 (U.S. survey) = 2 589 998 m^2
1 square yard (yd^2) = **9** ft^2 = **1 296** in^2 = **0.836 127 36** m^2
1 tablespoon = **3** teaspoon = **16^{-1}** cup = 14.786 76 mL
1 ton [long] (t) = **2 240** lb = 1 016.047 kg
1 ton [metric or tonne] (t) = **1 000 000** g = **1 000** kg = 1.102 311 ton (short) = 2 204. 623 lb av
1 ton [short] (t) = **2 000** lb = 907.184 7 kg = 0.907 184 7 ton (metric) or tonne
1 ton (short) per cubic yard (t/yd^3) = 1.186 553 tonne/m^3 = 1 186.553 kg/m^3
1 yard (yd) = **36** in = **3** ft = **0.914 4** m
1 watt (W) = 0.001 341 022 hp (550 ft·lbf/s) = 0.737 562 ft·lbf/s = 0.001 340 483 hp (electric) = **1** joule/sec

SI PREFIXES

TABLE 6.5 SI prefixes

Factor	Prefix	Symbol	Factor	Prefix	Symbol
10^1	deka	da	10^{-1}	deci	d
10^2	hecto	h	10^{-2}	centi	c
10^3	kilo	k	10^{-3}	milli	m
10^6	mega	M	10^{-6}	micro	µ
10^9	giga	G	10^{-9}	nano	n
10^{12}	tera	T	10^{-12}	pico	p
10^{15}	peta	P	10^{-15}	femto	f
10^{18}	exa	E	10^{-18}	atto	a
10^{21}	zetta	Z	10^{-21}	zepto	z
10^{24}	yotta	Y	10^{-24}	yocto	y

Source: Taylor 1995.

GREEK ALPHABET

TABLE 6.6 Greek alphabet

Name	Letters	Name	Letters
alpha	A α	nu	N ν
beta	B β	xi	Ξ ξ
gamma	Γ γ	omicron	O o
delta	Δ δ	pi	Π π
epsilon	E ε	rho	P ρ
zeta	Z ζ	sigma	Σ σ
eta	H η	tau	T τ
theta	Θ θ	upsilon	Y υ
iota	I ι	phi	Φ φ
kappa	K κ	chi	X χ
lambda	Λ λ	psi	Ψ ψ
mu	M µ	omega	Ω ω

REFERENCES

Lide, D.R., ed. 1997. *CRC Handbook of Chemistry and Physics.* 78th ed. Boca Raton, FL: CRC Press.

Liepman, D.S. 1972. Mathematical constants. In *Handbook of Mathematical Functions with Formulas, Graphs, and Mathematical Tables.* Applied Mathematics Series 55, 10th printing. Edited by M. Abramowitz and I.A. Stegun. National Bureau of Standards. Washington, DC: U.S. Government Printing Office. 1–3.

Mohr, P.J., and B.N. Taylor. 1999. CODATA Recommended Values of the Fundamental Physical Constants: 1998. *Journal of Physical and Chemical Reference Data* 28(6): 1713–1852. Washington, DC: American Chemical Society and the American Institute of Physics for NIST.

NIST Web site. Accessed August 2001. <http://www.physics.nist.gov/cuu/index.html>.

Ramani, R.V. 1992. Mine ventilation. In *SME Mining Engineering Handbook.* 2nd ed., Vol. 1. Edited by H.L. Hartman. Littleton, CO: Society for Mining, Metallurgy, and Exploration, Inc. 1052–1092.

Sytchev, V.V., A.A.Vasserman, A.D. Kozlov, G.A. Spiridonov, and V.A. Tsymarny. 1987. *Thermodynamic Properties of Air.* New York: Hemisphere Publishing.

Taylor, B.N. 1995. *Guide for the Use of the International System of Units (SI).* NIST Special Publication 811. Washington, DC: U.S. Government Printing Office.

U.S. Naval Observatory. 1994. *The Astronomical Almanac for the Year 1995.* Washington, DC: U.S. Government Printing Office. K6.

CHAPTER 7
Sampling and Analysis

Marcus A. Wiley, P.E.

SAMPLING

Sampling Theory
Sampling is a technique for obtaining objective, reliable information about a population. Because it is usually impractical to measure the entire population, samples that will represent the whole population must be selected. The sample contains only the facts that are available on the subject slice of the population. Therefore, any conclusions drawn about the total population are merely inferences and are subject to verification. Conclusions about a population are only as good as the data on which they are based.

Geologic Data Collection and Recording

TABLE 7.1　Geologic data collection: key features

Location Data
Sample, map, mine, or drill location on each sheet. May include geographic data such as state, county, section, township, range, latitude, longitude, coordinates, elevation, mining district, mine, pit, bench, level, working, claim, claim corner, or any and all information that will clearly identify the unique location of the geologic data points. Data cannot be used if the geologist does not know where they originated.
Lithologic Data
Typical data to describe rock, sample, or unit. Should include color, texture, mineralogic characteristics, lithology, and rock type. Appropriate descriptive modifiers, stratigraphic information (if known), top and bottom data, age relationships, and general gross features as hardness, competency, and bedding characteristics. Subjective generic terms should be avoided unless well-established or qualified to distinguish inference from observable facts. Primary sedimentary structure and sedimentological features such as bedding, laminations, casts, soft-sediment deformation, graded bedding, burrows, bioturbation, fossil content, or banding, foliation, and lineation, with appropriate attitudes, should be noted where possible.
Structural Data
Secondary structural features that post-date rock formation. Should include clear description and attitudes of joints, fractures, faults, breccias with quantitative description of selvages, gouge zones, fragment size, and healed or recemented character of breccias. Folds, dragfolds, crenulations, lineations, and foliation should be noted. Age relationships, mineralization association, and overall effect on rock mass are important. Weathering and oxidation intensity data are usually critical and commonly related to structure, but may be included with lithologic data. Where possible, quantification of structural data is extremely important because these data may play a key role in determining minability of a deposit.

continues next page

TABLE 7.1 Geologic data collection: key features (continued)

Alteration Data

Nature, mineralogy, intensity, and distribution of features. Should include color, texture, mineralogy, intensity, pervasiveness, fracture control, stages, mineralization association, and overall effect on rock mass. Weathering and oxidation intensity are important but may be included with lithologic data. Where possible, quantification is extremely useful, as is description of age relationships between various alteration features.

Mineralization Data

Nature, intensity, mineralogy, and distribution of the desired resource. Should include primary and secondary classification; estimates of specific and total quantity of various minerals; intensity; character of veinlet, vein, or disseminations; supergene features; weathering and oxidation intensity; and associated gangue mineralogy. Vein age relationships tied to mineralogy, alteration, or lithology offer important data for understanding both zoning and grade estimates and overall deposit genesis.

Coal Data

Logging of any and all features that aid in correlation and understanding the distribution of sedimentary facies, as well as construction of a depositional model of the coalbed(s) and coal-bearing sequence (in addition to the standard lithological and structural data listed above). Critical to include detailed descriptions of horizons immediately above and below the roof and floor and accurate measurements of depths and thickness of all units associated with the coal. Key features include abundance and type of marine or freshwater fossils, slickenside in roof or floor rocks, the presence of roots representing old soil horizons, pyrite bands, nodules or streaks, siderite or ironstone nodules, and plant debris. Description of individual coalbeds, either the banded or nonbanded groups, requires careful measurement or estimates of the banded lithotypes, vitrain, clarain, durain, and fusain content (Stopes 1919; Ward 1984). A more practical system (Schopf 1960) describes the thickness and amount or concentration of vitrain and fusain bands in a matrix of atrital coal. The latter is described by five luster levels that range from bright to dull. Nonbanded sapropelic coals, boghead, and cannel end-members descriptions rely on identification of these massive, faintly banded, fine-grained accumulations of algae or spores and usually require a microscope for adequate description. The nature of cleats, partings, bone, and shale layers needs description and careful thickness measurements to separate "net" from "gross" coalbed thicknesses. Coalbed description, although straightforward, requires some supervised training to ensure adequate data recording.

Other Features

Other features that may supply extremely important information with direct bearings on mining and/or metallurgy. This may be reasonably objective, such as fracture frequency, rock quality determination measurements, and longest and shortest unbroken core recovered in a run, or more subjective, such as an overall estimate of rock strength, friability, or competency. Metallurgically significant features such as hardness (which affects grindability), grain size (which controls grinding for particle liberation), or oxidation intensity (which affects flotation recoveries) should be noted. Added testing is almost always needed here; however, geologic data collections should indicate these and other potential problem areas requiring specialized study.

Source: Adapted from Erickson 1992.

Size, Frequency, or Density of Sample

For a sample to be representative of the whole population, it must be of the proper size and frequency. The larger the sample, and the greater the number of increments taken, the more likely that the sample will be representative of the whole. The following tables offer guidelines to assist in obtaining a representative sample.

Drill-hole size is an important consideration in obtaining a representative sample. A smaller drill-hole diameter can be used if the deposit is expected to be relatively homogeneous. For example, a 12-in. blast hole sample will contain 50 times the amount of sample that would be obtained in a BX-size core hole.

Table 7.3 furnishes diameter size data for commonly used wireline drills.

Sufficient material must be obtained and submitted to the testing laboratory to permit proper execution of all desired tests. The quantities given in Table 7.4 generally provide adequate material for routine grading and quality analysis. A rule of thumb for sample size is to obtain a 50-lb sample for each inch diameter of the coarsest particle size. A sample that is too

TABLE 7.2 Geologic data collection: useful equipment

Equipment routinely used in day-to-day geologic work:
- Compass
- Geologist's pick
- Hand lens
- Pocket tape
- Several cloth survey tapes or chains (nonmetallic for safety)
- Knife
- Protractor, or 6-in. (152-mm) plastic scales with imprinted protractor
- Dilute hydrochloric acid (HCl)
- Magnet and nail
- Survey spads
- Cement nails
- Sample bags and tags, plus a sample ring for holding sample bags during chip sampling
- Marking pens and aluminum sheet holder (preferably end-opening with an end-extractable pencil holder attached), pencils, pens, mapping sheets, or logging forms.

In addition, a field vest (with large pockets for samples) and clipboard is helpful. Access to brushes and water buckets is important for core logging, and spray bottles help. Extra core blocks are useful, as is a good binocular microscope.

Source: Adapted from Erickson 1992.

TABLE 7.3 Commonly used wireline core drilling sizes

Boyles Bros.*			CBC†			Longyear‡		
Designation	Core Diameter (in.)	Hole Diameter (in.)	Designation	Core Diameter (in.)	Hole Diameter (in.)	Designation	Core Diameter (in.)	Hole Diameter (in.)
AX	1.067	1.890	AXE	1.015	1.852	AQ	1.062	1.890
BX	1.432	2.360	BXE	1.437	2.385	BQ	1.432	2.360
		2.400						
		2.440						
NX	1.875	2.980	NXE	2.000	3.005	NQ	1.875	2.980
		3.032						
		3.125						
		3.250						
HX	2.400	3.650	NCE	2.406	3.685	NQ	2.500	3.782
		3.750						
		3.937						
CP	3.345	4.827	3-in.	3.000	3.915	PQ	3.345	4.827

* Personal communication, C. Hirschi, Boyles Bros., Reno, Nevada.
† Personal communication, N. Trujillo, CBC Drilling, Tucson, Arizona.
‡ Personal communication, A.O. Krause, Longyear Company, Peoria, Arizona.

Source: Metz 1992.

large can be reduced in the laboratory, but a sample that is too small is of no value and represents wasted effort.

For all in situ and non-in situ samples, the top size to which any sample is crushed is important for determining the weight of the sample required.

The spacing of the drill holes or other exploratory openings must be considered if a representative sample is to be obtained. Table 7.5 contains general guidelines for a horizontal deposit such as coal. A deposit with a complex structure would require spacing with a greater density.

The concept of error can be divided into two components that are referred to as precision (or repeatability) and accuracy (or lack of bias). Figure 7.1 illustrates the concepts of precision and accuracy by considering four targets. The shooting results shown on target A are both inaccurate and imprecise. Target B is imprecise but accurate. Target C is precise but inaccurate. And target D is both accurate and precise (Springett 1984).

The quality of a geologic deposit cannot be accurately represented by a single value. To obtain a representative sample, multiple increments must be taken. Three methods can be used: (1) systematic sampling, where increments are spaced evenly in time or position; (2) random sampling, where increments are spaced at random but a prerequisite number are taken; or (3) stratified random sampling, where the unit is divided by time or amount into a

TABLE 7.4 Required size of coal samples

	Mechanically Cleaned Coal		
Top size	5/8 in. (16 mm)	2 in. (50 mm)	6 in. (150 mm)
Minimum number of increments	15	15	15
Minimum weight per increment	2 lb (1 kg)	6 lb (3 kg)	15 lb (7 kg)
Total sample weight	30 lb (15 kg)	90 lb (45 kg)	225 lb (105 kg)
	Raw Uncleaned Coal		
Top size	5/8 in. (16 mm)	2 in. (50 mm)	6 in. (150 mm)
Minimum number of increments	35	35	35
Minimum weight per increment	2 lb (1 kg)	6 lb (3 kg)	15 lb (7 kg)
Total sample weight	70 lb (35 kg)	210 lb (105 kg)	525 (245 kg)

Source: Adapted from American Society for Testing and Materials (ASTM) D 2234-97a.

TABLE 7.5 Spacing of drill holes in exploration of coalfields

	Spacing (ft)*	
Type of Deposit	Measured Reserves	Indicated Reserves
Horizontal or gently sloping		
Uniform beds	3,000	6,000
Fairly consistent beds	1,500	3,000
Inconsistent beds	750	1,500
Deposits with simple first-order folding		
Uniform beds	1,000	2,000
Fairly consistent beds	750	1,500
Inconsistent beds	500	1,000
Deposits with complex folding and consequent faulting		
Uniform beds	750	1,500
Fairly consistent beds	400	750
Inconsistent beds	Explored during mining	—

* Grid dimensions for horizontal deposits; distances between exploratory profiles for folded and complex deposits.
Source: Adapted from Misaqi 1973.

number of equal strata and one or more increments are taken at random from each (Thomas 1992). Table 7.6 details certain minimum numbers of increments.

In Situ and Non-In Situ Sampling

In general, a better sample can be obtained by sampling the deposit or mineral commodity in its natural state (in situ). Channel samples and core drilling are examples of in situ sample techniques. Non-in situ sampling consists of sampling the material after it has been removed from its original deposit. An example of this would be a sample taken from a truck, boat, conveyor belt, or stockpile.

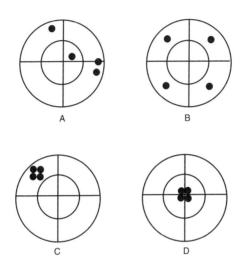

Source: Springett 1984.

FIGURE 7.1 Accuracy and precision

TABLE 7.6 Minimum numbers of increments required for a coal shipment up to 1,000 tonnes

Sampling situation	Common Sample for Total Moisture and General Analysis			Total Moisture Sample		General Analysis Sample		Size Analysis Sample
	Sized coals: dry cleaned or washed	Washed smalls (50 mm)	Blended, part treated, untreated, run of mine and "unknown" coals	Sized coals: dry cleaned or washed; unwashed dry coals	Washed smalls (50 mm), blended, part treated, untreated, run of mine, and "unknown" coals	Sized coals: dry cleaned, or washed or unwashed dry coals	Blended, part treated, untreated, run of mine, and "unknown" coals	All coals
Moving streams	20	35	35	20	35	20	35	40
Wagons and trucks, barges, grabs or conveyors unloading ships	25	35	50	20	35	25	50	40
Holds of ships, stockpiles	35	35	65	20	35	35	65	40

Source: Osborne 1988 *(reprinted with permission of Kluwer Law International).*

Grab Sampling   Randomly picking up a sample of the material can easily lead to a bias in the selection of the sample and inconclusive results. Grab sampling can, however, be used to obtain generalized information about the material sampled.

In Situ Sampling Methods

Channel Sampling   Channel sampling is recommended for obtaining a representative sample of a laminated deposit such as coal. Clean the area to be sampled to expose a fresh section. Normally, the material is sampled perpendicular to the bedding. Cut a channel of uniform width and depth into the seam and collect the material on a plastic sheet or metal plate placed on the floor. Figure 7.2 shows a composite method in which the entire cross section is collected as one sample, and Figure 7.3 illustrates a ply method in which each individual layer is sampled and recorded individually to make up the total cross section. This method allows additional information to be obtained—for example, the top of the seam might be high in sulfur and could be removed in a mining process before the remainder of the seam is recovered.

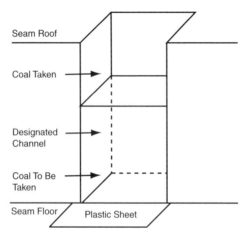

Source: Robertson Research 1987 (reprinted with permission of RobSearch Australia Pty. Ltd.).

FIGURE 7.2   Composite channel sample procedure

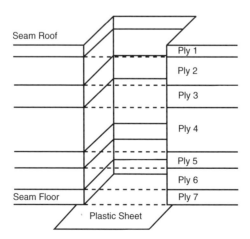

Source: Thomas 1992 (reprinted with permission of John Wiley & Sons Ltd.).

FIGURE 7.3   Ply channel sample procedure

Core Sampling Core sampling is the preferred method for obtaining a nonweathered sample of the deposit, along with roof and floor material that could make up seam dilution in mining operations. The sample can be recorded as a whole, or in plies (individual components), as Figure 7.4 shows for a coal seam. Bag, label, and record individual samples.

Cuttings Sampling This method is considerably less accurate than core sampling. As drilling progresses down-hole, samples of bit cuttings are collected over specified intervals. Although this method is better than no information at all, it gives only a generalized representation of the material sampled.

Bulk Sampling Larger volume samples taken from an outcrop or small pit or shaft are called bulk samples. They offer a larger representation of the material to be sampled than is available through a borehole. This method is generally used when size consistency, washability, or variability data are needed.

Non-In Situ Sampling Methods

Collecting samples after the mineral has been mined is required to determine the "as-mined" or "as-shipped" quality. These data will vary from the in situ quality because of mining dilution or processing after mining. The sample taken should be random and should consist of the entire cross section of material. Conduct the sampling in accordance with ASTM guidelines. The following procedures are a quick guide to obtaining accurate samples.

Stockpile Sample Procedure Mined minerals are commonly stored in a stockpile. Figure 7.5 illustrates the problem of obtaining a representative sample because of segregation and material crumbling within the pile. Avoid taking stockpile samples if possible.

But if sampling a stockpile is the only method possible, insert a sheet of thin metal or plywood (approximately 2 ft by 2 ft) corner first, and horizontally to a depth of at least 1 ft into the stockpile and one-third of the aboveground total height of the stockpile. The sheet should be inserted directly above the area to be sampled and at several locations around the stockpile. Obtain the sample from the stockpile directly below the sheet of metal or plywood. Discard material to a depth of approximately 6 in. by removing it with a shovel. Take samples until a sufficient amount of the material is obtained (Johnson and Forst 2000).

Conveyor Belt Sample Procedure A representative sample can be obtained from a conveyor belt. Take a minimum of three increments and combine them into one field sample. Use a sweep-type sampler if available. Stop the belt and clear the material at least 6 in. ahead of

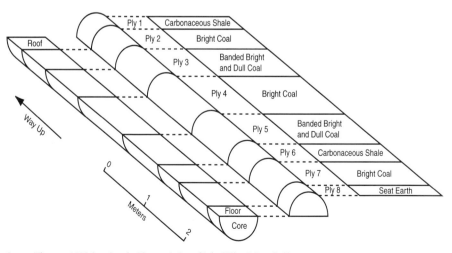

Source: Thomas 1992 (reprinted with permission of John Wiley & Sons Ltd.).

FIGURE 7.4 Ply sampling of borehole core

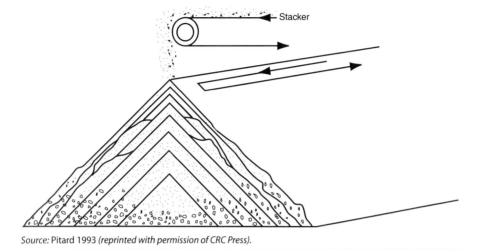

Source: Pitard 1993 *(reprinted with permission of CRC Press).*

FIGURE 7.5 Stockpile cross section

and behind the sample. Carefully scoop all the material between the cleared areas into a sample bag, then collect the fines on the belt with a brush and add them to the sample bag (Johnson and Forst 2000).

Auger Sampling Procedure Augers are generally used to obtain samples of material from transportation units such as trucks, rail cars, and barges. Take all increments at the full depth of the material, and be careful to exclude any segregation or stratification of the material. Collect a minimum of three secondary crushed increments per auger increment, evenly spaced throughout (Johnson and Forst 2000).

Sample Splitting

To analyze sampled material, it must first be reduced in size. Figure 7.6 is a sample reduction diagram that illustrates requirements for a gold operation.

The best method for reducing the size of a sample is to use a true mechanical splitting device, which is typically available in a laboratory. In the field, however, to reduce the amount of material to send to a lab while maintaining representation, use the procedures that follow.

Coning and Quartering Coning and quartering, illustrated in Figure 7.7, is a very old method. The material to be sampled is mixed and carefully piled on top of a conical heap. Flatten the cone into a flat circular cake that maintains the cone's symmetry, and then divide this cake into four identical quarters. Take care to divide the cake with a 90-degree cross-angle so that each section is strictly equivalent, then take the sample from one of the quarters or a combination of two quarters. Although this method is useful for splitting small samples, the alternate shovel method is an easier method for obtaining larger samples.

Alternate Shoveling Figure 7.8 illustrates the alternate shoveling method. Extract the material to be sampled one shovelful at a time and place it in two alternating distinct heaps. All shovelfuls should be approximately the same size, and each heap should consist of the same number of shovelfuls. One heap should contain only odd increments and the other only even increments. When the sample is selected at random, the sampling equity is preserved in the reduced sample.

Fractional Shoveling Another method of sampling is shown in Figure 7.9. In this method, place the material to be sampled, one shovelful at a time, in several distinct heaps. All shovelfuls are approximately the same size and each heap contains the same number of shovel increments. Take a sample at random from one or a combination of the individual heaps. The illustration shows this method with five individual heaps.

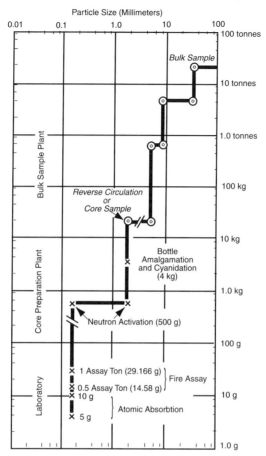

Source: Springett 1984.

FIGURE 7.6 Sample reduction diagram

Sample Preparation

To maintain the integrity of a sample after it has been obtained in the field, yet to reduce it for analysis in a laboratory, use the procedures that follow.

Table 7.7 outlines a procedure for a relatively homogeneous material with small ore particles; Table 7.8 describes a procedure for a nuggety-type ore such as gold.

The usual method for reducing a core sample is to split the core and then crush one-half of it to ¼-in. After this ¼-in. material is blended, take a subsample of 1 or 2 lb, then grind this material to 100 mesh, blend again, and keep a few grams for analysis.

Figure 7.10 is a diagram illustrating sample preparation and reduction. Each point represents a given quantity of material crushed to a given size. Sample preparation is a succession of reductions in weight and grain size. At each vertical line of the stairway, error is involved.

Figure 7.11 shows a sample preparation procedure for reducing a bulk sample of steam coal.

Sampling Error

Table 7.9 contains a formula for calculating minimum sampling error.

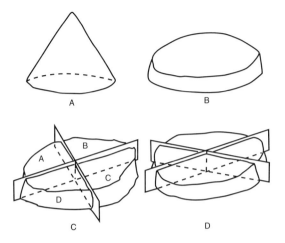

(A) Coning. (B) Flattening. (C) Correct quartering. (D) Incorrect quartering.

Source: Pitard 1993 *(reprinted with permission of CRC Press).*

FIGURE 7.7 Coning and quartering

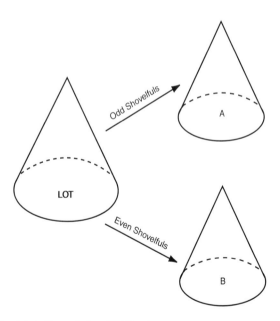

Source: Pitard 1993 *(reprinted with permission of CRC Press).*

FIGURE 7.8 Alternate shoveling

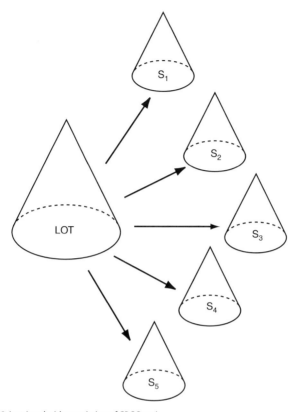

Source: Pitard 1993 (reprinted with permission of CRC Press).

FIGURE 7.9   Fractional shoveling

ANALYSIS

The following list shows instruments and abbreviations now being used for analyzing geologic materials (Cardwell 1984).

- Atomic absorption (AA, AAS)
  - Hydride generation
  - Graphite furnace
  - Cold vapor generation
- Colorimetry
- Fluorimetry
- Emission spectrometry (ES, E-Spec)
- Inductively coupled plasma (ICP, DCICP, ICAP)
- X-ray fluorescence (XRF)
- Neutron activation (INAA, NA, PGAA)
- Electron microprobe (EM)
- Ion chromatography (IC).

Table 7.10 gives guidelines for the instrument or method to be used, listed by element of interest.

TABLE 7.7   Representative sample preparation procedure

Dry Sample
(Typical weight 2 to 10 lb, or 1.0 to 4.5 kg)
Crush
(Typical product 8 to 10 mesh, or 2.4 to 2.0 mm)
Riffle Split
(Retain 1/2 lb, or 250 g)
Pulverize
(Typical product 100 to150 mesh, or 150 to 100 µm)

Source: Gumble, Post, and Hill 1992.

TABLE 7.8   Sample preparation procedure for nuggety materials

Dry Sample
(Typical weight 8 to 20 lb, or 3.6 to 9.0 kg)
Crush
(Typical product 8 to 10 mesh, or 2.4 to 2.0 mm)
Riffle Split
(Retain 4 to 10 lb, or 1.8 to 4.5 kg)
Plate Pulverize
(Product 40 mesh, 420 µm, or finer)
Riffle Split
(Retain 1 to 2 lb, or 450 to 900 g)
Pulverize in Ring Mill
(Typical product 100 to 150 mesh, or 150 to 100 µm)

Source: Gumble, Post, and Hill 1992.

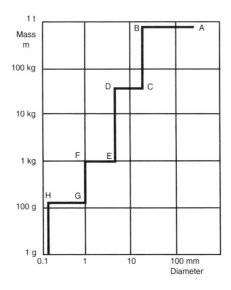

Source: David 1977.

FIGURE 7.10   Schematic representation of sample preparation

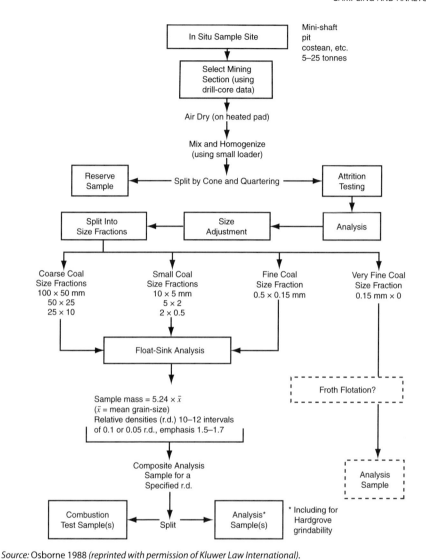

FIGURE 7.11 Coal sample preparation diagram

Source: Osborne 1988 *(reprinted with permission of Kluwer Law International).*

TABLE 7.9 Pierre Gy's formula for calculating minimum sampling error

$$\sigma^2 FE = \left(\frac{1}{m_s} - \frac{1}{m_L}\right)\frac{1-a_l}{a_l}[(1-a_l)\lambda_c + (a_l + \lambda_g)]lfgd^3$$

$\sigma^2 FE$*	(Sampling error in %)²
m_s	Sample weight in grams
m_L	Total sample weight in grams
a_l	Mineral content expressed in decimal %
λ_c	Specific gravity of mineral g/cm³
λ_g	Specific gravity of gangue g/cm³
l	Liberation factor
f	Particle shape factor
d	Maximum particle size in sample expressed in cm
g	Granulometric factor

* Total sampling error = $\sqrt{\Sigma \sigma^2 FE}$.
Source: Schwarz, Weber, and Erickson 1992.

TABLE 7.10 Commonly used analysis methods

Element	Digestion	Analysis by
Au	HCl, HNO₃ Fusion	AA Fire assay
Ag	HCl, HNO₃ Fusion	AA Fire assay
Hg	HNO₃ Thermal	AA cold vapor Au film
As	HClO₄ + other various acids	AA hydride AA furnace
Sb	Various acids Fusion	AA hydride AA
Mo	HClO₄, H₂SO₄	AA Colorimetric
Cu	HClO₄ and various acids	AA Titrametric
Pb	HClO₄ and various acids	AA Titrametric
Zn	HClO₄ and various acids	AA Titrametric
F	Fusion	Specific ion electrode
Tl	Various acids	AA Colorimetric
Te	Various acids	AA Colorimetric
W	Fusion	Colorimetric
Ba	Fusion	AA ICP
Sn	Fusion	AA
S		Leco induction furnace
Mn	Various Acids	AA Colorimetric ICP
B		ICP

Source: Cardwell 1984.

REFERENCES

American Society for Testing and Materials (ASTM). 1997. *Standard Practice for Collection of a Gross Sample of Coal.* Designation: D 2234-97a. West Conshohocken, PA: ASTM.

Cardwell, G.J. 1984. Analytical methods for applied geology. In *Applied Mining Geology.* Edited by A.J. Erickson Jr. New York: Society of Mining Engineers of the American Institute of Mining, Metallurgical, and Petroleum Engineers, Inc. 197–203.

David, M. 1977. *Geostatistical Ore Reserve Estimation.* Amsterdam: Elsevier Scientific Publishing.

Erickson, A.J. Jr. 1992. Geologic data collection and recording. In *SME Mining Engineering Handbook.* 2nd ed., Vol. 1. Edited by H.L. Hartman. Littleton, CO: Society for Mining, Metallurgy, and Exploration, Inc. (SME). 288–313.

Gumble, G.E, E.V. Post, and W.E. Hill Jr. 1992. Sample preparation and assaying. In *SME Mining Engineering Handbook.* 2nd ed., Vol. 1. Edited by H.L. Hartman. Littleton, CO: SME. 327–332.

Johnson, R.A., and M.D. Forst. 2000. Unpublished presentation notes from annual sampling seminar, Construction Consulting and Testing, Inc., Waterville, Ohio.

Metz, R.A. 1992. Sample collection. In *SME Mining Engineering Handbook.* 2nd ed., Vol. 2. Edited by H.L. Hartman. Littleton, CO: SME. 314–326.

Misaqi, F.L. 1973. Special exploration techniques—coal and petroliferous solids. In *SME Mining Engineering Handbook.* Vol. 1. Edited by A.B. Cummins and I.A. Given. New York: Society of Mining Engineers of the American Institute of Mining, Metallurgical, and Petroleum Engineers, Inc. 5-50–5-53.

Osborne, D. 1988. *Coal Preparation Technology.* 2 Vols. London: Graham & Trotman.

Pitard, F.F. 1993. *Pierre Gy's Sampling Theory and Sampling Practice.* Boca Raton, FL: CRC Press.

Robertson Research (Australia) Pty. Ltd. 1987. *Coal Geologist's Manual.* Edited by P.G. Strauss and C.M. Atkinson. North Sydney, Australia: Robertson Research.

Schwarz, F.P. Jr., S.M. Weber, and A.J. Erickson Jr. 1992. Quality control of sample preparation at the Mount Hope Molybdenum Prospect, Eureka Country, Nevada. In *Applied Mining Geology.* Edited by A.J. Erickson Jr. New York: Society of Mining Engineers of the American Institute of Mining, Metallurgical, and Petroleum Engineers, Inc. 175–187.

Springett, M.W. 1984. Sampling practices and problems. In *Applied Mining Geology.* Edited by A.J. Erickson Jr. New York: Society of Mining Engineers of the American Institute of Mining, Metallurgical, and Petroleum Engineers, Inc. 189–195.

Thomas, L. 1992. *Handbook of Practical Coal Geology.* West Sussex, England: John Wiley & Sons.

CHAPTER 8

Economics and Costing

Marcus A. Wiley, P.E.

INTEREST FORMULAS (AFTER STERMOLE AND STERMOLE 1996)

Discrete Interest End-of-Period Dollar Values

Single payment compound amount factor: $F/P_{i,n} = (1 + i)^n$ (EQ 8.1)

Single payment present worth factor: $P/F_{i,n} = 1/(1 + i)^n$ (EQ 8.2)

Uniform series compound amount factor: $F/A_{i,n} = [(1 + i)^n - 1]/i$ (EQ 8.3)

Sinking fund deposit factor: $A/F_{i,n} = i/[(1 + i)^n - 1]$ (EQ 8.4)

Capital recovery factor: $A/P_{i,n} = i(1 + i)^n/[(1 + i)^n - 1]$ (EQ 8.5)

Uniform series present worth factor: $P/A_{i,n} = [(1 + i)^n - 1]/i(1 + i)^n$ (EQ 8.6)

Arithmetic gradient series factor: $A/G_{i,n} = (1/i) - \{n/[(1 + i)^n - 1]\}$ (EQ 8.7)

Effective annual interest rate: $E = (1 + i)^m - 1$ (EQ 8.8)

Continuous Interest Lump Sum End-of-Period Dollar Values

$$F/P_{r,n} = e^{nr}$$ (EQ 8.9)

$$P/F_{r,n} = 1/e^{nr}$$ (EQ 8.10)

$$F/A_{r,n} = (e^{nr} - 1)/(e^r - 1)$$ (EQ 8.11)

$$A/F_{r,n} = (e^r - 1)/(e^{nr} - 1)$$ (EQ 8.12)

$$P/A_{r,n} = (e^{nr} - 1)/(e^r - 1)e^{nr}$$ (EQ 8.13)

$$A/P_{r,n} = (e^r - 1)e^{nr}/(e^{nr} - 1)$$ (EQ 8.14)

where:
- A = uniform amount per interest period
- E = effective annual interest
- e = base of natural log (ln) = 2.7182818...
- F = future worth, value, or amount
- G = uniform gradient amount per interest period
- i = interest rate per period
- m = number of compounding periods per year
- n = number of compounding periods
- r = nominal interest rate compounded continuously
- P = present worth, value, or amount.

DISCRETE COMPOUND INTEREST TABLES

TABLE 8.1 Single payment compound amount factor ($F/P_{i,n}$)

	1%	2%	3%	4%	5%	6%	7%	8%
1	1.01000	1.02000	1.03000	1.04000	1.05000	1.06000	1.07000	1.08000
2	1.02010	1.04040	1.06090	1.08160	1.10250	1.12360	1.14490	1.16640
3	1.03030	1.06121	1.09273	1.12486	1.15763	1.19102	1.22504	1.25971
4	1.04060	1.08243	1.12551	1.16986	1.21551	1.26248	1.31080	1.36049
5	1.05101	1.10408	1.15927	1.21665	1.27628	1.33823	1.40255	1.46933
6	1.06152	1.12616	1.19405	1.26532	1.34010	1.41852	1.50073	1.58687
7	1.07214	1.14869	1.22987	1.31593	1.40710	1.50363	1.60578	1.71382
8	1.08286	1.17166	1.26677	1.36857	1.47746	1.59385	1.71819	1.85093
9	1.09369	1.19509	1.30477	1.42331	1.55133	1.68948	1.83846	1.99900
10	1.10462	1.21899	1.34392	1.48024	1.62889	1.79085	1.96715	2.15892
11	1.11567	1.24337	1.38423	1.53945	1.71034	1.89830	2.10485	2.33164
12	1.12683	1.26824	1.42576	1.60103	1.79586	2.01220	2.25219	2.51817
13	1.13809	1.29361	1.46853	1.66507	1.88565	2.13293	2.40985	2.71962
14	1.14947	1.31948	1.51259	1.73168	1.97993	2.26090	2.57853	2.93719
15	1.16097	1.34587	1.55797	1.80094	2.07893	2.39656	2.75903	3.17217
16	1.17258	1.37279	1.60471	1.87298	2.18287	2.54035	2.95216	3.42594
17	1.18430	1.40024	1.65285	1.94790	2.29202	2.69277	3.15882	3.70002
18	1.19615	1.42825	1.70243	2.02582	2.40662	2.85434	3.37993	3.99602
19	1.20811	1.45681	1.75351	2.10685	2.52695	3.02560	3.61653	4.31570
20	1.22019	1.48595	1.80611	2.19112	2.65330	3.20714	3.86968	4.66096
21	1.23239	1.51567	1.86029	2.27877	2.78596	3.39956	4.14056	5.03383
22	1.24472	1.54598	1.91610	2.36992	2.92526	3.60354	4.43040	5.43654
23	1.25716	1.57690	1.97359	2.46472	3.07152	3.81975	4.74053	5.87146
24	1.26973	1.60844	2.03279	2.56330	3.22510	4.04893	5.07237	6.34118
25	1.28243	1.64061	2.09378	2.66584	3.38635	4.29187	5.42743	6.84848

	9%	10%	12%	15%	20%	25%	30%	40%
1	1.090	1.100	1.120	1.150	1.200	1.250	1.300	1.400
2	1.188	1.210	1.254	1.323	1.440	1.563	1.690	1.960
3	1.295	1.331	1.405	1.521	1.728	1.953	2.197	2.744
4	1.412	1.464	1.574	1.749	2.074	2.441	2.856	3.842
5	1.539	1.611	1.762	2.011	2.488	3.052	3.713	5.378
6	1.677	1.772	1.974	2.313	2.986	3.815	4.827	7.530
7	1.828	1.949	2.211	2.660	3.583	4.768	6.275	10.541
8	1.993	2.144	2.476	3.059	4.300	5.960	8.157	14.758
9	2.172	2.358	2.773	3.518	5.160	7.451	10.604	20.661
10	2.367	2.594	3.106	4.046	6.192	9.313	13.786	28.925
11	2.580	2.853	3.479	4.652	7.430	11.642	17.922	40.496
12	2.813	3.138	3.896	5.350	8.916	14.552	23.298	56.694
13	3.066	3.452	4.363	6.153	10.699	18.190	30.288	79.371
14	3.342	3.797	4.887	7.076	12.839	22.737	39.374	111.120
15	3.642	4.177	5.474	8.137	15.407	28.422	51.186	155.568
16	3.970	4.595	6.130	9.358	18.488	35.527	66.542	217.795
17	4.328	5.054	6.866	10.761	22.186	44.409	86.504	304.913
18	4.717	5.560	7.690	12.375	26.623	55.511	112.455	426.879
19	5.142	6.116	8.613	14.232	31.948	69.389	146.192	597.630
20	5.604	6.727	9.646	16.367	38.338	86.736	190.050	836.683
21	6.109	7.400	10.804	18.822	46.005	108.420	247.065	1,171.356
22	6.659	8.140	12.100	21.645	55.206	135.525	321.184	1,639.898
23	7.258	8.954	13.552	24.891	66.247	169.407	417.539	2,295.857
24	7.911	9.850	15.179	28.625	79.497	211.758	542.801	3,214.200
25	8.623	10.835	17.000	32.919	95.396	264.698	705.641	4,499.880

ECONOMICS AND COSTING | 163

TABLE 8.2 Single payment present worth factor ($P/F_{i,n}$)

n	1%	2%	3%	4%	5%	6%	7%	8%
1	0.99010	0.98039	0.97087	0.96154	0.95238	0.94340	0.93458	0.92593
2	0.98030	0.96117	0.94260	0.92456	0.90703	0.89000	0.87344	0.85734
3	0.97059	0.94232	0.91514	0.88900	0.86384	0.83962	0.81630	0.79383
4	0.96098	0.92385	0.88849	0.85480	0.82270	0.79209	0.76290	0.73503
5	0.95147	0.90573	0.86261	0.82193	0.78353	0.74726	0.71299	0.68058
6	0.94205	0.88797	0.83748	0.79031	0.74622	0.70496	0.66634	0.63017
7	0.93272	0.87056	0.81309	0.75992	0.71068	0.66506	0.62275	0.58349
8	0.92348	0.85349	0.78941	0.73069	0.67684	0.62741	0.58201	0.54027
9	0.91434	0.83676	0.76642	0.70259	0.64461	0.59190	0.54393	0.50025
10	0.90529	0.82035	0.74409	0.67556	0.61391	0.55839	0.50835	0.46319
11	0.89632	0.80426	0.72242	0.64958	0.58468	0.52679	0.47509	0.42888
12	0.88745	0.78849	0.70138	0.62460	0.55684	0.49697	0.44401	0.39711
13	0.87866	0.77303	0.68095	0.60057	0.53032	0.46884	0.41496	0.36770
14	0.86996	0.75788	0.66112	0.57748	0.50507	0.44230	0.38782	0.34046
15	0.86135	0.74301	0.64186	0.55526	0.48102	0.41727	0.36245	0.31524
16	0.85282	0.72845	0.62317	0.53391	0.45811	0.39365	0.33873	0.29189
17	0.84438	0.71416	0.60502	0.51337	0.43630	0.37136	0.31657	0.27027
18	0.83602	0.70016	0.58739	0.49363	0.41552	0.35034	0.29586	0.25025
19	0.82774	0.68643	0.57029	0.47464	0.39573	0.33051	0.27651	0.23171
20	0.81954	0.67297	0.55368	0.45639	0.37689	0.31180	0.25842	0.21455
21	0.81143	0.65978	0.53755	0.43883	0.35894	0.29416	0.24151	0.19866
22	0.80340	0.64684	0.52189	0.42196	0.34185	0.27751	0.22571	0.18394
23	0.79544	0.63416	0.50669	0.40573	0.32557	0.26180	0.21095	0.17032
24	0.78757	0.62172	0.49193	0.39012	0.31007	0.24698	0.19715	0.15770
25	0.77977	0.60953	0.47761	0.37512	0.29530	0.23300	0.18425	0.14602
n	9%	10%	12%	15%	20%	25%	30%	40%
1	0.91743	0.90909	0.89286	0.86957	0.83333	0.80000	0.76923	0.71429
2	0.84168	0.82645	0.79719	0.75614	0.69444	0.64000	0.59172	0.51020
3	0.77218	0.75131	0.71178	0.65752	0.57870	0.51200	0.45517	0.36443
4	0.70843	0.68301	0.63552	0.57175	0.48225	0.40960	0.35013	0.26031
5	0.64993	0.62092	0.56743	0.49718	0.40188	0.32768	0.26933	0.18593
6	0.59627	0.56447	0.50663	0.43233	0.33490	0.26214	0.20718	0.13281
7	0.54703	0.51316	0.45235	0.37594	0.27908	0.20972	0.15937	0.09486
8	0.50187	0.46651	0.40388	0.32690	0.23257	0.16777	0.12259	0.06776
9	0.46043	0.42410	0.36061	0.28426	0.19381	0.13422	0.09430	0.04840
10	0.42241	0.38554	0.32197	0.24718	0.16151	0.10737	0.07254	0.03457
11	0.38753	0.35049	0.28748	0.21494	0.13459	0.08590	0.05580	0.02469
12	0.35553	0.31863	0.25668	0.18691	0.11216	0.06872	0.04292	0.01764
13	0.32618	0.28966	0.22917	0.16253	0.09346	0.05498	0.03302	0.01260
14	0.29925	0.26333	0.20462	0.14133	0.07789	0.04398	0.02540	0.00900
15	0.27454	0.23939	0.18270	0.12289	0.06491	0.03518	0.01954	0.00643
16	0.25187	0.21763	0.16312	0.10686	0.05409	0.02815	0.01503	0.00459
17	0.23107	0.19784	0.14564	0.09293	0.04507	0.02252	0.01156	0.00328
18	0.21199	0.17986	0.13004	0.08081	0.03756	0.01801	0.00889	0.00234
19	0.19449	0.16351	0.11611	0.07027	0.03130	0.01441	0.00684	0.00167
20	0.17843	0.14864	0.10367	0.06110	0.02608	0.01153	0.00526	0.00120
21	0.16370	0.13513	0.09256	0.05313	0.02174	0.00922	0.00405	0.00085
22	0.15018	0.12285	0.08264	0.04620	0.01811	0.00738	0.00311	0.00061
23	0.13778	0.11168	0.07379	0.04017	0.01509	0.00590	0.00239	0.00044
24	0.12640	0.10153	0.06588	0.03493	0.01258	0.00472	0.00184	0.00031
25	0.11597	0.09230	0.05882	0.03038	0.01048	0.00378	0.00142	0.00022

TABLE 8.3 Uniform series compound amount factor ($F/A_{i,n}$)

n	1%	2%	3%	4%	5%	6%	7%	8%
1	1.00000	1.00000	1.00000	1.00000	1.00000	1.00000	1.00000	1.00000
2	2.01000	2.02000	2.03000	2.04000	2.05000	2.06000	2.07000	2.08000
3	3.03010	3.06040	3.09090	3.12160	3.15250	3.18360	3.21490	3.24640
4	4.06040	4.12161	4.18363	4.24646	4.31013	4.37462	4.43994	4.50611
5	5.10101	5.20404	5.30914	5.41632	5.52563	5.63709	5.75074	5.86660
6	6.15202	6.30812	6.46841	6.63298	6.80191	6.97532	7.15329	7.33593
7	7.21354	7.43428	7.66246	7.89829	8.14201	8.39384	8.65402	8.92280
8	8.28567	8.58297	8.89234	9.21423	9.54911	9.89747	10.25980	10.63663
9	9.36853	9.75463	10.15911	10.58280	11.02656	11.49132	11.97799	12.48756
10	10.46221	10.94972	11.46388	12.00611	12.57789	13.18079	13.81645	14.48656
11	11.56683	12.16872	12.80780	13.48635	14.20679	14.97164	15.78360	16.64549
12	12.68250	13.41209	14.19203	15.02581	15.91713	16.86994	17.88845	18.97713
13	13.80933	14.68033	15.61779	16.62684	17.71298	18.88214	20.14064	21.49530
14	14.94742	15.97394	17.08632	18.29191	19.59863	21.01507	22.55049	24.21492
15	16.09690	17.29342	18.59891	20.02359	21.57856	23.27597	25.12902	27.15211
16	17.25786	18.63929	20.15688	21.82453	23.65749	25.67253	27.88805	30.32428
17	18.43044	20.01207	21.76159	23.69751	25.84037	28.21288	30.84022	33.75023
18	19.61475	21.41231	23.41444	25.64541	28.13238	30.90565	33.99903	37.45024
19	20.81090	22.84056	25.11687	27.67123	30.53900	33.75999	37.37896	41.44626
20	22.01900	24.29737	26.87037	29.77808	33.06595	36.78559	40.99549	45.76196
21	23.23919	25.78332	28.67649	31.96920	35.71925	39.99273	44.86518	50.42292
22	24.47159	27.29898	30.53678	34.24797	38.50521	43.39229	49.00574	55.45676
23	25.71630	28.84496	32.45288	36.61789	41.43048	46.99583	53.43614	60.89330
24	26.97346	30.42186	34.42647	39.08260	44.50200	50.81558	58.17667	66.76476
25	28.24320	32.03030	36.45926	41.64591	47.72710	54.86451	63.24904	73.10594

n	9%	10%	12%	15%	20%	25%	30%	40%
1	1.0000	1.0000	1.0000	1.0000	1.0000	1.0000	1.0000	1.0000
2	2.0900	2.1000	2.1200	2.1500	2.2000	2.2500	2.3000	2.4000
3	3.2781	3.3100	3.3744	3.4725	3.6400	3.8125	3.9900	4.3600
4	4.5731	4.6410	4.7793	4.9934	5.3680	5.7656	6.1870	7.1040
5	5.9847	6.1051	6.3528	6.7424	7.4416	8.2070	9.0431	10.9456
6	7.5233	7.7156	8.1152	8.7537	9.9299	11.2588	12.7560	16.3238
7	9.2004	9.4872	10.0890	11.0668	12.9159	15.0735	17.5828	23.8534
8	11.0285	11.4359	12.2997	13.7268	16.4991	19.8419	23.8577	34.3947
9	13.0210	13.5795	14.7757	16.7858	20.7989	25.8023	32.0150	49.1526
10	15.1929	15.9374	17.5487	20.3037	25.9587	33.2529	42.6195	69.8137
11	17.5603	18.5312	20.6546	24.3493	32.1504	42.5661	56.4053	98.7391
12	20.1407	21.3843	24.1331	29.0017	39.5805	54.2077	74.3270	139.2348
13	22.9534	24.5227	28.0291	34.3519	48.4966	68.7596	97.6250	195.9287
14	26.0192	27.9750	32.3926	40.5047	59.1959	86.9495	127.9125	275.3002
15	29.3609	31.7725	37.2797	47.5804	72.0351	109.6868	167.2863	386.4202
16	33.0034	35.9497	42.7533	55.7175	87.4421	138.1085	218.4722	541.9883
17	36.9737	40.5447	48.8837	65.0751	105.9306	173.6357	285.0139	759.7837
18	41.3013	45.5992	55.7497	75.8364	128.1167	218.0446	371.5180	1,064.6971
19	46.0185	51.1591	63.4397	88.2118	154.7400	273.5558	483.9734	1,491.5760
20	51.1601	57.2750	72.0524	102.4436	186.6880	342.9447	630.1655	2,089.2064
21	56.7645	64.0025	81.6987	118.8101	225.0256	429.6809	820.2151	2,925.8889
22	62.8733	71.4027	92.5026	137.6316	271.0307	538.1011	1,067.2796	4,097.2445
23	69.5319	79.5430	104.6029	159.2764	326.2369	673.6264	1,388.4635	5,737.1423
24	76.7898	88.4973	118.1552	184.1678	392.4842	843.0329	1,806.0026	8,032.9993
25	84.7009	98.3471	133.3339	212.7930	471.9811	1,054.7912	2,348.8033	11,247.1990

TABLE 8.4 Uniform series present worth factor ($P/A_{i,n}$)

n	1%	2%	3%	4%	5%	6%	7%	8%
1	0.99010	0.98039	0.97087	0.96154	0.95238	0.94340	0.93458	0.92593
2	1.97040	1.94156	1.91347	1.88609	1.85941	1.83339	1.80802	1.78326
3	2.94099	2.88388	2.82861	2.77509	2.72325	2.67301	2.62432	2.57710
4	3.90197	3.80773	3.71710	3.62990	3.54595	3.46511	3.38721	3.31213
5	4.85343	4.71346	4.57971	4.45182	4.32948	4.21236	4.10020	3.99271
6	5.79548	5.60143	5.41719	5.24214	5.07569	4.91732	4.76654	4.62288
7	6.72819	6.47199	6.23028	6.00205	5.78637	5.58238	5.38929	5.20637
8	7.65168	7.32548	7.01969	6.73274	6.46321	6.20979	5.97130	5.74664
9	8.56602	8.16224	7.78611	7.43533	7.10782	6.80169	6.51523	6.24689
10	9.47130	8.98259	8.53020	8.11090	7.72173	7.36009	7.02358	6.71008
11	10.36763	9.78685	9.25262	8.76048	8.30641	7.88687	7.49867	7.13896
12	11.25508	10.57534	9.95400	9.38507	8.86325	8.38384	7.94269	7.53608
13	12.13374	11.34837	10.63496	9.98565	9.39357	8.85268	8.35765	7.90378
14	13.00370	12.10625	11.29607	10.56312	9.89864	9.29498	8.74547	8.24424
15	13.86505	12.84926	11.93794	11.11839	10.37966	9.71225	9.10791	8.55948
16	14.71787	13.57771	12.56110	11.65230	10.83777	10.10590	9.44665	8.85137
17	15.56225	14.29187	13.16612	12.16567	11.27407	10.47726	9.76322	9.12164
18	16.39827	14.99203	13.75351	12.65930	11.68959	10.82760	10.05909	9.37189
19	17.22601	15.67846	14.32380	13.13394	12.08532	11.15812	10.33560	9.60360
20	18.04555	16.35143	14.87747	13.59033	12.46221	11.46992	10.59401	9.81815
21	18.85698	17.01121	15.41502	14.02916	12.82115	11.76408	10.83553	10.01680
22	19.66038	17.65805	15.93692	14.45112	13.16300	12.04158	11.06124	10.20074
23	20.45582	18.29220	16.44361	14.85684	13.48857	12.30338	11.27219	10.37106
24	21.24339	18.91393	16.93554	15.24696	13.79864	12.55036	11.46933	10.52876
25	22.02316	19.52346	17.41315	15.62208	14.09394	12.78336	11.65358	10.67478
n	9%	10%	12%	15%	20%	25%	30%	40%
1	0.91743	0.90909	0.89286	0.86957	0.83333	0.80000	0.76923	0.71429
2	1.75911	1.73554	1.69005	1.62571	1.52778	1.44000	1.36095	1.22449
3	2.53129	2.48685	2.40183	2.28323	2.10648	1.95200	1.81611	1.58892
4	3.23972	3.16987	3.03735	2.85498	2.58873	2.36160	2.16624	1.84923
5	3.88965	3.79079	3.60478	3.35216	2.99061	2.68928	2.43557	2.03516
6	4.48592	4.35526	4.11141	3.78448	3.32551	2.95142	2.64275	2.16797
7	5.03295	4.86842	4.56376	4.16042	3.60459	3.16114	2.80211	2.26284
8	5.53482	5.33493	4.96764	4.48732	3.83716	3.32891	2.92470	2.33060
9	5.99525	5.75902	5.32825	4.77158	4.03097	3.46313	3.01900	2.37900
10	6.41766	6.14457	5.65022	5.01877	4.19247	3.57050	3.09154	2.41357
11	6.80519	6.49506	5.93770	5.23371	4.32706	3.65640	3.14734	2.43826
12	7.16073	6.81369	6.19437	5.42062	4.43922	3.72512	3.19026	2.45590
13	7.48690	7.10336	6.42355	5.58315	4.53268	3.78010	3.22328	2.46850
14	7.78615	7.36669	6.62817	5.72448	4.61057	3.82408	3.24867	2.47750
15	8.06069	7.60608	6.81086	5.84737	4.67547	3.85926	3.26821	2.48393
16	8.31256	7.82371	6.97399	5.95423	4.72956	3.88741	3.28324	2.48852
17	8.54363	8.02155	7.11963	6.04716	4.77463	3.90993	3.29480	2.49180
18	8.75563	8.20141	7.24967	6.12797	4.81219	3.92794	3.30369	2.49414
19	8.95011	8.36492	7.36578	6.19823	4.84350	3.94235	3.31053	2.49582
20	9.12855	8.51356	7.46944	6.25933	4.86958	3.95388	3.31579	2.49701
21	9.29224	8.64869	7.56200	6.31246	4.89132	3.96311	3.31984	2.49787
22	9.44243	8.77154	7.64465	6.35866	4.90943	3.97049	3.32296	2.49848
23	9.58021	8.88322	7.71843	6.39884	4.92453	3.97639	3.32535	2.49891
24	9.70661	8.98474	7.78432	6.43377	4.93710	3.98111	3.32719	2.49922
25	9.82258	9.07704	7.84314	6.46415	4.94759	3.98489	3.32861	2.49944

CAPITALIZED COSTS

$$P = A/i \qquad (EQ\ 8.15)$$

DEPRECIATION

Straight Line

$$D_j = (C - S_n)/n \qquad (EQ\ 8.16)$$

where:
- D_j = depreciation in year j
- C = cost
- S_n = salvage value in year n
- n = recovery period in years

Accelerated Cost Recovery System (ACRS)

$$D_j = (\text{Modified ACRS depreciation rate})(C) \qquad (EQ\ 8.17)$$

Book Value

$$BV = \text{initial cost} - \Sigma D_j \qquad (EQ\ 8.18)$$

TABLE 8.5 Modified ACRS depreciation rates, percent

Year	Recovery Period					
	3-Year	5-Year	7-Year	10-Year	15-Year	20-Year
1	33.33	20.00	14.29	10.00	5.00	3.750
2	44.45	32.00	24.49	18.00	9.50	7.219
3	14.81	19.20	17.49	14.40	8.55	6.677
4	7.41	11.52	12.49	11.52	7.70	6.177
5		11.52	8.93	9.22	6.93	5.713
6		5.76	8.92	7.37	6.23	5.285
7			8.93	6.55	5.90	4.888
8			4.46	6.55	5.90	4.522
9				6.56	5.91	4.462
10				6.55	5.90	4.461
11				3.28	5.91	4.462
12					5.90	4.461
13					5.91	4.462
14					5.90	4.462
15					5.91	4.462
16					2.95	4.461
17						4.462
18						4.461
19						4.462
20						4.461
21						2.231

DEPLETION ALLOWANCE

Cost Depletion (after Whitney and Sibbald 1992)

$$CD_n = (CB - \Sigma D_i)\,[U_n/(U_n + U_r)] \quad \text{(EQ 8.19)}$$

where:
- CD_n = cost depletion allowance in year n
- CB = original cost basis
- ΣD_i = accumulated depletion in preceding years (both cost and percentage)
- U_n = units of product sold during year n
- U_r = units of product remaining at year end (units in reserves)

Percentage Depletion

TABLE 8.6 Depletion percentages for minerals

(1) 22%
  a. sulphur and uranium
  b. if from deposits in the United States—anorthosite, clay, laterite, nephelite syenite (to the extent that alumina and aluminum compounds are extracted therefrom), asbestos, bauxite, celestite, chromite, corundum, fluorspar, graphite, ilimenite, kyanite, mica, olivine, quartz crystals (radio grade), rutile, block steatite talc, zircon, and ores of the following metals: antimony, beryllium, bismuth, cadmium, cobalt, columbium, lead, lithium, manganese, mercury, molybdenum, nickel, platinum and platinum group metals, tantalum, thorium, tin, titanium, tungsten, vanadium, and zinc

(2) 15%—if from deposits in the United States.
  a. gold, silver, copper, and iron ore
  b. oil shale [except shale described in (5)]

(3) 14%
  a. metal mines [if (1) b. or (2) a. do not apply], rock asphalt, and vermiculite
  b. if (1)b., (5), or (6)b. do not apply—ball clay, bentonite, china clay, sagger clay, and clay used or sold for use for purposes dependent on its refractory properties

(4) 10%
  a. asbestos [if (1)b. does not apply], brucite, coal, lignite, perlite, sodium chloride, and wollastonite

(5) 7½%
  a. clay and shale used or sold for use in the manufacture of sewer pipe or brick, and clay, shale, and slate used or sold for use as sintered or burned lightweight aggregates

(6) 5%
  a. gravel, peat, pumice, sand, scoria, shale [except shale described in (2)b. or (5)], and stone [except stone described in (7)]
  b. clay use, or sold for use, in the manufacture of drainage and roofing tile, flower pots, and kindred products
  c. if from brine wells—bromine, calcium chloride, and magnesium chloride

(7) 14%
  a. all other minerals, including, but not limited to, aplite, barite, borax, calcium carbonates, diatomaceous earth, dolomite, feldspar, fullers earth, garnet, gilsonite, granite, limestone, magnesite, magnesium carbonates, marble, mollusk shells (including clam shells and oyster shells), phosphate rock, potash, quartzite, slate, soapstone, stone (used or sold for use by the mine owner or operator as dimension stone or ornamental stone), thenardite, tripoli, trona, and [if (1)b. does not apply] bauxite, flake graphite, fluorspar, lepidolite, mica, spodumene, and talc (including pyrophyllite), except that, unless sold on bid in direct competition with a bona fide bid to sell a mineral listed in (3), the percentage shall be 5% for any such other mineral [other than slate to which (5) applies] when used, or sold for use, by the mine owner or operator as rip rap, ballast, road material, rubble, concrete aggregates, or for similar purposes. The term "all other minerals" does not include: (a) soil, sod, dirt, turf, water, or mosses; (b) minerals from sea water, the air, or similar inexhaustible sources; or (c) oil and gas wells. Minerals (other than sodium chloride) extracted from brines pumped from a saline perennial lake within the United States shall not be considered minerals from an inexhaustible source.

Source: Adapted from Internal Revenue Service, Internal Revenue Code Title 26 Section 613, 2001.

EFFECTIVE TAX RATE

$$\text{Effective tax rate} = s + f(1 - s) \quad \text{(EQ 8.20)}$$

where:
- s = marginal state tax rate, decimal
- f = marginal federal tax rate, decimal

BENEFIT COST RATIO (BCR) (ALSO KNOWN AS PROFITABILITY INDEX)

BCR = present worth net positive cash flow at i^* / | present worth net negative cash flow at i^* | (EQ 8.21)

where i^* = discount rate.

PRESENT VALUE RATIO (PVR)

Present Value Ratio = net present value at i^* / | present worth net negative cash flow at i^* | (EQ 8.22)

$$\text{PVR} = \text{BCR} - 1 \quad \text{(EQ 8.23)}$$

INFLATION

$$i = (1 + f)(1 + i') - 1 \quad \text{(EQ 8.24)}$$

where:
- i = escalated dollar rate of return
- i' = constant dollar rate of return
- f = inflation rate

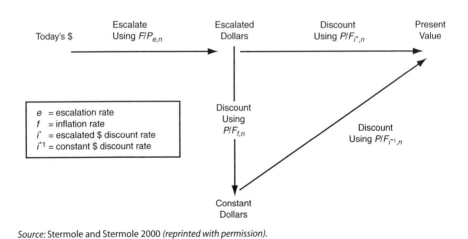

Source: Stermole and Stermole 2000 (reprinted with permission).

FIGURE 8.1 Equivalent escalated dollar and constant dollar present value calculations

CONSUMER PRICE INDEX

TABLE 8.7 Consumer price index

Consumer Price Index (Major Expenditures—All Items: 1982–1984 = 100)									
Year	Index	Year	Index	Year	Index	Year	Index	Year	Index
1960	29.6	1970	38.8	1980	82.4	1990	130.7	2000	172.2
1961	29.9	1971	40.5	1981	90.9	1991	136.2	2001	177.1
1962	30.2	1972	41.8	1982	96.5	1992	140.3		
1963	30.6	1973	44.4	1983	99.6	1993	144.5		
1964	31.0	1974	49.3	1984	103.9	1994	148.2		
1965	31.5	1975	53.8	1985	107.6	1995	152.4		
1966	32.4	1976	56.9	1986	109.6	1996	156.9		
1967	33.4	1977	60.6	1987	113.6	1997	160.5		
1968	34.8	1978	65.2	1988	118.3	1998	163.0		
1969	36.7	1979	72.6	1989	124.0	1999	166.6		

Source: Council of Economic Advisers 2001.

CAPITAL STRUCTURE

TABLE 8.8 Components of capital structure

Assets =	Liabilities +	Equity
Current assets	Current liabilities	Common shares
Cash	Accounts payable	Preferred shares
Marketable securities	Notes payable	Share premium above par
Accounts receivable	Current portion of long-term debt	Retained earnings
Inventory	Taxes	
	Lease payments	
Long-term assets	Long-term liabilities	Subordinated debt?
Mine/plant	Bank loans	
Buildings	Bonds	
Equipment	Debentures	
Land	Long-term notes	
Mineral rights	Leases	
Ore reserves	Subordinated debt	

Source: Tinsley 1992.

CASH FLOW

TABLE 8.9 Factors for consideration in cash-flow analysis

Preproduction Period	
Exploration expenses	Land and mineral rights
Water rights	Environmental costs
Mine and plant capital requirements	Development costs
Sunk costs	Financial structure
Working capital	Administration
Production Period	
Price	Capital investment—replacement and expansions
Processing costs	Royalty
Recovery	Mining cost
Postconcentrate cost	Development cost
Reserves and percent removable	Exploration cost
Grade	General and administration
Investment tax credit	Insurance
State taxes	Production rate in tons per year
Federal taxes	Financial year production begins
Depletion rate	Percent production not sent to processing plant
Depreciation	Operating days per year
Postproduction Period	
Salvage value	Contractual and reclamation expenditures

Source: Laing 1977.

TABLE 8.10 Components for developing cash flows

Calculation	Component
	Revenue
Less	Royalties
Equal	Gross income from mining
Less	Operating costs
Equal	Net operating income
Less	Depreciation and amortization allowance
Equal	Net income after depreciation and amortization
Less	Depletion allowance
Equal	Net taxable income
Less	State income tax
Equal	Net federal taxable income
Less	Federal income tax
Equal	Net profit after taxes
Add	Depreciation and amortization allowances
Add	Depletion allowance
Equal	Operating cash flow
Less	Capital expenditures
Less	Working capital
Equal	Net annual cash flow

Source: Gentry and O'Neil 1984.

COSTING

Six-Tenths Rule (after Katell and Wellman 1968)

$$C_a = C_b \, (A/B)^{0.6} \qquad \text{(EQ 8.25)}$$

where:
- C_a = cost of item of size or capacity A
- C_b = cost of similar item of size or capacity B
- A = size or capacity of item A
- B = size or capacity of item B

The exponential factor varies from 0.1 to greater than 1. For mining projects typical values range from 0.5 to 0.9 (Gentry and O'Neil 1984).

Cost Reports

TABLE 8.11 Example of a monthly cost report for a mine and a mill

	Actual ($)	Budget ($)	Variance Amount	Variance %
Mine department				
Production				
Ore tons mined	1,120,000	1,250,000	–130,000	90
Waste tons mined	2,017,000	2,005,000	12,000	101
Ore grade (% Cu)	0.80	0.78	0.02	103
Cost ($)				
Operating labor	920,000	876,000	(44,000)	105
Mechanical labor	756,000	779,000	23,000	97
Salary labor	105,000	104,000	(1,000)	101
Fuel	947,000	976,000	29,000	97
Operating supplies	876,000	894,000	18,000	98
Mechanical supplies	1,002,000	808,000	(194,000)	124
Miscellaneous and administrative	14,000	28,000	14,000	50
Allocated overhead	420,000	393,000	(27,000)	107
Total	5,040,000	4,858,000	(182,000)	104
Mill department				
Production				
Recovery	87.8%	88.3%	–0.5%	99
Product grade	96.4%	96.3%	0.1%	100
Copper produced (lb)	15,733,760	17,218,500	–1,484,740	91
Cost ($)				
Operating labor	187,000	189,000	2,000	99
Mechanical labor	205,000	212,000	7,000	97
Salary labor	52,100	49,600	(2,500)	105
Power	530,000	565,000	35,000	94
Operating supplies	164,000	176,000	12,000	93
Mechanical supplies	107,500	105,000	(2,500)	102
Miscellaneous and administrative	15,000	21,000	6,000	71
Allocated overhead	116,000	113,000	(3,000)	103
Total	1,376,600	1,430,600	54,000	96

Source: Cavender 1999.

TABLE 8.12 Example of a monthly income statement: full absorption costing basis

(Full absorption costing basis; $000)	Actual ($)	Budget ($)	Variance Amount ($)	%
Sales (at $1/lb Cu)	15,734	17,219	1,485	91
Less: Cost of goods sold				
Mining cost	5,040	4,858	(182)	104
Milling cost	1,377	1,431	54	96
Gross margin	9,317	10,930	1,613	85
Less: Selling, general, and administrative costs	8,143	7,725	(418)	105
Operating profit	1,174	3,205	2,031	37

Source: Cavender 1999.

TABLE 8.13 Example of monthly income statement: variable costing basis

(Variable costing basis; $000)	Actual ($)	Budget ($)	Variance Amount ($)	%
Sales (at $1/lb Cu)	15,734	17,219	1,485	91
Less: Variable costs				
Mining cost	4,217	4,073	(144)	104
Milling cost	1,097	1,138	41	96
Variable selling, general, and administrative costs	6,514	6,180	(334)	105
Contribution margin	3,905	5,827	1,922	67
Less: Fixed costs				
Mining cost	823	785	(38)	105
Milling cost	280	292	13	96
Fixed selling, general, and administrative costs	1,629	1,545	(84)	105
Operating profit	1,174	3,205	2,031	37

Source: Cavender 1999.

REFERENCES

Cavender, B. 1999. *Mineral Production Costs: Analysis and Management*. Littleton, CO: Society for Mining, Metallurgy, and Exploration, Inc. (SME). 97–98.

Council of Economic Advisers. 2000. *Economic Report of the President*. Washington, DC: U.S. Government Printing Office.

Gentry, D.W., and T.J. O'Neil. 1984. *Mine Investment Analysis*. Littleton, CO: SME. 502 pp.

Internal Revenue Service (IRS). 2001. *Internal Revenue Code Title 26 Section 613*. Washington, DC: IRS. Available online at http://tns-www.lcs.mit.edu/uscode/TITLE_26/toc.html.

Katell, S., and P. Wellman. 1968. *Process Evaluation Cost Analysis Seminar*. Morgantown, WV: U.S. Bureau of Mines. II-31.

Laing, G.J. 1977. Effects of state taxation on the mining industry in the Rocky Mountain states. *Colorado School of Mines Quarterly*. 72(1): 126 pp.

Stermole, F.J., and J.M. Stermole. 2000. *Economic Evaluation and Investment Decision Methods*. 10th ed. Golden, CO: Investment Evaluations Corporation. 258.

Tinsley, C.R. 1992. Mine financing. In *SME Mining Engineering Handbook*. 2nd ed., Vol. 1. Edited by H.L. Hartman. Littleton, CO: SME. 471.

Whitney, J.W., and G.H. Sibbald. 1992. Taxation and depletion. In *SME Mining Engineering Handbook*. 2nd ed., Vol. 1. Edited by H.L. Hartman. Littleton, CO: SME. 90.

CHAPTER 9

Quality and Specifications of Products

Keith E. Dyas (retired P.E.)

ABRASIVES

TABLE 9.1 Classification of abrasives

Natural Abrasives				Manufactured Abrasives		Types of Abrasive Products
Superior Hardness (above 7 on Mohs scale)	Intermediate Hardness (H-5.5 to 7)		Inferior Hardness (H-under 5.5)			
	Silica Abrasives	Other Rocks and Minerals				
Diamond H-10	Buhrstone	Argillaceous limestone	Apatite	Boron carbide	Lampblack	Abrasive grains and powders, loose
Corundum H-9	Chalcedony	Basalt	Calcite	Boron nitride	Lime	Abrasive grains bonded into wheels, blocks and special shapes
Emery H-7 to 9	Chert	Feldspar	Chalk	Calcium carbonate (pptd.)	Magnesia (pptd.)	Coated abrasives; grains bonded to paper and cloth
Garnet H-6.5 to 7.5	Flint	Granite	Clay	Calcium phosphate	Manganese dioxide	Abrasive grains and powders; paste form; oil or water vehicles
Staurolite H-7.0 to 7.5	Novaculite	Mica schist	Diatomite	Cerium oxide	Periclase (artif.)	Abrasive grains and powders; brick and stick form; grease, glue and wax binders
	Quartz	Perlite	Dolomite	Chromium oxide	Silicon carbide	Natural rocks shaped into grindstones, pulpstones, chaser stones, millstones, etc.
	Quartzite	Pumice and pumicite	Iron oxides	Clay (hard burned)	Tantalum carbide	Natural rocks shaped into sharpening stones, such as oil stones, whetstones, scythe stones, razor hones, etc.
	Sandstone	Quartz conglomerate	Limestone	Diamond	Tin oxide	Natural stones shaped into rubbing and polishing stones such as holystones and pumice scouring blocks
	Silica sand		Rottenstone	Fused alumina	Titanium carbide	Natural stones shaped into blocks for tube-mill and pebble-mill liners
			Siliceous shale	Glass	Tungsten carbide	Pebbles, natural and manufactured, for grinding mills
			Silt	Iron oxides	Zirconium oxide	
			Talc		Zirconium silicate	

continues next page

TABLE 9.1 Classification of abrasives (continued)

Natural Abrasives				Manufactured Abrasives	Types of Abrasive Products
Superior Hardness (above 7 on Mohs scale)	Intermediate Hardness (H-5.5 to 7)		Inferior Hardness (H-under 5.5)		
	Silica Abrasives	Other Rocks and Minerals			
			Tripoli	Metallic abrasives, including steel wool, steel shot, angular steel grit, brass wool, and copper wool	
			Whiting	Porcelain blocks for mill liners and grinding pebbles	

Source: Adapted from Hight 1983 in Wellborn 1994.

AGGREGATE

TABLE 9.2 Functional uses and related specifications for mineral aggregates

Code	Use	Specification
304	Concrete aggregate—nominal gradations from 88.9 mm (3.5 in.) maximum to 12.7 mm (0.5 in.) maximum	American Society for Testing and Materials (ASTM) specification C33
305	Bituminous aggregate—nominal gradations from 88.9 mm maximum to 12.7 mm maximum	ASTM specification D602
306	Macadam aggregate—nominal gradations from 88.9 mm maximum to 12.7 mm maximum	ASTM specification D693
307	Sense graded roadbase aggregate—nominal gradations from 38.1 mm (1.5 in.) maximum	ASTM specification D2940
308	Surface treatment aggregate—nominal gradations from 25.4 mm (1 in.) maximum	ASTM specification D1139
309	Unspecified construction aggregate and roadstone	Not applicable
310	Riprap and jetty stone—heavy, irregular rock chunks used for riverbank, harbor and shoreline stabilization	Specifications of Corps of Engineers (formerly U.S. Army Corps of Engineers) as appropriate
311	Railroad ballast—nominal gradations from 63.5 mm (2.5 in.) to No. 4 sieve (4.75 mm)	Specifications of American Railway Engineering Association (AREA) or equivalent ASTM specifications
312	Filter aggregate—any porous aggregate through which water is filtered	Federal (GSA) specification SS744a
313	Manufactured fine aggregate—(stone sand or crushed gravel) 100% passing 9.5 mm (3/8 in.)	ASTM specification C33
314	Terazzo and exposed aggregate—small chips or pieces of crushed stone or other aggregate 12.7 mm (0.5 in.) to 19.1 mm (0.75 in.)	Not applicable

Source: Adapted from Schenck and Torries 1983 in McCarl 1994.

QUALITY AND SPECIFICATIONS OF PRODUCTS | **175**

TABLE 9.3 Grading requirements for lightweight aggregates (weight percentages passing sieves with square openings)

Size Designation	25.0 mm	19.0 mm	12.5 mm	9.5 mm	4.75 mm	2.36 mm	1.18 mm	0.29 mm	0.149 mm
Lightweight Aggregates for Structural Concrete (ASTM C330)									
Fine aggregate									
4.75 to 0 mm	—	—	—	100	85–100	—	40–80	10–35	5–25
Coarse aggregate									
25.0 to 4.75 mm	95–100	—	25–60	—	0–10	—	—	—	—
19.0 to 4.75 mm	100	90–100	—	10–50	0–15	—	—	—	—
12.5 to 4.75 mm	—	100	90–100	40–80	0–20	0–10	—	—	—
9.5 to 2.36 mm	—	—	100	80–100	5–40	0–20	0–10	—	—
Combined fine and coarse aggregate									
12.5 to 0 mm	—	100	95–100	—	50–80	—	—	5–20	2–15
9.5 to 0 mm	—	—	100	90–100	65–90	35–65	—	10–25	5–15
Lightweight Aggregates for Concrete Masonry Units (ASTM C331)									
Fine aggregate									
4.75 to 0 mm	—	—	—	100	85–100	—	40–80	10–35	5–25
Coarse aggregate									
12.5 to 4.75 mm	—	100	90–100	40–80	0–20	0–10	—	—	—
9.5 to 2.36 mm	—	—	100	80–100	5–40	0–20	0–10	—	—
Combined fine and coarse aggregate									
12.5 to 0 mm	—	100	95–100	—	50–80	—	—	5–20	2–15
9.5 to 0 mm	—	—	100	90–100	65–90	35–65	—	10–25	5–15

Source: Mason 1994.

ALUMINUM

TABLE 9.4 Typical specifications for grades of bauxite (weight-percent, maximum content unless otherwise specified)

Constituent	Metal Grade (Dried Jamaican Type)	Refractory Grade (Calcined)	Abrasive Grade (Calcined)
Al_2O_3	47.0*	86.5*	83.0*
SiO	3.0	7.0	6.0
Fe_2O_3	22.0	2.5	8.0
TiO_2	3.0	3.75	3.0–4.5†
$K_2O + Na_2O$	NS‡	0.2	0.7
$MgO + CaO$	NS	0.3	NS
CaO	NS	NS	0.2
MgO	NS	NS	0.4
$MnO_2 + Cr_2O_3 + V_2O_5$	2.0	1.0	1.0
P_2O_5	1.5	NS	0.5
Loss on ignition	NS	0.5	1.0

* Minimum.
† Range.
‡ No specification.
Source: McCawley and Baumgardner 1985.

BARITE

Material for weighting muds must be finely ground, heavy, and chemically inert; consequently, barite for this purpose must have a specific gravity of 4.2 or higher, it must be free of soluble salts, and 90% to 95% of the material must be able to pass through a 325-mesh screen. In chemical manufacturing, purity is the principal concern, and a maximum of 1% each of Fe_2O_3 and $SrSO_4$ and a trace of fluorine usually are set, with a minimum of 94% $BaSO_4$. Most chemical manufacturers specify a size range of 4 to 20 mesh. Glass manufacturers usually require a minimum of 95% $BaSO_4$ with a maximum of 2.5% SiO_2 and 0.15% Fe_2O_3. Particle size that is generally preferred is minus 30 to plus 140 mesh (Ampian 1985a).

CHROMIUM

Grades of chromite containing not more than 40% Cr_2O_3 are used in the refractory industry. Grades containing more than 40% Cr_2O_3 but less than 46% are used by the refractory, chemical, and metallurgical industries, and grades with 46% or more Cr_2O_3 are used by the metallurgical and chemical industries. The quantity of minerals, other than chromite, contained in the chromite ore or concentrate is an important factor in end uses. Silica content is important to both the refractory and chemical industries, and alumina content is important to the refractory industry for some applications (Papp 1985).

CLAYS

Clays are classified in six groups—kaolin, ball clay, fireclay, bentonite, fuller's earth, and common clay and shale. Many producers and consumers rely on their own tests and specifications applicable to their specific needs (Ampian 1985b).

Bentonite

Many companies use a yield specification for bentonite. Yield is a term used in an earlier American Petroleum Institute specification for the number of barrels of 15-centipoise viscosity mud that can be made from a ton of bentonite. The yield requirement is ordinarily 90 bbl/ton minimum.

Kaolin

TABLE 9.5 Definitions of kaolinitic raw materials

Ball clay	A plastic, kaolinitic clay with minor to abundant organic matter producing high green strength and fired shrinkage in ceramic bodies and usually firing to white. The term comes from the old English practice of rolling clays in ~25 cm balls.
Fireclay	A plastic, kaolinitic clay with sufficient Al_2O_3 to be refractory. The clay usually occurs as an underclay.
Flint clay	A hard, nonplastic, kaolinitic claystone that breaks with conchoidal fracture and does not disperse in water. "Burley flint clay" is a claystone in which aluminum-rich minerals such as diaspore raise the Al_2O_3 above that of kaolinite (~40%).
Kaolin	A soft, white, relatively nonplastic but dispersible, kaolinitic clay. In primary and sandy kaolins that are less than one-half kaolinite, the term kaolin reflects the purity and properties of the fine-fraction separate.
Underclay	A stratum of soft, dispersible clay or claystone that typically underlies coal. There are three facies: (1) shale-type or illitic, (2) soil-type or fireclay-type (kaolinitic), and (3) gley-type (rich in mixed-layered illite/smectite).

Source: Adapted from Burst and Hughes 1994.

QUALITY AND SPECIFICATIONS OF PRODUCTS | 177

TABLE 9.6 Kaolin quality factors

• Brightness (TAPPI, ISO, etc.)	• Color (X,Y, Z; L, a, b)
• Reduction leach brightness	• Oxidation brightness
• Magnet brightness	• Flotation brightness
• Brookfield viscosity	• Hercules viscosity
• Crystal particle size	• Crystal aspect ratio
• Delamination response	• Calcination response
• Dispersant demand	• Fired brightness
• Cation exchange capacity	• Einlehner abrasivity
• Pyrometric cone temperature	• Ink receptivity
• Sheet opacity, brightness	• Water release rate
• Differential thermal analysis	• Water retention rate
• Fe, Ti, K, Ca, Mg, Na content	• Oil absorption
• Bulk density, packing	• Filterability
• Bacteria and fungus content	• Resistivity

Source: Pickering and Murray 1994.

COAL

Classification of coals by rank is in Table 2.11 in the Coal section of Chapter 2, which covers material properties.

TABLE 9.7 Significance of coal characteristics for combustion performance[*]

	Stokers				Pulverized Firing	Cyclone	Fluidized Bed
	Single-retort	Multiple-retort	Traveling Grate	Spreader Stoker			
1. Size consist (as fired)	V	I	I	V	V	V	I
2. Moisture[†]	M	M	N	M	V	M	M
3. Caking index[‡]	I	I	V	M	N	N	N
4. Ash fusibility	I	I	M	M	I	V	M
5. Grindability	N	N	N	N	V	N	N
6. Friability	M	M	M	M	N	N	M
7. Volatile matter	M	M	M	M	I	M	M
8. Fixed carbon	N	N	N	N	M	N	N
9. Ash content	M	M	M	M	M	M	M
10. Calorific value	N	N	N	N	N	N	N
11. Ash viscosity	M	M	M	N	I	V	M
12. Ash composition[§]							
13. Sulfur[**]							

Rating code: V = Very important
            I = Important
            M = Minor importance
            N = Little or no importance

[*] Degree of fineness is a better term for pulverized firing.
[†] Surface moisture is more critical than inherent moisture. Moisture is very important from the standpoint of plant flowability.
[‡] Some engineers are attempting to use the FSI as an index of the degree of caking.
[§] Ash composition is very important as it affects fireside fouling, but not important to combustion.
[**] Sulfur is important from a corrosive standpoint, but not important to combustion.

Source: Buttermore and Leonard 1991.

TABLE 9.8 Rating of coking coals for blending

	Coal classification								
	High-volatile A			Medium-volatile			Low-volatile		
Property	Good	Medium	Poor	Good	Medium	Poor	Good	Medium	Poor
Volatile matter, %	31.0–33.0	33.0–36.0	+36	21.0–24.0	24.0–27.0	27.0–31.0	18.0–21.0	15.0–18.0	<15.0
Vitrinite Reflectance, %	0.92–1.09	0.85–0.95	0.68–0.85	1.40–1.50	1.20–1.40	1.10–1.20	1.51–1.70	1.70–1.85	>1.85
Fluidity, ddpm	+20,000	5,000–20,000	<5,000	500–8,000	300–20,000	<300–>20,000	100–300	30–1,000	<30–>1,000
Free-swelling index	9	6–8	<6	9	7–8	<7	9	7–8	<7
Hardgrove grindability index	48–75		32–70	80–135		60–90	90–120		85–105
Composition balance index	0.40–0.80	0.80–1.40	>1.4	1.0–1.50	1.50–2.00	>2.0	2.00–3.50	3.50–5.00	>5.00
Rank index	3.4–4.3	3.0–3.4	2.2–3.0	6.0–6.5	4.3–5.5	<4.3	>6.8	6.0–7.5	<7.5
Other Characteristics for Good Coking Coals:									
Sulfur				<0.7%			<1.0%		
P_2O_5 (whole-coal basis)				0.05%–0.06% maximum			<0.03% preferred		
Ash				<6.0%			<8.0%		
Moisture				6%–8% maximum					
K_2O and Na_2O (ash basis)				<1.0%			<3.0%		
Ash-softening temperature				<2,300°F			2,300–2,500°F		
Limits on Cu, Ni, Co, Mo, Sn, Cr, V, Zn, Pb, TiO_2, As, Sb, Cl, F, SiO_2, Al_2O_3, Mn									

Source: Gray, Goscinski, and Schoeberger 1978.

COBALT

National Defense Stockpile purchase specifications designate three grades of cobalt—electrolytic broken cathodes (grades A and B) and granules. Material designated grade A must contain 99.9% cobalt; grade B, 99.65%; and granules, 99.5% (Kirk 1985).

COPPER

There are about 370 types of copper and copper alloys, which are divided into the broad categories of wrought and cast metals. Within these two categories, the metals are further subdivided into classes as follows:

- Coppers—metals containing at least 99.3% copper. There are 44 numbered coppers, including oxygen-free, tough-pitch, and deoxidized varieties.
- High-copper alloys—copper content of cast alloys is at least 94%; copper content of wrought alloys is 96% to 99.3%. This class includes the cadmium, beryllium, and chromium copper alloys.
- Brasses—copper alloys that contain zinc as the principal alloying element. There are three families of wrought brasses and five families of cast brasses.
- Bronzes—copper alloys in which the main alloying element is usually tin, and which contain other metals such as aluminum, lead, phosphorus, and silicon (but not zinc or nickel).
- Copper nickels—copper alloys with nickel as the principal alloy metal.
- Copper-nickel-zinc alloys—copper alloys that contain nickel and zinc as the principal and secondary alloying elements, commonly known as nickel silver.
- Leaded coppers—cast copper alloys that contain 20% or more lead, usually a small amount of silver, but no zinc or tin.
- Special alloys—copper alloys not covered by the above descriptions (Jolly, J.L.W. 1985).

DIAMONDS, INDUSTRIAL

TABLE 9.9 Standard grade numbers for diamond powders

Grade Nos.	μ Range	Approximate Mesh Equivalent
1/2	0–1	50,000
1	1–2	14,000
3	2–4	8,000
6	4–8	3,000
9	8–12	1,800
15	12–22	1,200
30	22–36	800
45	36–54	500/600
60	54–80	400/500

Source: Reckling et al. 1994.

DIATOMITE AND PERLITE

TABLE 9.10 Key properties for diatomite and perlite used as filters

Mineral/Grade	Color	Density (wet)	kg/m³ (dry)	Relative Flow Rate*	Ignition Loss (%)	Medium Pore Size (μ)
Diatomite						
Flux calcined	White	330	220	700–2,300	0.2	15.0
Calcined	Pink	375	140	100–430	0.5	3.5
Natural	Gray	260	105	<100	2.5	2.0
Perlite	White	300	120	170–930	3.0	—

* Water permeability flow ratio, Grefco's dicalite 215 as 100.
Source: Grefco Inc. and Celite Corp. in Van Kouteren 1994.

FLUORSPAR

Three principal grades of fluorspar are available commercially—acid, ceramic, and metallurgical. Acid-grade fluorspar (acidspar) contains at least 97% CaF_2. Some manufacturers of hydrofluoric acid in the United States and Europe can use 96% CaF_2 (or slightly lower) if the remaining impurities are acceptable. User specifications may impose limits on silica, calcium carbonate, sulfide or free sulfur, calcite, beryllium, arsenic, lead, phosphates, and other constituents. Moisture content of the dried material is preferably 0.1% or less. Particle size and distribution are sometimes specified for proper control of the rate of chemical reaction and stack losses. Ceramic-grade fluorspar is generally marketed as No. 1 ceramic, containing 95% to 96% CaF_2. An intermediate grade of about 93% to 94% CaF_2 is also available. Specifications on impurities vary but may allow a maximum of 2.5% to 3.0% silica, 1.0% to 1.5% calcite, 0.12% ferric oxide, and trace quantities of lead and zinc. Metallurgical-grade fluorspar (metspar) contains a minimum of 60% effective CaF_2. In the United States, metspar is usually quoted in terms of effective CaF_2 units, obtained by subtracting 2.5 times the silica content of the ore from its total CaF_2 content. The term metspar is usually used to refer to material with a maximum CaF_2 content of 85% but is sometimes used for material as high as 96% (Pelham 1985).

GOLD

Fineness refers to the weight proportion of pure gold in an alloy, expressed in parts per thousand; 1000 "fine gold" is 100% pure gold. Commercially traded gold bullion is usually 995 fine, or higher. The term fine gold may also be used to designate the particle size of gold in its native state; for example, a placer deposit with gold particles ranging from 0.015 to 0.03 in. in diameter contains fine gold; a similar deposit with particles over 0.06 in. in diameter contains coarse gold. The term "karat," like fineness, refers to purity, but is expressed in 24ths; thus, 24-karat (24k) gold is 1000 fine or pure gold, and 14k gold is 14/24 or 58.3% gold. Several alloys can be designated by a given karat number, differing from each other in the number, identity, and proportions of their nongold constituent metals (Lucas 1985).

GRAPHITE

Sri Lankan lump graphite is classified either as amorphous or crystalline. Each type is divided into a number of grades, depending on the particle size (lump, ranging from the size of walnuts to that of peas; chip, from the size of peas to about that of wheat grains; dust, finer than 60 mesh), graphite carbon content, and degree of consolidation. Amorphous graphite is graded primarily on graphitic carbon content. Commercial products contain from 50% to 94% carbon. Crystalline flake graphite from the Malagasy Republic is divided into two main grades, flake (coarse flake) and fines (fine flake). Malagasy crucible flake must have a minimum of 85% carbon and be essentially all minus 8 to plus 60 mesh. Other crystalline flake graphite is graded according to graphitic carbon content and particle size (Taylor 1994).

IRON ORE

U.S. Lake Superior iron ores are graded and priced on the basis of chemical composition and physical structure. Bessemer ores contain 0.045% phosphorus or less; non-Bessemer ores contain more than 0.045% phosphorus; high-phosphorus ores contain 0.18% phosphorus or more. Manganiferous ores contain 2% or more manganese. Siliceous ore contains 18% or more of silica (Klinger 1985).

LEAD

Minimum purity ranges from 99.85% to 99.9999% for "zone-refined" lead. Most soft pig lead consumed in the world is specified at the London Metal Exchange (LME) grade pure lead minimum of 99.97%. In the United States, there are four grades: corroding, 99.94% minimum with 0.0025% silver plus copper maximum; common, 99.94% minimum with 0.005% silver maximum and 0.0015% copper maximum; and undesilverized chemical and copper bearing, 99.90% minimum with up to 0.1% silver and copper. Maximums of arsenic, zinc, iron, and bismuth are specified for these grades (Woodbury 1985).

LIMESTONE AND DOLOMITE

TABLE 9.11 Physical and chemical specifications for glass-grade limestone

Size (mm)	Typical Sieve Analysis		
	% Retained	% Retained Cumulative	% Passing Cumulative
1.68 (12 mesh)	0.00	0.00	100.00
1.19 (16 mesh)	0.35	0.17	99.83
0.84 (20 mesh)	5.06	5.20	94.80
0.30 (50 mesh)	57.05	62.25	37.75
0.15 (100 mesh)	26.26	88.90	11.10
0.07 (200 mesh)	9.98	98.40	1.60
PAN	1.60	100.00	0.00

Moisture Content 0.09%

Chemical	Typical Chemical Analysis	
	Reported as	%
Calcium carbonate	$CaCO_3$	97.80
Magnesium carbonate	$MgCO_3$	1.25
Iron oxide	Fe_2O_3	0.095
Silica	SiO_2	0.56
Alumina	Al_2O_3	0.23
Nickel	Ni	<0.002
Chromium	Cr_2O_3	<0.001
Strontium oxide	SrO	0.03
Manganese oxide	MnO	<0.01

Source: Carr, Rooney, and Freas 1994.

TABLE 9.12 Typical analyses of limestone, dolomite, and lime fluxes

Component	Limestone		Dolomite		Burnt Lime	
	Blast Furnace*	Sinter Plant	Blast Furnace*	Sinter Plant	High-Calcium	Dolomitic
	(One-year average, % by weight)				(Calculated)	
$CaCO_3$	95.3	93.8	54.5	52.4		
$MgCO_3$	3.1	3.6	42.0	40.0		
CaO	53.4	52.5	30.6	29.4	88.5	56.0
MgO	1.5	1.7	20.1	19.1	2.5	36.8
R_2O_3†	0.3	0.4	0.3	0.4	0.5	0.5
SiO_2	0.7	1.8	2.6	6.8	1.2	4.7
	(typical, % by weight)					
Fe_2O_3	0.20	0.30	0.2	0.3		
Al_2O_3	0.30	0.20	0.3	0.3		
Mn	0.01	0.02	0.01	0.03		
P	0.01	0.01	0.01	0.01		
S	0.03	0.04	0.05	0.03		
K_2O	0.10	0.10	0.10	0.10		
Na_2O	0.02	0.02	0.02	0.02		
LOI‡	43.40	46.10	6.00	1.00	1.0	2.0

* Also typical of flux used in fluxed pellets and as stone for lime production.
† $R_2O_3 = Fe_2O_3 + Al_2O_3 + Cr_2O_3 + TiO_2$.
‡ Loss on ignition.
Source: Kokal and Ranade 1994.

TABLE 9.13 Size analyses of typical flux stone for blast furnace sinter plant, lime plant, and fluxed pellet plant

	(% Passing)						
	Blast Furnace or Lime Plant Stone		Mixed Blend 50:50*	Sinter Flux		Pellet Flux 50:50†
Size	Limestone	Dolomite		Limestone	Dolomite	
cm						
7.620			99.6			
6.350	87.6		99.0			
5.080	46.2	88.7	91.0			
4.445	18.7	72.7				
3.810	9.1	50.5	64.6			
3.175	4.4	26.4				
2.540	2.6	14.8	15.7			
1.905	1.7	2.0				
1.270			2.5			
0.952				99.8	100.0	
0.635			1.2			
mm						
4.699				97.5	97.8	
3.327				90.6	86.1	
1.651				66.0	58.6	
0.589				20.3	20.0	
0.295				8.2	10.0	99.9
0.208						99.3
0.147				2.4	4.7	90.4
0.074				0.9	2.4	65.9
0.053						58.2
0.043						52.2
0.038						47.2
0.026						42.0

* Feed to crusher at fluxed pellet plant (50% limestone, 50% dolomite).
† Ball mill grinding.
Source: Kokal and Ranade 1994.

MANGANESE

Manganese content of the more commonly used and traded ores, concentrates, nodules, and sinter for metallurgical purposes is in the approximate range of 38% to 55%. A manganese content of 48% is considered standard as a pricing basis. Besides manganese content, quality also depends on the manganese-to-iron ratio and the concentration of frequently encountered impurities such as alumina, silica, and lime. For metallurgical ore the manganese-to-iron ratio is ideally about 7.5 to 1 for manufacture of standard ferromanganese, which contains 78% manganese (Jones 1985).

MERCURY

Mercury produced from mining operations is called prime virgin mercury or virgin metal, and is usually more than 99.9% pure. Virgin metal with a clean and bright appearance contains less than 1 part per million (ppm) of any base metal and is acceptable for nearly all end uses. Mercury is packaged in cast iron, wrought iron, or spun steel bottles, flasks, or metric ton containers, which can vary in diameter, height, and weight. Mercury is sold and priced on the basis of a flask containing 76 lb, and market quotations cover prime virgin mercury only (Carrico 1985).

MICA

TABLE 9.14 Typical chemical analysis and physical properties of different forms of ground muscovite mica (British standards)

Chemical Analysis (approximate)	Dry Ground (wt %)	Micronized (wt %)	Wet Ground (wt %)
SiO_2	45.57	46.27	48.65
Al_2O_3	36.10	35.24	32.04
K_2O	9.87	9.87	9.87
Fe_2O_3	2.48	2.48	3.68
Na_2O	0.62	0.60	0.28
TiO_2	0.20	0.20	0.20
CaO	0.21	0.21	0.21
MgO	0.15	0.16	0.16
H_2O	0.10	0.20	0.20
P_2O_5	0.03	0.02	0.018
S	0.01	0.01	0.010
C	0.44	0.44	0.44
LOI	4.30	4.30	4.30
Total	100.08	100.00	100.05
Physical properties			
Index of refraction	1.58	1.58	1.58
Hardness (Mohs scale)	2.5	2.5	2.5
pH in distilled water	6.2	5.2	5.2
Oil absorption (B.S. 3483)	60.75%	60.75%	60.75%
Water soluble (B.S. 1765)	<0.3%	<0.3%	<0.3%
Brightness	66–75	75	75
Effect by common acids	Slight	Slight	Slight
Phericity factor	0.01	0.01	0.01
Softening point	1,538°C	1,538°C	1,538°C
Apparent density (kg/m³)	1,920–256	160–224	160–224

Source: Rajgarhia 1987 in Tanner 1994.

MOLYBDENUM

Molybdenite concentrate generally contains about 90% molybdenite (MoS_2); the grade may be somewhat lower, particularly if produced at copper by-product concentrating plants. Technical-grade molybdic oxide (MoO_2) is produced by roasting molybdenite concentrate. Typically, the oxide has a MoO_2 content of 85% to 90%, or a minimum of 57% contained molybdenum. Other raw materials, including ferromolybdenum, purified molybdic oxide, ammonium and sodium molybdate, and molybdenum metal powder are produced from technical-grade oxide (Blossom 1985).

NICKEL

TABLE 9.15 Commercial forms of primary nickel

	Composition (%)								
	Ni	C	Cu	Fe	S	Co	O	Si	Cr
Pure unwrought nickel									
Cathode	>99.9	0.01	0.005	0.002	0.001	—	—	—	—
Pellets	>99.97	<0.01	0.0001	0.0015	0.0003	0.00005	—	—	—
Powder	99.74	<0.1	—	<0.01	<0.001	—	<0.15	—	—
Briquets	99.9	0.01	0.001	0.002	0.0035	0.03	—	—	—
Ferronickel*	20–50†	1.5–1.8	—	Balance	<0.3	†	—	1.8–4.0	1.2–1.8
Nickel oxide	76.0	—	0.75	0.3	>0.006	1.0	Balance	—	—
Nickel salts‡									
Nickel chloride	24.70	—	—	—	—	—	—	—	—
Nickel nitrate	20.19	—	—	—	—	—	—	—	—
Nickel sulfate	20.90	—	—	—	—	—	—	—	—

* Ranges used to denote variable grades produced.
† Cobalt (1% to 2%) included with nickel.
‡ Theoretical nickel content.
Source: Sibley 1985.

PHOSPHATE ROCK

The phosphate content or grade of phosphate rock is normally reported as P_2O_5. It may also be expressed as bone phosphate of lime (BPL), reminiscent of the time when bones comprised the principal source of phosphate in fertilizer manufacture (percent BPL = 2.1853 × percent P_2O_5). It is the basis on which phosphate rock is sold, almost always as a beneficiated concentrate. Chemical analysis of pebble, the plus-16-mesh washer product, and flotation concentrates, 1 mm by 0.1 mm, ranges from 25% to 34% P_2O_5 in Florida mines. The percentages of P_2O_5, CaO, Fe_2O_3, Al_2O_3, and MgO are of interest (Stowasser 1985).

PLATINUM GROUP METALS

Commercial-grade platinum normally must be at least 99.95% pure and palladium, 99.9% pure. Platinum at least 99.999% pure is considered chemically pure and is the grade required for thermocouples and resistance thermometers. According to federal voluntary product standards, articles made wholly or partially of platinum must contain a minimum of 95% platinum to be called platinum. Special stamping provisions cover some alloys developed for the jewelry trade. In the United Kingdom, all platinum jewelry sold must have 95% platinum content to be hallmarked as platinum (Loebenstein 1985).

POTASH

TABLE 9.16 Potash product specifications

Grade	Minimum K$_2$O Equivalent	Approximate Particle Size Range*		Type of Potash
		Mesh†	mm	
Granular	60, 50, 22	6–30	3.34–0.85	Muriate and sulfates
Coarse	60	8–28	2.4–0.6	Muriate
Standard	60, 50, 22	14–65	1.2–0.21	Muriate and sulfates
Special standard	60	35–150	0.4–0.11	Muriate and sulfates
Soluble	62	35–150	0.4–0.11	Muriate
Chemical	63	Not applicable	Not applicable	Muriate

* From approximately 2 to 98% by weight percent cumulative.
† Tyler standard.
Source: Williams-Stroud, Searles, and Hite 1994.

PUMICE

Specifications for ground pumice sizes used for abrasives range from minus 6 mesh for cleaning to minus 300 mesh for polishing. Pumice for abrasive use should have thin vesicle walls, a maximum of grains approaching the optimum cubic shape, and uniformity in composition. In addition, it must be free of impurities. Size gradations for pumice aggregate are best determined by tests for each pumice source, but conform to specifications for lightweight aggregates. Key factors to be considered for pumice aggregate use in structural concrete, building blocks, and plaster are bulk density, compressive strength and modulus of elasticity, fire resistance, sound transmission, and thermal conductivity (Meisinger 1985).

QUARTZ CRYSTAL

TABLE 9.17 Arkansas lascas* grades and properties

Grade Physical Properties	Chemical Analysis (typical; in ppm)								
	Al	Fe	Ca	Mg	Na	K	Li	Ti	Total
1. 90% clear to the unaided eye and essentially free of crystal faces	15	2	1	1	2	2	1	1	25
2. 50% to 60% clear to the unaided eye; contains minor air and water inclusions, but essentially free of crystal faces	15	2	1	1	2	2	1	1	25
3. Translucent to light	15	3	1	1	15	7	1	1	44
4. Opaque quartz of milky white appearance	20	5	1	1	25	10	1	1	64

* Lascas is the SiO$_2$ feedstock material needed for cultured quartz crystal production.
Source: Ferrell 1985.

SAND AND GRAVEL

Sand is defined throughout the industry and by ASTM as naturally occurring unconsolidated or poorly consolidated rock particles that pass through a No. 4 mesh (0.187-in.) U.S. Standard sieve and are retained on a No. 200 mesh (0.0029-in.) U.S. Standard sieve. Gravel is naturally occurring unconsolidated or poorly consolidated rock particles that pass through a sieve with 3-in. square openings and are retained on a No. 4 mesh U.S. Standard sieve. Sand and gravel is made up of varying amounts of different rock types and is therefore of varying chemical composition. Silica is the major constituent of most commercial sands, and lesser amounts of feldspar, mica, iron oxides, and heavy minerals are common. Most applications of sand and gravel have specifications for size, physical characteristics, and chemical composition. Specifications for sand and gravel used in road building and concrete construction are often rigid in terms of particle size gradation and shape and include physical as well as some chemical properties (Davis and Tepordei 1985).

SILVER

Purity of silver in bullion, coinage, jewelry, or other items is usually expressed by its "fineness," or parts per thousand. Pure silver or fine silver is 1,000 parts fine, or 100.0% silver. Sterling silver is 925 fine, or 925 parts silver and 75 parts copper. Domestic coin silver is an alloy that was used in minting coinage until 1964 and contains 900 parts silver and 100 parts copper. Commercial silver bullion ranges from a minimum of 999 to 999.9 fine. For any fineness of silver bullion, the principal impurities are gold or copper. Dore silver is unrefined silver bullion generally containing a variable percentage of gold as an impurity. Silver for the National Defense Stockpile is required to be 999 fine, free of slag, dirt, or other foreign material, and in bars of about 1,000 troy ounces (Reese 1985).

SODA ASH

A British standard specifies not less than 57.25% Na_2O and not more than 0.005% Fe_2O_3; ASTM, a minimum of 99.16% Na_2CO_3. Sodium sulfate made from natural brine usually contains less than 0.5% impurities, but that produced as a by-product of other manufacturing may contain much larger quantities. The material meeting U.S. Pharmacopia (USP) specifications and that intended for glass making must contain at least 99% sodium sulfate. In addition, glassmakers' grade must be low in iron and heavy metals. Technical grades of sodium sulfate may have from 2% to 6% impurities. Purchases of detergent- or rayon-grade sodium sulfate are based primarily on whiteness. Salt content may be between 1.5% and 2.0%, and iron content between 60 and 100 ppm (Kostick 1985).

STONE, CRUSHED

TABLE 9.18   Classification of rocks commonly used for crushed stone

Group	General Classification	Rock	Approximate Specific Gravity	Los Angeles Abrasion Test* Average	Mid-Range[†]
Igneous	Intrusive	Gabbro	2.9	18	
		Granite	2.6	38	27–49
		Syenite	2.7	24	15–27
		Diorite	2.8		
		Peridotite	2.9		
	Extrusive	Felsite[‡]	2.6	18	
		Traprocks:			
		Basalt	2.8	14	10–17
		Diabase	2.9	18	13–21
Sedimentary	Calcareous	Dolomite	2.7	25	18–31
		Limestone	2.6	26	19–30
		Coquina			
		Coral	<2.6		
		Shell			
	Siliceous	Chert	2.5	26	
		Sandstone	2.6	38	24–48
		Graywacke	2.6	17	14–20[§]
	Foliated	Amphibolite	3.0	35	22–46
		Gneiss	2.7	45	33–57
		Schist	2.8	38	26–49

continues next page

TABLE 9.18 Classification of rocks commonly used for crushed stone (continued)

Group	General Classification	Rock	Approximate Specific Gravity	Los Angeles Abrasion Test*	
				Average	Mid-Range†
Metamorphic					
	Nonfoliated	Marble	2.7	47	26–64
		Quartzite	2.7	28	20–35
		Serpentine	2.3–2.6		

* Source: Woolf 1953, modified by Schenck and Torries 1983.
† Highest and lowest value after excluding top one-fifth and lowest one-fifth of tests reported by Woolf (1953).
‡ Including andesite, dacite, rhyolite, and trachyte.
§ Private correspondence.
Source: Herrick 1994.

SULFUR

Types of sulfur can be categorized as follows:
- Native sulfur—sulfur occurring in nature in the elemental form
- Pyrites—iron sulfide minerals that include pyrite, marcasite, and pyrrhotite
- Sulfur ore—unprocessed ore containing native sulfur
- Elemental sulfur—processed sulfur in the elemental form produced from native sulfur or combined sulfur sources, generally with a minimum sulfur content of 99.5%
- Frasch sulfur—elemental sulfur produced from native sulfur by the Frasch mining process
- Recovered sulfur—elemental sulfur produced from combined sulfur by any method
- Crude sulfur—commercial nomenclature for elemental sulfur
- Brimstone—synonymous with crude sulfur
- Broken sulfur—solid crude sulfur crushed to minus 8-in. size
- Slated sulfur—solid crude sulfur in the form of slate-like lumps produced by allowing molten sulfur to solidify on a water-cooled moving belt
- Prilled sulfur—solid crude sulfur in the form of pellets produced by cooling molten sulfur with air or water
- Bright sulfur—crude sulfur free of discoloring impurities and bright yellow in color
- Dark sulfur—crude sulfur discolored by minor quantities of hydrocarbons ranging up to 0.3% carbon content
- Sulfuric acid—sulfuric acid of commerce produced from all source of sulfur, generally reported in terms of 100% H_2SO_4 with a 32.69% sulfur content (Morse 1985).

TALC

TABLE 9.19 Talc properties important in specific markets

Automotive industry—lubricants, body putty, and asphaltic undercoating	**Cosmetics industry**
Free from grit (pure platy talc for lubricants)	Quartz content (1.0% maximum; may be lowered to 0.1% maximum)
Chemically inert	Tremolite content (0.1% maximum)
Nonwicking (undercoating)	Loss on ignition (6.0% maximum at 1,000°C)
Ceramics industry	Neutral pH
Uniform chemical composition	Acid-soluble substances (2.0% maximum)
Constant amount of shrinkage on firing	Water-soluble substances (0.1% maximum)
Fired color	Arsenic content (3 ppm maximum)
Particle size distribution	Lead content (20 ppm maximum)
	Odor, slip (lubricity), particle size, fragrance retention, and whiteness according to customer specification

continues next page

TABLE 9.19 Talc properties important in specific markets (continued)

Paint industry	Plastics industry
Whiteness	Platy particle shape (reinforcing ability)
Platy particle shape	Whiteness
Low oil absorption	Chemical inertness
Opacifying power	Low iron content
Fine particle size (Hegman gauge)	Electrical resistivity
Paper industry	Superfine particle size
Paper filler:	Low abrasion
Whiteness	Powder bulk density
Brightness (GE 78 minimum)	**Roofing industry**—asphalt backing and surfacing
Controlled top size (50 µm maximum)	Platy particle shape
Platy particle shape	Particle size distribution and its consistency
Opacifying power	Low oil absorption
Low abrasion	Minimal dust content
Particle size (8 to 12 µm median)	**Rubber industry**
Pitch control:	Particle size (<2 µm median; controlled top size)
Surface area (12 m^2/g minimum)	Platy particle shape
Whiteness	Chemical inertness
Brightness (GE 78 minimum)	Electrical resistivity
Low abrasion	Good lubricity
Particle size (2 to 5 µm median)	
Good dispersion without surfactants	

Source: Piniazkiewicz, McCarthy, and Genco 1994.

TIN

Primary or virgin tin metal is cast and sold as bars, ingots, pigs, and slabs in weights of 50 kg or less. Most of the tin metal imported into the United States is in the form of 45-kg pigs. Commercially pure tin, often designated "Straits" or "grade A" tin, has a minimum tin content of 99.8%. Higher grades, such as electrolytic, have a minimum tin content of 99.95% or even 99.98%. Hard tin contains 99.6% tin, and a still lower grade, common tin, has a 99% minimum tin content (Carlin 1985).

TITANIUM

TABLE 9.20 Typical titanium ore specifications for chloride pigment feedstocks and chemical analyses of commercial ores

Compound	Typical Specification	RBM Slag RSA	Synthetic Rutile Kerr-McGee USA	Ishahara Japan	Natural Rutile
Total Ti as TiO_2	85% minimum	86.3%	90.5–93.0	96.1	95.2
Ti_3+	?	—	—	—	—
CaO	0.2% maximum	.15	<0.35	0.01	0.02
MgO*	1%–2.0% maximum	1.20	0.05	0.07	0.07
MnO*	0.1%–1.0% maximum	1.78	0.10	0.03	0.01
Al_2O_3	1.0% maximum	1.13	<1.0	0.46	0.20
SiO_2	Varies	1.79	1.5–2.5	0.50	1.0
Th and U	100 ppm maximum	22 ppm	†	†	†
Particle size	−30+200 mesh	†	†	†	†

* Some companies specify that the combination of these two compounds should not exceed 1%.
† Not available.
Source: Garnar and Stanaway 1994.

TABLE 9.21 Composition of typical commercial titanium concentrates (weight percent)

	Ilmenite				Slag		Rutile	Synthetic Rutile		
	United States					Republic of South Africa, Richards Bay	Australia, East Coast	United States Kerr-McGee	Australia, Western Titanium	Japan, Isihara
	New York	Florida	Norway, Tellnes	Australia, Bunbury	Canada, Sorel					
TiO_2 (total)	46.1	64.00	45.0	54.4	80	85.0*	95.2	94.15	92.0	96.1
Ti_2O_3	—	—	—	—	16	25.0†	—	—	10.0	—
Fe (total)	—	—	—	—	—	—	—	—	3.6	—
Fe (metallic)	—	—	—	—	—	—	—	—	0.2	—
FeO	39.3	1.33	34.0	19.8	9.0	—	0.9	—	—	—
Fe_2O_3	6.7	28.48	12.5	19.0	—	—	1.0	2.6	—	1.3
SiO_2	1.5	0.28	2.8	0.7	2.4	—	0.2	1.3	0.7	0.5
Al_2O_3	1.4	1.23	0.6	1.5	2.9	—	0.02	0.48	0.7	0.46
CaO	0.5	0.007	0.25	0.04	0.6	0.15†	0.07	0.003	0.03	0.01
MgO	1.9	0.20	5.0	0.45	5.0	1.3†	0.18	0.2	0.15	0.07
Cr_2O_3	0.009	—	0.076	0.02	0.17	0.3†	0.6	0.16	—	0.15
V_2O_5	0.05	—	0.16	0.12	0.57	0.6†	0.01	0.16	0.12	0.20
MnO	0.5	—	0.25	1.4	0.25	2.5†	0.008	0.04	2.0	0.03
S	0.6	—	0.05	0.01	0.06	—	0.1	—	0.15	—
Na_2O	—	—	—	—	—	—	0.04	—	—	—
C	0.22	—	0.055	—	0.05	—	0.03	—	0.15	—
P_2O_5	0.008	0.12	0.04	0.02	—	—	0.8	—	—	0.17
ZrO_2	0.01	—	—	—	—	—	0.2	—	—	0.15
Nb_2O_5	0.01	0.10	0.01	0.14	—	—	0.03	—	—	0.25
Ignition loss	1.3	—	—	0.4	—	—	0.1	0.6	—	—
Sources	(‡)	(§)	(16)	(17)	(**)	(28)	(‡)	(††)	(3,29)	(30)

* Minimum.
† Maximum.
‡ NL Industries.
§ E.I. du Pont de Nemours & Co.
** QIT-Fer et Titane Inc.
†† Kerr-McGee Chemical Corp.
Source: Lynd 1985.

VANADIUM

TABLE 9.22 Typical chemical specifications for commercial forms of ferrovanadium

Alloy	Composition, percent							
	Vanadium	Carbon	Nitrogen	Aluminum	Silicon	Phosphorus	Sulfur	Manganese
50%–60% ferrovanadium	50.0–60.0	0.2 maximum	—	2.0 maximum	1.0 maximum	0.05 maximum	0.05 maximum	—
70%–80% ferrovanadium	70.0–80.0	—	—	1.0 maximum	2.5 maximum	0.05 maximum	0.10 maximum	—
80% ferrovanadium	77.0–83.0	0.50 maximum	—	0.50 maximum	1.25 max.	0.05 maximum	0.05 maximum	0.5 maximum
Proprietary alloys								
Carvan (Umetco Minerals Corporation)	82.0–86.0	10.5–14.5	—	0.10 maximum	0.10 maximum	0.05 maximum	0.10 maximum	0.05 maximum
Ferrovanadium Carbide* (Reading Alloys Inc.)	70–73	10.0–12.0	—	—	0.50 maximum	0.05 maximum	0.05 maximum	0.05 maximum
Ferovan (Foote Mineral Company)	42 minimum	0.85 maximum	—	—	7.0 maximum	—	—	4.5 maximum
Nitrovan (Umetco Minerals Corporation)	78.0–82.0	10.0–12.0	6.0 minimum	0.10 maximum	0.10 maximum	0.05 maximum	0.10 maximum	0.05 maximum

* Iron (Fe) 14.0% to 19.0%.
Source: Kuck 1985.

ZEOLITES

TABLE 9.23 Classification of zeolites (after Breck 1974)
(Figures in parentheses refer to types of zeolite frameworks, e.g., D4R = double 4-ring, T_5O_{10} = a unit of 5 tetrahedons.)

Group 1 (S4R)
Analcime
Harmotome
Phillipsite
Gismondine
Paulingite
Laumontite
Yugawaralite

Group 2 (S6R)
Erionite
Offretite
Levynite
Sodalite

Group 3 (D4R)
A-Type Zeolites

Group 4 (D6R)
Faujasite
X
Y
Chabazite
Gmelinite

Group 5 (T_5O_{10})
Natrolite
Scolecite
Mesolite
Thomsonite
Gonnardite
Edingtonite

Group 6 (T_8O_{16})
Mordenite
Dachiardite
Ferrierite
Epistibite
Bikitaite

Group 7 ($T_{10}O_{20}$)
Heulandite
Clinoptilolite
Stilbite
Brewsterite

Source: Holmes 1994.

ZINC

TABLE 9.24 Standard specifications for zinc

Grade (UNS)	Composition (%)							
	Lead	Iron, Maximum	Cadmium, Maximum	Aluminum, Maximum	Copper, Maximum	Tin, Maximum	Total Non-Zinc, Maximum	Zinc, Minimum by Difference
Special High Grade (Z13001)	0.003, maximum	0.003	0.003	0.002	0.002	0.001	0.010	99.990
High Grade (Z15001)	0.03, maximum	0.02	0.02	0.01	—	—	0.10	99.90
Prime Western (Z19001)	0.5–1.4	0.05	0.20	0.01	0.20	—	2.0	98.0

Source: ASTM 1998 *(reprinted with permission).*

Two other grades of zinc supplied to customer specifications and used for galvanizing have gained acceptance. Continuous galvanizing grade contains up to 0.35% lead and some aluminum, whereas controlled lead grade contains less than 0.18% lead and no aluminum (Jolly, J.H. 1985).

REFERENCES

Ampian, S.G. 1985a. Barite. In *Mineral Facts and Problems*. Bulletin 675. Edited by A.W. Knoerr. Washington, DC: U.S. Bureau of Mines. 65–74.

Ampian, S.G. 1985b. Clays. In *Mineral Facts and Problems*. Bulletin 675. Edited by A.W. Knoerr. Washington, DC: U.S. Bureau of Mines. 157–169.

ASTM. 1998. Standard specification for zinc. In *Chemical Requirements*. B6. Vol. 02.04. West Conshohocken, PA: ASTM.

Blossom, J.W. 1985. Molybdenum. In *Mineral Facts and Problems*. Bulletin 675. Edited by A.W. Knoerr. Washington, DC: U.S. Bureau of Mines. 521–534.

Breck, D.W. 1974. *Zeolite Molecular Sieves*. New York: Wiley-Interscience.

Burst, J.F., and R.E. Hughes. 1994. Clay-Based Ceramic Raw Materials. In *Industrial Minerals and Rocks*. 6th edition. Edited by D.D. Carr. Littleton, CO: Society for Mining, Metallurgy, and Exploration, Inc. (SME). 317–324.

Buttermore, W.H., and J.W. Leonard III. 1991. Utilization. In *Coal Preparation*. 5th ed. Edited by J.W. Leonard III. Littleton, CO: SME. 907–951.

Carlin, J.J. 1985. Tin. In *Mineral Facts and Problems*. Bulletin 675. Edited by A.W. Knoerr. Washington, DC: U.S. Bureau of Mines. 847–858.

Carr, D.D., L.F. Rooney, and R.C. Freas. 1994. Limestone and dolomite. In *Industrial Minerals and Rocks*. 6th ed. Edited by D.D. Carr. Littleton, CO: SME. 605–629.

Carrico, L.C. 1985. Mercury. In *Mineral Facts and Problems*. Bulletin 675. Edited by A.W. Knoerr. Washington, DC: U.S. Bureau of Mines. 499–508.

Davis, L.L., and V.V. Tepordei. 1985. Sand and gravel. In *Mineral Facts and Problems*. Bulletin 675. Edited by A.W. Knoerr. Washington, DC: U.S. Bureau of Mines. 689–703.

Ferrell, J.E. 1985. Quartz crystal. In *Mineral Facts and Problems*. Bulletin 675. Edited by A.W. Knoerr. Washington, DC: U.S. Bureau of Mines. 641–646.

Garnar, T.E., and K.J. Stanaway. 1994. Titanium minerals. In *Industrial Minerals and Rocks*. 6th ed. Edited by D.D. Carr. Littleton, CO: SME. 1071–1089.

Gray, R.J., J.S. Goscinski, and R.W. Schoeberger. 1978. Selection of coals for coke making. *Conference of Iron and Steel Society*. Pittsburgh, PA: The Iron and Steel Society.

Herrick, D.H. 1994. Crushed stone. In *Industrial Minerals and Rocks*. 6th ed. Edited by D.D. Carr. Littleton, CO: SME. 975–986.

Hight, R.P. 1983. Abrasives. In *Industrial Minerals and Rocks*. 5th ed. Edited by S.J. Lefond. New York: American Institute of Mining, Metallurgical, and Petroleum Engineers (AIME). 11–32.

Holmes, D.A. 1994. Zeolites. In *Industrial Minerals and Rocks*. 6th ed. Edited by D.D. Carr. Littleton, CO: SME. 1129–1158.

Jolly, J.H. 1985. Zinc. In *Mineral Facts and Problems*. Bulletin 675. Edited by A.W. Knoerr. Washington, DC: U.S. Bureau of Mines. 923–940.

Jolly, J.L.W. 1985. Copper. In *Mineral Facts and Problems*. Bulletin 675. Edited by A.W. Knoerr. Washington, DC: U.S. Bureau of Mines. 197–221.

Jones, T.S. 1985. Manganese. In *Mineral Facts and Problems*. Bulletin 675. Edited by A.W. Knoerr. Washington, DC: U.S. Bureau of Mines. 483–498.

Kirk, W.S. 1985. Cobalt. In *Mineral Facts and Problems*. Bulletin 675. Edited by A.W. Knoerr. Washington, DC: U.S. Bureau of Mines. 171–183.

Klinger, F.L. 1985. Iron ore. In *Mineral Facts and Problems*. Bulletin 675. Edited by A.W. Knoerr. Washington, DC: U.S. Bureau of Mines. 385–403.

Kokal, H.R., and M.G. Ranade. 1994. Fluxes for metallurgy. In *Industrial Minerals and Rocks*. 6th ed. Edited by D.D. Carr. Littleton, CO: SME. 661–675.

Kostick, D.S. 1985. Soda ash. In *Mineral Facts and Problems*. Bulletin 675. Edited by A.W. Knoerr. Washington, DC: U.S. Bureau of Mines. 741–755.

Kuck, P.H. 1985. Vanadium. In *Mineral Facts and Problems*. Bulletin 675. Edited by A.W. Knoerr. Washington, DC: U.S. Bureau of Mines. 895–915.

Loebenstein, J.R. 1985. Platinum-group metals. In *Mineral Facts and Problems*. Bulletin 675. Edited by A.W. Knoerr. Washington, DC: U.S. Bureau of Mines. 595–616.

Lucas, J.M. 1985. Gold. In *Mineral Facts and Problems*. Bulletin 675. Edited by A.W. Knoerr. Washington, DC: U.S. Bureau of Mines. 323–337.

Lynd, L.E. 1985. Titanium. In *Mineral Facts and Problems*. Bulletin 675. Edited by A.W. Knoerr. Washington, DC: U.S. Bureau of Mines. 859–879.

Mason, B.H. 1994. Lightweight aggregates. In *Industrial Minerals and Rocks*. 6th ed. Edited by D.D. Carr. Littleton, CO: SME. 343–350.

McCarl, H.N. 1994. Aggregates: markets and uses. In *Industrial Minerals and Rocks*. 6th ed. Edited by D.D. Carr. Littleton, CO: SME. 287–293.

McCawley, F.X., and L.H. Baumgardner. 1985. Aluminum. In *Mineral Facts and Problems*. Bulletin 675. Edited by A.W. Knoerr. Washington, DC: U.S. Bureau of Mines. 9–31.

Meisinger, A.C. 1985. Pumice. In *Mineral Facts and Problems*. Bulletin 675. Edited by A.W. Knoerr. Washington, DC: U.S. Bureau of Mines. 635–640.

Morse, D.E. 1985. Sulfur. In *Mineral Facts and Problems*. Bulletin 675. Edited by A.W. Knoerr. Washington, DC: U.S. Bureau of Mines. 783–797.

Papp, J.F. 1985. Chromium. In *Mineral Facts and Problems*. Bulletin 675. Edited by A.W. Knoerr. Washington, DC: U.S. Bureau of Mines. 139–156.

Pelham, L. 1985. Fluorspar. In *Mineral Facts and Problems*. Bulletin 675. Edited by A.W. Knoerr. Washington, DC: U.S. Bureau of Mines. 277–290.

Pickering, S.M. Jr., and H.H. Murray. 1994. Kaolin. In *Industrial Minerals and Rocks*. 6th ed. Edited by D.D. Carr. Littleton, CO: SME. 255–277.

Piniazkiewicz, R.J., E.F. McCarthy, and N.A. Genco. 1994. Talc. In *Industrial Minerals and Rocks*. 6th ed. Edited by D.D. Carr. Littleton, CO: SME. 1049–1069.

Rajgarhia, M.L. 1987. *Ground Mica.* Mica Manufacturing Co. Pvt. Ltd. 30 pp.

Reckling, K., R.B. Hoy, S.J. Lefond, D.G. Fullerton, and U.H. Rowell. 1994. Industrial diamonds. In *Industrial Minerals and Rocks*. 6th ed. Edited by D.D. Carr. Littleton, CO: SME. 379–395.

Reese, R.G. 1985. Silver. In *Mineral Facts and Problems*. Bulletin 675. Edited by A.W. Knoerr. Washington, DC: U.S. Bureau of Mines. 729–739.

Schenck, G.H.K., and T.F. Torries. 1983. Construction materials: aggregates—crushed stone. In *Industrial Minerals and Rocks.* 5th ed. Edited by S.J. Lefond. New York: AIME. 60–61.

Sibley, S.F. 1985. Nickel. In *Mineral Facts and Problems*. Bulletin 675. Edited by A.W. Knoerr. Washington, DC: U.S. Bureau of Mines. 535–551.

Stowasser, W.F. 1985. Phosphate rock. In *Mineral Facts and Problems*. Bulletin 675. Edited by A.W. Knoerr. Washington, DC: U.S. Bureau of Mines. 579–594.

Tanner, J.T. Jr. 1994. Mica. In *Industrial Minerals and Rocks*. 6th ed. Edited by D.D. Carr. Littleton, CO: SME. 693–710.

Taylor, H.A. Jr. 1994. Graphite. In *Industrial Minerals and Rocks*. 6th ed. Edited by D.D. Carr. Littleton, CO: SME. 561–570.

Van Kouteren, S. 1994. Filters and absorbents. In *Industrial Minerals and Rocks*. 6th ed. Edited by D.D. Carr. Littleton, CO: SME. 497–507.

Wellborn, W.W. 1994. Abrasives. In *Industrial Minerals and Rocks*. 6th edition. Edited by D.D. Carr. Littleton, CO: SME. 67–79.

Williams-Stroud, S.C., J.P. Searls, and R.J. Hite. 1994. Potash resources. In *Industrial Minerals and Rocks*. 6th ed. Edited by D.D. Carr. Littleton, CO: SME. 783–802.

Woodbury, W.D. 1985. Lead. In *Mineral Facts and Problems*. Bulletin 675. Edited by A.W. Knoerr. Washington, DC: U.S. Bureau of Mines. 433–452.

Woolf, D.O. 1953. *Results of Physical Tests of Road Building Aggregates*. Bureau of Public Roads, Washington, DC: U.S. Department of Commerce.

CHAPTER 10

Haul Roads

John E. Feddock, P.E.

HAUL TRUCK SPECIFICATIONS

Haul road design parameters are based on the specifications of the largest vehicle that uses the road on a regular schedule. In most cases, this is an "off-highway truck."

TABLE 10.1 Summary of off-highway truck specifications by gross vehicle weight (GVW)

(Pounds)	100,000<GVW<200,000		200,000<GVW<400,000		GVW >400,000			
(Kilograms)	45,400<GVW<90,700		90,700<GVW<181,400		GVW >181,400			
Capacity								
(Tons)	40	44	58–60	100	150	190–200	240	310
(Tonnes)	36	40	53–54	91	136	172–181	218	281
Tires								
Front & Dual Rear	18.00R33	18.00R33	24.00R35	27.00R49	33.00R51	37.00R57	40.00–57	48/95–R57
Wheel Load								
(Pounds)	23,250	27,000	33,375	55,825	85,000	117–120,000	135,000	165–176,000
(Kilograms)	10,500	12,200	15,100	25,300	38,600	53–54,400	61,200	75–79,800
Turning Radius								
(Feet)	28–32.5	32.5	34.5–39.4	32.5–42.8	44.4–49.5	40–49.5	46.5–49.5	45.5
(Meters)	8.5–10.5	9.9	10.5–12.0	9.9–13.0	13.5–15.1	12.2–15.1	14.2–15.1	13.9
Width								
(Feet)	12–16.5	16.4	14.6–16.7	17–20	21.75	21.7–26.8	23.9–29.3	26.5
(Meters)	3.7–5.0	5	4.4–5.1	5.2–6.1	8.08	6.6–8.2	7.3–8.9	8.1
Max. Speed								
(Miles per hour)	35.6–47	35	28–35	37–42	33–35	34	30–34	40.1
(Kilometers/hour)	57.3–75.6	56.3	45.1–56.3	59.5–67.6	53.1–56.3	54.7	48.3–54.7	64.5
Weight Empty								
(Pounds)	69,000	74,900	88–90,300	142–149,000	210–212,500	250–269,000	323–333,000	397,800
(Kilograms)	31,300	34,000	40–40,900	64–67,500	95–96,000	113–122,000	146–151,000	180,441
Gross Vehicle Weight								
(Pounds)	149,000	163,000	204–210,000	350–355,000	510–550,000	630–700,000	830,000	1,017,800
(Kilograms)	67,600	73,900	92–95,300	158–161,000	231–249,500	286–317,500	376,486	461,671

Sources: Feddock 2000; Caterpillar, Inc.1997; Komatsu, Inc.1997.

STOPPING DISTANCE

Stopping distance is the sum of the distance traveled during brake reaction time and the distance required to decelerate the vehicle. A vehicle should be capable of stopping within the sight distance. When designing a road segment, use Figures 10.1, 10.2, and 10.3 to determine the stopping distance for a vehicle with a particular GVW.

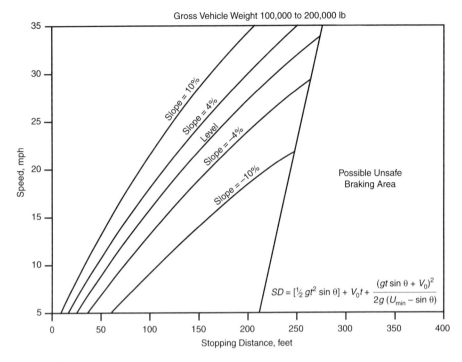

where:
SD = stopping distance, feet
g = gravitational pull (32.2 fps^2)
t = the sum of the brake reaction time (1.5) and the lag time of reaction (1.5), seconds
θ = angle of descent
V_0 = speed at time of perception, feet per second
U_{min} = coefficient of friction at the tire–road contact area, 0.30, dimensionless

Sources: Feddock 2000; Kaufman and Ault 1977.

FIGURE 10.1 Stopping distance for GVW 100,000 to 200,000 lb

NUMBER OF LANES

Haul roads from a pit to an external location may require more than a single lane per direction. The number of lanes may be determined by the following equations (Atkinson 1992):

$$n = (t)(d_b) / (550)(v) \quad \text{(EQ 10.1a)}$$

$$n = (t)(d_b) / (100)(v) \quad \text{(EQ 10.1b)}$$

where:
n = number of lanes for unidirectional travel
v = vehicle speed in miles per hour or kilometers per hour
t = traffic density in vehicles per hour
d_b = normal safe distance between trucks in feet or meters

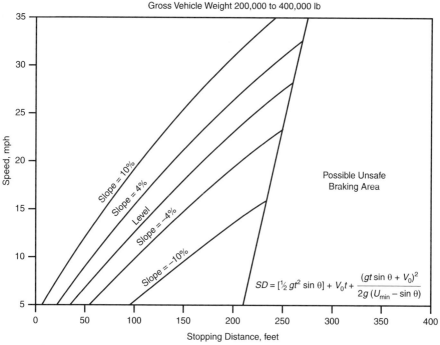

FIGURE 10.2 Stopping distance for GVW 200,000 to 400,000 lb

where:
SD = stopping distance, feet
g = gravitational pull (32.2 fps^2)
t = the sum of the brake reaction time (2.75) and the lag time of reaction (1.5), seconds
θ = angle of descent
V_0 = speed at time of perception, feet per second
U_{min} = coefficient of friction at the tire–road contact area, 0.30, dimensionless

Sources: Feddock 2000; Kaufman and Ault 1977.

SAFE DISTANCE BETWEEN TRUCKS

Safe distance depends upon driver reaction time (usually estimated at 2.0 s), gradient, and the road surface, plus an allowance (usually 16.5 ft or 5 m). It can be determined from the following equations (Atkinson 1992):

$$d_b = (v / 1.08) + [v^2 / (91.5)(C_t \pm i)] \quad \text{(EQ 10.2a)}$$

$$d_b = [(2.0)(v) / 3.6] + [v^2 / (254)(C_t \pm i)] + 5.0 \quad \text{(EQ 10.2b)}$$

where:
C_t = coefficient of traction
i = steepest haul road gradient as a fraction

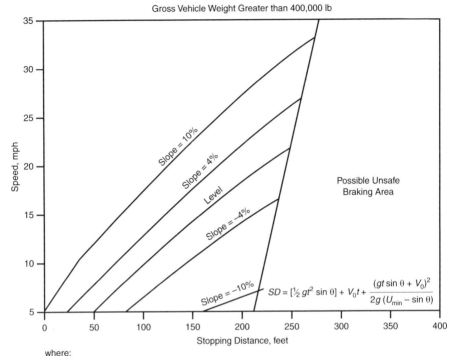

where:
SD = stopping distance, feet
g = gravitational pull (32.2 fps^2)
t = the sum of the brake reaction time (t_1) and the lag time of reaction (t_2), seconds
θ = angle of descent
V_0 = speed at time of perception, feet per second
U_{min} = coefficient of friction at the tire–road contact area, 0.30, dimensionless

Sources: Feddock 2000; Kaufman and Ault 1977.

FIGURE 10.3 Stopping distance for GVW greater than 400,000 lb

ROAD WIDTH

TABLE 10.2 Width for straight regular grade haul roads

Number of Lanes	Factor × Maximum Vehicle Width
1	2.0
2	3.5
3	5.0
4	6.5

Source: Atkinson 1992.

TABLE 10.3 Width for haul roads with sharp curves

	% of Table 8.2 (inside radius)*		
Vehicle	20 ft (6 m)	150 ft (45 m)	200 ft (60 m)
Rear-dump and unitized trucks	125	118	110
Articulated trucks	155	135	115

* Percentages for other radii may be interpolated.
Source: Atkinson 1992.

SIGHT DISTANCE

Sight distance must be sufficient to enable a vehicle traveling at a given speed to stop before reaching a hazard that is 0.5 ft in height.

TABLE 10.4 Minimum height of driver's eye

GVW	Articulated	Single Unit
100,000 to 200,000 lb	8 ft–6 in.	8 ft–6 in.
200,000 to 400,000 lb	9 ft–0 in.	11–0 in.
> 400,000 lb	11 ft–0 in.	13 ft–7 in.

Sources: Feddock 2000; Caterpillar, Inc. 1997; Komatsu, Inc. 1997.

Sight distance on vertical curves is measured by the chord length between points of tangency. Vertical curve lengths necessary to provide adequate stopping distance (i.e., equal to sight distance) may be computed as follows:

If S is greater than L_e:

$$L_s = 2S - 200\,[(H_1)^{1/2} + (H_2)^{1/2}]^2 / A \tag{EQ 10.3}$$

If S is less than L_e:

$$L_s = (A)(S^2) / 100\,[(2H_1)^{1/2} + (2H_2)^{1/2}]^2 \tag{EQ 10.4}$$

where:
- L_e = existing curve length
- L_S = safe curve length
- A = algebraic difference in grades
- S = stopping distance
- H_1 = height of eye of driver
- H_2 = height of object above road surface (0.5 ft)

GRADIENT

TABLE 10.5 Haul road gradients for most situations*

Sustained maximum	8% to 15%
Usually optimum	≈ 8%
For trolley-assisted trucks	≤12%

* 150-ft sections at ≤ 2% gradient should be included for every 1,500 to 1,800 ft of severe gradient.
Source: Adapted from Atkinson 1992.

SUPER ELEVATION

Haul road curves for vehicle speeds at or above 10 mph should be super elevated.

TABLE 10.6 Super elevation for haul road curves, inches per yard

Radius (yd)	Truck Speed (mph)					
	10	15	20	25	30	>35
5	1.5	1.5	—	—	—	—
10	1.5	1.5	1.5	—	—	—
15	1.5	1.5	1.5	1.8	—	—
25	1.5	1.5	1.5	1.5	2.2	—
30	1.5	1.5	1.5	1.5	1.8	2.2
60	1.5	1.5	1.5	1.5	1.5	1.8
100	1.5	1.5	1.5	1.5	1.5	1.5

Source: Atkinson 1992.

The portion of haulageway for transforming a normal cross-slope section into a superelevated section is the runout length. One-third should be in the curve and two-thirds on the tangent.

TABLE 10.7 Maximum rates of super elevation change on tangents, inches per yard

Truck Speed, mph	15	20	25	30	<40
Super elevation change/100 yd of tangent (in./yd)	9.35	9.00	8.28	7.20	5.76

Source: Atkinson 1992.

CONSTRUCTION MATERIAL FACTORS

TABLE 10.8 Adhesion and rolling resistance coefficients

Description	Coefficient of Adhesion	Rolling Resistance % GVW
Concrete of blacktop	0.9	2.0
Rock base, dry	0.7	3.0
Rock base, wet	0.65	3.5
Pit floor, rock	0.55	5.0
Partly compacted gravel	0.45	5.0
Gravel, unmaintained, wet	0.4	7.5
Weak materials, flexing considerably	0.35	8.0
Soft, muddy, rutted road	0.3	15.0–20.00
Ice	0.1	1.0

Source: Atkinson 1992.

The coefficient of rolling resistance may be related to tire penetration by the following equation (Atkinson 1992):

$$CRR = 0.02 + 0.0007 \, (tp) \tag{EQ 10.5}$$

where:
 CRR = coefficient of rolling resistance
 tp = tire penetration in mm

OTHER CONSIDERATIONS

- Avoid introducing sharp horizontal curvature at or near the crest of a hill.
- Design sections of haulage road with long tangents and constant grades.
- Avoid intersections near crest verticals and sharp horizontal curvatures.
- Adequate drainage should have a cross slope of ¼-in. to ½-in. drop for each foot of width.

See Figure 10.4 on page 201 for subbase design, English units.
See Figure 10.5 on page 202 for subbase design, SI units.

REFERENCES

Atkinson, T. 1992. Design and layout of haul roads. In *SME Mining Engineering Handbook*. 2nd ed., Vol. 1. Edited by H.L. Hartman. Littleton, CO: Society for Mining, Metallurgy, and Exploration, Inc. 1334–1342.
Caterpillar, Inc. 1997. *Caterpillar Performance Handbook*. 28th ed. Peoria, IL: Caterpillar.
Feddock, J.E. 2000. Internal report. Bluefield, VA: Marshall Miller & Associates.
Komatsu, Inc. 1997. *Specifications and Application Handbook*. 18th ed. Vernon Hills, IL: Komatsu, Inc.
Kaufman, W.W., and J.C. Ault. 1977. *Design of Surface Mine Haulage Roads—A Manual*. Information Circular 8758. Washington, DC: U.S. Bureau of Mines.

HAUL ROADS | **201**

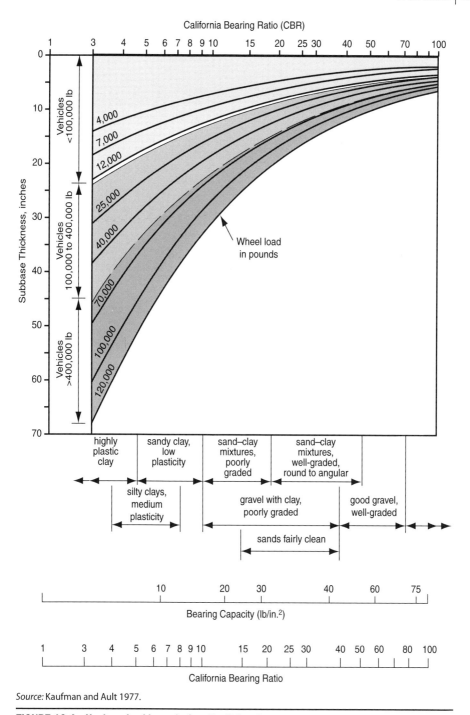

Source: Kaufman and Ault 1977.

FIGURE 10.4 Haul road subbase design, English units

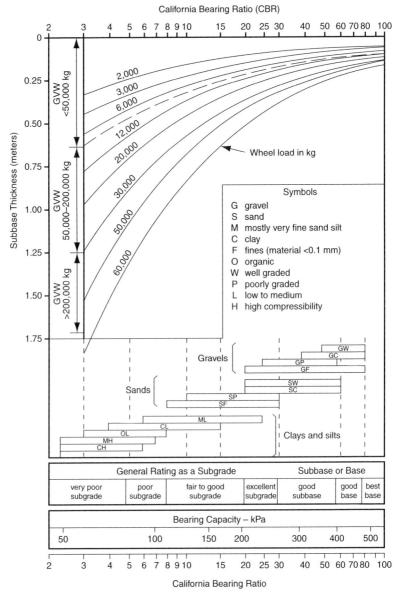

Source: Kaufman and Ault 1977.

FIGURE 10.5 Haul road subbase design, SI units

CHAPTER 11

Blasting and Explosives

Larry C. Schneider, P.E.

CHARACTERISTICS OF EXPLOSIVES

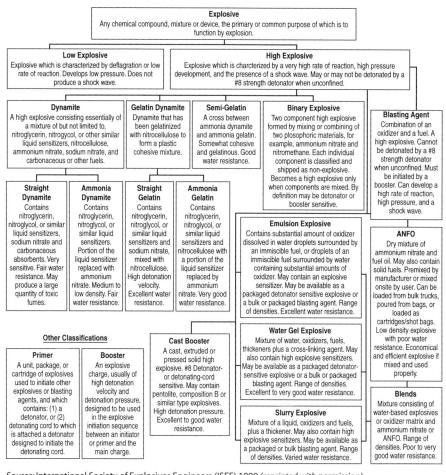

Source: International Society of Explosives Engineers (ISEE) 1999 (reprinted with permission).

FIGURE 11.1 Descriptive classification of explosives

TABLE 11.1 Typical characteristics of commercial explosives

Explosive Type	Grade	Specific Gravity	Velocity of Detonation (ft/s)	Relative Bulk Strength (ANFO* = 100)	Water Resistance
Straight dynamite	50% ditching dynamite	1.32	17,400		Good
Ammonia dynamite	40%	1.30–1.32	9,800–12,500	146	Fair
	60%	1.28–1.30	12,470–15,300	152	Very good
Semi-gelatin		1.29	13,100		Good
		1.16	12,600		Very good
		1.26	17,500	149	Very good
		0.94	11,300		Good
Straight gelatin	80%	135–1.43	18,700–22,000	201	Excellent
Ammonia gelatin	40%	1.30–1.50	11,000–18,500	146–171	Excellent
	60%	1.34–1.43	17,400–18,900	164	Excellent
	80%	1.34	20,200	199	Excellent
Water gels	Packaged 2-in.–9-in. diameter	1.18–1.52	13,230–19,000	128–183	Very good
Emulsions	Bulk	0.95–1.28	13,100–19,000	103–156	Excellent
	Package high explosive	1.10–1.20	13,500–17,000	102–167	Excellent
ANFO	Poured	0.77–0.85	9,100–15,100	100	None
	Pneumatic loading	0.85–1.10	9,000–11,000	100–122	None
ANFO/emulsion blends	20:80	1.05	16,400	138	Poor
	30:70	1.2	17,400	166	Fair
	40:60	1.3	18,100	185	Good
	50:50	1.3	17,900	179	Excellent
	60:40	1.3	17,500	175	Excellent
	70:30	1.3	17,100	169	Excellent
	80:20	1.3	16,700	165	Excellent

* Ammonium nitrate and fuel oil.

Classification of Explosives by Fume Characteristics

The Mine Safety and Health Administration (MSHA) designates an explosive as "permissible" for use in underground coal mines if the carbon monoxide (CO) produced by detonation does not exceed 2.5 ft^3/lb of explosive. In this approval method, other toxic gases are measured and their volumes converted to the equivalent volume of CO.

Explosives not designated as permissible are classified by the Institute of Makers of Explosives (IME) in accordance with Table 11.2.

Detonation Pressure

The detonation pressure of an explosive, which is an important factor in determining an explosive's effectiveness as a primer charge or a booster, can be calculated as follows:

$$P = (2.32 \times 10^{-7}) \rho D^2 \qquad \text{(EQ 11.1)}$$

where:
    P = detonation pressure (kilobars)
    ρ = specific gravity of explosive
    D = detonation velocity (ft/s)

TABLE 11.2 IME fume classification

Fume Class	Amount of Poisonous Gases per 1¼ in. × 8 in. (32 mm × 203 mm) Cartridge of Explosive Material
1	less than 0.16 ft^3 (4.53 L)
2	0.16 to 0.33 ft^3 (4.53 to 9.35 L)
3	0.33 to 0.67 ft^3 (9.35 to 18.98 L)

Source: IME 1997 (reprinted with permission).

TABLE 11.3 Loading factor table (pounds of explosives per foot of depth)

Explosives Diameter (in.)	Explosives Specific Gravity									
	0.80	0.85	0.90	0.95	1.00	1.05	1.10	1.20	1.25	1.30	1.40
1	0.27	0.29	0.31	0.32	0.34	0.36	0.37	0.41	0.43	0.44	0.48
1¼	0.42	0.45	0.48	0.50	0.53	0.56	0.58	0.64	0.67	0.69	0.74
1½	0.61	0.65	0.69	0.73	0.77	0.80	0.84	0.92	0.96	0.99	1.07
1¾	0.83	0.89	0.94	0.99	1.05	1.10	1.15	1.25	1.31	1.36	1.46
2	1.09	1.16	1.23	1.30	1.36	1.43	1.50	1.64	1.70	1.77	1.91
2¼	1.38	1.47	1.55	1.64	1.73	1.81	1.90	2.07	2.16	2.77	2.98
2½	1.70	1.81	1.91	2.02	2.13	2.23	2.34	2.55	2.66	2.77	2.98
2¾	2.06	2.19	2.32	2.44	2.57	2.70	2.83	3.09	3.22	3.34	3.60
3	2.45	2.60	2.76	2.91	3.06	3.22	3.37	3.68	3.83	3.98	4.29
3¼	3.33	3.54	3.75	3.96	4.16	4.37	4.58	5.00	5.20	5.42	5.83
4	4.35	4.62	4.89	5.16	5.44	5.71	5.98	6.52	6.80	7.07	7.61
4½	5.51	5.85	6.19	6.54	6.88	7.23	7.57	8.26	8.60	8.95	9.63
5	6.81	7.22	7.65	8.07	8.50	8.93	9.35	10.20	10.62	11.05	11.90
5½	8.23	8.74	9.25	9.77	10.28	10.80	11.31	12.34	12.85	13.37	14.39
6	9.81	10.40	11.01	11.62	12.24	12.85	13.46	14.68	15.30	15.91	17.13
6¼	10.63	11.29	11.95	12.62	13.28	13.95	14.61	15.94	16.60	17.27	18.59
6½	11.49	12.21	12.93	13.65	14.36	15.08	15.80	17.24	17.95	18.67	20.11
6¾	12.39	13.17	13.94	14.72	14.49	16.27	17.04	18.59	19.36	20.14	21.69
7	13.33	14.16	15.00	15.83	16.66	17.50	18.33	20.00	20.83	21.66	23.33
7½	15.30	16.26	17.21	18.17	19.13	20.08	21.04	22.95	23.91	24.87	26.78
7⅞	16.87	17.92	18.97	20.03	21.08	22.14	23.19	25.30	26.35	27.41	29.51
8	17.41	18.50	19.59	20.68	21.76	22.85	23.94	26.12	27.20	28.29	30.47
9	22.03	23.41	24.78	26.16	27.54	28.91	30.29	33.04	34.42	35.80	38.55
9⅞	26.52	28.18	29.84	31.50	33.15	34.81	36.47	39.79	41.44	43.10	46.42
10	27.20	28.90	30.60	32.30	34.00	35.70	37.40	40.80	42.50	44.20	47.60
10⅝	30.71	32.62	34.54	36.46	38.38	40.30	42.22	46.06	47.98	49.90	53.73
12¼	40.81	43.37	45.92	48.47	51.02	53.57	56.12	61.22	63.77	66.32	71.43
15	61.20	65.03	68.85	72.68	76.50	80.33	84.15	91.80	95.63	99.45	107.1

Source: Kentucky Department of Mines and Minerals 1996.

Values not contained in Table 11.3 can be calculated as

$$LF = 0.3405 \, (d^2) \, \rho \qquad \text{(EQ 11.2)}$$

where:
- LF = loading factor in (lb/ft)
- d = explosives column diameter (in.)
- ρ = specific gravity of explosives

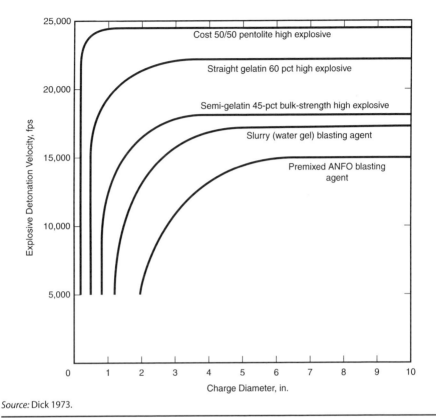

Source: Dick 1973.

FIGURE 11.2 Effects of charge diameter on detonation velocity

BLAST DESIGN

The following standards for calculating the parameters of a blast pattern are first approximations, or rules of thumb.

Method Based on Burden Ratios

To use this method of designing a blasting pattern, the explosive's diameter must be known, and a "burden ratio" corresponding to the type of rock to be shot and the type of explosive used must be selected. Equations 11.3 through 11.7 provide the necessary values for the blast pattern (Ash 1963, Konya and Walter 1990).

$$B = (K_b \times D_e)/12 \qquad \text{(EQ 11.3)}$$

where:
B = burden distance (ft)
D_e = diameter of explosives (in.)
K_b = burden ratio selected from Table 11.4

$$H = K_d \times B \qquad \text{(EQ 11.4)}$$

where:
H = borehole depth (ft)
B = burden distance (ft)
K_d = depth ratio, which ranges from 1.5 to 4.0

TABLE 11.4 Values of the burden ratio, K_b

	Rock Type		
Explosive Energy	Soft	Average	Hard
Low	30	25	20
Average	35	30	25
High	40	35	30

Source: Ash 1963.

$$J = 0.3 \times B \qquad \text{(EQ 11.5)}$$

where:
J = borehole subdrilling (ft)
B = burden distance (ft)

$$T = 0.7 \times B \qquad \text{(EQ 11.6)}$$

where:
T = stemming length (ft)
B = burden distance (ft)

$$S = K_s \times B \qquad \text{(EQ 11.7)}$$

where:
S = spacing (ft)
B = burden (ft)
K_s = 2.0 for simultaneous initiation
K_s = 1.0 for long delay intervals between holes in the same row
$1.2 < K_s < 1.8$ for short interval delays between holes in the same row

Method Based on Known Powder Factor

This method requires the use of a known powder factor (pounds of explosives per cubic yard of rock) and calculates the geometry of the blast necessary to attain this powder factor.

$$T = 1.7 \times D_e \qquad \text{(EQ 11.8)}$$

where:
T = stemming length (ft)
D_e = explosive diameter (in.)

$$J = 0.5 \times D_e \qquad \text{(EQ 11.9)}$$

where:
J = borehole subdrilling (ft)
D_e = explosive diameter (in.)

$$C = H + J - T \qquad \text{(EQ 11.10)}$$

where:
C = powder column (ft)
H = bench height (ft)
J = subdrilling (ft)
T = stemming (ft)

$$W = C \times L_f \quad \text{(EQ 11.11)}$$

where:
- W = weight of explosive per borehole (lb)
- C = powder column (ft)
- L_f = loading factor from Table 11.3 (lb/ft)

$$Y = W / P_f \quad \text{(EQ 11.12)}$$

where:
- Y = volume of rock broken per borehole (yd^3)
- W = weight of explosive per borehole (lb)
- P_f = powder factor (lb/yd^3)

For a rectangular pattern:

$$B = \sqrt{\frac{(18 \cdot Y)}{H}} \quad \text{(EQ 11.13)}$$

$$S = 1.5 \times B \quad \text{(EQ 11.14)}$$

For a square pattern:

$$B = \sqrt{\frac{(27 \cdot Y)}{H}} \quad \text{(EQ 11.15)}$$

$$S = B \quad \text{(EQ 11.16)}$$

where:
- B = burden (ft)
- S = spacing (ft)
- Y = volume of rock broken per borehole (yd^3)
- H = bench height (ft)

Determining Amount of Rock Broken per Borehole

$$V = \frac{B \times S \times H}{27} \quad \text{(EQ 11.17)}$$

where:
- V = volume of shot rock per borehole (yd^3)
- B = burden distance (ft)
- S = spacing distance (ft)
- H = bench height (ft)

NOTE: bench height is the borehole depth minus any subdrilling.

Powder Factor—Ratio of Explosives Used per Unit Weight/Volume of Rock

As commonly in surface coal mining (pounds of explosive per cubic yard of rock):

$$P_f = \frac{W_e}{V_r} \quad \text{(EQ 11.18)}$$

where:
- P_f = powder factor (lb/yd^3)
- W_e = weight of explosives used (lb)
- V_r = volume of rock broken (yd^3)

As commonly used in the quarry industry (tons of rock per pound of explosive):

$$P_f = \frac{W_r}{W_e} \quad \text{(EQ 11.19)}$$

where
P_f = powder factor (ton/lb)
W_r = weight of rock broken (ton)
W_e = weight of explosives used (lb)

Powder Factor for Underground Blasting

A heading in an underground mine has only one free face; this type of blasting is much more confined and rock movement is constricted. As shown in Figure 11.3, a higher powder factor is required, which relates the powder factor to the cross-sectional area of the heading.

Controlled Blasting

The following equations give some first approximations for spacing of pre-split shots and the burden for cushion blasting. Two rules of thumb for spacing in controlled blasting indicate that the spacing of a pre-split pattern should be 10 times the borehole diameter, and the spacing for cushion blasting should be 16 times the borehole diameter. These simple rules agree well with Equations 11.20 and 11.22 (Konya and Walter 1990).

TABLE 11.5 Typical powder factors for surface coal mines

Primary Excavating Equipment or Mining Method	Type of Rock Blasted	Typical Powder Factors (lb/yd³)
Large dragline	Shale	0.5–0.8
	Sandstone	0.7–1.0
Shovel	Shale	0.6–0.9
	Sandstone	0.7–1.2
Front-end loader	Shale	0.7–1.1
	Sandstone	0.8–1.3
Cast blasting	Shale	0.9–1.3
	Sandstone	1.1–1.8

TABLE 11.6 Typical powder factors for surface metal mines

Geology	Compressive Seismic Velocity Range (ft/s)	Powder Factor (lb/yd³)
Weathered limestone	1,000–2,000	0.5
Weathered porphyry	2,000–3,000	0.6
Rhyolite breccia	3,000–4,000	0.75
Monzonite porphyry	4,000–5,000	1.0
Quartz sericite porphyry	5,000–6,000	1.0
Fresh limestone	5,000–6,000	1.0
Massive jasperoid	6,000–12,000	1.3

Source: Heinen and Dimock 1976.

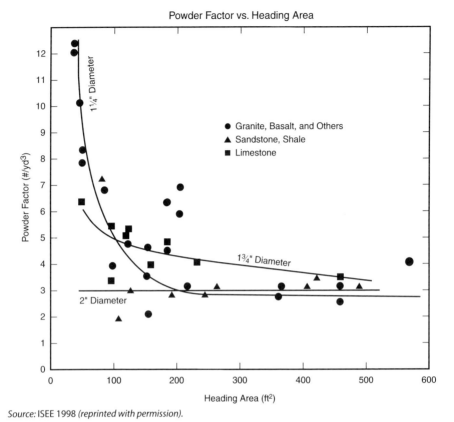

Source: ISEE 1998 *(reprinted with permission).*

FIGURE 11.3 Relationship of powder factor versus heading area

Pre-splitting: $$S = 1.1 \times d \times \left(\frac{1}{\rho_r}\right)^{1/3}$$ (EQ 11.20)

$$LF = \frac{d^2}{28}$$ (EQ 11.21)

Cushion blasting: $$S = 1.1 \times d \times \left(\frac{1}{\rho_r}\right)^{1/3}$$ (EQ 11.22)

$$B = 1.33 \times S$$ (EQ 11.23)

where:
- S = spacing between holes (ft)
- d = diameter of borehole (in.)
- ρ_r = specific gravity of rock
- LF = loading factor (lb of explosive/ft of borehole)
- B = minimum burden distance for cushion blasting only (ft)

Height of Water-Resistant Explosives to Build Out of Water in a Borehole

When loading a bulk non-water-resistant blasting agent into a borehole containing water, it is possible to use cartridged water-resistant explosives to build out of the water. The final height of such cartridges should be equal to the final height of water (Atlas Powder Company 1987).

$$h_f = \frac{h_o \times d_b^2}{d_b^2 - d_e^2} \quad \text{(EQ 11.24)}$$

where:
- h_f = final height of water in borehole (ft)
- h_o = original height of water in borehole (ft)
- d_b = diameter of borehole (in.)
- d_e = diameter of explosive cartridge (in.)

GROUND VIBRATIONS AND AIR CONCUSSION FROM BLASTING

Scaled Distance Equations

The "scaled distance," sometimes called the "scale factor," is a parameter useful for comparing the seismic effects of blasts of varying charge weights per delay and distances. It is also useful for predicting the magnitude of peak particle velocity (PPV).

$$S = \frac{D}{\sqrt{W}} \quad \text{(EQ 11.25)}$$

For compliance purposes, the allowable weight per delay can be calculated from the scaled distance equation. The use of specific scaled distances, such as 50, 55, and 65, has been shown statistically to limit peak particle vibrations to specified levels. A scaled distance of 50 will maintain PPV below 2.0 in./s with greater than 95% certainty. Likewise, a scaled distance of 55 will limit the PPV to less than 1.00 in./s, and a scaled distance of 65 to less than 0.75 in./s, also with greater than 95% confidence. Where the variables are as above, the form used in this case is

$$W = \left(\frac{D}{S}\right)^2 \quad \text{(EQ 11.26)}$$

where:
- S = scaled distance (ft/lb$^{1/2}$)
- D = distance from blast to structure (ft)
- W = maximum weight of explosives detonated per 8 ms delay (lb)

VIBRATION ATTENUATION AND PEAK PARTICLE VELOCITY PREDICTIONS

If seismic data are plotted on a log-log graph with PPV on the y axis and scaled distance on the x axis, an equation can be developed to describe the seismic characteristics of a particular site. Figure 11.4 shows a typical plot, and the equation for the lines is in the form:

$$V = kS^m \quad \text{(EQ 11.27)}$$

where:
- V = peak particle velocity (in./s)
- k = a site constant equal to the y intercept of the line, where scaled distance = 1
- S = scaled distance, (lb/ft$^{1/2}$)
- m = site constant equal to the slope of the line

The data in Figure 11.4 are from a number of surface coal mines and show the three components of ground vibration. The wide range of data scatter typical of such plots is illustrated. The regression lines are in the form $V = kS^m$.

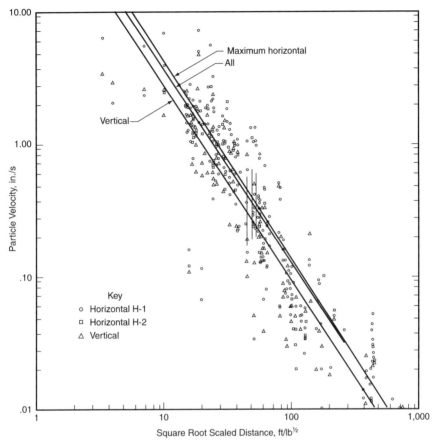

Source: Siskind et al. 1980.

FIGURE 11.4 Log-log plot of PPV versus scaled distance

TABLE 11.7 Typical values of site constants

Direction	Range of k	Average k	Range of m values	Average m
Radial	44 to 135	106	−1.729 to −1.324	−1.454
Transverse	40 to 106	61	−1.562 to −1.234	−1.368
Vertical	56 to 335	156	−1.825 to −1.551	−1.642

Source: Siskind et al. 1980.

Vector Sum

$$R = \sqrt{V^2 + L^2 + T^2} \qquad \text{(EQ 11.28)}$$

where:
- R = resultant vector sum velocity (in./s)
- V = particle velocity in vertical direction (in./s)
- L = particle velocity in longitudinal direction (in./s)
- T = particle velocity in transverse direction (in./s)

NOTE: Values of V, L, and T must occur at the same instant of time.

Vibration Level Criteria

The safe blasting vibration criteria in Figure 11.5, which were developed by the U.S. Bureau of Mines, have been generally accepted as the most reliable standards available. This curve has been adopted in many regulatory schemes on the federal, state, and local levels.

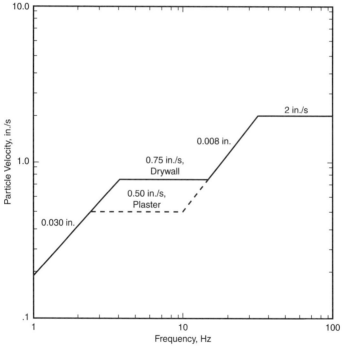

Source: Siskind et al. 1980.

FIGURE 11.5 Safe levels of blasting vibrations for residential structures based on both peak particle velocity and frequency

REFERENCES

Ash, R.L. 1963. The mechanics of rock breakage. *Pit and Quarry.* August: 98–112.

Atlas Powder Company. 1987. *Explosives and Rock Blasting.* Dallas: Atlas Powder Company.

Dick, R.A. 1973. Explosives and borehole loading. In *SME Mining Engineering Handbook.* Vol. 1. Edited by A.B. Cummins and I.A. Given. New York: Society of Mining Engineers of The American Institute of Mining, Metallurgical, and Petroleum Engineers, Inc. 11-78–11-99.

Heinen, R.H., and R.R. Dimock. 1976. The use of seismic measurements to determine the blastability of rock. *Proceedings of the Second Conference of Explosives and Blasting Techniques:* 234–248. Morgantown, WV: ISEE.

IME. 1997. *Glossary of Commercial Explosives Industry Terms.* Safety Library Publication No. 12. Washington, DC: IME.

ISEE. 1998. *Blaster's Handbook.* 17th ed. Cleveland, OH: ISEE.

ISEE. 1999. *Membership Directory and Desk Reference 1999–2000 Edition.* Cleveland, OH: ISEE.

Kentucky Department of Mines and Minerals. 1996. *Study Guide for the General Blasters Examination.* Lexington, KY: Kentucky Department of Mines and Minerals.

Konya, C.J., and E.J. Walter. 1990. *Surface Blast Design.* Englewood Cliffs, NJ: Prentice-Hall.

Siskind, D., M.S. Stagg, J.W. Kopp, and C.H. Dowding. 1980. *Structure Response and Damage Produced by Ground Vibration from Surface Mine Blasting.* Report of Investigation 8507. Washington, DC: U.S. Bureau of Mines.

CHAPTER 12

Excavation, Loading, and Material Transport

Frank J. Filas, P.E.

EQUIPMENT PERFORMANCE

General definitions, equations, and factors for determining the performance of crawler and rubber-tired equipment are presented below. Pulling and resistive forces are solved in kilograms and pounds. Convert kilograms to newtons by multiplying by 9.81 m/s².

Drawbar Pull

Drawbar pull is the horizontal force available at the drawbar of tracked equipment and is typically determined in field tests conducted on firm, level ground.

$$DBP = (EBP \times EFF \times UCF)/V_{tr} \qquad \text{(EQ 12.1)}$$

where:
DBP = drawbar pull, kg (lb)
EBP = engine brake power, kW (hp)
EFF = efficiency in converting engine power to drawbar power, decimal
UCF = unit conversion factor, 102 (m-kg)/(kW-s) or 33,000 (ft-lb)/(hp-min)
V_{tr} = tractor speed, m/s (ft/min)

Rim Pull

Rim pull is the maximum pulling force that the engine can deliver to the tires of the driving wheels of rubber-tired equipment at the point of contact with the ground.

$$RP = (EBP \times EFF \times UCF)/V_{eq} \qquad \text{(EQ 12.2)}$$

where:
RP = rim pull, kg (lb)
EBP = engine brake power, kW (hp)
EFF = efficiency in converting engine power to rim pull power, decimal (typically about 0.85 for mechanical drives)
UCF = unit conversion factor, 367 (km-kg)/(kW-h) or 375 (mi-lb)/(hp-h)
V_{eq} = equipment speed, km/h (mi/h)

Traction

Usable drawbar pull and rim pull are limited by the traction of the tracks and drive wheels, respectively, on the road surface.

$$UDBP = CT \times GWT \qquad (EQ\ 12.3)$$

$$URP = CT \times GWD \qquad (EQ\ 12.4)$$

where:
$UDBP$ = usable drawbar pull, kg (lb)
GWT = gross weight of tractor, kg (lb)
URP = usable rim pull, kg (lb)
GWD = gross weight on drive wheels, kg (lb)
CT = coefficient of traction, decimal (see Table 12.1)

TABLE 12.1 Coefficient of traction values

Road Surface	Coefficient of Traction	
	Rubber Tires	Tracks
Concrete, new	0.80–1.00	0.45
Concrete, old	0.60–0.80	
Concrete, wet	0.45–0.80	
Asphalt, new	0.80–1.00	
Asphalt, old	0.60–0.80	
Asphalt, wet	0.30–0.80	
Gravel, packed and oiled	0.55–0.85	
Gravel, loose	0.35–0.70	0.50
Gravel, wet	0.35–0.80	
Rock, crushed	0.55–0.75	0.55
Rock, wet	0.55–0.75	
Cinders, packed	0.50–0.70	
Cinders, wet	0.65–0.75	
Earth, firm	0.55–0.70	0.90
Earth, loose	0.45	0.60
Sand, dry	0.20	0.30
Sand, wet	0.40	0.50–0.55
Snow, packed	0.20–0.55	0.25
Snow, loose	0.10–0.25	
Snow, wet	0.30–0.60	
Ice, smooth	0.10–0.25	0.12
Ice, wet	0.05–0.10	
Coal, stockpiled	0.45	0.60

Source: Hays 1990.

Resistance

Rolling resistance is the sum of the forces opposing motion over level terrain for rubber-tired equipment. Grade resistance is the gravitational force that resists movement up a slope (+) and assists movement down a slope (–). Total resistance is the sum of rolling and grade resistance and represents the resisting force that the usable rim pull must exceed before the equipment can move.

$$RR = RRF \times GMW \qquad (EQ\ 12.5)$$

$$GRF = GCF \times \%Grade \qquad (EQ\ 12.6)$$

$$GR = GRF \times GMW \qquad (EQ\ 12.7)$$

$$TR = RR + GR \qquad (EQ\ 12.8)$$

where:
- RR = rolling resistance, kg (lb)
- RRF = rolling resistance factor, kg/t (lb/ton) (see Table 12.2)
- GMW = gross machine weight, t (ton)
- GRF = grade resistance factor, kg/t (lb/ton)
- GCF = grade conversion factor = 10 kg/t or 20 lb/ton
- $\%Grade$ = (± vertical distance/horizontal distance) × 100
- GR = grade resistance, kg (lb)
- TR = total resistance, kg (lb)

Total resistance may also be expressed as effective grade in percent of vehicle weight.

$$EG = Grade + (RRF/GCF)\% \qquad (EQ\ 12.9)$$

where:
- EG = effective grade (%)
- Grade = (± vertical distance/horizontal distance) × 100%
- RRF = rolling resistance factor, kg/t (lb/ton) (see Table 12.2)
- GCF = grade conversion factor = 10 kg/t or 20 lb/ton

TABLE 12.2 Rolling resistance factors

Surface	High-Pressure Tires (radial)		Low-Pressure Tires (bias ply)	
	kg/t (lb/ton)	% of Weight	kg/t (lb/ton)	% of Weight
Concrete	15 (30)	1.5	20 (40)	2.0
Asphalt	18 (36)	1.8	24 (48)	2.4
Packed gravel	22.5 (45)	2.3	30 (60)	3.0
Packed earth	30 (60)	3.0	40 (80)	4.0
Unplowed earth terrain	75 (150)	7.5	50 (100)	5.0
Rutted and uneven earth	105 (210)	10.5	90 (180)	9.0
Loose sand and gravel	140 (280)	14.0	120 (240)	12.0
Soft, muddy, deeply rutted	175 (350)	17.5	160 (320)	16.0
Snow-packed	25 (50)	2.5	35 (70)	3.5

Source: Adapted from Drevdahl 1963.

Maximum and Average Speed

The maximum speed that rubber-tired haulage equipment can achieve for a section of road with a specified total resistance can be determined from the manufacturer's specifications (i.e., performance and retarding curves). Calculating an average speed requires allowance for acceleration, deceleration, shifting, and braking, as well as safely negotiating curves, steep downhill grades, and congested areas. Use Table 12.3 to estimate an average speed for a section of road. This table is based on dividing the haul road into segments, with each segment having a relatively uniform total resistance or effective grade. The maximum speed of the truck is determined for each segment based on the manufacturer's specifications and then derated using the factors given in the table.

TABLE 12.3 Factors for converting maximum speed to average speed

Length of Haul Road Section		Factors for Converting Maximum Speed to Average Speed		
m	ft	Short, Level Hauls 500–1,000 ft Total Length	Unit Starting from Stop	Unit in Motion When Entering Road Section
0–107	0–350	0.20	0.25–0.50	0.50–2.00
107–229	350–750	0.30	0.35–0.60	0.60–0.75
229–457	750–1,500	0.40	0.50–0.65	0.70–0.80
457–762	1,500–2,500		0.60–0.75	0.75–0.80
762–1,067	2,500–3,500		0.65–0.75	0.80–0.85
over 1,067	over 3,500		0.70–0.85	0.80–0.90

NOTES:
1. The average speed may be above the maximum speed for a short section if the haulage unit enters the section at high speed.
2. The return time is generally governed by job conditions and safety precautions. If no steep downgrades or operating hazards are present, the following factors apply to top speed, empty: under 500 feet (favorable = 0.65, average = 0.60, unfavorable = 0.55); over 500 feet (favorable = 0.85, average = 0.80, unfavorable = 0.75).
3. Recommended average speeds in loading areas for favorable, average, and unfavorable conditions are 16 km/h (10 mi/h), 11.2 km/h (7 mi/h), and 6.4 km/h (4 mi/h), respectively.
4. Recommended maximum downgrade speeds include 40–56 km/h (25–35 mi/h) for 0%–6% grades, 33 to 40 km/h (21–25 mi/h) for 7%–8% grades, 27–32 km/h (17–20 mi/h) for 9%–10% grades, 21–26 km/h (13–16 mi/h) for 11%–12% grades, and less than 21 km/h (13 mi/h) for grades over 12%.
5. Safety considerations may require the use of slower speeds than listed in this table.

Source: Bishop 1968.

Altitude Correction

With increasing altitude, there is a corresponding loss in engine power (and drawbar and rim pull) because the air becomes thinner. For naturally aspirated, four-cycle gasoline and diesel engines, a rule-of-thumb reduction of 3% is used for each 305 m (1,000 ft) above the first 305 m (1,000 ft). For two-cycle diesel engines, a reduction of 1% is used for each 305 m (1,000 ft) above the first 305 m (1,000 ft). These rule-of-thumb reductions do not apply to turbocharged engines, which typically do not experience a loss in power until they reach about 1,500 m (about 5,000 ft) or higher. For high-elevation work, using the altitude derating tables supplied in equipment performance manuals is recommended.

PRODUCTION CALCULATIONS

General definitions, equations, and factors for determining the production of excavation, loading, and transportation equipment are presented below.

Density and Swell

- LCM = loose cubic meter
- LCY = loose cubic yard
- BCM = bank cubic meter
- BCY = bank cubic yard

Density is the mass per unit volume (see Table 2.1 in Chapter 2, which covers material properties, for densities.)

$$\text{Loose density } (LD) = \text{mass (kg)}/LCM \text{ or mass (lb)}/LCY \qquad \text{(EQ 12.10)}$$

$$\text{Bank density } (BD) = \text{mass (kg)}/BCM \text{ or mass (lb)}/BCY \qquad \text{(EQ 12.11)}$$

Swell is the percentage increase in volume that a material exhibits when removed from its natural state (see Table 2.1 in Chapter 2 for swell factors.)

$$\text{Swell} = (100\%)(LCM - BCM)/BCM = (100\%)(LCY - BCY)/BCY \qquad \text{(EQ 12.12)}$$

$$\text{Swell factor } (SF) = LD/BD = 100/(100 + \text{swell}) \qquad \text{(EQ 12.13)}$$

NOTE: Equation 12.13 yields swell factors of less than one. Some references invert this equation, which results in swell factors greater than one. Either method is correct as long as the derived swell factors are used with the appropriate equation.

Operating Efficiency

The maximum estimated equipment production is adjusted down to account for the time periods when equipment is not operating to its full potential. For primary mining and excavating equipment operating under relatively uniform conditions, the efficiency is commonly calculated as

$$E = A \times U \tag{EQ 12.14}$$

where:
- E = operating efficiency, decimal
- A = availability factor—the portion of scheduled time that the equipment is mechanically and electrically ready to be operated, decimal
- U = utilization factor—the portion of available time that the equipment is being operated at full potential, decimal

For equipment operating under more variable conditions or as part of an integrated system, the operating efficiency may be estimated as a function of previous experience with similar job and management conditions.

$$E = JC \times MC \tag{EQ 12.15}$$

where:
- E = operating efficiency, decimal
- JC = job conditions factor, decimal
- MC = management conditions factor, decimal

Bucket and Dipper Fill Factor

The fill factor for excavating equipment is the percentage of the bucket or dipper's heaped capacity that actually fills with material, expressed as

$$FF = LCM\ (\text{actual})/LCM\ (\text{capacity}) = LCY\ (\text{actual})/LCY\ (\text{capacity}) \tag{EQ 12.16}$$

Load Cycle Time

The load cycle time (T_{cl}) is the time required for excavating and loading equipment to complete one cycle of filling the bucket or dipper, swinging or tramming to the dump point, dumping, swinging or tramming to the excavation point, and positioning the bucket or dipper for filling.

Loading Production

Use Equation 12.17 to calculate production for excavating and loading equipment. This equation includes propel (or move) time between production locations as a separate factor. Propel time may also be included as part of the operating efficiency or omitted if it is not a significant consideration. To determine total production, multiply the hourly production rate by the scheduled work hours.

$$P_l = 3{,}600(Cb)(SF)(E)(FF)(PT)/(T_{cl}) \tag{EQ 12.17}$$

where:
- P_l = loading production, BCM/h (BCY/h)
- $3{,}600$ = seconds in 1 h (s/h)
- Cb = bucket or dipper heaped capacity, LCM (LCY)
- SF = swell factor, ratio of BCM/LCM (ratio of BCY/LCY)
- E = operating efficiency, decimal

FF = fill factor, decimal
PT = propel time factor, decimal
T_{cl} = load cycle time, s

Travel Time
Use the formula that follows to calculate equipment travel time.

$$TT = D/(V_{ave} \times UCF) \qquad \text{(EQ 12.18)}$$

where:
TT = travel time, min
D = distance, m (ft)
V_{ave} = average speed, km/h (mi/h)
UCF = units conversion factor = 16.7 (m-h)/(km-min) or 88 (ft-h)/(mi-min)

Haulage Cycle Time
The theoretical cycle time for haulage equipment is the sum of the load time, travel time to the dump point, dump or spread time, and return time. The actual or corrected cycle time also includes waiting and expected delays (if these are not already included in the operating efficiency factor).

$$T_{ch} = T_l + TT_o + T_{dp} + TT_r \qquad \text{(EQ 12.19)}$$

$$TC_{ch} = T_{ch} + T_w + T_d \qquad \text{(EQ 12.20)}$$

where:
T_{ch} = theoretical haulage cycle time, min
T_l = equipment load time, min
TT_o = travel time to dump point, min
T_{dp} = dump or spread time, min
TT_r = travel return time, min
TC_{ch} = corrected haulage cycle time, min
T_w = wait time, min
T_d = delay time, min

Haulage Production
Use Equation 12.21 to estimate haulage production. To determine the production in tonnes per hour or tons per hour, multiply by the bank density (t/BCM or ton/BCY).

$$P_h = 60(N_h)(L_h)(E)/(TC_{ch}) \qquad \text{(EQ 12.21)}$$

where:
P_h = haulage production, BCM/h (BCY/h)
60 = minutes in 1 h, min/h
N_h = number of haulage units, integer
L_h = haul load, BCM (BCY)
E = operating efficiency, decimal
TC_{ch} = corrected cycle time, min

EQUIPMENT OPERATING COSTS

Equipment operating costs include labor, electrical power or fuel, preventive maintenance, repairs, and tire replacement (if applicable). General guidelines for estimating some of these costs follow.

Maintenance and Repair

Most large, nonmobile excavation and haulage equipment have operating lives of 20 to 30 years with annual maintenance and repair costs ranging between 5% and 15% of installed cost. Use Equation 12.22 and the repair factors listed in Table 12.4 to estimate repair costs. Repair factors vary depending on operating conditions and the age of the equipment. Table 12.4 also presents typical operating lives for mobile equipment. Preventive maintenance costs for mobile equipment—including the labor required for equipment servicing—average between 15% and 25% of the fuel costs.

$$Rc = (Fr)(Vd)/10{,}000 \qquad (EQ\ 12.22)$$

where:
- Rc = hourly repair cost, \$/h
- Fr = repair factor, decimal (see Table 12.4)
- Vd = depreciable new value of the equipment, \$ (exclude tire costs)
- $10{,}000$ = conversion factor, h

For more information, see chapter 21 covering Maintenance and Inventory.

TABLE 12.4 Mobile equipment operating lives and repair factors

Equipment	Normal Operating Life (h)	Repair Factor
Wheel loader	7,000–12,000	0.30–1.00
Rear dump truck, mechanical drive	20,000–30,000	0.30–1.00
Rear dump truck, electric drive	30,000–40,000	0.20–0.65
Tractor trailer truck	20,000–30,000	0.25–0.80
Conventional scraper	12,000–16,000	0.30–1.40
Push-pull scraper	12,000–16,000	0.35–1.45
Elevating scraper	12,000–16,000	0.40–1.60
Track dozer	8,000–18,000	0.16–0.50
Wheel dozer	8,000–12,000	0.20–0.40

Source: Compiled from Hays 1990 and Atkinson 1992.

Tires

Tire life is affected by tire type and construction, equipment loads and velocity, operating surface, and tire load and position. Table 12.5 lists the average life of tires used in their correct applications. Estimate hourly tire costs by dividing the cost of the tires by the life of the tires, then adding 15% for repairs.

TABLE 12.5 Tire life

	Job Conditions			
	A	B	C	D
Average life, operating hours	4,000–5,000	3,300–3,500	2,000–2,500	400–1,500

NOTES:
A = Good road surface with operation over soft, nonabrasive soil or rock.
B = Average road surface over soft soil or rock.
C = Average road surface over abrasive, medium-hard rock.
D = Road surface consisting of hard, angularly fragmented rock.
Source: Atkinson 1992.

Fuel and Power Consumption

Use Equation 12.23 to calculate fuel consumption for diesel equipment. This equation is based on an engine fuel consumption of 0.26 kg/kW·h (0.42 lb/hp-h) and a diesel fuel density of 0.85 kg/L (7.1 lb/gal). Engine load factors typically range between 0.25 and 0.75, depending on the equipment type and use level.

$$Cf = (EBP)(Fl)(UCF) \qquad \text{(EQ 12.23)}$$

where:
- Cf = fuel consumption, L/h (gal/h)
- EBP = engine brake power, kW (hp)
- Fl = engine load factor, decimal
- UCF = unit conversion factor, 0.3 L/kW·h (0.06 gal/hp-h)

The average power consumption for electrically powered equipment can be found in the manufacturer's equipment specifications. If this is not available, average consumption may be estimated to be about 35% of the total continuous power ratings of the motors. The power consumption per unit of bank material moved generally ranges between 0.45 and 1.2 kW·h/*BCM* (0.35 and 0.93 kW·h/*BCY*) for excavating equipment. Generally, larger equipment is more efficient.

DRAGLINES

Dragline size selection is based on the maximum volume of overburden to be removed per unit time and the required maximum reach of the machine. Given the required stripping rate and reach, estimate the cycle time and determine an approximate bucket capacity by using Equation 12.17 and Tables 12.6 through 12.9. Propel time is included in the utilization factor given in Table 12.9.

CLAMSHELLS

Clamshells are useful in lifting loose materials vertically from one level to another. They are used to clean ditches and settlement ponds, to load materials from a stockpile into trucks or a hopper, or to unload barges. Use Equation 12.17 to calculate clamshell production.

SHOVELS

When determining the required number and size of shovels for a mine, consider these factors: (1) ground-bearing pressure, (2) truck size, (3) bench height, (4) tonnage required, (5) blending required, (6) number of ore and waste faces required, (7) fragmentation expected, (8) maintenance facilities, (9) material weights, (10) pit geometry, (11) dipper weight required, and (12) infrastructure requirements (Sargent 1990). Use Tables 12.10 through 12.12, along with Equation 12.17, to estimate shovel production. Average propel time factors are 0.75 for strip mines, 0.85 for open-pit mines, 0.90 for sand and gravel pits, and 0.95 for high-face quarries (Atkinson 1992).

BACKHOES

Because of its configuration, a backhoe can efficiently excavate material located below its base. Larger backhoes occasionally find use in surface mining applications where the backhoe sits on top of the bench to be loaded. This includes pits having a wet or incompetent floor and narrow pits where trucks can be loaded by backing up to the front and slightly to either side of the backhoe (i.e., a minimal boom swing is required). Calculate backhoe production in the same manner used for shovels.

WHEEL LOADERS

Wheel loaders are used for both loading and load-and-carry applications as well as in other mine support functions. Table 12.13 gives truck-loading cycle times for wheel loaders operating in different types of digging conditions. The wheel loader is usually not used in hard digging conditions.

EXCAVATION, LOADING, AND MATERIAL TRANSPORT | 223

TABLE 12.6 Approximate specifications for walking draglines*

Bucket Size		Dumping Radius		Dumping Height		Width Over Shoes		Machine Mass (t)	Ballast (t)
yd³	m³	m	ft	m	ft	m	ft		
15	11.5	50	165	21	70	14	45	470	115
20	15	58	190	23	75	16	52	550	160
40	31	67	220	26	85	22	72	1,250	200
50	38	79	260	30	100	23	76	1,950	250
80	46	84	275	37	120	27	90	2,900	320
90	70	92	300	41	135	32	105	4,200	340
110	85	92	300	44	145	35	115	5,700	370

* Conversion factor, 1 ton = 0.9072 t.
Source: Atkinson 1992.

TABLE 12.7 Swell and bucket fill factors for walking draglines

Overburden Conditions	Swell Factor	Bucket Fill Factor
Light blasting	0.81	0.85–0.90
Medium blasting	0.75	0.80–0.90
Heavy blasting	0.71	0.75–0.85
Poor fragmentation	0.69	0.70–0.75

Source: Adapted from Atkinson 1992.

TABLE 12.8 Theoretical cycle times for draglines (s)

Bucket Size		Swing Angle (degrees)			
yd³	m³	90	120	150	180
Up to 19	Up to 15	55	62	69	77
20–34	16–26	56	63	70	78
35–59	27–44	57	64	71	79
60–74	45–57	59	65	72	80
75–120	58–92	60	66	73	81
121–200	93–150	62	69	76	84

Source: Atkinson 1992.

TABLE 12.9 Dragline operating efficiency

Availability	Utilization			
	Excellent	Good	Fair	Poor
Excellent	0.84	0.81	0.76	0.70
Good	0.78	0.76	0.71	0.64
Fair	0.72	0.69	0.65	0.60
Poor	0.63	0.61	0.57	0.52

Source: Humphrey 1990.

TABLE 12.10 Shovel cycle times and fill factors

Bucket Capacity		Average Cycle Time (s)			
m³	yd³	E	M	M-H	H
3	4	18	23	28	32
4	5	20	25	29	33
5	6	21	26	30	34
5.5	7	21	26	30	34
6	8	22	27	31	35
8	10	23	28	32	36
9	12	24	29	32	37
11.5	15	26	30	33	38
15	20	27	32	35	40
19	25	29	34	37	42
35	45	30	36	40	45
Average Fill Factor		0.95–1.0	0.85–0.90	0.80–0.85	0.75–0.80

NOTES:
E (Easy Digging) = loose free running material (e.g., sand, small gravel, loose earth, ashes, bituminous coal).
M (Medium Digging) = partially consolidated materials (e.g., clayey gravel, packed earth, clay, anthracite coal).
M-H (Med-Hard Digging) = materials requiring some blasting (e.g., weaker rock, gravel w/ boulders, heavy clay).
H (Hard Digging) = materials requiring heavy blasting (e.g., hard, competent rock).
Source: Adapted from Atkinson 1992.

TABLE 12.11 Bench and swing angle corrections

Optimum digging depth, %	40	60	80	100		
Cycle time correction factor	1.25	1.10	1.02	1.00		
Angle of swing, degrees	45	60	75	90	120	150	180
Cycle time correction factor	0.83	0.91	0.95	1.00	1.10	1.19	1.30

Source: Adapted from Atkinson 1992.

TABLE 12.12 Shovel and loader operating efficiency

	Management Conditions			
Job Conditions	Excellent	Good	Fair	Poor
Excellent	0.83	0.80	0.77	0.70
Good	0.76	0.73	0.70	0.64
Fair	0.72	0.69	0.66	0.60
Poor	0.63	0.61	0.59	0.54

Source: Atkinson 1992.

EXCAVATION, LOADING, AND MATERIAL TRANSPORT

TABLE 12.13 Wheel-loader cycle times

Bucket Capacity		Cycle Time (s)		
yd³	m³	E	M	M-H
5	4.0	32	33	41
6	4.5	33	34	42
7	5.5	33	35	44
10	7.5	37	39	51
12	9.0	39	42	56
15	11.5	41	44	60

NOTES:
E (Easy Digging) = loose free running material (e.g., sand, small gravel, loose earth, ashes, bituminous coal).
M (Medium Digging) = partially consolidated materials (e.g., clayey gravel, packed earth, clay, anthracite coal).
M-H (Med-Hard Digging) = materials requiring some blasting (e.g., weaker rock, gravel w/ boulders, heavy clay).
H (Hard Digging) = materials requiring heavy blasting (e.g., hard, competent rock).
Source: Adapted from Atkinson 1992.

Wheel loaders are used to load and transport materials for short distances up to about 150 m (490 ft) in various production, stockpiling, and process feed applications. The fixed portion of the load-and-carry cycle time, which consists of positioning, loading, and dumping, typically requires between 30 and 40 seconds per cycle depending on digging conditions, amount of maneuvering required, and dumping procedures. Average tram speeds for load-and-carry applications typically vary from 3 km/h (1.8 mi/h) for short distances to more than 15 km/h (9.0 mi/h) for longer distances. Use the manufacturer's performance charts and Table 12.3 to estimate tram speed.

To estimate loader productivity, use Equation 12.17 with propel time (PT) omitted. Fill factors are similar to those listed in Table 12.10 for shovels (i.e., 0.95 to 1.0 for easy digging, 0.85 to 0.90 for medium digging, and 0.80 to 0.85 for medium-hard digging). The maximum bucket capacity of a wheel loader may also be limited by its maximum rated payload (kilograms or pounds). Use Table 12.12 to estimate operating efficiency (E). For load-and-carry operations, the load cycle time (T_{cl}) in Equation 12.17 is replaced with the load-and-carry cycle time.

CRAWLER-MOUNTED LOADERS
Because of their reduced mobility, crawler- or track-mounted loaders find fewer appropriate applications in the mining industry than wheel loaders. They do provide an advantage over wheel loaders where the ground is very abrasive, less competent (i.e., soft or wet), or steeply dipping (e.g., decline development). Production calculations for a crawler-mounted unit are similar to those for a wheel unit with longer loading and tram times.

Trucks
Surface haul trucks include (Hays 1990):
- Conventional rear dump trucks—used in and around open pits, strip mines, and quarries where traction and maneuverability are needed
- Tractor-trailer trucks—used for high-speed, long-distance haulage
- Integral bottom dump trucks—used for haulage of coal and other soft, free-flowing materials.

Truck Loading
Trucks and loading equipment are typically matched in size so that between three and six passes by the loading equipment are required to load each truck. To determine the number of passes required to load a truck to its weight capacity, use Equation 12.24a. For less-dense materials, it may also be necessary to determine the number of passes required to load a truck

to its heaped volume capacity (see Equation 12.24b). For estimating purposes, the lower of the two Np values is rounded to the next whole integer (NP). Equation 12.25 can then be used to estimate the actual load per truck.

$$Np = Ctw/(Cb \times FF \times SF \times BD) \quad \text{(EQ 12.24a)}$$

$$Np = (Ctv \times SF)/(Cb \times FF \times SF) = Ctv/(Cb \times FF) \quad \text{(EQ 12.24b)}$$

$$L_h = NP \times Cb \times FF \times SF \quad \text{(EQ 12.25)}$$

where:
- Np = number of passes, decimal number
- NP = number of passes, integer
- Ctw = truck capacity, t (tons)
- Cb = loading bucket heaped capacity, LCM (LCY)
- FF = bucket fill factor, decimal
- SF = material swell factor, ratio of BCM/LCM (ratio of BCY/LCY)
- BD = bank density, t/BCM (ton/BCY)
- Ctv = truck heaped capacity, LCM (LCY)
- L_h = truck load, BCM (BCY)

Truck Spot and Load Time

Spot time is the time needed for the truck to maneuver into position for loading, and load time is the time required for the loading machine to make the number of passes required to load the truck. These times may overlap because the loading machine may perform parts of its work cycle while the truck is being spotted. If the spot time is less than the loading machine cycle time, use Equation 12.26a to determine the combined spot and load time. If the spot time is greater than the loading machine cycle time, use Equation 12.26b (Hays 1990). Spot times for rear dump trucks typically range between 0.3 and 0.6 min, with smaller trucks generally being faster. Spot times for tractor-trailer units may range from 0.15 to 1.0 min, depending on the loading method and the need for backing (Bishop 1968).

$$T_l = NP(T_{cl}) \quad \text{(EQ 12.26a)}$$

$$T_l = T_s + (NP - 1)T_{cl} \quad \text{(EQ 12.26b)}$$

where:
- T_l = truck spot and load time, min
- NP = number of passes by the loading machine, integer
- T_{cl} = loading machine cycle time, min
- T_s = truck spot time, min

Truck Dump Time

Dump time (T_{dp}) for a rear dump truck is the time required to enter the dump area, turn, back up, raise the body to dump, and lower the body. For a side dump or bottom dump tractor-trailer, the dump time is the time required to pull through or over the dump area, sometimes dumping without stopping. Approximate dump times are 1.0 min for rear dump units and 0.5 min for side and bottom dump units (Hays 1990).

Truck Cycle Time and Production

Calculate truck travel time using the truck's specifications, Table 12.3, and Equation 12.18. To calculate the theoretical truck cycle time (T_{ch}) and actual or corrected cycle time (TC_{ch}), use Equations 12.19 and 12.20, respectively. Use Equation 12.21 to calculate truck fleet production.

EXCAVATION, LOADING, AND MATERIAL TRANSPORT

Number of Trucks

Use Equation 12.27 to determine the estimated number of trucks needed to service a shovel or loader. Most operations also have spare trucks available in case of breakdowns.

$$N_h = T_{ch}/T_l \qquad \text{(EQ 12.27)}$$

where:
N_h = number of trucks
T_{ch} = theoretical truck cycle time (min)
T_l = truck spot and load time (min)

SCRAPERS

Scrapers commonly in use include (Hays 1990):
- Conventional scrapers (both single and dual engines)—used to load a wide range of materials in conjunction with a pusher dozer
- Push-pull scrapers (dual engines)—used to load a wide range of materials with the ability to load in tandem without a pusher dozer
- Elevating scrapers (both single and dual engines)—used to self-load easily excavated, fine-grained materials.

Scraper Loading and Spreading

Scrapers are generally filled to their heaped capacity with the resulting load calculated using Equation 12.28. The scraper load time (T_l) and spread time (T_{dp}) depend on the type of scraper used, material characteristics, pushing or pulling force, and operating conditions. Typical load and spread times for various conditions are presented in Tables 12.14 and 12.15, respectively.

TABLE 12.14 Scraper load times

	Load Time (min)		
Scraper Type	Favorable	Average	Unfavorable
Conventional, push loaded			
Single engine	0.40	0.70	1.00
Dual engines	0.35	0.60	0.90
Conventional, push-pull			
Dual engines*	0.60	0.90	1.50
Elevating			
Single engine	0.60	0.90	1.30
Dual engines	0.45	0.70	1.00

* Time for loading both scrapers.
Source: Hays 1990.

TABLE 12.15 Scraper spread times

	Spread Time (min)		
Scraper Type	Favorable	Average	Unfavorable
Conventional			
Single engine	0.30	0.60	1.00
Dual engines	0.30	0.50	0.90
Elevating			
Single engine	0.40	0.70	1.10
Dual engines	0.30	0.60	1.00

Source: Hays 1990.

$$Lh = Cs \times SF \qquad \text{(EQ 12.28)}$$

where:
- Lh = scraper load, BCM (BCY)
- Cs = scraper heaped capacity, LCM (LCY)
- SF = swell factor, ratio of BCM/LCM (ratio of BCY/LCY)

Scraper Cycle Time and Production

To calculate scraper travel time, use the scraper's performance specifications, Table 12.3, and Equation 12.18. Use Equations 12.19 and 12.20, respectively, to determine the theoretical scraper cycle time (T_{ch}) and the actual or corrected cycle time (TC_{ch}). Equation 12.21 may be used to calculate scraper fleet production.

Pusher Dozers

Pusher cycle time (T_{pc}) is the time required for the pusher dozer to push load the scraper, boost the scraper out of the cut, move to the next scraper, and position the push blade against the scraper. It also includes the average wait time, if any. The pusher cycle time depends on the type of push loading pattern employed, as shown in Table 12.16.

TABLE 12.16 Pusher cycle times with no waiting

	Pusher Cycle Time (min)*		
Pusher Pattern	Favorable	Average	Unfavorable
Back track loading	0.75	1.25	1.80
Chain loading	0.60	0.95	1.40
Shuttle loading	0.60	0.95	1.40

* For tandem pushers, multiply pusher cycle time by 1.20.
Source: Hays 1990.

Equation 12.29 gives the theoretical number of scrapers that a pusher dozer can service. Times for other tasks such as ripping, clearing, or salvaging of topsoil need to be included in the pusher cycle time if the dozer is required to perform these operations in addition to pushing.

$$N_h = T_{ch}/T_{pc} \qquad \text{(EQ 12.29)}$$

where:
- N_h = number of scrapers, integer
- T_{ch} = theoretical scraper cycle time, min
- T_{pc} = pusher cycle time with no waiting, min

BULLDOZERS

The track-mounted bulldozer is used in mine production for pushing overburden or product (e.g., ore, coal) up to about 80 m (260 ft), as well as for ripping unconsolidated soil and rock. Track and wheel bulldozers are also used for a wide variety of mine support applications.

Dozer Blades

Common dozer blades include straight, universal, semiuniversal, angling, and cushion blades. The first three blades are used in production dozing with the straight blade having the best penetration and the universal blade the greatest capacity. The blade capacity for dozers is provided in the equipment specifications as LCM or LCY and may be converted to bank measure for production calculations using Equation 12.30. For side-by-side or slot dozing, increase the capacity by 20%.

$$L_h = Cbd \times SF \qquad \text{(EQ 12.30)}$$

where:
- L_h = dozer load, *BCM* (*BCY*)
- Cbd = blade capacity, *LCM* (*LCY*)
- SF = swell factor, ratio of *BCM/LCM* (ratio of *BCY/LCY*)

Bulldozer Cycle Time and Production

Use Equation 12.21 to calculate bulldozer production. The bulldozer cycle time (TC_{ch}) consists of the time necessary to excavate and push the material the desired distance, spread the material, back up to the starting point, and position the blade for the next pass. When excavating, cut lengths range from 7.6 to 23 m (25 to 75 ft) and load times range from 0.15 to 0.45 min, depending on the type of material involved. Pushing the material is typically done in first or second gear at speeds between 0.8 and 1.7 m/s (2.6 and 5.6 ft/s). Spread times usually range from 0.08 to 0.12 min. Backing up to the beginning of the cut in third gear is typically limited to about 2.2 m/s (7.2 ft/s) for track-mounted units and 4.4 m/s (14.4 ft/s) for wheel units (Hays 1990).

Ripping

To estimate ripper production, use Equation 12.31. The ripper cycle time (T_{cr}) consists of the time needed to make a single pass while ripping, including the time necessary to turn the dozer to begin the next pass. During ripping, the dozer is typically operated in first gear at velocities of 25 to 40 m/min (82 to 131 ft/min). Turn times are typically between 0.20 and 0.35 min (Hays 1990). If depth of penetration (*d*) is left out of the equation, the area ripped per hour (square meter per hour or square yard per hour) can be calculated.

$$P_r = (l)(w)(d)(E)(60)/T_{cr} \qquad \text{(EQ 12.31)}$$

where:
- P_r = ripper production, *BCM*/h (*BCY*/h)
- l = length of the area being ripped, m (yd)
- w = width between ripper passes, m (yd)
- d = depth of ripper penetration, m (yd)
- E = efficiency, decimal
- 60 = minutes in 1 h, min/h
- T_{cr} = ripper cycle time, min

RAIL HAULAGE

Rail haulage is employed in a variety of underground mines and in some large surface mines with long (i.e., more than 4 km), moderately sloped haulage requirements.

Haulage grades are usually limited to a maximum of 3% uphill and 4% downhill (Brauns and Orr 1968). In the equations presented below, pulling and resistive forces are solved for in kilograms and pounds. Convert kilograms to newtons by multiplying by 9.81 m/s².

Tractive Resistance

The factors influencing a train's resistance to movement include (a) frictional resistance, (b) grade resistance, (c) track curvature resistance, and (d) acceleration resistance. Frictional resistance for locomotives is usually estimated as 7.5 kg/t (15 lb/ton) for antifriction bearings and 10 kg/t (20 lb/ton) for bushed bearings. Use Table 12.17 to estimate the frictional resistance of the mine cars, and employ the equations presented below to estimate grade curve and acceleration resistances. Typical locomotive speeds are between 16 and 32 km/h (10 and 20 mi/h/s) with starting accelerations between 0.16 and 0.32 km/h/s (0.1 to 0.2 mi/h/s; Brantner 1973). Wind resistance is negligible for the speeds employed in most mining applications.

TABLE 12.17 Rolling frictional resistance of mine cars

Resistance		Type of Car	Track Condition
kg/t	lb/ton		
3–5	6–10	Large, modern railroad type with excellent bearings, 30 tons and over	Excellent
5.0–7.5	10–15	Large, mine type with good bearings, 8-wheel, 15 to 30 tons	Good
10.0–12.5	20–25	Medium size, roller bearings, 5 to 10 tons	Fair to Good
15.0–17.5	30–35	Small, bronze bearings, under 5 tons	Fair

Source: Adapted from Brantner 1973.

$$CRF = CCF \times C \qquad (EQ\ 12.32)$$

where:
 CRF = curvature resistance factor, kg/t (lb/ton)
 CCF = curvature conversion factor, 0.4 kg/t-degree (0.8 lb/ton-degree)
 C = track curvature (degrees)

$$ARF = ACF \times ACC \qquad (EQ\ 12.33)$$

where:
 ARF = acceleration resistance factor, kg/t (lb/ton)
 ACF = acceleration conversion factor, 31 kg-h-s/t-km (100 lb-h-s/ton-mi)
 ACC = acceleration, km/h/s (mi/h/s)

$$TR = LW(FRF_L) + TW(FRF_C) + GTW(GRF + CRF + ARF) \qquad (EQ\ 12.34)$$

where:
 TR = total tractive resistance, kg (lb)
 LW = locomotive weight, t (ton)
 FRF_L = locomotive frictional resistance factor, kg/t (lb/ton)
 TW = gross trailing weight of the cars, t (ton)
 FRF_C = car frictional resistance factor, kg/t (lb/ton; see Table 12.17)
 GTW = gross train weight (locomotive and cars), t (ton)
 GRF = grade resistant factor, kg/t (lb/ton; see Equation 12.6)
 CRF = curvature resistance factor, kg/t (lb/ton; see Equation 12.32)
 ARF = acceleration resistance factor, kg/t (lb/ton; see Equation 12.33)

Tractive Effort

Tractive effort, defined as the total force delivered by the locomotive to its drive wheels, is a function of its engine power and speed. The net tractive effort (i.e., tractive effort minus tractive resistance) must be greater than zero before the train can move.

$$TE = (EP - EPa)(EFF)(UCF)/V \qquad (EQ\ 12.35)$$

$$NTE = TE - TR \qquad (EQ\ 12.36)$$

where:
 TE = tractive effort of locomotive, kg (lb)
 EP = engine power of locomotive, kW (hp)
 EPa = engine power to the auxiliaries, kW (hp)
 EFF = efficiency in converting engine power to tractive effort, decimal (typically 0.80 to 0.85)
 UCF = unit conversion factor, 367 (km-kg)/kW-h or 375 (mi-lb)/hp-h
 V = train speed, km/h (mi/h)
 NTE = net tractive effort, kg (lb)
 TR = total tractive resistance, kg (lb; see Equation 12.34)

Adhesion

Adhesion for locomotives is analogous to traction for crawler and rubber-tired equipment. When the tractive effort exceeds adhesion, the wheels of the locomotive will slip (i.e., the available tractive effort is limited by adhesion). Table 12.18 supplies coefficient of adhesion values for various rail conditions.

TABLE 12.18   Coefficient of adhesion values

Description	Unsanded Rails	Sanded Rails
Clean dry rails, starting and accelerating	0.30	0.40
Clean dry rails, continuous running	0.25	0.35
Clean dry rails, locomotive braking	0.20	0.30
Thoroughly wet rails	0.18	0.25
Greasy, moist rails	0.15	0.20
Dry, snow-covered rails	0.11	0.15

NOTE: For cast-iron wheels on steel rails, reduce listed values by 20%.
Source: Adapted from Brantner 1973.

$$TE_A = CA \times LW \times UCF \quad \text{(EQ 12.37)}$$

where:
TE_A = available tractive effort, kg (lb)
CA = coefficient of adhesion, decimal
LW = locomotive weight, t (ton)
UCF = unit conversion factor, 1,000 kg/t (2,000 lb/ton)

Acceleration

Acceleration of the train depends on the available net tractive effort.

$$T_a = (TCF/NTE)(V_2 - V_1) = (V_2 - V_1)/ACC \quad \text{(EQ 12.38)}$$

$$D_a = (DCF/NTE)(V_2 - V_1) = (UCF)(T_a)(V_2 - V_1)/2 \quad \text{(EQ 12.39)}$$

where:
T_a = acceleration time, s
D_a = acceleration distance, m (ft)
TCF = time conversion factor, 27 s-h-kg/km (95.6 sec-h-lb/mi)
DCF = distance conversion factor, 6 m-kg-h/km (70 ft-lb-h/mi)
NTE = net tractive effort, kg (lb)
V_1 = initial velocity, km/h (mi/h)
V_2 = final velocity, km/h (mi/h)
ACC = acceleration, km/h/s (mi/h/s)
UCF = units conversion factor, 0.278 m-h/km-s (1.47 ft-h/mi-s)

Braking

A deceleration rate (D) of 0.5 km/h per s (0.3 mi/h/s) is normally acceptable for bringing a train to a stop, although a higher rate may be needed for emergencies (Brantner 1973). Braking time and distance depends on the available retarding force (see Equation 12.40). To find stopping time and distance, substitute deceleration for acceleration or available retarding force (F_r) for NTE in Equations 12.38 and 12.39, respectively (Ramani 1990). Consistent with other equations in this chapter, uphill grades are considered positive and downhill grades, negative in Equation 12.40.

$$F_r = BE + TW(FRF_C) + GTW(GRF) \qquad \text{(EQ 12.40)}$$

where:
- F_r = available retarding force, kg (lb)
- BE = braking effort, kg (lb)
- TW = gross trailing weight of the cars, t (ton)
- FRF_C = car frictional resistance factor, kg/t (lb/ton; see Table 12.17)
- GTW = gross train weight (locomotive and cars), t (ton)
- GRF = grade resistant factor, kg/t (lb/ton; see Equation 12.6)

Rail Haulage Cycle Time and Production

To calculate haulage travel time, use Equation 12.18 and an estimated average locomotive velocity. The average velocity will depend on the *NTE* available, as well as on acceleration/deceleration requirements. Use Equations 12.19 and 12.20, respectively, to calculate the theoretical haulage cycle time and corrected cycle time. Because loading and dumping times vary substantially, they must be estimated on a case-by-case basis. Use Equation 12.21 to calculate rail haulage production.

Track

Rail weight depends on the weight per wheel of the locomotive and cars, with the minimum rail weight equal to about 5.5 kg/m for each tonne (10 lb/yd for each ton) of weight on a wheel. The track gauge may vary from 46 to 168 cm (18 to 66 in.) with 107 cm (42 in.) being considered standard gauge in the United States. Wood ties should have a length of about twice the track gauge. Spacing of ties is dependent on rail use and ground conditions and may vary from 40 cm (16 in.) on main lines, with soft beds to 1 m (3.3 ft) or more in production areas.

BELT CONVEYORS

Belt conveyors, which are used in a wide variety of mine and plant applications, are the most common type of conveyor. Maximum belt inclinations vary from 15 to 25 degrees, depending on the material characteristics.

Conveyor Production Capacity

Use Equation 12.41 to calculate the production capacity of conveyors. Conveyors are typically designed for peak production requirements. For belt conveyors, the average cross-sectional area depends on belt width, surcharge angle, lump size, and troughing angle. Belt speed is dependent on belt width and material properties. To estimate the optimum belt width and speed for a given production capacity and material density, use Tables 12.19 through 12.21 and Figures 12.1 and 12.2 in an iterative process. These tables and figures apply to conveyors up to about 600 m (about 2,000 ft) in length with 20- or 35-degree troughing angles and horizontal or inclined configurations (Duncan and Levitt 1990).

$$P_c = (A)(LD)(V_c)(UCF) \qquad \text{(EQ 12.41)}$$

where:
- P_c = conveyor production capacity, t/h (ton/h)
- A = average cross-sectional area of material on the conveyor, m² (ft²)
- LD = density of material (loose), kg/m³ (lb/ft³)
- V_c = conveyor speed, m/s (ft/min)
- UCF = unit conversion factor, 3.6 t-s/h-kg (0.03 ton-min/h-lb)

EXCAVATION, LOADING, AND MATERIAL TRANSPORT | 233

TABLE 12.19 Recommended maximum belt speeds

Material Being Conveyed	Belt Speeds		Belt Width	
	m/s	(ft/min)	mm	(in.)
Grain or other free-flowing, nonabrasive material	2.5	(500)	450	(18)
	3.6	(700)	600–750	(24–30)
	4.1	(800)	900–1,050	(36–42)
	5.1	(1,000)	1,200–2,400	(48–96)
Coal, damp clay, soft ores, overburden and earth, fine-crushed stone	2.0	(400)	450	(18)
	3.0	(600)	600–900	(24–36)
	4.1	(800)	1,050–1,500	(42–60)
	5.1	(1,000)	1,800–2,400	(72–96)
Heavy, hard, sharp-edged ore, coarse-crushed stone	1.8	(350)	450	(18)
	2.5	(500)	600–900	(24–36)
	3.0	(600)	Over 900	Over (36)
Feeder belts, flat or troughed, for feeding fine, nonabrasive, or mildly abrasive materials from hoppers and bins	0.3–0.5	(50–100)	Any width	

Source: Adapted from Conveyor Equipment Manufacturers Association 1997 (shown with permission).

TABLE 12.20 Cross-sectional area of 20° troughed belt

Belt Width		A Cross Section of Load (m^2)						A Cross Section of Load (ft^2)					
		Surcharge Angle						Surcharge Angle					
mm	(in.)	0°	5°	10°	15°	20°	25°	30°	0°	5°	10°	15°	20°	25°	30°
450	(18)	0.008	0.010	0.012	0.014	0.016	0.017	0.019	0.089	0.108	0.128	0.147	0.167	0.188	0.209
600	(24)	0.016	0.019	0.023	0.026	0.030	0.033	0.037	0.173	0.209	0.246	0.283	0.320	0.359	0.399
750	(30)	0.026	0.032	0.037	0.043	0.048	0.054	0.060	0.284	0.343	0.402	0.462	0.522	0.585	0.649
900	(36)	0.039	0.047	0.055	0.064	0.072	0.080	0.089	0.423	0.509	0.596	0.684	0.774	0.866	0.960
1,050	(42)	0.055	0.066	0.077	0.088	0.100	0.112	0.124	0.588	0.708	0.828	0.950	1.074	1.201	1.332
1,200	(48)	0.073	0.087	0.102	0.117	0.132	0.148	0.164	0.781	0.940	1.099	1.260	1.424	1.592	1.765
1,350	(54)	0.093	0.112	0.131	0.150	0.169	0.189	0.210	1.002	1.204	1.407	1.613	1.822	2.037	2.258
1,500	(60)	0.116	0.139	0.163	0.187	0.211	0.236	0.261	1.249	1.501	1.753	2.009	2.270	2.537	2.812
1,800	(72)	0.170	0.204	0.238	0.272	0.308	0.344	0.381	1.826	2.192	2.560	2.933	3.312	3.701	4.102
2,100	(84)	0.233	0.280	0.327	0.374	0.423	0.472	0.524	2.513	3.014	3.519	4.030	4.551	5.085	5.635
2,400	(96)	0.307	0.369	0.430	0.493	0.556	0.621	0.689	3.308	3.967	4.631	5.302	5.986	6.687	7.411

NOTE: Three equal rolls standard edge distance = 0.055 b + 0.9 in. (b = belt width).
Source: Adapted from Conveyor Equipment Manufacturers Association 1997 (shown with permission).

TABLE 12.21 Cross-sectional area of 35° troughed belt

Belt Width		A Cross Section of Load (m²)							A Cross Section of Load (ft²)						
		Surcharge Angle							Surcharge Angle						
mm	(in.)	0°	5°	10°	15°	20°	25°	30°	0°	5°	10°	15°	20°	25°	30°
450	(18)	0.013	0.015	0.016	0.018	0.020	0.021	0.023	0.144	0.160	0.177	0.194	0.212	0.230	0.248
600	(24)	0.026	0.029	0.032	0.035	0.038	0.041	0.044	0.278	0.309	0.341	0.373	0.406	0.440	0.474
750	(30)	0.042	0.047	0.052	0.057	0.062	0.067	0.072	0.455	0.506	0.557	0.609	0.662	0.716	0.772
900	(36)	0.063	0.070	0.077	0.084	0.091	0.098	0.106	0.676	0.751	0.826	0.903	0.980	1.060	1.142
1,050	(42)	0.087	0.097	0.107	0.117	0.126	0.137	0.147	0.940	1.044	1.148	1.254	1.361	1.471	1.585
1,200	(48)	0.116	0.129	0.141	0.154	0.168	0.181	0.195	1.248	1.385	1.523	1.662	1.804	1.949	2.099
1,350	(54)	0.149	0.165	0.181	0.198	0.215	0.232	0.250	1.599	1.774	1.950	2.128	2.309	2.494	2.686
1,500	(60)	0.185	0.205	0.226	0.246	0.267	0.289	0.311	1.994	2.211	2.429	2.651	2.876	3.107	3.345
1,800	(72)	0.271	0.300	0.330	0.359	0.390	0.421	0.453	2.913	3.229	3.547	3.869	4.197	4.532	4.879
2,100	(84)	0.372	0.412	0.453	0.494	0.536	0.578	0.623	4.007	4.440	4.876	5.317	5.766	6.226	6.701
2,400	(96)	0.490	0.543	0.596	0.650	0.705	0.761	0.819	5.274	5.842	6.415	6.994	7.584	8.189	8.812

Note: Three equal rolls standard edge distance = 0.055 b + 0.9 in. (b = belt width).
Source: Adapted from Conveyor Equipment Manufacturers Association 1997 (shown with permission).

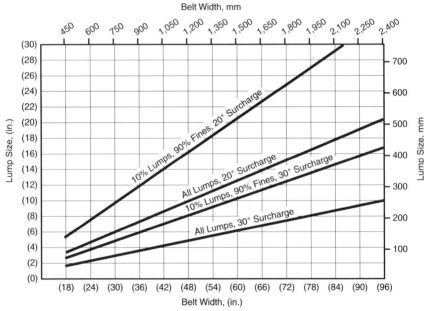

Source: Adapted from Conveyor Equipment Manufacturers Association 1997 (shown with permission).

FIGURE 12.1 Belt width necessary for a given lump size

EXCAVATION, LOADING, AND MATERIAL TRANSPORT | 235

Very Free Flowing	Free Flowing	Average Flowing		Sluggish	
5° Angle of Surcharge	10° Angle of Surcharge	20° Angle of Surcharge	25° Angle of Surcharge	30° Angle of Surcharge	
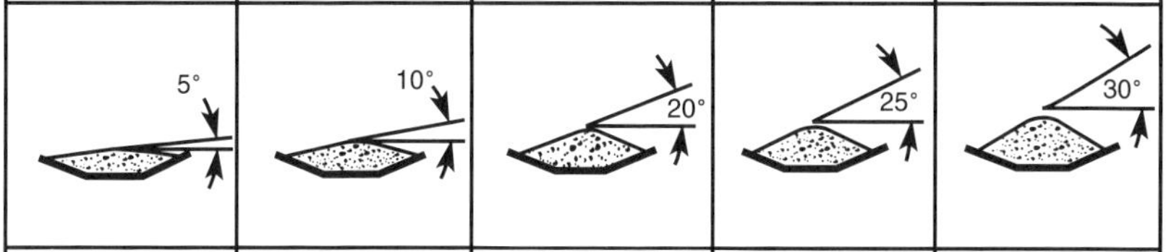 5°	10°	20°	25°	30°	
0°–19° Angle of Repose	20°–29° Angle of Repose	30°–34° Angle of Repose	35°–39° Angle of Repose	40°–up Angle of Repose	
Material Characteristics					
Uniform size, very small rounded particle, either very wet or very dry, such as dry silica sand, cement, wet concrete, etc.	Rounded, dry polished particles, of medium weight, such as whole grain and beans.	Irregular, granular or lumpy materials of medium weight, such as anthracite coal, cottonseed meal, clay, etc.	Typical common materials such as bituminous coal, stone, most ores, etc.	Irregular, stringy, fibrous, interlocking material, such as wood chips, bagasse, tempered foundry sand, etc.	

Source: Adapted from Conveyor Equipment Manufacturers Association 1997 *(shown with permission).*

FIGURE 12.2 Flowability, angle of surcharge, and angle of repose

Belt Conveyor Power Requirements

Belt conveyors require power to: (1) drive the empty conveyor, (2) elevate the material, and (3) convey the material horizontally. Use Equation 12.42, Table 12.22, and Figures 12.3–12.5 to estimate power requirements.

$$PR_c = (R1)(V_c) + (R2)(H) + (R3)(P_c) \qquad \text{(EQ 12.42)}$$

where:
- PR_c = total conveyor power required, W (hp)
- $R1$ = power rate to drive empty conveyor, W/0.5 m/s (hp/100 ft/min; see Table 12.22 and Figure 12.3)
- V_c = conveyor belt speed, m/s (ft/min)
- $R2$ = power rate to elevate material, W/m (hp/ft; see Figure 12.4)
- H = height of lift, m (ft)
- $R3$ = power rate to convey material horizontally, W/25 kg/s (hp/100 t/h; see Figure 12.5)
- P_c = conveyor production capacity, kg/s (ton/h; see Figure 12.5)

TABLE 12.22 Weight per unit length of belt and idlers, kilograms (pounds)

Belt Width		Material Density kg/m³ (lb/ft³)							
mm	(in.)	800	(50)	1,600	(100)	2,400	(150)	3,200	(200)
450	(18)	17.9	(12)	20.8	(14)	25.3	(17)	25.3	(17)
600	(24)	23.8	(16)	28.3	(19)	34.2	(23)	34.2	(23)
750	(30)	29.8	(20)	35.7	(24)	43.2	(29)	43.2	(29)
900	(36)	41.7	(28)	52.1	(35)	61.0	(41)	71.4	(48)
1,050	(42)	50.6	(34)	62.5	(42)	72.9	(49)	87.8	(59)
1,200	(48)	61.0	(41)	75.9	(51)	102.7	(69)	114.6	(77)
1,350	(54)	71.4	(48)	86.3	(58)	116.1	(78)	132.4	(89)
1,500	(60)	89.3	(60)	104.2	(70)	129.5	(87)	147.3	(99)
1,800	(72)	110.1	(74)	123.5	(83)	168.2	(113)	193.5	(130)
2,100	(84)	151.8	(102)	189.0	(127)	221.7	(149)	245.5	(165)
2,400	(96)	174.1	(117)	212.8	(143)	269.4	(181)	269.4	(181)

Source: Adapted from Conveyor Equipment Manufacturers Association 1997 (shown with permission).

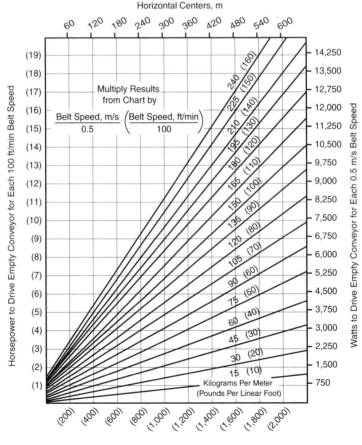

Source: Adapted from Conveyor Equipment Manufacturers Association 1997 (shown with permission).

FIGURE 12.3 Power required to drive empty conveyor

EXCAVATION, LOADING, AND MATERIAL TRANSPORT | 237

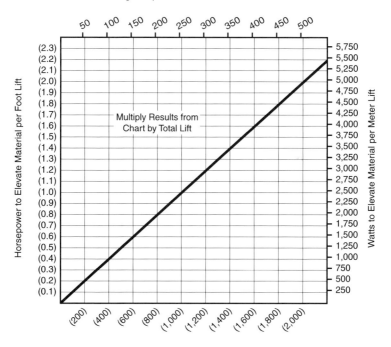

Source: Adapted from Conveyor Equipment Manufacturers Association 1997 *(shown with permission).*

FIGURE 12.4 Power required to elevate material

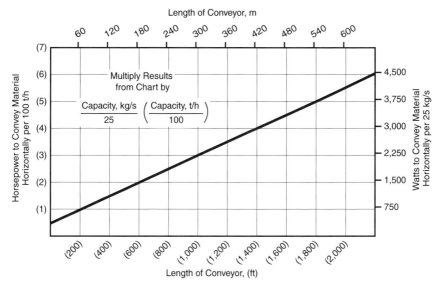

Source: Adapted from Conveyor Equipment Manufacturers Association 1997 *(shown with permission).*

FIGURE 12.5 Power required to convey material horizontally

HOISTING

The two types of hoists in common use are the drum hoist and the friction or Koepe hoist. Drum hoists consist of one or two steel or cast-iron hoisting drums on which the hoisting rope is stored. The other end of the hoisting rope is connected to the conveyance (i.e., either a skip or cage). In friction hoists, the hoist rope passes over a wheel or sheave with a conveyance on each end of the rope or with a conveyance on one end and a counterweight on the other end.

Hoist Cycle Time

Use Equation 12.43 to calculate the cycle time for loading, hoisting, and dumping (Edwards 1992). Maximum hoist speeds are typically between 2 and 20 m/s (6.6 and 66 ft/s); acceleration and deceleration rates can be as high as 1.2 m/s² (4 ft/s²), but are typically closer to 0.75 m/s² (2.5 ft/s²). Mining laws and regulations may also place limits on maximum accelerations and speeds. Typical combined dump and load times are about 20 s.

$$TC_H = (V_{max}/ACC) + (H/V_{max}) - [(V_{max}/2)(1/ACC + 1/RET)] + (V_{max}/RET) + T_{dl} \quad \text{(EQ 12.43)}$$

where:
TC_H = hoist cycle time, s (sec)
H = hoisting distance, m (ft)
V_{max} = maximum hoisting speed, m/s (ft/sec)
ACC = skip acceleration, m/s² (ft/sec²)
RET = skip retardation or deceleration, m/s² (ft/sec²)
T_{dl} = combined time to load and dump the skip, s (sec)

NOTE: For a single skip system, calculate the time to lower the skip by excluding T_{dl} in Equation 12.43.

Skips

Mine skips include the Kimberley or overturning skip, the swing-out body skip, and the bottom-dump skip. To estimate the approximate weight of the empty skip, use Equation 12.44 (Edwards 1992).

$$WS = 0.5(L_h) + SWC \quad \text{(EQ 12.44)}$$

where:
WS = weight of empty skip, kg (lb)
L_h = skip payload, kg (lb)
SWC = skip weight constant, 680 kg (1,500 lb)

Hoisting Ropes

Hoisting ropes include round-strand, flattened-strand, and locked-coil ropes. Each has different internal construction, weight per foot, and breaking strength characteristics. Table 12.23 lists the properties for the most common rope dimensions. Mining laws and regulations typically require a minimum load strength for the hoisting rope(s) equal to a safety factor multiplied by the maximum suspended load. Safety factors between four and ten are common.

Hoisting Capacity

Use Equation 12.21 to estimate the hoisting capacity of production skips. Convert cycle time to minutes. The haul load is typically given in tonnes (tons) rather than bank measure. The operating efficiency (E) for hoists averages about 0.7.

EXCAVATION, LOADING, AND MATERIAL TRANSPORT

TABLE 12.23 Wire rope data

Rope Diameter		18 × 7 I.W.R.C. Sinking (Round)				6 × 27 F.C. Flattened Strand (Triangular)				Locked Coil			
		Weight		Breaking Strength		Weight		Breaking Strength		Weight		Breaking Strength	
in.	cm	lb/ft	k/m	1,000 lb	1,000 kg	lb/ft	k/m	1,000 lb	1,000 kg	lb/ft	kg/m	1,000 lb	1,000 kg
¾	1.9	1.11	1.65	55.23	25.05	0.97	1.44	57	25.9	1.31	1.95	69.4	31.5
⅞	2.2	1.42	2.11	74.04	33.58	1.33	1.98	78	35.4	1.85	2.75	94.0	42.6
1	2.5	1.84	2.74	95.85	43.48	1.74	2.59	103	46.7	2.45	3.65	123.2	55.9
1⅛	2.9	2.33	3.47	121.21	54.98	2.22	3.30	132	59.9	3.08	4.58	156.8	71.1
1¼	3.2	2.80	4.17	148.65	67.43	2.74	4.08	163	73.9	3.75	5.58	192.6	87.4
1⅜	3.5	3.38	5.03	179.44	81.39	3.32	4.94	197	89.4	4.53	6.74	233.0	105.7
1½	3.8	4.02	5.98	212.53	96.40	3.95	5.88	235	106.6	5.25	7.81	277.8	126.0
1⅝	4.1	4.69	6.98	249.00	112.95	4.66	6.93	276	125.2	6.24	9.29	324.8	147.3
1¾	4.4	5.62	8.36	298.47	135.39	5.37	7.99	319	144.7	7.29	10.85	376.4	170.7
1⅞	4.8	6.18	9.20	331.19	150.23					8.46	12.59	432.4	196.1
2	5.1	7.08	10.54	375.79	170.46					9.67	14.39	492.8	223.5
2⅛	5.4	8.19	12.19	434.85	197.25								
2¼	5.7	9.15	13.62	485.54	220.24								

NOTES: Data is for standard tensile wire with a nominal breaking load of 120 long tons/square inch.
I.W.R.C. = Independent wire rope core.
F.C. = Fiber core.
Source: Adapted from Edwards 1992 (shown with permission of Wire Rope Industries, Ltd.).

Hoist Power Requirements

The running power requirements (i.e., hoisting at constant velocity) for a hoist motor are given in Equation 12.45. When designing hoist power systems, the power is normally calculated for each phase of hoisting (e.g., acceleration, constant speed, deceleration) to determine the peak accelerating power and root-mean-square power required.

$$PR_h = (TWS \times V_{max})/(UCF \times EFF) \qquad (EQ\ 12.45)$$

where:
- PR_h = approximate hoist motor power required, kW (hp)
- TWS = total weight of loaded skip including rope, kg (lb)
- V_{max} = maximum hoisting speed, m/s (ft/s)
- UCF = unit conversion factor, 102 (m-kg)/(kW-s) or 550 (ft-lb)/(hp-s)
- EFF = gear efficiency in converting engine power, decimal (typically about 0.85)

SELECTED UNDERGROUND EQUIPMENT

The common types of equipment used in underground coal and hardrock production operations are discussed below.

Continuous Miners

Continuous miners are used to excavate coal and other soft materials such as trona and potash in underground room and pillar operations. Cutting heads are typically between 2.4 to 3.7 m (8 to 12 ft) in width. Most rooms and cross-cuts are between 14 and 24 ft in width (i.e., two passes of the machine are required in each face). Although the continuous miner is capable of excavating between 13.6 and 32.7 t/min (15 to 36 st/min), production is controlled primarily by the mine plan, face dimensions, roof conditions, amount of reject material, efficiency in moving between production faces, the number of shuttle cars and average haulage cycle time, equipment availability (including main line haulage) and scheduled production hours per shift. Depending on these factors, average production rates of clean coal can vary between 20 and 130 t/h (22 and 143 st/h).

Shuttle Cars

Shuttle cars are wheel-mounted units used to haul coal from the continuous miner to a loading point (typically a belt feeder). They may be powered by cable-reel electric, battery, or diesel, and their capacities typically range between 6.8 and 10.9 t (7.5 and 12 tons). Shuttle cycle times vary according to haul road conditions and the length of haul. A typical cycle time for a 91 to 122 m (300 to 400 ft) haul consists of: loading time = 60 s, hauling to feeder = 75 s, unloading time = 30 s, tramming to face = 75 s (Breithaupt 1982).

Longwall Systems

To estimate productions for shearer-type longwall systems, use Equation 12.46 (adapted from Peng and Chiang 1992). Haulage speeds of shearers typically range between 12 and 18 m/min (40 to 60 ft/min). The listed values for operating efficiency are for the shearer only. These values include stopping or reversing at each end of the face and some stoppage during cutting operations. If production delays are also expected as a result of conveyor, roof support, or other ancillary operations, the operating efficiency should be reduced accordingly. Moving the longwall system between panels typically requires 1–2 weeks and must be included in any long-term production calculations. Longwall panels are typically 120 to 300 m (400 to 1,000 ft) in width and 900 to 5,500 m (3,000 to 18,000 ft) in length.

$$P_{she} = 60(Hc)(S)(V_s)(BD)(LC)(E) \qquad (EQ\ 12.46)$$

where:
- P_{she} = shearer production, t/h (ton/h)
- 60 = minutes in one hour, min/h
- Hc = mining height, m (ft)
- S = cutting web or width, m (ft)
- V_s = shearer haulage speed, m/min (ft/min)
- BD = bank density of coal, 1.2 to 1.4 t/m³ (0.037 to 0.044 ton/ft³)
- LC = loading coefficient of the shearer, decimal (typically 0.90 to 0.95)
- E = operating efficiency of the shearer, decimal (typically 0.70 to 0.90)

Slushers

A slusher consists of a scraper pulled by wire ropes over a series of rotating drums. Slushers are used to load and transport ore over a short distance, generally between 7.5 and 150 m (25 and 500 ft). Typical scraper capacities range from 0.25 *LCM* (0.33 *LCY*) for 0.7-m (28-in.) widths up to 2.8 *LCM* (3.7 *LCY*) for 2.1-m (84-in.) widths. Rope speeds are typically between 46 and 107 m/min (150 to 350 ft/min) with the higher speeds used for finer materials, smooth bottoms, and longer hauls (Rhoades 1982).

Load-Haul-Dump Units

A load-haul-dump (LHD) unit is a wheel-mounted, bidirectional, articulated machine equipped with four-wheel drive and a bucket. Load-haul-dump units are typically powered by diesel, with a low and narrow profile for underground use. Capacities range from about 1 t (1.1 ton) to more than 20 t (22 tons). Use Equation 12.21 to estimate productivity for the load, haul, and dump operation with the operating efficiency (E) ranging from 0.75 (severe job conditions) to 0.92 (excellent job conditions). Use Equations 12.18–12.20 to estimate cycle times, with the combined time for loading, maneuvering, and dumping ranging between 0.8 to 1.4 min (depending on operating conditions). Tram speeds range from 5 to 16 km/h (3 to 10 mi/h) depending on haulageway clearances (Stevens and Acuna 1982).

Mine Trucks

Underground mine trucks include tip dumpers, telescoping dumpers, and push-plate dumpers (Stevens 1982). Tip dumpers require headroom to raise the truck bed during dumping; telescoping and push-plate dumpers dump out the back. Underground trucks are usually articulated and may be two- or four-wheel drive. Use Table 12.3 to estimate haulage speeds, but underground safety considerations dictate that maximum speeds be reduced by at least one gear and that average speeds be estimated at the low end of the range of values. To estimate travel and haulage cycle times, use Equations 12.18 through 12.20, and employ Equation 12.21 to estimate haulage production. Operating efficiencies and load, maneuver, and dump times are approximately the same as for LHD units discussed above.

REFERENCES

Atkinson, T. 1992. Selection and sizing of excavating equipment. In *SME Mining Engineering Handbook*. Edited by H.L. Hartman. Littleton, CO: Society for Mining, Metallurgy, and Exploration, Inc. (SME). 1311–1333.

Bishop, T.S. 1968. Trucks. In *Surface Mining*. Edited by E.P. Pfleider. New York: The American Institute of Mining, Metallurgical, and Petroleum Engineers, Inc. 553–588.

Brantner, J.W. 1973. Mine haulage locomotive calculations. In *SME Mining Engineering Handbook*. Edited by A.B. Cummins and I.A. Given. New York: Society of Mining Engineers of the American Institute of Mining, Metallurgical, and Petroleum Engineers, Inc. 14-7–14-19.

Brauns, J.W., and D.H. Orr Jr. 1968. Railroad. In *Surface Mining*. Edited by E.P. Pfleider. New York: The American Institute of Mining, Metallurgical, and Petroleum Engineers, Inc. 531–552.

Breithaupt, R.L. 1982. Shuttle cars. In *Underground Mining Methods Handbook*. Edited by W.A. Hustrulid. New York: Society of Mining Engineers of the American Institute of Mining, Metallurgical, and Petroleum Engineers, Inc. 1223–1226.

Conveyor Equipment Manufacturers Association. 1997. *Belt Conveyors for Bulk Materials*. 5th ed. Naples, FL: Conveyor Equipment Manufacturers Association.

Drevdahl, E.R. Jr. 1963. *Fundamentals of Excavation Equipment for Engineering and Technology*. Tucson, AZ: Roadrunner Technical Publications Desert Laboratories, Inc.

Duncan, L.D., and B.J. Levitt. 1990. Belt conveyors. In *Surface Mining*. 2nd ed. Edited by B.A. Kennedy. Littleton, CO: SME. 692–705.

Edwards, F.A. 1992. Hoisting systems. In *SME Mining Engineering Handbook*. Edited by H.L. Hartman. Littleton, CO: SME. 1646–1673.

Hays, R.M. 1990. Trucks. Scrapers. Wheel loaders. Dozers. Four chapters in *Surface Mining*. 2nd ed. Edited by B.A. Kennedy. Littleton, CO: SME. 672–691, 709–723.

Humphrey, J.D. 1990. Walking draglines. In *Surface Mining*. 2nd ed. Edited by B.A. Kennedy. Littleton, CO: SME. 638–655.

Peng, S.S., and H.S. Chiang. 1992. Longwall mining. In *SME Mining Engineering Handbook*. Edited by H.L. Hartman. Littleton, CO: SME. 1780–1788.

Ramani, R.V. 1990. Rail haulage. In *Surface Mining*. 2nd ed. Edited by B.A. Kennedy. Littleton, CO: SME. 658–671.

Rhoades, W.A. 1982. Slushers. In *Underground Mining Methods Handbook*. Edited by W.A. Hustrulid. New York: Society of Mining Engineers of the American Institute of Mining, Metallurgical, and Petroleum Engineers, Inc. 1172–1178.

Sargent, F.R. 1990. Mining and quarry shovels. In *Surface Mining*. 2nd ed. Edited by B.A. Kennedy. Littleton, CO: SME. 626–633.

Stevens, R.M. 1982. Mine trucks. In *Underground Mining Methods Handbook*. Edited by W.A. Hustrulid. New York: Society of Mining Engineers of the American Institute of Mining, Metallurgical, and Petroleum Engineers, Inc. 1201–1219.

Stevens, R.M , and A. Acuna. 1982. Load-haul-dump units. In *Underground Mining Methods Handbook*. Edited by W.A. Hustrulid. New York: Society of Mining Engineers of the American Institute of Mining, Metallurgical, and Petroleum Engineers, Inc. 1179–1197.

CHAPTER 13

Ground Control/Support

Daniel F. Kump, P.E. and Alan A. Campoli, P.E.

INTRODUCTION

Rock masses are generally not homogenous or isotropic, but they are jointed; for this reason, the ground control requirements can vary considerably throughout a mine. While relying heavily on trial and error, ground control programs should be tempered with common sense and experience. For example, in a track drift one might start with 4-in. sections of steel with which to build sets. If that proves inadequate, 6-in. or 8-in. sections may have to be used to hold the ground. Similarly, in one part of a mine, 6-ft-long rock bolts may be used to hold the back, but a change to 8-ft-long bolts may be required to hold the ground in another part of the mine.

Table 13.1 shows some important properties that can be used to design effective ground control programs. See Chapter 2 on material properties for additional information.

TABLE 13.1 Strength properties of rock and coal

Rock or Coal Type*	Uniaxial Compressive Strength (psi)			Uniaxial Tensile Strength (psi)			Modulus of Elasticity (ksi)†			Poisson's Ratio		
	From	To	Mean	From	To	Mean	From	To	Mean	From	To	Mean
Basalt	6,090	51,475	21,750	290	4,060	1,885	2,320	14,645	7,685	0.13	0.38	0.22
Dolerite	32,915	46,255	40,600	1,740	3,770	2,900	8,700	13,050	10,150	0.15	0.29	0.20
Gneiss	10,585	49,300	23,055	435	3,045	2,030	2,320	14,935	8,410	0.10	0.40	0.22
Granite	4,350	46,980	24,070	435	5,655	1,740	1,450	10,730	6,525	0.10	0.39	0.23
Limestone	6,960	30,450	14,790	290	5,800	1,740	145	13,340	6,960	0.08	0.39	0.25
Norite	42,050	47,270	43,210	2,175	3,625	2,900	13,050	15,950	14,500	0.21	0.26	0.24
Quartzite	29,000	44,080	36,540	2,465	4,060	3,625	10,150	15,225	13,050	0.11	0.25	0.16
Sandstone	5,800	25,955	13,920	435	1,015	725	1,450	6,670	3,190	0.10	0.40	0.24
Shale	5,220	24,940	13,775	290	725	435	1,450	6,380	4,060	0.10	0.19	0.14
Pittsburgh coal	2,088	4,307	3,219	276	464	363	218	537	464	—	—	0.37
Pocahontas No. 3 coal	2,639	2,828	2,741	—	—	—	348	392	377	—	—	—
Herrin No. 6 coal	1,450	2,045	1,653	—	—	—	450	551	508	—	—	0.42

* Rock specimens were 4.25 in. high and 2.125 in. in diameter; coal specimens were 3-in. cubes.
† One ksi equals 1,000 psi.
Source: Bise 1986.

TABLE 13.2 Rock mass rating (RMR) system (after Bieniawski 1989)

A. Classification Parameters and Their Ratings

Parameter			Range of Values				
1 Strength of intact rock material	Point-load strength index	>10 MPa	4–10 MPa	2–4 MPa	1–2 MPa	For this low range, uniaxial compressive test is preferred	
	Uniaxial comp. strength	>250 MPa	100–250 MPa	50–100 MPa	25–50 MPa	5–25 MPa	1–5 MPa	<1 MPa
Rating		15	12	7	4	2	1	0
2 Drill core Quality RQD		90%–100%	75%–90%	50%–75%	25%–50%	<25%		
Rating		20	17	13	8	3		
3 Spacing of discontinuities		>2 m	0.6–2 m	200–600 mm	60–200 mm	<60 mm		
Rating		20	15	10	8	5		
4 Condition of discontinuities (See E)		Very rough surfaces; not continuous; no separation; unweathered wall rock	Slightly rough surfaces; separation <1 mm; slightly weathered walls	Slightly rough surfaces; separation <1 mm; highly weathered walls	Slickensided surfaces, or gouge <5 mm thick, or separation 1–5 mm continuous	Soft gouge >5 mm thick, or separation >5 mm, continuous		
Rating		30	25	20	10	0		
5 Groundwater	Inflow per 10 m tunnel length (v/m)	None	<10	10–25	25–125	>125		
	(Joint water press)/(Major principal σ)	0	<0.1	0.1–0.2	0.2–0.5	>0.5		
	General conditions	Completely dry	Damp	Wet	Dripping	Flowing		
Rating		15	10	7	4	0		

B. Rating Adjustment for Discontinuity Orientations (See F)

Strike and dip orientations		Very favorable	Favorable	Fair	Unfavorable	Very Unfavorable
Ratings	Tunnels and mines	0	-2	-5	-10	-12
	Foundations	0	-2	-7	-15	-25
	Slopes	0	-5	-25	-50	

C. Rock Mass Classes Determined from Total Ratings

Rating	100←81	80←61	60←41	40←21	<21
Class number	I	II	III	IV	V
Description	Very good rock	Good rock	Fair rock	Poor rock	Very poor rock

D. Meaning of Rock Classes

Class number	I	II	III	IV	V
Average stand-up time	20 yrs for 15-m span	1 year for 10-m span	1 week for 5-m span	10 hrs for 2.5-m span	30 min for 1-m span
Cohesion of rock mass (kPa)	>400	300–400	200–300	100–200	<100
Friction angle of rock mass (deg)	>45	35–45	25–35	15–25	<15

E. Guidelines for Classification of Discontinuity Conditions*

Discontinuity length (persistence)	<1 m	1–3 m	3–10 m	10–20 m	>20 m
Rating	6	4	2	1	0
Separation (aperture)	None	<0.1 mm	0.1–1.0 mm	1–5 mm	>5 mm
Rating	6	5	4	1	0
Roughness	Very rough	Rough	Slightly rough	Smooth	Slickensided
Rating	6	5	3	1	0
Infilling (gouge)	None	Hard filling <5 mm	Hard filling >5 mm	Soft filling <5 mm	Soft filling >5 mm
Rating	6	4	2	2	0
Weathering	Unweathered	Slightly weathered	Moderately weathered	Highly weathered	Decomposed
Rating	6	5	3	1	0

continues next page

GROUND CONTROL/SUPPORT | 245

TABLE 13.2 Rock mass rating (RMR) system (after Bieniawski 1989) (continued)

F. Effect of Discontinuity Strike and Dip Orientation in Tunneling†			
Strike perpendicular to tunnel axis		Strike parallel to tunnel axis	
Drive with dip—dip 45–90°	Drive with dip—dip 20–45°	Dip 45–90°	Dip 20–45°
Very favorable	Favorable	Very favorable	Fair
Drive against dip—dip 45–90°	Drive against dip—dip 20–45°	Dip 0–20°—irrespective of strike	
Fair	Unfavorable	Fair	

* Some conditions are mutually exclusive. For example, if infilling is present, the roughness of the surface will be overshadowed by the influence of the gouge. In such cases, use A.4 directly.
† Modified after Wickham et al. (1972).
Source: Hoek, Kaiser, and Bawden 1997 *(reprinted with permission of A.A. Balkema).*

TABLE 13.3 RMR system guidelines for excavation and support in rock tunnels

Shape: Horseshoe; Width: 10 m; Vertical Stress: Below 25 MPa; Construction: Drilling and Blasting

Rock Mass Class	Excavation	Support		
		Rock bolts (20 mm diameter, fully grouted)	Shotcrete	Steel sets
Very good rock I RMR: 80–100	Full face; 3 m advance	Generally no support required except for occasional spot bolting		
Good rock II RMR: 61–80	Full face; 1.0–1.5 m advance; complete support 20 m from face	Locally bolts in crown 3 m long, spaced 2.5 m with occasional wire mesh	50 mm in crown where required	None
Fair rock III RMR: 41–60	Top heading and bench 1.5–3 m advance in top heading; begin support after each blast; complete support 10 m from face.	Systematic bolts 4 m long, spaced 1.5–2 m in crown and walls with wire mesh in crown	50–100 mm in crown and 30 mm in sides	None
Poor rock IV RMR: 21–40	Top heading and bench 1.0–1.5 m advance in top heading; install support concurrently with excavation 10 m from face	Systematic bolts 4–5 m long, spaced 1–1.5 m in crown and walls with wire mesh	100–150 mm in crown and 100 mm in sides	Light to medium ribs spaced 1.5 m apart where required
Very poor rock V RMR: <20	Multiple drifts; 0.5–1.5 m advance in top heading; install support concurrently with excavation; shotcrete as soon as possible after blasting	Systematic bolts 5–6 m long, spaced 1–1.5 m in crown and walls with wire mesh; bolt invert	150–200 mm in crown, 150 mm in sides, and 50 mm on face	Medium to heavy ribs spaced 0.75 m apart with steel lagging and forepoling if required; close invert

Source: Bieniawski 1979.

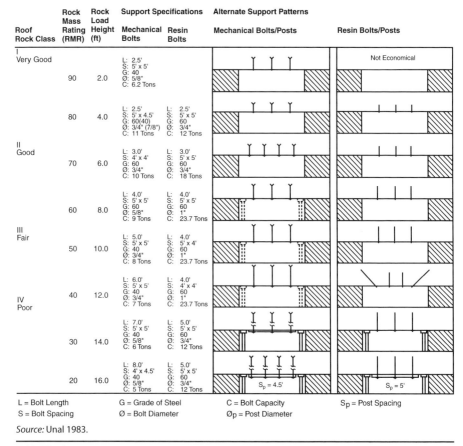

FIGURE 13.1 Rock support selection chart for coal entries 20 ft wide

SAFETY FACTOR

Safety factors are based on the required longevity of openings. The value of a safety factor considered acceptable for a design is usually established from previous experience with successful designs. A safety factor of 1.3 would generally be considered adequate for a temporary mine opening while a value of 1.5 to 2.0 may be required for a "permanent" excavation, such as an underground crusher station (Hoek, Kaiser, and Bawden 1997).

ROCK BOLTING

Empirical Rules

The empirical rules for rock bolting that follow are only general guidelines (Bieniawski 1992).
- Minimum bolt length
  - Greatest of:
    - Twice the bolt spacing
    - Three times the width of critical and potentially unstable rock blocks defined by the average discontinuity spacing in the rock mass
    - For spans less than 20 ft (6 m), use a bolt length of one-half the span. For spans from 20 ft (6 m) to 60 ft (18 m), interpolate between 10 ft (3 m) and 15 ft (5 m) lengths, respectively. For excavations higher than 60 ft (18 m), sidewall bolts are one-fifth of wall height.

GROUND CONTROL/SUPPORT | 247

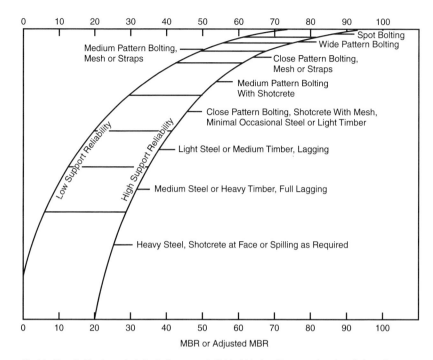

Spot bolting: Bolting to restrain limited areas or individual blocks of loose rock, primarily for safety.
Wide pattern bolting: Bolt spacing on 1.5 m to 1.8 m, or wider in very large openings.
Medium pattern bolting, with or without mesh or straps: Bolts spaced 0.9 m to 1.5 m, 23 cm wide straps or 100 mm welded wire mesh.
Close pattern bolting, mesh or straps: Bolt spacing less than 0.9 m, 100 mm welded wire mesh, 0.3 m straps, or chain link.
Medium pattern bolting with shotcrete: Bolts spaced 0.9 m to 1.5 m and 80 mm (nominal) of shotcrete. Light mesh for wet rock to alleviate shotcrete adherence problems.
Close pattern bolting, shotcrete with mesh, minimal occasional steel or light timber: Bolt spacing less than 0.9 m with 100 mm welded wire mesh or chain link throughout, and nominal 100 mm of shotcrete. Localized conditions may require light wide flange steel sets or timber sets.
Light steel, medium timber, lagging: Bolting as required for safety at the face—full contact (grouted or split set) bolts only. Light wide flange steel sets or 0.25 m timber sets spaced 1.5 m, with full crown lagging and rib lagging in squeezing areas.
Medium steel, heavy timber, full lagging: Medium wide flange steel sets or 0.3 m timber sets spaced 1.5 m, fully lagged across the crown and ribs. Support to be installed as close to the face as possible.
Heavy steel, shotcrete at face or spilling as required: Heavy wide flange steel sets spaced 1.2 m, fully lagged on crown and ribs, carried directly to face. Spilling or shotcreting of face as necessary.
General: Bolting: bolts in spot bolting through close pattern bolting are considered to be 19 mm in diameter, fully grouted or resin anchored standard rockbolts; mechanical anchors are acceptable in material of MBR > 60. Split set use is at the discretion of the operator.

Source: Bieniawski 1992.

FIGURE 13.2 Support recommendations for mine drifts

- Maximum bolt spacing
  - Least of:
    - One-half the bolt length
    - One and one-half the width of critical and potentially unstable rock blocks
    - Six ft (2 m); spacing greater than 6 ft (2 m) makes attachment of wire mesh difficult
- Minimum bolt spacing: 3 ft (0.9 m).

NOTE: Where discontinuity spacing is close and the span is relatively large, the superposition of two bolting patterns may be appropriate. For example, long, heavy bolts on wide centers may be necessary to support the span, and shorter, thinner bolts on closer centers may be needed to stabilize the surface against raveling caused by close jointing.

PILLARS

Tributary Area Method (Bieniawski 1992)

The tributary area method is a simple and conservative approach for calculating pillar load. Several assumptions must be made:
- Loading is only by vertical pressure.
- Each pillar supports the column of rock over an area that is the sum of the cross-sectional area of the pillar plus a portion of the room area.
- Load is uniformly distributed over the cross-sectional area of the pillar. Research has shown:
  - Stress is not evenly distributed over the cross-section of a pillar. The maximum stress occurs at the corners formed by the intersection of the three orthogonal planes, namely two sidewalls of the pillar and the roof or the floor.
  - Stress on the pillars increases with percentage extraction.
  - Stress distribution in pillars depends on the ratio of pillar width to pillar height.

For rectangular pillars, find the pillar load using

$$S_p = [1.1\, H(w + B)(L + B)] / (w \times L) \qquad \text{(EQ 13.1)}$$

where:
- S_p = pillar load, psi
- H = depth below surface, ft
- w = pillar width, ft
- L = pillar length, ft
- B = entry width, ft

For extraction ratio of the mined-out area to the total area, use

$$e = 1 - [(w / (w + B)][(L / (L + B)] \qquad \text{(EQ 13.2)}$$

where e = extraction ratio, decimal.

GROUND CONTROL/SUPPORT | 249

TABLE 13.4 Roof bolt characteristics by type

Type	Bolt Material	Anchoring Mechanism	Advantages	Disadvantages
Grouted rebar	Deformed bar	Resin	High pullout resistance, corrosion resistance	Dual components, cost
Conventional	Smooth bar	Expansion shell	Single component, cost	Anchor slippage, corrosion, tension loss
Cable	Wrapped wire	Cement or resin	High pullout resistance, elasticity	Cost, dual components
Friction	Split or expanding tube	Tube deformations	Ease of installation	Low pullout resistance, corrosion

Source: Personal communication with Mike L. Thomson, Fosroc, Inc., Georgetown, Kentucky.

TABLE 13.5 Typical bolt, hole, and resin cartridge combinations

Series	Cartridge diameter (mm)	Hole diameter (in.)	Bar size	Expected pullout strength (tons/in.)
A	23	1	No. 6	2
B	23	1	No. 5	1.5
E	32	1 ³⁄₈	No. 7	1.7

Source: Personal communication with Peter Mills of Fosroc, Inc., Georgetown, Kentucky.

TABLE 13.6 Common coal mine roof bolt bar capacity

Diameter (in.)	Designation	Grade (ksi)	Capacity (tons)	Typical Anchorage
⅝	No. 5	40	6	Expansion shell
⅝	No. 5	55	8	Resin
⅝	No. 5	60	9	Resin
⅝	No. 5	75	12	Resin
¾	No. 6	40	9	Resin
¾	No. 6	55	12	Resin
¾	No. 6	60	13	Resin
¾	No. 6	75	17	Resin
⅞	No. 7	55	17	Resin
⅞	No. 7	60	18	Resin
⅞	No. 7	75	23	Resin
1	No. 8	55	22	Resin
1	No. 8	60	24	Resin
1	No. 8	75	29	Resin

NOTE: Rock anchorage may limit bolt capacity.

Analysis of Retreat Mining Pillar Stability (ARMPS)

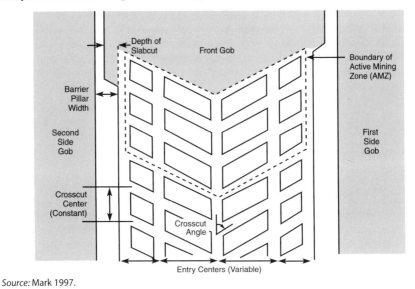

Source: Mark 1997.

FIGURE 13.3 Section layout parameters used in ARMPS

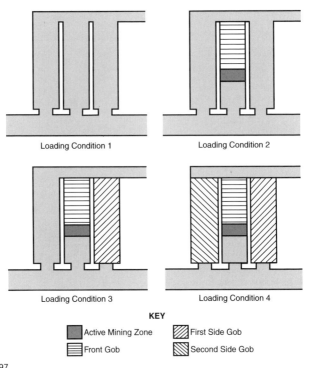

Source: Mark 1997.

FIGURE 13.4 Loading conditions evaluated with ARMPS

GROUND CONTROL/SUPPORT | 251

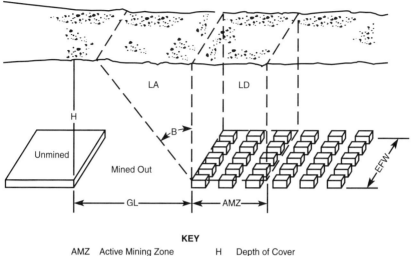

KEY

AMZ	Active Mining Zone	H	Depth of Cover
B	Abutment Angle	LA	Abutment Load
EFW	Extraction Front Width	LD	Development Load
GL	Mined Out Area		

Source: Mark 1997.

FIGURE 13.5 Schematic of loads

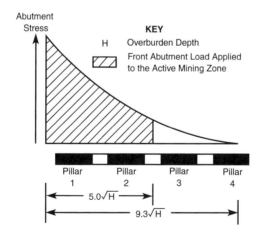

Source: Mark 1997.

FIGURE 13.6 Distribution of abutment stress

TABLE 13.7 Satisfactory ARMPS cases for coal mines

State	Seam	Stability Factor	Seam Thickness (ft)	Cover Thickness (ft)	Entry Width (ft)	Crosscut Angle (deg)	Entry Centers (ft)	Crosscut Centers (ft)	Loading Condition
AL	Blue Creek	1.96	6	1,150	20	90	130	130	2
CO	Cameo	1.11	7	700	18	90	100	50	2
CO	Cameo	1.14	7	600	18	90	100	50	3
CO	Cameo	1.86	7	400	18	90	100	50	3
CO	D seam	1.23	9	850	20	90	140	70	2
CO	D seam	1.44	9	700	20	90	70	130	2
IL	Herrin 6	1.14	8	700	14	90	60	60	3
KY	Harlan	1.94	6.5	300	20	90	60	60	3
KY	Hazard No. 4	1.36	4.4	420	20	60	55	55	3
KY	Kellioka	0.45	5	1,450	18	75	55	65	3
KY	Kellioka	1.18	5	675	18	90	50	85	3
KY	Kellioka	1.41	5	860	18	90	55	85	2
KY	Kellioka	1.61	5	600	18	75	55	85	3
KY	L. Elkhorn (No. 2 Gas)	1.64	13	400	20	90	80	80	3
KY	Pond Creek	1.2	5.5	700	19	90	60	60	2
KY	Pond Creek	1.69	5.5	450	19	90	50	80	3
KY	Pond Creek	1.98	5.5	450	19	90	50 or 60	70	3
KY	Pond Creek	2.22	5.5	450	19	90	50 or 60	80	2
OH	Freeport (L)	1.49	5	630	18	60	43.3	50	1
OH	Freeport (L)	1.66	5	530	18	60	43.3	50	1
OH	Freeport (L)	1.72	5	510	18	60	43.3	50	1
OH	Freeport (L)	2.6	5	600	18	60	60	70	1
OH	Mahoning	2.5	3	250	20	60	25	60	3
PA	Freeport (L)	2.06	6	400	20	90	60	90	3
PA	Kittanning (L)	1.62	6.5	340	17	60	70	50	3
PA	Kittanning (L)	1.92	6.5	330	17	60	75	50	2
PA	Kittanning (L)	2.02	6.5	330	17	60	75	50	3
PA	Lower Kittanning	1.92	6.5	340	17	60	100	50	3
PA	Lower Kittanning	1.96	6.5	350	17	60	100	50	3
PA	Lower Kittanning	2.08	6.5	340	17	60	100	50	2
PA	Lower Kittanning	2.09	6.5	315	17	60	100	50	3
PA	Lower Kittanning	2.14	5	550	17	90	54	114	3
PA	Pittsburgh	2.1	7	900	18	90	80		3
PA	Pittsburgh	2.78	7.2	853	17	90	87	97	2
PA	Sewickley	1.7	5.25	600	17	90	70	70	3
PA	Sewickley	1.7	5.25	600	17	90	70	70	3
PA	Sewickley	2.32	5.25	600	17	90	70	70	2
PA	Upper Freeport	2.68	5	250	20	60	50	50	1
PA	Upper Freeport	1.88	4.2	210	18	90	35	35	1
TN	Beach Grove	0.98	2.5	1,026	20	60	55	50	2
UT	Gilson	0.5	9	2,000	20	90	80	80	2
VA	Blair	1.65	3.8	600	20	90	60	60	3
VA	Glamorgan	2.31	6	400	20	90	70	70	3
VA	Jawbone	1.46	4.6	500	20	90	60		3
VA	Jawbone	1.97	4.6	400	20	90	60	60	3
VA	Jawbone	2.15	4.2	500	20	90	70	70	3
VA	Jawbone	2.86	4.2	450	20	90	70	70	2
VA	Mossy-Haggy	2.05	3	500	20	90	60	60	3
VA	Pocahontas No. 3	0.92	5.5	1,700	20	90	80	80	2
VA	Pocahontas No. 3	1.21	5.5	1,700	20	90	80	80	3
VA	Pocahontas No. 3	1.89	5	500	20	90	60	70	2

continues next page

TABLE 13.7 Satisfactory ARMPS cases for coal mines (continued)

State	Seam	Stability Factor	Seam Thickness (ft)	Cover Thickness (ft)	Entry Width (ft)	Crosscut Angle (deg)	Entry Centers (ft)	Crosscut Centers (ft)	Loading Condition
VA	Pocahontas No. 4	0.76	6.5	1,450	20	90	75	90	3
VA	Pocahontas No. 4	0.91	6	1,200	18	90	70	95	3
VA	Pocahontas No. 4	2.77	3	300	20	90	50	50	2
VA	Red Ash	2.44	3	500	20	90	70	70	2
VA	Red Ash	2.44	3	700	20	90	70	70	3
VA	Tiller	2.22	4	500	20	90	70	70	3
WV	Beckley	0.9	6.5	900	19	90	75	80	3
WV	Beckley	1.17	7	700	19	90	70	90	3
WV	Beckley	1.17	7	700	19	90	70	90	3
WV	Cedar Grove (L)	0.88	6	800	20		70	90	3
WV	Coalburg	0.53	8	425	20	90	30	60	3
WV	Coalburg	0.94	8	425	20	90	60	60	2
WV	Coalburg	0.96	9.2	500	20	90	60	60	3
WV	Coalburg	1.22	10	350	20	90	60	60	3
WV	Coalburg	1.23	9.7	350	20	90	60	60	3
WV	Coalburg	1.28	10	425	20	90	70	60	3
WV	Coalburg	1.3	9.8	400	20	90	60	60	3
WV	Coalburg	1.3	8	425	20	90	60	60	2
WV	Coalburg	1.37	7.5	380	20	90	60	60	3
WV	Coalburg	1.39	9.2	400	20	90	60	80	3
WV	Coalburg	1.39	9.8	375	20	90	60	70	3
WV	Coalburg	1.45	10	300	20	60	60	60	3
WV	Dorothy (Winifrede)	1.32	10	287	20	90	60	60	2
WV	Dorothy (Winifrede)	1.46	10	255	20	90	50	50	2
WV	Dorothy (Winifrede)	1.49	10	325	20	90	60	60	2
WV	Dorothy (Winifrede)	1.72	10	255	20	90	50	50	2
WV	Dorothy (Winifrede)	1.76	10	325	20	90	60	60	2
WV	Dorothy (Winifrede)	2.05	10	287	20	90	60	60	2
WV	Dorothy (Winifrede)	2.1	11	225	20	90	60	60	2
WV	Fire Creek	1.24	4.5	850	20	90	80	60	2
WV	Peerless	1.56	4.75	700	20	90	60	80	2
WV	Sewell	2.55	4	350	20	60	60	60	2
WV	Stockton	1.56	10	220	20	90	50	50	2
WV	Stockton	1.99	10	245	20	90	60	60	2
WV	Winifrede	1.73	6.5	600	18.5	90	53.5	78.5	2
WV	Winifrede	1.75	6.5	600	20	90	70	90	2

Source: Personal communication with Chris Mark, NIOSH.

SUBSIDENCE

TABLE 13.8 Typical angles of draw

Coalfield/Country	Reference	Angle of Draw (degrees)
Limburg/Netherlands	Brauner (1973)	35–45
Limburg/Netherlands	Pottgens (1978)	45
Northern France	Brauner (1973)	35
USSR	Brauner (1973)	30
Rurh/Germany	Brauner (1973)	30–45
Rurh/Germany	Kratzsch (1983)	55
Saar/Germany	Kratzsch (1983)	40
United Kingdom	ICE (Anon., 1977)	25–35
Midlands/United Kingdom	Orchard (1957), Wardell (1969), NCB (1975)	35
United States		
East—Anthracite	Montz and Norris (1930)	25
Southwestern Pennsylvania	Newhall and Plein (1936)	10–25
Appalachian	Cortis (1969)	15–27
Appalachian	Peng and Chyan (1981)	22–38
Northern Appalachian	Adamek and Jeran (1981)	12–17
Central—Illinois	Wade and Conroy (1977)	23–29
Illinois	Conroy (1979)	15–30
Illinois	Bauer and Hunt (1981)	12–26
Illinois	Hood (1981)	17–18 (long.)
Illinois		42–44 (trans.)
West—Raton, New Mexico	Gentry and Abel (1977)	16
Deer Creek, Emery, Utah	Allgaier (1988)	30
Somerset, Gunnison, Colorado	Dunrud (1984)	15–25
Salina, Utah	Dunrud (1984)	8–20
Sheridan, Wyoming	Dunrud (1984)	6–9

Source: Singh 1992.

TABLE 13.9 Residual subsidence duration over longwall mines

Reference	Country/Coalfield	Residual Subsidence Duration
Institution of Municipal Engineers (Anon., 1947)	United Kingdom	2 to 10 years
Orchard and Allen (1974)	United Kingdom	Several months to 3 to 6 years (strong overburden)
Collins (1977)	United Kingdom	2 to 4.5 years
Grard (1969)	France	6 to 12 months
Brauner (1973)	Germany	1 year (Cretaceous overburden); 2 years (sandstone overburden)
Brauner (1973)	USSR	2 years (shallow mines); 4 to 5 years (deep mines, >1300 ft or 400 m)
Shadrin and Zamotin (1977)	USSR	2 to 25 months
Gray et al. (1977)	U.S./Appalachian	Few months to few years
Hood et al. (1981)	U.S./Illinois	12 months

Source: Singh 1992.

GROUND CONTROL/SUPPORT | 255

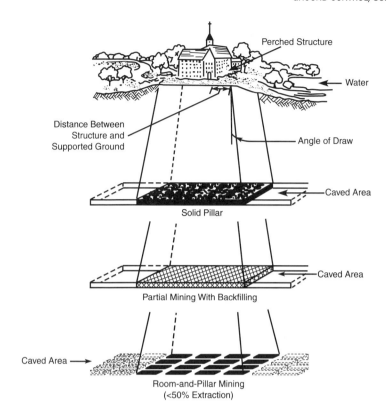

Source: Singh 1992.

FIGURE 13.7 Protective zones for surface structures

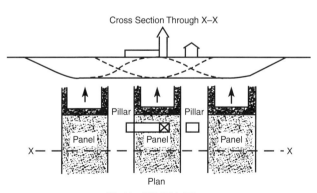

Source: Singh 1992.

FIGURE 13.8 Sized pillars for protecting surface structures

TABLE 13.10 Minimum cover for total extraction under water bodies*

Seam Thickness (t)		Minimum Cover Thickness		
Feet	Meters	In terms of t	Feet	Meters
3	0.9	117 t	351	107.0
4	1.2	95 t	380	115.8
5	1.5	80 t	400	121.9
6	1.8	71 t	426	129.8
7	2.1	63 t	441	134.4
7.5	2.3	60 t	450	137.2
>7.5	>2.3	60 t	450	137.2

* Potential for causing catastrophic damage.
Source: Singh 1992.

SLOPE STABILITY

Usually a safety factor (F) of 1.3 is considered adequate for slopes that are required to stand for a short time; for long-term stability, a value of 1.5 is desirable (Bise 1986).

The most common form of slope failure in coal and hardrock mines is planar failure. Planar failure can result when a competent block of rock lies along a planar discontinuity that dips with regard to the slope. When the dip angle is greater than the peak friction angle for the discontinuity surface, the block tends to slide.

The safety factor for planar failure can be approximated with the following equation (see Figure 13.9 for illustration of variables). Assume a 1-ft wide slice of slide block (Bise 1986):

$$F = \{c(A) + [W(\cos \psi_p) - U - V(\sin \psi_p)] \tan \varphi\} / [W(\sin \psi_p) + V(\cos \psi_p)] \quad \text{(EQ 13.3)}$$

where:
- c = cohesion, lb/ft^2
- A = length of slip surface = $(H - z)(\csc \psi_p)$, ft
- U = water pressure uplift = $0.5(\gamma_w)(z_w)(H - z)(\csc \psi_p)$, lb
- V = force from water in tension crack = $0.5(\gamma_w)(z_w^2)(\sin \psi_p)$, lb
- W = weight of block, lb
- ψ_p = angle of discontinuity plane, degrees
- φ = friction angle, degrees
- γ_w = density of water (62.4 lb per ft^3)

GROUND CONTROL/SUPPORT | **257**

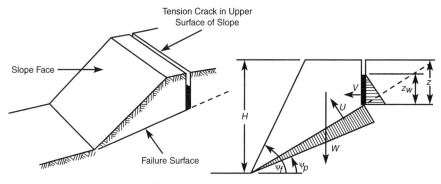

Geometry of Slope With Tension Crack in Upper Slope Surface

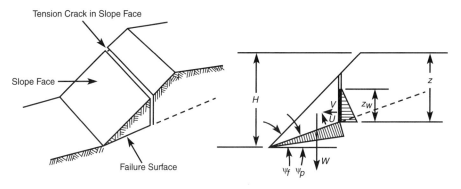

Geometry of Slope With Tension Crack in Slope Face

Source: Bise 1986.

FIGURE 13.9 Planer failure

REFERENCES

Bieniawski, Z.T. 1979. The geomechanics classification in rock engineering applications. *Proceedings 4th International Congress on Rock Mechanics*. Vol. 2. Rotterdam, The Netherlands: A.A. Balkema Publishers.

Bieniawski, Z.T. 1989. Engineering Rock Mass Classifications. New York: Wiley. 251 pp.

Bieniawski, Z.T. 1992. Ground control. In *SME Mining Engineering Handbook*. 2nd ed., Vol. 1. Edited by H.L. Hartman. Littleton, CO: Society for Mining, Metallurgy, and Exploration, Inc. (SME). 897–937.

Bise, C.J. 1986. *Mining Engineering Analysis*. Littleton, CO: Society of Mining Engineers, Inc.

Hoek, E., P.K. Kaiser, and W.F. Bawden. 1997. *Support of Underground Excavations in Hard Rock*. Rotterdam, The Netherlands: A.A. Balkema Publishers.

Kendorski, F., et al. 1983. Rock mass classification for block caving mine drift support. *Proceedings 5th International Congress on Rock Mechanics*. Melbourne, Australia: International Society for Rock Mechanics. B51–B63.

Mark, C. 1997. Analysis of retreat mining pillar stability. *Proceedings: New Technology for Ground Control in Retreat Mining*. IC 9446. Washington, DC: National Institute of Occupational Safety and Health (NIOSH).

Singh, M.M. 1992. Mine subsidence. In *SME Mining Engineering Handbook.* 2nd ed., Vol. 1. Edited by H.L. Hartman. Littleton, CO: SME. 938–971.

Unal, E. 1983. *Design Guidelines and Roof Control Standards for Coal Mine Roofs.* Ph.D. diss. University Park, PA: Pennsylvania State University.

Wickham, G.E., H.R. Tiedemann, and E.H. Skinner, 1972. "Support Determination Based on Geologic Predictions." *Proceedings Rapid Excavation and Tunneling Conference,* AIME, New York, pp. 43–64.

CHAPTER 14

Ventilation

William J. Francart, P.E. and Kelvin K. Wu, P.E.

AIR COMPOSITION, DENSITY, AND PSYCHROMETRY

TABLE 14.1 Air composition

Gas	Percent by Volume	Percent by Weight
Nitrogen	78.09	75.55
Oxygen	20.95	23.13
Carbon dioxide	0.03	0.05
Argon, other rare gases	0.93	1.27

Source: Hartman, Mutmansky, Ramani and Wang 1997 *(reprinted with permission of John Wiley & Sons).*

TABLE 14.2 Barometric pressure, temperature, and air density at different altitudes

	Barometric Pressure P_b		At Constant $t = 70°F$		At Varying t and Z	
Altitude Above or Below Sea Level, Z (ft)	Pounds per square inch	Inches mercury	Relative Air Density	Air Density, w (lb/ft³)	Air Temperature (°F)	Air Density (lb/ft³)
−1,000	15.23	31.02	1.037	0.0778	73.8	0.0771
−500	14.94	30.47	1.018	0.0764	71.9	0.0761
0	14.70	29.92	1.000	0.0750	70.0	0.0750
500	14.42	29.38	0.981	0.0736	68.1	0.0740
1,000	14.16	28.86	0.964	0.0723	66.1	0.0730
1,500	13.91	28.33	0.947	0.0710	64.2	0.0719
2,000	13.66	27.82	0.930	0.0698	62.3	0.0709
2,500	13.41	27.31	0.913	0.0685	60.4	0.0698
3,000	13.16	26.81	0.896	0.0672	58.4	0.0687
3,500	12.92	26.32	0.880	0.0660	56.5	0.0676
4,000	12.68	25.84	0.864	0.0648	54.6	0.0666
4,500	12.45	25.36	0.848	0.0636	52.6	0.0657
5,000	12.22	24.89	0.832	0.0624	50.7	0.0648
5,500	11.99	24.43	0.816	0.0612	48.8	0.0638
6,000	11.77	23.98	0.799	0.0599	46.9	0.6628
6,500	11.55	23.53	0.786	0.0590	45.0	0.0619
7,000	11.33	23.09	0.774	0.0580	43.0	0.0610
7,500	11.12	22.65	0.758	0.0568	41.0	0.0600

continues next page

TABLE 14.2 Barometric pressure, temperature, and air density at different altitudes (continued)

Altitude Above or Below Sea Level, Z (ft)	Barometric Pressure Pb		At Constant t = 70°F		At Varying t and Z	
	Pounds per square inch	Inches mercury	Relative Air Density	Air Density, w (lb/ft³)	Air Temperature (°F)	Air Density (lb/ft³)
8,000	10.91	22.22	0.739	0.0554	39.0	0.0590
8,500	10.70	21.80	0.728	0.0546	37.1	0.0581
9,000	10.50	21.38	0.715	0.0536	35.2	0.0573
9,500	10.30	20.98	0.701	0.0526	33.3	0.0564
10,000	10.10	20.58	0.687	0.0515	31.3	0.0555
10,500	9.90	20.18	0.674	0.0506	29.4	0.0546
11,000	9.71	19.75	0.661	0.0496	27.5	0.0538
11,500	9.52	19.40	0.648	0.0486	25.5	0.0529
12,000	9.34	19.03	0.636	0.0477	23.6	0.0521
12,500	9.15	18.65	0.624	0.0468	21.6	0.0513
13,000	8.97	18.29	0.611	0.0458	19.7	0.0505
13,500	8.80	17.93	0.599	0.0449	17.7	0.0496
14,000	8.62	17.57	0.587	0.0440	15.8	0.0488
14,500	8.45	17.22	0.576	0.0432	13.9	0.0480
15,000	8.28	16.88	0.564	0.0423	12.0	0.0473

Source: Madison 1949 (reprinted with permission of Howden Buffalo Inc., Camden, South Carolina [formerly Buffalo Forge, New York]).

TABLE 14.3 Temperature of dew point in degrees Fahrenheit (pressure = 29.0 in.)

Air Temperature (t)	Vapor Pressure (e)	Depression of Wet-Bulb Thermometer (t − t′)												
		1.0	2.0	3.0	4.0	5.0	6.0	7.0	8.0	9.0	10.0	11.0	12.0	13.0	14.0
20	0.103	17	13	8	2	−5	−18								
21	0.108	18	14	10	4	−3	−14	−42							
22	0.113	19	15	11	6	−1	−10	−29							
23	0.118	20	16	12	8	+1	−7	−22							
24	0.124	21	18	14	9	3	−4	−17							
25	0.130	22	19	15	11	5	−2	−12	−36						
26	0.136	23	20	16	12	7	±0	−9	−26						
27	0.143	24	21	18	14	9	+3	−5	−19						
28	0.150	25	22	19	15	11	5	−2	−14	−45					
29	0.157	26	24	20	17	12	7	±0	−9	−29					
30	0.164	27	25	22	18	14	9	+3	−5	−20					
31	0.172	29	26	23	20	16	11	5	−2	−14	−50				
32	0.180	30	27	24	21	17	13	8	+1	−9	−29				
33	0.187	31	28	25	22	19	15	10	8	−5	−20				
34	0.195	32	29	27	24	20	16	12	6	−2	−14	−50			
35	0.203	33	30	28	25	22	18	14	8	+1	−8	−28			
36	0.211	34	31	29	26	23	20	15	11	4	−4	−19			
37	0.219	35	32	30	27	24	21	17	13	7	−1	−12	−44		
38	0.228	36	33	31	28	26	23	19	14	9	+3	−7	−25		
39	0.237	37	34	32	29	27	24	21	16	12	6	−3	−16		

continues next page

TABLE 14.3 Temperature of dew point in degrees Fahrenheit (pressure = 29.0 in.) (continued)

Air Temperature (t)	Vapor Pressure (e)	\multicolumn{14}{c}{Depression of Wet-Bulb Thermometer (t − t′)}
		1.0	2.0	3.0	4.0	5.0	6.0	7.0	8.0	9.0	10.0	11.0	12.0	13.0	14.0
40	0.247	38	35	33	31	28	25	22	18	14	8	+1	−10	−35	
41	0.256	39	37	34	32	29	26	23	20	16	11	4	−5	−21	
42	0.266	40	38	35	33	30	28	25	21	17	13	7	−1	−13	−59
43	0.277	41	39	36	34	31	29	26	23	19	15	10	+3	−7	−28
44	0.287	42	40	38	35	32	30	27	24	21	17	12	6	−2	−17
45	0.298	43	41	39	36	34	31	29	26	22	19	14	8	+2	−9
46	0.310	44	42	40	37	35	32	30	27	24	20	16	11	5	−4
47	0.322	45	43	41	39	36	34	31	28	25	22	18	13	8	±0
48	0.334	46	44	42	40	37	35	32	30	27	23	20	15	10	+4
49	0.347	47	45	43	41	39	36	34	31	28	25	21	17	13	7
50	0.360	48	46	44	42	40	37	35	32	29	27	23	19	15	9
51	0.373	49	47	45	43	41	39	36	34	31	28	25	21	17	12
52	0.387	50	48	46	44	42	40	37	35	32	29	26	23	19	14
53	0.402	51	49	47	45	43	41	39	36	34	31	28	24	21	16
54	0.417	52	50	49	47	44	42	40	38	35	32	29	26	23	19
55	0.432	53	52	50	48	46	43	41	39	36	34	31	28	24	21
56	0.448	54	53	51	49	47	45	43	40	38	35	32	29	26	23
57	0.465	55	54	52	50	48	46	44	42	39	36	34	31	28	24
58	0.482	56	55	53	51	49	47	45	43	40	38	35	32	29	26
59	0.499	57	56	54	52	50	48	46	44	42	39	37	34	31	28
60	0.517	58	57	55	53	51	49	48	45	43	41	38	35	32	29
61	0.536	59	58	56	54	52	51	49	46	44	42	39	37	34	31
62	0.555	60	59	57	55	54	52	50	48	46	43	41	38	35	32
63	0.575	61	60	58	56	55	53	51	49	47	45	42	40	37	34
64	0.595	62	61	59	58	56	54	52	50	48	46	44	41	38	36
65	0.616	63	62	60	59	57	55	53	51	49	47	45	43	40	37
66	0.638	64	63	61	60	58	56	54	53	51	48	46	44	42	39
67	0.661	65	64	62	61	59	57	56	54	52	50	48	45	43	40
68	0.684	67	65	63	62	60	58	57	55	53	51	49	47	44	42
69	0.707	68	66	64	63	61	60	58	56	54	52	50	48	46	43
70	0.732	69	67	66	64	62	61	59	57	55	53	51	49	47	45
71	0.757	70	68	67	65	63	62	60	58	57	55	53	51	49	46
72	0.783	71	69	68	66	65	63	61	60	58	56	54	52	50	48
73	0.810	72	70	69	67	66	64	62	61	59	57	55	53	51	49
74	0.838	73	71	70	68	67	65	64	62	60	58	56	54	53	50
75	0.866	74	72	71	69	68	66	65	63	61	60	58	56	54	52
76	0.896	75	73	72	70	69	67	66	64	62	61	59	57	55	53
77	0.926	76	74	73	71	70	68	67	65	64	62	60	58	56	54
78	0.957	77	75	74	72	71	69	68	66	65	63	61	59	58	56
79	0.989	78	76	75	73	72	70	69	67	66	64	62	61	59	57
80	1.022	79	77	76	75	73	72	70	69	67	65	64	62	60	58

Source: Marvin 1915.

262 | SME MINING REFERENCE HANDBOOK

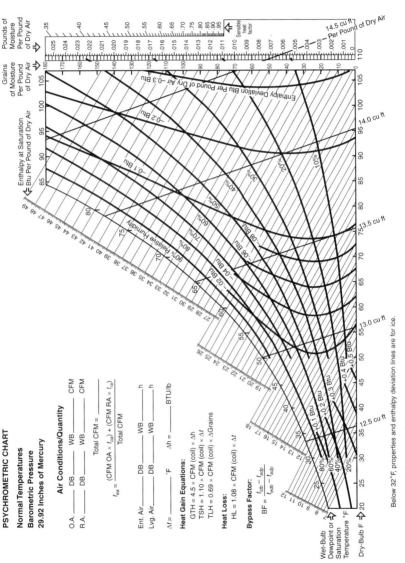

Source: Carrier Corporation 1959 *(reprinted with permission).*

FIGURE 14.1 Psychrometric chart, sea level

$$w = 1.327B / (460 + t_d) \qquad \text{(EQ 14.1a; after Ramani 1992)}$$

$$w = B / 0.287(273 + t_d) \qquad \text{(EQ 14.1b; SI units)}$$

where:
- w = density in pounds per cubic foot (kilograms per cubic meter)
- B = barometric pressure in inches of mercury (Pa)
- t_d = dry-bulb temperature in degrees F (degrees C)

Water vapor correction:

$$w = 1.327 (B - 0.378f) / (460 + t_d) \qquad \text{(EQ 14.2a; after Ramani 1992)}$$

$$w = (B - 0.378f) / 0.287 (273 + t_d) \qquad \text{(EQ 14.2b; SI units)}$$

where f = vapor pressure at the dew point temperature in inches.

GAS LAWS (AFTER HARTMAN, MUTMANSKY, AND WANG 1997)

See Chapter 4 on physical science and engineering for more information.

$$p_1 v_1 = p_2 v_2 \text{ constant temperature} \qquad \text{(EQ 14.3)}$$

$$v_1 / v_2 = T_1 / T_2 \text{ constant pressure} \qquad \text{(EQ 14.4)}$$

$$p_1 v_1 / T_1 = p_2 v_2 / T_2 \qquad \text{(EQ 14.5)}$$

where:
- p = absolute pressure
- v = specific volume
- T = absolute temperature

AIRFLOW

See Chapter 4 for more information.

Quantity

$$Q = VA \qquad \text{(EQ 14.6; after Hartman, Mutmansky, and Wang 1997)}$$

where:
- Q = quantity in cubic feet per minute (cubic meters per second)
- V = velocity in feet per minute (meters per second)
- A = area in square feet (square meters)

Velocity Head (after Ramani 1992)

$$H_v = w (V/1098)^2 \qquad \text{(EQ 14.7a)}$$

$$H_v = wV^2 / 2g \qquad \text{(EQ 14.7b; SI units)}$$

$$(w/g = kg/m^3)$$

where:
- H_v = velocity head in inches of water (Pa)
- V = velocity in feet per minute (meters per second)
- W = density in pounds per cubic foot (kilograms per cubic meter)

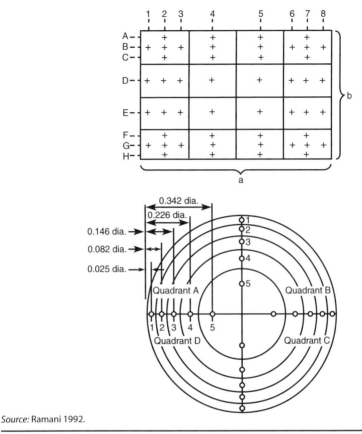

Source: Ramani 1992.

FIGURE 14.2 Measurement points for fixed traversing in rectangular and circular airways

Friction Loss (after Ramani 1992)

$$H_f = KLOV^2 / 5.2 A \qquad \text{(EQ 14.8a)}$$

$$H_f = KLOV^2 / A \qquad \text{(EQ 14.8b; SI units)}$$

where:
- H_f = friction loss in airway in inches of water (Pa)
- L = length of airway in feet (meters)
- O = perimeter of airway in feet (meters)
- V = velocity of air in feet per minute (meters per second)
- A = cross-sectional area of airway in square feet (square meters)
- K = friction factor in lb-min^2/ft^4 (kg/m^3; from Table 14.4)

$$H_f = RQ^2 \qquad \text{(EQ 14.9)}$$

where:
- R = resistance in inch-min^2/ft^6 (N-s^2/m^8)
- Q = quantity of air in cubic feet per minute (cubic meters per second)

TABLE 14.4 Friction factors for mine airways

Type of Airway	Irregularities of Surfaces, Areas, and Alignment	Straight			Sinuous or Curved								
					Slightly			Moderately			High Degree		
		Clean (basic values)	Slightly Obstructed	Moderately Obstructed	Clean	Slightly Obstructed	Moderately Obstructed	Clean	Slightly Obstructed	Moderately Obstructed	Clean	Slightly Obstructed	Moderately Obstructed
Smooth lined	Minimum	10	15	25	20	25	35	25	30	40	35	40	50
	Average	15	20	30	25	30	40	30	35	45	40	45	55
	Maximum	20	25	35	30	35	45	35	40	50	45	50	60
Sedimentary rock (or coal)	Minimum	30	35	45	40	45	55	45	50	60	55	60	70
	Average	55	60	70	65	70	80	70	75	85	80	85	95
	Maximum	70	75	85	80	85	95	85	95	100	95	100	110
Timbered (5-ft centers)	Minimum	80	85	95	90	95	105	95	100	110	105	110	120
	Average	95	100	110	105	110	120	110	115	125	120	125	135
	Maximum	105	110	120	115	120	130	120	125	135	130	135	145
Igneous rock	Minimum	90	95	105	100	105	115	105	110	120	115	120	130
	Average	145	150	160	155	160	165	160	165	175	170	175	195
	Maximum	195	200	210	205	210	220	210	215	225	220	225	235

NOTE: All values of K are for air weighing 0.0750 lb/ft^3. Values in the table are expressed in whole numbers but must be multiplied by 10^{-10} to obtain the proper K value.
Conversion factors: 1 lb-min^2/ft^4 = 1.855 × 10^6 kg/m^3, 1 lb/ft^3 = 16.018 kg/m^3.
Source: McElroy 1935.

Values of $K \times 10^{10}$, lb-min^2/ft^4

Shock Losses

TABLE 14.5 Equivalent lengths for various sources of shock loss

Source	Feet	(Meters)	Source	Feet	(Meters)
Bend, acute, round	3	(1)	Contraction, gradual	1	(1)
Bend, acute, sharp	150	(45)	Contraction, abrupt	10	(3)
Bend, right, round	1	(1)	Expansion, gradual	1	(1)
Bend, right, sharp	70	(20)	Expansion, abrupt	20	(6)
Bend, obtuse, round	1	(1)	Splitting, straight branch	30	(10)
Bend, obtuse, sharp	15	(5)	Splitting, deflected branch (90°)	200	(60)
Doorway	70	(20)	Junction, straight branch	60	(20)
Overcast	65	(20)	Junction, deflected branch (90°)	30	(10)
Inlet	20	(6)	Mine car or skip (20% of airway area)	100	(30)
Discharge	65	(20)	Mine car or skip (40% of airway area)	500	(150)

Source: Hartman, Mutmansky, Ramani, and Wang 1997 *(reprinted with permission of John Wiley & Sons).*

FANS

TABLE 14.6 Fan laws

Variance in Performance Characteristics	Law 1, with Speed Change, n (D and w constant)	Law 2, with Size Change, D (w and Dn constant)	Law 3, with Specific Weight Change, w (n and D constant)
Quantity, Q	Directly	As square	Constant
Head, H_s or H_t	As square	Constant	Directly
Power, P_a or P_m	As cube	As square	Directly
Efficiency, η	Constant	Constant	Constant

Source: Hartman, Mutmansky, Ramani, and Wang 1997 *(reprinted with permission of John Wiley & Sons).*

For speed change

$$Q_2 = (n_2/n_1)Q_1 \quad \text{(EQ 14.10)}$$

$$H_2 = (n_2/n_1)^2 H_1 \quad \text{(EQ 14.12)}$$

$$P_2 = (n_2/n_1)^3 P_1 \quad \text{(EQ 14.14)}$$

For air density change

$$Q_2 = Q_1 \quad \text{(EQ 14.11)}$$

$$H_2 = (w_2/w_1) H_1 \quad \text{(EQ 14.13)}$$

$$P_2 = (w_2/w_1) P_1 \quad \text{(EQ 14.15)}$$

where:
- Q = air quantity in cubic feet per minute (cubic meters per second)
- n = fan speed in revolutions per minute
- H = fan head in inches of water (Pa)
- P = horsepower (kW)
- w = density of air in pounds per cubic foot (kilograms per cubic meter)

Horsepower Requirements (after Ramani 1992)

$$AHP = H_L Q / 6{,}350 \quad \text{(EQ 14.16a)}$$

$$AHP = H_L Q / 1{,}000 \quad \text{(EQ 14.16b; SI units)}$$

where:
- AHP = air horsepower (kW)
- H_L = total head loss in inches of water (Pa)

NATURAL VENTILATING PRESSURE (AFTER HARTMAN, MUTMANSKY, RAMANI, AND WANG 1997)

$$H_n = 0.255\, B\, L\, (1/T_1 - 1/T_2)$$ (EQ 14.17)

where:
- H_n = natural ventilating pressure in inches of water (Pa)
- B = average absolute pressure in inches of mercury (Pa)
- L = vertical height of air columns in feet (meters)
- T_1 and T_2 = the average absolute temperatures of the air columns in °F (°C)

PARALLEL FLOW

See Chapter 4 for more information.

$$Q_n = Q(R_{eq}/R_n)^{1/2}$$ (EQ 14.18; after Hartman, Mutmansky, Ramani, and Wang 1997)

where:
- Q_n = quantity in particular entry n
- Q = total quantity of common entries
- R_{eq} = equivalent resistance of common entries
- R_n = resistance of entry n

$$R_2 = R_1 / N^2$$ (EQ 14.19; after Kingery 1960)

where:
- R_1 = the original resistance for n_1 number of entries
- R_2 = the resistance for n_2 number of entries
- N = the ratio n_2/n_1

NATURAL SPLITTING (AFTER KINGERY 1960)

Formula for Potential Air Splitting

$$F_n = A_n (A_n/L_n O_n)^{1/2}$$ (EQ 14.20)

where:
- A_n = cross-sectional area of airway in ft²
- L_n = length of airway n in 1,000-ft units
- O_n = perimeter of airway n in feet
- F_n = the potential (no units) for split n

and

$$Q_n = [F_n/(F_1 + F_2 + \ldots F_n)]\, Q_{total}$$ (EQ 14.21)

where Q_n = quantity in the nth split in cubic feet per minute.

CONTROLLED SPLITTING (AFTER KINGERY 1960)

Regulator Formula

$$A = 40\, Q / H^{1/2}$$ (EQ 14.22)

where:
- A = area of the opening in ft²
- Q = quantity of air in the split in 100,000 cfm
- H = head loss through regulator in inches of water

MINE GASES

TABLE 14.7 Properties of mine gases

Gas	Chemical Symbol	Specific Gravity	Explosive Range	Health Hazards	Solubility	Color	Odor	Taste
Air	—	1.000	—	—	—	—	—	—
Oxygen	O_2	1.1054	Supports combustion	Oxygen deficiency: 17%—panting; 15%—dizziness and headache; 9%—unconsciousness; 6%—death	Moderate	—	—	—
Nitrogen	N_2	0.9674	—	Asphyxiant (oxygen depletion)	Slight	—	—	—
Carbon dioxide	CO_2	1.5291	—	Increases breathing rate; may cause death in high concentrations	Soluble	—	—	Acid in high concentrations
Carbon monoxide	CO	0.9672	12.5%–74.2%	Highly toxic; can be an asphyxiant.	Slight	—	—	—
Nitrogen dioxide	NO_2 N_2O_4	1.5894	—	Highly toxic; corrosive effect on lungs; can be an asphyxiant.	Only slight	Reddish brown	Blasting powder fumes	Blasting powder fumes
Hydrogen	H_2	0.0695	4.0%–74.2%; highly explosive	Asphyxiant (oxygen depletion)	—	—	—	—
Hydrogen sulfide	H_2S	1.1906	4.3%–45.5%	Highly toxic; can be an asphyxiant.	Soluble	—	Rotten eggs	Sweetish
Sulfur dioxide	SO_2	2.2638	—	Highly toxic; can be an asphyxiant.	Highly	—	Sulfurous	Acid (bitter)
Methane	CH_4	0.5545	5%–15%	Asphyxiant (rare)	Slight	—	—	—
Ethane	C_2H_6	1.0493	3.0%–12.5%	Asphyxiant (rare)	Slight	—	—	—
Propane	C_3H_8	1.5625	2.12%–9.35%	Asphyxiant (rare)	Slight	—	"Gassy" in high concentrations	—
Butane	C_4H_{10}	2.0100	1.86%–8.41%	Asphyxiant (rare)	Slight	—	"Gassy" in high concentrations	—
Acetylene	C_2H_2	0.9107	2.5%–80%	Only slightly toxic. Asphyxiant (rare).	Only slight	—	—	Garlic
Radon	Rn	7.526	—	Exposure to radiation	Highly	—	—	—

Source: Mine Safety and Health Administration (MSHA) undated.

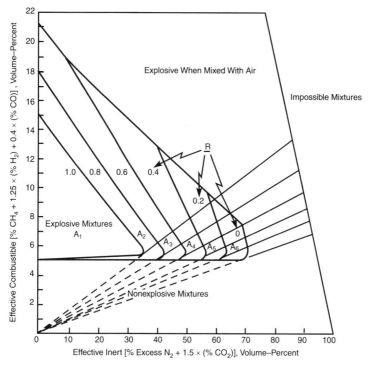

Source: Zabetakis, Stahl, and Watson 1959.

FIGURE 14.3 U.S. Bureau of Mines explosibility diagram

Percent excess nitrogen = %N_2 − 3.8% O_2 (EQ 14.23; Zabetakis, Stahl, and Watson 1959)

Effective inert = %N_2 + 1.5% CO_2 (EQ 14.24; Zabetakis, Stahl, and Watson 1959)

Effective combustible = %CH_2 + 0.4 (%CO) + 1.25 (%H_2) (EQ 14.25; Zabetakis, Stahl, and Watson 1959)

R = %CH_4/% effective combustible (EQ 14.26; Zabetakis, Stahl, and Watson 1959)

REGULATORY REQUIREMENTS

TABLE 14.8 Key ventilation requirements for underground coal mines

Last open crosscut, minimum	9,000 cfm
Each working face, minimum	3,000 cfm
Longwall intake, minimum	30,000 cfm
Mean entry air velocity, minimum	60 fpm
Trolley entry air velocity, maximum	250 fpm
Belt entry air velocity when using atmospheric monitoring system (fire detection), minimum	50 fpm
Ventilation of diesel engine exhausts	Nameplate quantity
Ventilation of multiple diesel exhausts	Sum of nameplate quantities
Gas quantities	
Oxygen, minimum	19.5%
Carbon dioxide, maximum	0.5%
No harmful quantities of noxious or poisonous gases	Threshold limited values
Actions for excessive methane in working places and intake airways:	
1.0% CH_4 or more	Changes or adjustments to less than 1.0%. Cut off power to energized face equipment.
1.5% CH_4 or more	Withdraw miners (except those adjusting ventilation). Cut off power to endangered area of the mine.

Source: Adapted from Code of Federal Regulations, Title 30, 75.300 to 75.389.

TUBING

$$H = C(V/4{,}005)^2 \qquad (EQ\ 14.27)$$

where:

H = head loss for fitting in inches of water
C = loss coefficient (from tables)
V = velocity in feet per minute

Type	Illustration	Conditions	Loss Coefficient
Gradual Contraction		θ	C_2
		30°	0.02
		45°	0.04
		60°	0.07
Equal Area Transformation	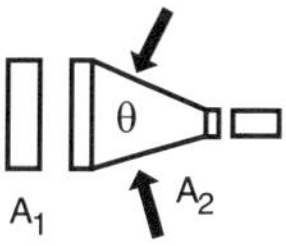	$A_1 = A_2$ $\theta \leq 14°$	C 0.15
Flanged Entrance	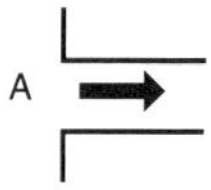	$A = \infty$	C 0.34
Duct Entrance	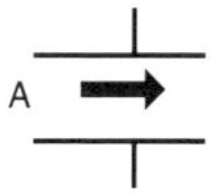	$A = \infty$	C 0.85
Formed Entrance		$A = \infty$	C 0.03
Gradual Expansion		θ	C_1
		5°	0.17
		7°	0.22
		10°	0.28
		20°	0.45
		30°	0.59
		40°	0.73
Abrupt Exit	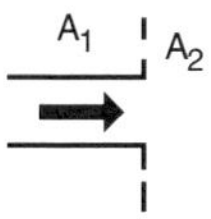	$A_2 = \infty$ $A_1/A_2 = 0.0$	1.00

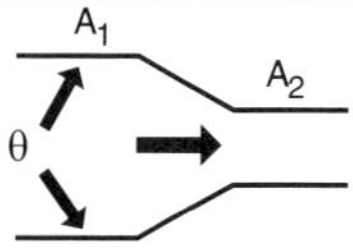

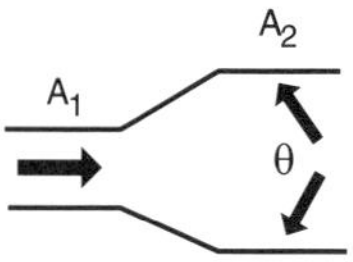

Note: A "C" with a subscript indicates the cross-section at which velocity is calculated.

Source: Schauenburg Flexadux Corporation *(undated; reprinted with permission)*.

FIGURE 14.4 Loss coefficients for area changes in ducts

Type	Illustration	Conditions	Pressure Loss Loss Coefficient	Pressure Loss L/D Ratio
N°	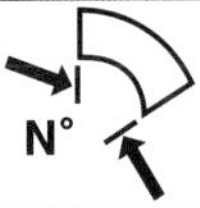	Rectangular or round, with or without vanes	(N/90) times value for similar 90° elbow	
90° Round Section		Miter	1.30	65
		R/D = 0.5	0.90	
		0.75	0.45	23
		1.0	0.33	17
		1.5	0.24	12
		2.0	0.19	10

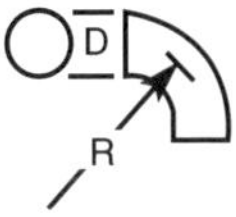

Source: Schauenburg Flexadux Corporation *(undated; reprinted with permission)*.

FIGURE 14.5 Loss coefficients for elbows in ducts

TABLE 14.9 Equivalent lengths for vent pipe and tubing

Source	Equivalent Length, Feet (Meters)
Entrance loss*	35 (10.7)
Exit loss*	100 (30.5)
Bend (generic), 90°	35 (10.7)
Bend (generic), 60°	24.5 (7.5)
Bend (generic), 45°	17.5 (5.3)
Coupling, spiral-reinforced duct	8 (2.4)
Fiberglass duct	
90° round sharp† bend	23D (7D)
90° round smooth‡ bend	12.5D (3.8D)
90° oval sharp† vertical bend	20D (6.1D)
90° oval smooth‡ vertical bend	10D (3.0D)
90° oval sharp† horizontal bend	29D (8.8D)
90° oval smooth‡ horizontal bend	14D (4.3D)
45° bend	50% of 90° value
60° bend	70% of 90° value
Lay-flat tubing	
Coupling	6 (2)
90° bend§	150 (46)
45° bend	50% of 90° value
60° bend	70% of 90° value

* Used only if fan is not located at the point of loss.
† Sharp bend: duct centerline radius = 0.75D.
‡ Smooth bend: duct centerline radius = 1.25D.
§ Duct centerline radius = D.
NOTES: (1) Most values derived from ABC Industries, Inc., laboratory measurements. (2) The table variable D refers to duct diameter.
Source: Hartman, Mutmansky, Ramani, and Wang 1997 *(reprinted with permission of John Wiley & Sons)*.

VELOCITIES FOR MINERAL DUSTS AND GASES

TABLE 14.10 Capture velocities for mineral dusts

Location	Toxicity	Capture Velocity (fpm)
Low speed conveyor transfer	Low	100
	High	200
Welding	Low	100
	High	200
Conveyor Loading	Low	200
	High	500
Grinding	Low	500
	High	2,000
Degreasing	Low	50
	High	100

Source: American Conference of Governmental Industrial Hygienists, Committee on Industrial Ventilation 1995.

TABLE 14.11 Conveying velocities of dusts and gases (in feet per minute)

Gases	1,000 to 1,200
Light dust	2,000 to 2,500
Mineral dusts	3,500 to 4,000

Source: American Conference of Governmental Industrial Hygienists, Committee on Industrial Ventilation 1995.

EXPLOSION-RESISTANT SEALS (APPROVED BY THE MINE SAFETY AND HEALTH ADMINISTRATION [MSHA])

- OMEGA 384 Block Seal
- CELUSEAL
- MICON 550
- MESHBLOCK
- Solid Concrete Block Seal
- RIBFILL
- ROCKFAST M-FGL (Hydroseal)
- TEKSEAL
- TEKGROUT
- SUPERBLOCK
- Wood Crib Block
- Packsetter Bags.

For more information on specific seal construction requirements, call the Mine Safety and Health Administration's Technical Support line at (412) 386-6936.

REFERENCES

American Conference of Governmental Industrial Hygienists, Committee on Industrial Ventilation. 1995. Lansing, Michigan.

Carrier Corporation. 1959. Company literature. Syracuse, NY.

Code of Federal Regulations, Title 30. 1997. Mineral Resources. Part 75–Mandatory Safety Standards–Underground Coal Mines. Subpart D–Ventilation. Washington, DC: U.S. Government Printing Office. 460–494.

Hartman, H.L., J.M. Mutmansky, R.V. Ramani, and Y.J. Wang. 1997. *Mine Ventilation and Air Conditioning.* 3rd ed. New York: John Wiley & Sons.

Kingery, D.S. 1960. *Introduction to Mine Ventilating Principles and Practices.* Bulletin 589. Washington, DC: U.S. Bureau of Mines.

Madison, R.D. 1949. *Fan Engineering.* 5th ed. Buffalo, NY: Buffalo Forge (now Howden Buffalo, Inc., Camden, SC).

Marvin, C.F. 1915. *Psychrometric Tables.* Bulletin 235. Washington, DC: U.S. Weather Bureau.

McElroy, G.E. 1935. *Engineering Factors in the Ventilation of Metal Mines.* Bulletin 385. Washington, DC: U.S. Bureau of Mines.

MSHA. Undated. *Mine Rescue Training Handbook, Visual 9.* Washington, DC: U.S. Department of Labor.

Ramani, R.V. 1992. Mine ventilation. In *SME Mining Engineering Handbook.* 2nd ed., Vol. 1. Edited by H.L. Hartman. Littleton, CO: Society for Mining, Metallurgy, and Exploration, Inc. 1052–1092.

Schauenburg Flexadux Corporation. Undated. *Ventilation, Designing a Mine Auxiliary Ventilation System.* Grand Junction, CO: Schauenburg Flexadux Corporation.

Zabetakis, M.G., R.W. Stahl, and H.A. Watson. 1959. *Determining the Explosibility of Mine Atmospheres.* Information Circular 7901. Washington, DC: U.S. Bureau of Mines.

CHAPTER 15

Pumping

Daniel F. Kump, P.E.

INTRODUCTION

Pumps move water out of the way and supply water for mining and milling operations. The material presented here applies to centrifugal pumps, which are probably the most popular type of pump used in mining.

CENTRIFUGAL PUMPS

A single-stage (single-impeller) pump is normally appropriate for high volume and low to moderate heads. For heads greater than about 250 ft (76.2 m), multistage pumps are generally used. A multistage pump has two or more stages and is essentially for high-head pumping (Warner 1992).

CALCULATING DYNAMIC HEAD

To select a pump, determine the flow rate in gallons per minute (gpm) (liters per second) and the total dynamic head in feet (meters). Because the dynamic head depends on the flow rate, this is an iterative process. Generally, only the elevation and friction loss components are needed to calculate head. The elevation head is the difference in elevation of the pump's discharge and the delivery point. The friction head is the amount of head used in overcoming friction in the pump's discharge pipeline.

$$\text{Head} = H_{el} + H_f \quad \text{(EQ 15.1)}$$

where:
Head = total dynamic head in feet (meters)
H_{el} = elevation head in feet (meters)
H_f = friction head in feet (meters)

In most cases, head to create velocity of flow is negligible and thus is often ignored (Bise 1986). Properly designed pipelines limit friction loss. The Hazen-Williams Equation (15.2) is commonly used to calculate pipeline friction losses (Warner 1992).

$$H_{f100} = \frac{K(Q/C)^{1.852}}{D^{4.87}} \quad \text{(EQ 15.2)}$$

where:
H_{f100} = friction loss per 100 ft (100 m) of pipe
K = constant = 1,045 for English units or 1.22×10^{12} for SI units

Q = flow rate in gallons per minute (liters per second)
C = retardation coefficient = 120 for coated steel or 150 for polyvinyl chloride (PVC)
D = inside pipe diameter in inches (millimeters)

Table 15.1 contains pipe friction loss data, and Table 15.2 lists equivalent lengths of pipe for some common fittings and valves. Generally, friction losses caused by fittings and valves are negligible in long pipelines.

TABLE 15.1 Pipe friction loss

Flow Rate (gpm)*	Friction Loss							
	2-in. i.d.,[†] HDPE[‡]	2-in. i.d., steel	4-in. i.d., HDPE	4-in. i.d., steel	6-in. i.d., HDPE	6-in. i.d., steel	8-in. i.d., HDPE	8-in. i.d., steel
	(foot per 100 feet)							
25	1.3	2.0	0.0	0.1				
50	4.8	7.1	0.2	0.2				
75	10.1	15.1	0.3	0.5	0.0	0.1		
100	17.3	25.8	0.6	0.9	0.1	0.1		
125	26.1	39.1	0.9	1.3	0.1	0.2		
150	36.7	55.0	1.2	1.9	0.2	0.3	0.0	0.1
175	48.9	73.2	1.6	2.5	0.2	0.3	0.1	0.1
200	62.6	93.8	2.1	3.2	0.3	0.4	0.1	0.1
300			4.5	6.7	0.6	0.9	0.2	0.2
400			7.7	11.5	1.1	1.6	0.3	0.4
500			11.6	17.4	1.6	2.4	0.4	0.6
750			24.7	36.9	3.4	5.1	0.8	1.2
1,000			42.1	63.1	5.8	8.7	1.4	2.1
1,500					12.3	18.5	3.0	4.5
2,000					21.0	31.5	5.1	7.7
2,500					31.8	47.7	7.8	11.7
3,000					44.7	67.0	10.9	16.4
4,000					76.3		18.7	28.0
5,000							28.3	42.4

* gpm = gallons per minute.
† i.d. = inside diameter.
‡ HDPE = high-density polyethelene.

TABLE 15.2 Equivalent feet of straight pipe for fittings

Fitting	Inside Diameter of Fitting			
	2 in.	4 in.	6 in.	8 in.
90° elbow	5.0	10	15	20
45° elbow	2.5	5.0	8.0	10
Tee straight	3.5	7.0	10	14
Tee side	10	20	30	40
Gate valve, open	1.5	2.6	4.0	5.0
Swing check valve, open	9.0	17	25	35

Source: After Crane Company 1988.

NET POSITIVE SUCTION HEAD (NPSH)

Pumps do not pull fluid up a suction pipe; atmospheric or external pressure pushes it up the suction pipe of a pump (Bise 1986). Atmospheric pressure provides most of a pump's NPSH. At sea level, atmospheric pressure is 14.7 psi (34 ft of water head; some consider 23 ft to be the practical limit of suction lift; Warner 1992). Table 15.3 contains equivalent feet of suction head at various elevations.

PUMP CHARACTERISTIC CURVES

Pump manufacturers publish pump characteristic curves (pump curves) that show the relationship between pump discharge rate in gallons per minute (liters per second) and head in feet (meters). See Figure 15.1 for a typical pump curve. Frequently, manufacturers also show pump efficiency and horsepower on pump curves.

Series Pumps

By connecting pumps in series, the pump combination's head is increased. To plot the curve for a series pump arrangement, add the pumps' heads at a given flow rate.

Parallel Pumps

By connecting pumps in parallel, the pump combination's flow is increased. To plot the curve for a parallel pump arrangement, add the pumps' flow rates at a given head.

TABLE 15.3 Equivalent feet of suction head available at various elevations, water at 750°F

Elevation (feet)	Elevation (meters)	Head of water (feet)
0	0	34.0
2,000	609.6	31.6
4,000	1,219.2	29.4
6,000	1,028.8	27.3
8,000	2,438.4	25.2
10,000	3,048.0	23.4
15,000	4,572.0	19.1

Source: Adapted from Heald 1992.

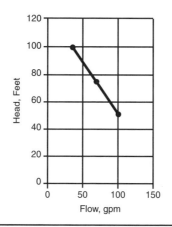

FIGURE 15.1 Typical pump characteristic curve

Best Efficiency Point (BEP)
A pump is designed to operate at its BEP, which is the flow–head point on its curve at which the pump operates at its greatest efficiency.

Allowable Operating Range
Although a pump is designed to operate at the BEP, pump manufacturers often provide greater flexibility by specifying an allowable working range for the pump.

System Head Curve
A system head curve is a plot of the expected head at which one plans to run a pump. The intersection of the system head curve and the pump curve is the operating point at which the pump is forecast to run.

PUMPING POWER FORMULAS

$$\text{Brake horsepower (BHP)} = \frac{Q \times H}{3,960 \times \text{Eff}_p} \quad \text{(EQ 15.3; Warner 1992)}$$

where:
- BHP = brake horsepower
- Q = flow rate in gallons per minute
- H = head in feet
- Eff_p = pump efficiency as decimal

Equation 15.3 can be rearranged to solve directly for Q and H.

$$\text{kw Input} = \frac{\text{BHP} \times 0.746}{\text{Eff}_m} \quad \text{(EQ 15.4)}$$

where:
- kw Input = kilowatt input to the motor
- BHP = brake horsepower
- 0.746 = conversion factor (1 hp = 0.746 kw)
- Eff_m = motor efficiency as decimal

Efficiency Values
Both pump and motor efficiency values are available from the manufacturers. Use the following values as rules of thumb.

- Eff_p = pump efficiency = 70% (use as 0.70 in the above formulas)
- Eff_m = motor efficiency = 90% for a 1,780-rpm motor (use as 0.90 in the formulas)
- Eff_m = motor efficiency = 85% for a 3,500-rpm motor (use as 0.85 in the formulas)

FLOW RATE FORMULA AND OTHER FACTS
See Chapter 4 on physical science and engineering for more information.

$$Q = V \times A \quad \text{(EQ 15.5)}$$

where:
- Q = flow rate in cubic feet per second
- V = velocity in feet per second
- A = area in square feet
- 1 cfs = 448.83 gpm
- 1 psi = 2.31 ft of water
- 1 ft^3 = 7.48 gal.

AFFINITY LAWS

When the speed of a centrifugal pump is changed, a pump's operation changes according to three fundamental laws (Bise 1986):

1. Q varies directly as speed.
2. H varies as the square of speed.
3. BHP varies as the cube of speed.

REFERENCES

Bise, C.J. 1986. *Mining Engineering Analysis.* Littleton, CO: Society of Mining Engineers, Inc. 59–70.

Crane Company. 1988. *Flow of Fluids*. Technical Paper No. 410. Stamford, CT: Crane Company.

Heald, C.C. 1992. *Cameron Hydraulic Data.* 17th ed., second printing. Woodcliff Lake, NJ: Ingersoll-Rand.

Warner, R.C. 1992. Design and management of water and sediment control systems. In *SME Mining Engineering Handbook*. 2nd ed., Vol. 1. Edited by H.L. Hartman. Littleton, CO: Society for Mining, Metallurgy, and Exploration, Inc. 1158–1169.

CHAPTER 16

Power: Electrical and Compressed Air

Daniel F. Kump, P.E.

ELECTRICAL POWER

See Chapter 4, which covers physical science and engineering, for more information on electricity.

Miscellaneous

- Synchronous motor efficiency: depending on their rating, unity power factor synchronous motors will have a full load efficiency of 93% to more than 97%. Motors with an 80% power factor will have lower efficiencies (Gibbs 1971).
- Values for an induction motor's efficiency:
  - Use 90% for a 1,780-rpm motor (use as 0.90 in the formulas).
  - Use 85% for a 3,500-rpm motor (use as 0.85 in the formulas).
- When sizing a back-up generator, allow one kilowatt per horsepower of load.
- When sizing a transformer, allow one kilovolt-ampere per horsepower of load.

TABLE 16.1 Allowable ampacities

Allowable Ampacities of Insulated Conductors Rated 0 Through 2000 Volts, 60°C Through 90°C (140°F Through 194°F) Not More Than Three Current-Carrying Conductors in Raceway, Cable, or Earth (Directly Buried), Based on Ambient Temperature of 30°C (86°F)							
Size	Temperature Rating of Conductor						Size
	60°C (140°F)	75°C (167°F)	90°C (194°F)	60°C (140°F)	75°C (167°F)	90°C (194°F)	
	Types TW, UF	Types FEPW, RH, RHW, THHW, THW, THWN, XHHW, USE, ZW	Types TBS, SA, SIS, FEP, FEPB, MI, RHH, RHW-2, THHN, THHW, THW-2, THWN-2, USE-2, XHH, XHHW, XHHW-2, ZW-2	Types TW, UF	Types RH, RHW, THHW, THW, THWN, XHHW, USE	Types TBS, SA, SIS, THHN, THHW, THW-2, THWN-2, RHH, RHW-2, USE-2, XHH, XHHW-2, ZW-2	
AWG or kcmil	Copper			Aluminum or Copper-clad Aluminum			AWG or kcmil
18	—	—	14	—	—	—	—
16	—	—	18	—	—	—	—
14	20	20	25	—	—	—	—
12	25	25	30	20	20	25	12
10	30	35	40	25	30	35	10
8	40	50	55	30	40	45	8

continues next page

TABLE 16.1 Allowable ampacities (continued)

Allowable Ampacities of Insulated Conductors Rated 0 Through 2000 Volts, 60°C Through 90°C (140°F Through 194°F) Not More Than Three Current-Carrying Conductors in Raceway, Cable, or Earth (Directly Buried), Based on Ambient Temperature of 30°C (86°F)

Size	Temperature Rating of Conductor						Size
	60°C (140°F)	75°C (167°F)	90°C (194°F)	60°C (140°F)	75°C (167°F)	90°C (194°F)	
	Types TW, UF	Types FEPW, RH, RHW, THHW, THW, THWN, XHHW, USE, ZW	Types TBS, SA, SIS, FEP, FEPB, MI, RHH, RHW-2, THHN, THHW, THW-2, THWN-2, USE-2, XHH, XHHW, XHHW-2, ZW-2	Types TW, UF	Types RH, RHW, THHW, THW, THWN, XHHW, USE	Types TBS, SA SIS, THHN, THHW, THW-2, THWN-2, RHH, RHW-2, USE-2, XHH, XHHW, XHHW-2, ZW-2	
AWG or kcmil	Copper			Aluminum or Copper-clad Aluminum			AWG or kcmil
6	55	65	75	40	50	60	6
4	70	85	95	55	65	75	4
3	85	100	110	65	75	85	3
2	95	115	130	75	90	100	2
1	110	130	150	85	100	115	1
1/0	125	150	170	100	120	135	1/0
2/0	145	175	195	115	135	150	2/0
3/0	165	200	225	130	155	175	3/0
4/0	195	230	260	150	180	205	4/0
250	215	255	290	170	205	230	250
300	240	285	320	190	230	255	300
350	260	310	350	210	250	280	350
400	280	335	380	225	270	305	400
500	320	380	430	260	310	350	500
600	355	420	475	285	340	385	600
700	385	460	520	310	375	420	700
750	400	475	535	320	385	435	750
800	410	490	555	330	395	450	800
900	435	520	585	355	425	480	900
1,000	455	545	615	375	445	500	1,000
1,250	495	590	665	405	485	545	1,250
1,500	520	625	705	435	520	585	1,500
1,750	545	650	735	455	545	615	1,750
2,000	560	665	750	470	560	630	2,000
Correction Factors							
Ambient Temp. (°C)	For ambient temperature other than 30°C (86°F), multiply the allowable ampacities shown above by the appropriate factor shown below.						Ambient Temp. (°F)
21–25	1.08	1.05	1.04	1.08	1.05	1.04	70–77
26–30	1.00	1.00	1.00	1.00	1.00	1.00	78–86
31–35	0.91	0.94	0.96	0.91	0.94	0.96	87–95
36–40	0.82	0.88	0.91	0.82	0.88	0.91	96–104
41–45	0.71	0.82	0.87	0.71	0.82	0.87	105–113
46–50	0.58	0.75	0.82	0.58	0.75	0.82	114–122
51–55	0.41	0.67	0.76	0.41	0.67	0.76	123–131
56–60	—	0.58	0.71	—	0.58	0.71	132–140
61–70	—	0.33	0.58	—	0.33	0.58	141–158
71–80	—	—	0.41	—	—	0.41	159–176

Source: National Fire Protection Association (NFPA) 1998. (Reprinted with permission from NFPA 70–1999, National Electric Code®, Copyright © 1998, National Fire Protection Association, Quincy, MA 02269. This reprinted material is not the complete and official position of the NFPA on the referenced subject, which is represented only by the standard in its entirety.)

TABLE 16.2 Power factor of typical alternating current loads

Load	Approximate power factor
Near-Unity Power Factor	
Incandescent lamps	1.0
Fluorescent lamps (with built-in capacitor)	0.95–0.97
Resistor heating apparatus	1.0
Synchronous motors (also built for leading power factor operation)	1.0
Rotary converters	1.0
Lagging Power Factor	
Induction motors (at rated load)	
Split-phase, below 1 hp	0.55–0.75
Split-phase, 1–10 hp	0.75–0.85
Polyphase, squirrel cage	
High-speed, 1–10 hp	0.75–0.90
High-speed, 10 hp and larger	0.85–0.92
Low-speed	0.70–0.85
Wound rotor	0.80–0.90
Groups of induction motors	0.50–0.85
Welders	
Motor-generator type	0.50–0.60
Transformer type	0.50–0.70
Arc furnaces	0.80–0.90
Induction furnaces	0.60–0.70
Leading Power Factor	
Synchronous motors	0.9, 0.8, 0.7, 0.6, etc. leading power factor depending on the rated leading power factor for which they are built
Synchronous condensers	Nearly zero leading power factor; output practically all leading reactive kva
Capacitors	Zero leading power factor; output practically all leading reactive kva

Source: Bise 1986.

COMPRESSED AIR

See Chapter 4, which covers physical science and engineering, for more information on gases; see particularly the section on thermodynamics and heat transfer.

Use the following equation to determine the pipe diameter for compressed air lines:

$$D = [(V^2 L)/(2{,}000(P_1^2 - P_2^2))]^{0.2} \qquad \text{(EQ 16.1)}$$

where:
    D = pipe diameter, in.
    V = volume of free air, cfm
    L = pipe length, ft
    P_1 = absolute pressure at beginning of pipe, psig
    P_2 = absolute pressure at end of pipe, psig

TABLE 16.3 Electrical formulas

Required	Direct Current	Alternating Current — Single-Phase	Alternating Current — 2-Phase, 4-Wire*	Alternating Current — 3-Phase
Amperes when horsepower is known	$\dfrac{746(hp)}{(E)(eff)}$	$\dfrac{746(hp)}{(E)(eff)(pf)}$	$\dfrac{746(hp)}{2(E)(eff)(pf)}$	$\dfrac{746(hp)}{1.73(E)(eff)(pf)}$
Amperes when kilowatts are known	$\dfrac{1{,}000(kw)}{E}$	$\dfrac{1{,}000(kw)}{(E)(pf)}$	$\dfrac{1{,}000(kw)}{2(E)(pf)}$	$\dfrac{1{,}000(kw)}{1.73(E)(pf)}$
Amperes when kilovolt-ampere is known		$\dfrac{1{,}000(kva)}{E}$	$\dfrac{1{,}000(kva)}{2(E)}$	$\dfrac{1{,}000(kva)}{1.73(E)}$
Power kilowatts	$\dfrac{(I)(E)}{1{,}000}$	$\dfrac{(I)(E)(pf)}{1{,}000}$	$\dfrac{2(I)(E)(pf)}{1{,}000}$	$\dfrac{1.73(I)(E)(pf)}{1{,}000}$
Power kilovolt-ampere		$\dfrac{(I)(E)}{1{,}000}$	$\dfrac{2(I)(E)}{1{,}000}$	$\dfrac{1.73(I)(E)}{1{,}000}$
Power output horsepower	$\dfrac{(I)(E)(eff)}{746}$	$\dfrac{(I)(E)(pf)(eff)}{746}$	$\dfrac{2(I)(E)(pf)(eff)}{746}$	$\dfrac{1.73(I)(E)(pf)(eff)}{746}$

I = amperes; E = volts; eff = efficiency (as a decimal); hp = horsepower; pf = power factor; kw = kilowatts; kva = kilovolt-amperes.
If actual voltage is above or below rated value, the correct value should be used in the above formulas.
* In 2-phase, 3-wire circuits the current in the common conductor is 1.41 times that in either of the other conductors.

TABLE 16.4 Air consumption multipliers for operation of rock drills (based on 80 to 100 psig air pressure)

Altitude (ft)	0	1,000	2,000	3,000	4,000	5,000	6,000	7,000	8,000	9,000	10,000	12,500	15,000
Multiplier	1.0	1.02	1.05	1.08	1.11	1.14	1.18	1.22	1.26	1.30	1.34	1.46	1.58

Source: Bise 1986.

TABLE 16.5 Multiplier for air consumption by number of rock drills

Number of drills	1	2	3	4	5	6	7	8	9	10	12	15	20	30	40	50	70
Multiplier	1.0	2.0	3.0	4.0	5.0	6.0	6.8	7.5	8.2	9.0	10.5	12.6	16.0	23.5	31.0	38.0	52.5

Source: Bise 1986.

TABLE 16.6 Factors for correcting capacity of single-stage compressors (based on 7% cylinder clearance)

Altitude (ft)	90 psig factor	100 psig factor
Sea level	1.000	1.000
1,000	0.988	0.987
2,000	0.972	0.972
3,000	0.959	0.957
4,000	0.944	0.942
5,000	0.931	0.925
6,000	0.917	0.908
7,000	0.902	0.890
8,000	0.886	0.873
9,000	0.868	0.857
1,0000	0.853	0.840
11,000	0.837	—
12,000	0.818	—

Source: Bise 1986.

POWER: ELECTRICAL AND COMPRESSED AIR | 285

TABLE 16.7 Pressure loss in hose

Hose Length (inside diameter)	Free Air (cfm)	Line Pressure (psig)						
		60	80	100	120	150	200	300
50 ft; ¾ in.	60	3.1	2.4	2.0				
	80	5.3	4.2	3.5	2.9	2.4	1.8	1.2
	100	8.1	6.4	5.2	4.5	3.6	2.8	1.9
	120		9.0	7.4	6.3	5.1	3.9	2.7
	140		12.0	9.9	8.4	6.9	5.3	3.6
	160			12.7	10.8	8.9	6.8	4.6
	180				13.6	11.1	8.5	5.8
	200				16.6	13.5	10.4	7.1
	220					16.2	12.4	8.4
50 ft; 1 in.	120	2.7	2.1					
	150	4.1	3.2	2.7	2.3			
	180	5.8	4.6	3.8	3.2	2.6	2.0	1.3
	210	7.7	6.1	5.0	4.3	3.5	2.7	1.8
	240		7.9	6.5	5.5	4.5	3.4	2.3
	270		9.8	8.1	6.9	5.6	4.3	2.9
	300		12.0	9.9	8.4	6.9	5.3	3.6
	330			11.8	10.0	8.2	6.3	4.3
	360			13.9	11.9	9.7	7.4	5.0
	390				13.8	11.3	8.7	5.9
	420				15.9	13.0	10.0	6.8
	450					14.8	11.4	7.7
50 ft; 1¼ in.	200	2.4						
	250	3.7	2.9	2.4	2.0			
	300	5.2	4.1	3.4	2.9	2.3	1.8	1.2
	350	7.0	5.5	4.5	3.8	3.1	2.4	1.6
	400	8.9	7.0	5.8	4.9	4.0	3.1	2.1
	450		8.8	7.3	6.2	5.0	3.9	2.6
	500		10.8	8.9	7.6	6.2	4.7	3.2
	550			10.7	9.1	7.4	5.7	3.9
	600			12.6	10.7	8.7	6.7	4.6
	650			14.6	12.4	10.2	7.8	5.3
	700				14.3	11.7	9.0	6.1
	750					13.3	10.2	6.9
	800					15.0	11.5	7.8
50 ft; 1½ in.	300	2.1						
	400	3.7	2.9	2.4	2.0			
	500	5.6	4.4	3.7	3.1	2.5	1.9	1.3
	600	8.0	6.3	5.2	4.4	3.6	2.8	1.9
	700		8.5	7.0	5.9	4.9	3.7	2.5
	800		10.9	9.0	7.7	6.3	4.8	3.2
	900			11.2	9.5	7.8	6.0	4.1
	1,000			13.6	11.6	9.5	7.3	4.9
	1,100				14.0	11.4	8.8	6.0
	1,200					13.6	10.4	7.1
	1,300					15.8	12.1	8.3

Source: Bise 1986.

TABLE 16.8   Air requirements of representative drilling machines

Type of Machine	Hammer Diameter (in.)	Free Air Required (cfm)
Sinker	2 3/8	70
Sinker	2 1/2	95
Sinker	2 5/8	110
Sinker	2 3/4	115
Stoper	2 9/16	140
Stoper	2 3/4	160
Drifter	2 3/4	130
Drifter	3	140
Drifter	3 1/2	180
Drifter	4 1/2	200

Source: Bise 1986.

See Chapter 14 on Ventilation for more information; see Table 14.2 for barometric pressure, temperature, and air density at different altitudes.

REFERENCES

Bise, C.J. 1986. *Mining Engineering Analysis*. Littleton, CO: Society of Mining Engineers, Inc.
Gibbs, C.W. 1971. *Compressed Air and Gas Data*. 2nd ed. Woodcliff Lake, NJ: Ingersoll-Rand.
NFPA. 1998. *National Electric Code 1999*. Quincy, MA: NFPA.

CHAPTER 17

Mineral Processing

Paul D. Chamberlin, P.E.

CRUSHING
by Paul D. Chamberlin, P.E.

Guidelines
All crushers work best with a constant feed rate, meaning that surge capacity may be needed between crushing stages. In addition, all crushers work best if fines (product size and less) are removed from feed.

Overall reduction ratio: $R_T = R_1$ (1st stage) $\times R_2$ (2nd stage) $\times R_3$ (3rd stage) ......

Change the shape of the crushing chamber in jaw and cone crushers to suit the size distribution of the ore and get better wear and energy usage rates.

$$\text{Circulating load (CL) in \% of original feed} = 100 / [(e/r) - 1] \qquad \text{(EQ 17.1)}$$

where:
    e = % screen efficiency
    r = % oversize in crusher product

This formula is more accurate for +12 mm crushing and less accurate for –1.7 mm crushing: CL in tonnes = (total feed to crusher) – (original crusher feed). Determine circulating load at steady state by stopping all conveyors at the same time and weighing ±2 m of ore taken from feed, product, and recycle conveyors. Calculate CL.

Rod mill feed should be finer than F_{80} = 33 mm. Primary ball mill feed should be finer than F_{80} = 12.7 mm.

Crusher discharge is typically 100%, passing two times the closed-side setting (CSS); e.g., if CSS = 12.7 mm, the discharge will be 100% –25.4 mm.

Types of Crushers
Jaw   A very general estimate of crusher capacity is given by:

$$Q = 1.5F \qquad \text{(EQ 17.2)}$$

where:
    Q = m³/h
    F = discharge area in cm² at open setting

The reduction ratio is typically 3 or 4 to 1 but can be 6 or 8 to 1; use the double toggle style if abrasion index is >0.7. The maximum feed size is ≈90% of feed opening for single toggle and ≈80% for double toggle; do not choke feed.

Cone/Gyratory   The gyratory reduction ratio is usually 3 or 4 to 1 but can be 8 or 10 to 1; the largest rock in the feed should be 80%–90% of the open-side setting (OSS) of the feed opening. Fifty percent of the feed passing the midpoint setting of the crusher chamber is ideal; problems occur when >80% of the feed passes the midpoint. The midpoint opening is about 0.5 times (closed-side receiving opening plus CSS); should be choke-fed for lowest steel consumption, greatest reduction ratio, and lowest kilowatt-hours/ton. Reduce the circulating load by setting the CSS slightly smaller than the screen's opening; normally 60% of the crusher discharge is <CSS.

WaterFlush Cone Crusher   Because this type of crusher can produce <4 mm ore, they may be able to replace a rod mill. WaterFlush crushers are good for wet sticky ores and they can eliminate the need for third-stage screening. The ore may need to be pre-wetted before entering crusher. The water needed may give a crusher discharge of 30% solids (30% to 50% solids is usual). Because of this, the downstream cyclone overflow may be 25% solids, which is too low for leach or float, and a thickener may be needed before leaching or flotation. Testing requires a 10-tonne sample.

Vertical Shaft Impactor (VSI)   The feed should be <5 cm to avoid high wear on rotor tips; the minimum product size is typically <3.3 to 1.7 mm. VSI should operate in a closed circuit with very efficient screening to remove as much of the fines and moisture as possible from crusher feed. Fines make spongy rock beds and reduce crushing efficiency and increase power consumption, and clay lumps and wood plug the crusher. Add water to control dusting. VSIs need constant feed rate. The Bond Work Index is not applicable to VSIs. A larger-diameter rotor creates more fines (attrition) and increases power draw; extra throughput can be obtained by cascading a portion of the feed. Typically, circulating loads are ±300%. Wear steel costs from ≈$0.04/T to $0.80/T. The abrasion index should be <0.15. For impact crusher testwork by the vendor, >500 kg of sample is needed.

Horizontal Shaft Impactors   These impactors are used for low abrasive or friable ore. Some machines have internal rock pockets so crushing is mostly autogenous; others depend on anvils (rock on steel). Reduction ratios are high (up to 35). Hammer mills work on wet, sticky ores.

Rolls   Single, dual, or quad rolls are available; ore should be <172,000 kN/m² (<25,000 psi) compressive strength.

Mining Machinery Developments, Ltd. (MMD)   Because this type of crusher breaks rock in tension, not compression, it is best for soft ores or sticky ores.

Maintenance

Avoid bearing damage in some cone crushers by following this procedure:
- Dissipate bearing heat by idling the crusher without feed for 10–15 min before shutdown.
- On cold startup, idle without feed for 10 min and then slowly step up the clamping pressure over a period of 20 more min.

Motors should have reduced voltage starters (65% of voltage and 42% of amps); otherwise, inrush amps may be 5.6 × rated amps.

Liner consumption for cone crushers is given by (A_i values are given in Table 17.1):

$$\text{lb/ton passing through crusher} = 0.043 \, (A_i)^{1.255} \, (\text{head diameter})^{0.254} \quad \text{(EQ 17.3)}$$

where lb = total weight of the liners including unused waste portion at end of liner life; head diameter is in inches.

The type of cone crusher is not a factor.

Other Considerations

- The coarse ore storage capacity should be about 2 days.
- The feed pocket ahead of the primary crusher should have a capacity two times that of the ore truck.

- A grizzly is needed ahead of the primary jaw crusher. Keep it fully loaded to avoid spills.
- The operator should be positioned to allow visibility into the primary feeder.
- Rock boxes must be installed in chutes and transfer points.
- Belt scrapers must discharge onto a conveyor to avoid a mess.
- All electrical equipment should be housed.
- Crushing force monitors should be installed to help select the minimum CSS.

GRINDING
by Paul D. Chamberlin, P.E.

Guidelines
The energy to break rock is most effectively applied by blasting, then by crushing. Grinding is the least effective method for breaking rock. When grinding is necessary, however:
- Remove fine material from a grinding machine as quickly as possible.
- Consider screens instead of cyclones on mill discharges.
- Replace constant-speed cyclone feed pump motors with frequency-controlled adjustable speed drives.
- Incline cyclones to reduce pumping head.
- Return collected spills to the circuit gradually.
- Consider frequency-controlled adjustable speed drives on semiautogenous grinding (SAG) mills to accommodate variable-feed hardness.
- Note that chemicals such as surfactants and dispersants may reduce grinding power.

TABLE 17.1 Work index and abrasion indices (averages)

Material	W_i	A_i
Amphibolite	16 ± 3	0.2–0.45
Andesite	16 ± 2	0.5 ± 0.1
Basalt	20 ± 4	0.2 ± 0.1
Diabase	19 ± 4	0.3 ± 0.1
Diorite	19 ± 4	0.4
Dolomite	12 ± 3	0.01–0.005
Hematite ore	11 ± 3	0.5 ± 0.3
Magnetite ore	8 ± 3	0.2 ± 0.1
FeSi	11 ± 2	0.25 ± 0.7
Gabbro	20 ± 3	0.4 ± 0.1
Granite	16 ± 6	0.55 ± 0.1
Hornfels	18 ± 3	0.7 ± 0.2
Limestone	12 ± 3	0.001–0.03
Marble	12 ± 3	0.001–0.03
Porphyry	18 ± 3	0.1–0.9
Pyrite ore	10 ± 3	0.6 ± 0.2
Quartzite	16 ± 3	0.75 ± 0.1
Sandstone	10 ± 3	0.1–0.9
Syenite	19 ± 4	0.4 ± 0.1

Source: Ottergren and Steer 1996.

Types of Grinding Mills
Rod and Ball Mills

$$W = [10W_i / P^{0.5}] - [10W_i / F^{0.5}] \quad \text{(EQ 17.4)}$$

where:
- W = kWh per short ton
- W_i = work index = kWh/st to reduce one ton from infinite size to 80% passing 100 μ
- P = 80% passing size of product
- F = 80% passing size of feed

$$\text{Actual grinding mill hp} = 1.341 \, (W) \, (t) \, (EF_t) \quad \text{(EQ 17.5)}$$

where:
- 1.341 = hp/kWh; t = short tons per hour; EF_t = product of all efficiency factors
- EF_1 = dry grinding factor (rod or ball mills) = 1.3
- EF_2 = open circuit factor (ball mill only) = 1.04 to 1.7 (average 1.2)
- EF_3 = mill diameter (ft) factor: 3 ft = 1.217, 4 ft = 1.149, 5 ft = 1.099, 6 ft = 1.059, 7 ft = 1.027, 8 ft = 1.000, 9 ft = 0.977, 10 ft = 0.956, 11 ft = 0.938, 12 ft = 0.922, 13+ ft = 0.907
- EF_4 = oversize feed factor (rod or ball mills) = 1.000 (except when W_i is >14.0 and/or F is >16,000 μ (rod) or 4,000 μ (ball); call the factory in these cases). Alternately, $EF_4 = R_r + [W_i - 7] \, [\, (F_{80} - F_o) / F_o]$ where F_o = 16,000 μ $(13 / W_i^{0.5})$ for rod mills and F_o = 4,000 μ $(13 / W_i^{0.5})$ for ball mills
- EF_5 = fineness factor (ball mill only when P is <70 μ) = $[P + 10.3] / 1.145P$
- EF_6 = reduction ratio (rod mill) = 1.0 when R_r is 10 to 20; otherwise, call the factory
- EF_7 = reduction ratio (ball mill) = 1.0 when R_r is <5.0; otherwise, call the factory
- EF_8 = feed preparation factor = 1.4 if only a rod mill and open circuit crushing; 1.2 for rod + ball mill circuit if open circuit crushing; 1.2 if only a rod mill and closed circuit crushing; 1.0 for rod + ball mill circuits if closed circuit crushing.

To approximate plant operating work indices, multiply laboratory work indices for rod and SAG mills by 1.4. Wet grinding power is about 25% lower than that required for equivalent dry grinding, but media wear is five to seven times as much as dry grinding. For dry grinding, media consumption is much less; moisture should not exceed 1% unless special handling equipment is installed.

The rod mill feed should be <19 mm, and the ball mill feed should be <19 mm for soft ores and <7 mm for hard ores.

$$\text{Critical speed, rpm} = C_s = 42.305 / D^{0.5} \quad \text{(EQ 17.6)}$$

where D = diameter of the inside liners in m.

Mill power is nearly proportional to revolutions per minute (rpm); new liner designs allow for greater mill rpm; e.g., 20-ft-diameter mill at 77% of C_s, thus more throughput. Older %C_s rules of thumb are shown in Table 17.2.

TABLE 17.2 Critical-speed rules of thumb

Inside Diameter of Liners		Average %C_s	
Meters	Feet	Rod Mills	Ball Mills
0.91–1.83	3–6	76–73	80–78
1.83–2.74	6–9	73–70	78–75
2.74–3.66	9–12	70–67	75–72
3.66–4.57	12–15	67–64	72–69
4.57–5.49	15–18		69–66

Source: Rowland and Kjos 1978.

Overall availability for large ball mills can be as high as 98%. Ninety-five percent is more typical for rod mills and ball mills, and 85% to 90% is common for SAG mills.

Charge volume (% of mill filled with rods/balls) = 113 − 126(H/D) (EQ 17.7)

where:
- H = distance from top of inside liners to top of charge
- D = diameter inside the liners

The charge volume is usually 45% for wet grinding and 40% for overflow ball mills. It ranges between 28% and 35% for dry grinding. The void space in a rod charge is 30%, and is 42% in a ball charge. The weight of rods is ≈5.5 tonnes/m³; balls weigh ≈4.6 T/m³.

Rowland and Kjos (1978) define the largest diameter of rods or balls that should be added as makeup media as

Rods

$$R = 25.4\{(F^{0.75} / 160)[(W_i \times S_g) / (D^{0.5} \times \%C_s)]^{0.5}\}$$ (EQ 17.8)

where:
- R = rod diameter in mm
- F = 80% passing feed size in μ
- W_i = work index
- S_g = specific gravity of ore
- C_s = % of critical speed
- D = diameter inside the liners in m

Balls

$$B = 25.4\{(F / k)^{0.5}[S_g W_i / ((3.281D)^{0.5} \times \%C_s)]\}$$ (EQ 17.9)

where:
- B = ball diameter in mm
- k = 350 for wet overflow mills, 330 for wet diaphragm mills, and 335 for dry diaphragm mills. All other terms are the same as for rod diameter; 125 mm diameter is generally the largest used in ball mills and SAG mills. If there is no pebble crusher, the SAG grates are probably ≈25 mm and 125-mm balls would peen them shut, causing a big problem.

$$B = 6 \times \log d_k \times d^{0.5}$$ (EQ 17.10)

where:
- d_k = largest particle in the product in μ
- d = largest particle in the feed in mm

Ball and Liner Wear This guide is based on forged steel (Rowland and Kjos 1978):
- Wet rod mill rods: kg/kWh = 0.175 (A_i − 0.020)$^{.2}$
- Wet rod mill liners: kg/kWh = 0.175 (A_i − 0.015)$^{.3}$
- Wet ball mill balls: kg/kWh = 0.175 (A_i − 0.015)$^{.333}$
- Wet ball mill liners: kg/kWh = 0.013 (A_i − 0.015)$^{.3}$
- Approximate wear = 1.0 kg of balls per 15 kWh of power to mill
- The typical wear rate of steel balls in wet ball mills is 0.4 kg/tonne (range = 0.15 to 0.75). For high-chrome balls it is 0.1 kg/T (0.05 to 0.3); 12% chrome balls are probably the most economic; slugs may be an economic alternative.
- Grinding mill liners: 12% chrome is probably the most economic metal liner, but rubber is probably most economical for secondary grinding and regrinding ball mills. Because the wear rate of rubber in primary grinding mills can be high, consider steel-capped rubber liners.

SAG Mills

- Conduct tests on each ore type in the deposit: (1) 18-in.-diameter Aerofall grindability test on 250 kg of ore crushed no finer than 3.175 cm top size (supply coarser sample to laboratory); (2) Bond ball mill work index (12 kg of sample); and (3) Bond rod mill work index (20 kg of sample).
- To obtain the SAG mill kW, ±5%, use the same formula as for a ball mill but substitute the autogenous work index for the ball mill work index (MacPherson 1989, Austin 1990, and Scott and Barratt 1987).
- Maintain charge volume in the 30%–35% range.
- Assay and test SAG mill cobbles; are they worth crushing in semiautogenous, ball milling, crushing (SABC) circuit?
- Use screens or a rubber covered trommel rather than cyclones on mill discharge.
- To prevent mill overload, utilize power draw as a predictive tool:
  - Rising power at a fixed mill load suggests lowering the charge set point to get more impact grinding; it may also mean that the ore feed does not have enough large "media" and that a coarser blend of coarse and fine ore is needed.
  - When mill load does not decrease at maximum water input and the power draw is high, stop the mill feed for about 30 minutes.
  - Install an expert system.
- If the initial set of mill liners are troublesome, the initial grates may need to be changed to include pebble ports; coordinate with the supplier for fast delivery.
- In SABC circuits, set the crusher CSS as tight as possible.
- Install a second magnet and a metal detector to reject steel from the pebble load.
- Install a variable-speed spare pump with a short acceleration time on SAG mill discharge.
- The operating availability for a well-run SAG mill circuit should be 90%–92% or more.
- With a compression-type bolt sealer, retighten liner bolts after 1–2 hours of full load operation, again after 6–8 hours, and again after 5–7 days.

TABLE 17.3 Torque table

Bolt diameter, mm	Torque (kg-m)		
	Grade 2	Grade 5	Grade 8
32	117	201	293
38	180	252	401
44	281		638
51	423		961

Source: Kjos 1986.

- Shutdown sequence: (1) shut off feed, (2) shut off water, (3) allow mill to turn five revolutions to pump out slurry, (4) shut off power, and (5) hose and bar down overhead muck and balls. Do not grind out mill to inspect/maintain liners or to determine ball charge volume (bad for liners and grates).
- Wash muck from pebbles before conveying them to the SABC cone crusher.

Dry Grinding Systems

- Run mill at average 24% charge load.
- Do not try to grind finer than 1.7 mm.
- Limit velocities to 1,500 m/min in classifier ductwork.
- Use clean air fans, not dirty air fans. Ceramic line all dirty air fans.

High-Pressure Grinding Rolls (HPGRs) HPGRs must be choke-fed from an overhead hopper, and generally require <3% moisture in the ore. Less than a full feed rate causes excessive and uneven wear on the rolls.

HPGRs generate many microfractures that make subsequent grinding easier and provide higher extractions in subsequent leaching operations.

High steel wear is overcome by segmented roll faces that are partially covered with studs of very hard metal between which ore gets jammed; i.e., much of the roll face is rock, and grinding is partially autogenous.

Stirred Media Mills (Vertical, Tower)   These types of mills are used for fine wet or dry grinding and for regrinding.

For F_{80} <0.15 mm, stirred media mill's energy usage is ≈25% < ball mill's.

Because Bond energy equations show 20%–50% more power than is actually needed when grinding in 10-μ range, do not apply them directly.

For stirred media mills:

- Circulating load ≈300%–1,000%; % solids in mill ≈10%–70%; % solids in overflow ≈2%–45%; feed size 5 mm or less.
- Makeup balls generally 25 mm or less; maximum size = 32 mm. Total metal wear is less than for ball mills in the same application.
- Availability of scheduled operating time ≈95%–98%.

Vibrating Mills, Fluid Energy Mills, Jet Mills, Colloid Mills, Roller, Bowl, and Ring-Ball Mills

CLASSIFICATION
by Paul D. Chamberlin, P.E.

Guidelines

The main controls for most classifiers are (1) varying the percentage of solids, (2) varying the input and output volumes, and (3) varying the viscosity of slurry systems.

For air separators and classification below 250 μ, a unit with a rotor is preferable, and variable-speed drives are needed.

Circulating loads in high-efficiency air separators are 50%–400%; in traditional units they are 200%–1,000%.

Minimize turbulence in the pool area of spiral, rake, or drag classifiers.

For hydrocyclone information, see the section on solid–liquid separations. The best cyclone separations are obtained when the feed is ≈20% solids by weight. The D_{50} size is generally proportional to $1/Q^{0.5}$, where Q = the flow rate of the feed slurry.

Two stages of classification in series invariably give clean separations.

Use intermittent full flow discharge of solids from settling cones rather than continuous slow flows.

TABLE 17.4   Classifiers

Classifier	Maximum Feed Size	Feed Rate (tonne/h)	Volume % Solids Feed/Overflow/ Underflow	Power (kW)	Notes
Hydrocyclones	1,400 μ–45 μ	To 20 m³/min	4–35/2–15/30–50	35–400 kN/m² pressure head	Relatively efficient separations
Air separators	2 mm–38 μ	To 2,100		4–500	
Spiral, rake, drag	25 mm	5–850	Variable/2–20/45–65	0.5–110	Gives relatively clean sands; may eliminate pump in grinding circuit
Log washer	100 mm	40–450		10–60	Breaks agglomerates
Hydraulic bowl	12 mm	5–225	Variable/2–15/50–65	5–25	Gives very clean sands
Rake clarifier	25 mm	1–150	Variable/0.4–15/20–35	1–11	Gives very clean sands
Cone	6 mm	2–100	Variable/5–30/35–60	0	Simple
Elutriator	7.5 mm	4–120	15–35/0.4–5.0/20–35	1	Simple

Source: Adapted from Kelly and Spottiswood 1982.

Measurements of Performance for Air Classifiers (Klumpar 1987a and 1987b)

- Yield, or recovery, is the amount of product per unit of feed.

$$\text{Yield of fines} = Y_f = 100C(A - B) / A(C - B) \quad \text{(EQ 17.11)}$$

where:
- A = cumulative mass fraction of feed passing screen 'x'
- B = cumulative mass fraction of coarse product passing screen 'x'
- C = cumulative mass fraction of fines product passing screen 'x'

- Efficiency, E, is the difference between fine and coarse product yields.

$$E = Y_f - Y_c = 100(C - A)(A - B) / A(1 - A)(C - B) \quad \text{(EQ 17.12)}$$

- Selectivity, S, is the weight percentage of all particles of a given diameter in the feed, D, that go to the coarse product. The particles of a given diameter are defined as the narrow fraction passing one screen (for example 200 mesh, 74µ) but retained on the next smaller screen (250 mesh, 63µ). The average diameter of that fraction is the arithmetic mean or 68.5µ. Selectivity for particles of average diameter, D, is

$$S = b(c - a) / a(c - b) \quad \text{(EQ 17.13)}$$

where:
- a = narrow mass fraction of feed passing screen "x" and retained on next smaller screen
- b = narrow mass fraction of coarse product passing screen "x" and retained on next smaller screen
- c = narrow mass fraction of fines product passing screen "x" and retained on next smaller screen

- Performance is best characterized by the plot of selectivity versus average particle size, D. Plot $(c - a)$ versus $(c - b)$ for each D, draw a straight line through the points, and measure the line's slope, s. Typically, the ordinate is selectivity on a probability scale; average particle size, D, is the abscissa on a logarithmic scale. Ideal classification would result in a vertical line intersecting the curve at 50% selectivity.

$$s = (c - a) / (c - b) \quad \text{(EQ 17.14)}$$

SCREENING
by Paul D. Chamberlin, P.E.

Materials of Construction
- Woven wire.
- Polyurethanes (many varieties); best for most abrasive situations; good for wet screening; available in modular deck configurations that are interchangeable with woven wire panels; be sure to install thicker panels for deck sections consistently impacted; protect the screens when welding nearby.
- Rubber (many varieties); best in dry and high-impact areas.
- Perforated plate can be metal or covered with rubber or polyurethanes.
- Wedge wire; bar decks; grizzly bars.

Polyurethane and rubber screens have much less open area than woven wire screens.

Types of Screens
Screens are available in many configurations: horizontal vibrating; inclined vibrating; sieve bends, which are stationary; circular; wet or dry; divergator; grizzly; stationary; dewatering; scalping; washing; banana; interstage carbon screens; trommel screens; rotating probability (for wet screening fine materials); electrically heated screens to reduce plugging in winter (these

do not work well with >2%–3% moisture); Liwell (alternate flexing and tensioning of the synthetic deck prevents plugging); Bradford breakers for coal; multiple deck screens. Note that the stroke of a vibrating screen can be circular, elliptical, linear.

Typical screen availability is 92% to 96%.

TABLE 17.5 International standard and U.S. sieve sizes

Standard	U.S. Sieve	Standard	U.S. Sieve	Standard	U.S. Sieve
125.0 mm	5.000 in.	11.2 mm	0.438 in.	600 µ	No. 30
106.0 mm	4.240 in.	9.5 mm	0.375 in.	500 µ	No. 35
100.0 mm	4.000 in.	8.0 mm	0.313 in.	425 µ	No. 40
90.0 mm	3.500 in.	6.7 mm	0.265 in.	355 µ	No. 45
75.0 mm	3.000 in.	6.3 mm	0.250 in.	300 µ	No. 50
63.0 mm	2.500 in.	5.6 mm	No. 3.5	250 µ	No. 60
53.0 mm	2.120 in.	4.75 mm	No. 4	212 µ	No. 70
50.0 mm	2.000 in.	4.00 mm	No. 5	180 µ	No. 80
45.0 mm	1.750 in.	3.35 mm	No. 6	150 µ	No. 100
37.5 mm	1.500 in.	2.80 mm	No. 7	125 µ	No. 120
31.5 mm	1.250 in.	2.36 mm	No. 8	106 µ	No. 140
26.5 mm	1.060 in.	2.00 mm	No. 10	90 µ	No. 170
25.0 mm	1.000 in.	1.70 mm	No. 12	75 µ	No. 200
22.4 mm	0.875 in.	1.40 mm	No. 14	63 µ	No. 230
19.0 mm	0.750 in.	1.18 mm	No. 16	53 µ	No. 270
16.0 mm	0.625 in.	1.00 mm	No. 18	45 µ	No. 325
13.2 mm	0.530 in.	850 µ	No. 20	38 µ	No. 400
12.5 mm	0.500 in.	710 µ	No. 25		

Source: Matthews 1985.

Troubleshooting

- Distribute feed across the full width of the screen so that the entire screen is used. Poor feed distribution is a major cause of poor screening.
- Present the feed to the screen in the direction of its centerline; feed that approaches at an angle will segregate by size, with coarse particles migrating to the far side and fines remaining in the center. Without coarse particles to scrub the deck, damp sticky fines tend to blind openings.
- To reduce blinding, make screens more efficient, and increase screen's capacity, use woven wire screen with a large percentage of open area; i.e., smaller wire diameters rather than larger diameters. Use backing wire.
- Prevent blinding by having the openings in the sizing screen slightly smaller than the openings in the backing underneath.
- Replace screens with holes in them.
- If screen cloth tears down the support ribs, it is undertensioned.
- Ripping screen cloth along the hook strip is caused by tightening end-draw bolts first.
- Ensure that resonant frequency of the screen support structure is at least 3 times the operating frequency of the screen.
- Replace broken steel coil springs as a complete set rather than individually.
- Replace screen bearings in sets rather than individually.
- Install access around the screen so maintenance won't be ignored.
- All coil springs should be of same height and should not sag.
- Provide ≈6 cm clearance between the vibrating unit and any stationary structure. Start/stop the unit with and without load; note clearances, especially when stopping.

- Motor must be properly aligned and on a pivoting motor base. Are sheaves clean?
- Determine vibrator speed and stroke and compare with specifications.
- Overheating or noisy? Check for proper lubrication fluids. Check for failed bearings.
- Erratic motion? Check that screen is level; that coil springs aren't broken or weak or jammed with rock; that feed is evenly presented to screen.
- Watch screen's discharge and how clean it is (establish baseline); then, if discharge depth increases or if fines piggyback on coarse, there is a problem.

Check Screen Sizing

The formula and factors on the following table are guides only. Contact manufacturer for details. A separate calculation is required for each deck of a multiple-deck screen.

Terms in the equation are: Area = ft^2; U = short tons/hour of material in the feed to the screen that is smaller than the screen opening size, i.e., tons per hour of undersize in the feed; A = short tons/hour that can pass through one ft^2 of screen deck when feed contains 25% oversize and 40% half size; G = factor that applies when the open area of screening surface is less than open area shown in Factor A chart.

TABLE 17.6 Factors for calculating screen area

Formula: Screening Area = $\dfrac{U}{A \times B \times C \times D \times E \times F \times G \times H \times J}$

Basic Operating Conditions:
Feed to screening deck contains 25% oversize and 40% halfsize
Feed is granular free-flowing material
Material weighs 100 lb/ft^3
Operating slope of screen is: Inclined Screen 18°–20° with flow rotation; Horizontal Screen 0°
Objective screening efficiency—95%

FACTOR "A"		
Surface Square Opening (in.)	% Open Area	STPH Passing a Square Foot
4	75	7.69
3½	77	7.03
3	74	6.17
2¾	74	5.85
2½	72	5.52
2	71	4.90
1¾	68	4.51
1½	69	4.20
1¼	66	3.89
1	64	3.56
⅞	63	3.38
¾	61	3.08
⅝	59	2.82
½	54	2.47
⅜	51	2.08
¼	46	1.60
3/16	45	1.27
⅛	40	0.95
3/32	45	0.76
1/16	37	0.58
1/32	41	0.39

continues next page

TABLE 17.6 Factors for calculating screen area (continued)

FACTOR "B" (Percent of Oversize in Feed Deck)							
% oversize	5	10	15	20	25	30	35
Factor B	1.21	1.13	1.08	1.02	1.00	.96	.92
% oversize	40	45	50	55	60	65	70
Factor B	.88	.84	.79	.75	.70	.66	.62
% oversize	75	80	85	90	95		
Factor B	.58	.53	.50	.46	.33		

FACTOR "C" (Percent of Halfsize in Feed Deck)							
% half size	0	5	10	15	20	25	30
Factor C	.40	.45	.50	.55	.60	.70	.80
% half size	35	40	45	50	55	60	65
Factor C	.90	1.00	1.10	1.20	1.30	1.40	1.55
% half size	70	75	80	85	90		
Factor C	1.70	1.85	2.00	2.20	2.40		

FACTOR "D" (Deck Location)			
Deck	Top	Second	Third
Factor D	1.00	.90	.80

FACTOR "E" (Wet Screening)									
Opening	1/32 in.	1/16 in.	1/8 in.	3/16 in.	1/4 in.	3/8 in.	1/2 in.	3/4 in.	1 in.
Factor E	1.00	1.25	2.00	2.50	2.00	1.75	1.40	1.30	1.25

FACTOR "F" (Material Weight)									
lb/ft³	150	125	100	90	80	75	70	60	50	30
Factor F	1.50	1.25	1.00	.90	.80	.75	.70	.60	.50	.30

FACTOR "G"
(Screen Surface Open Area)

$$\text{Factor ``G''} = \frac{\text{\% open area of surface being used}}{\text{\% open area indicated in capacity}}$$

FACTOR "H" (Shape of Surface Opening)	
Square	1.00
Short slot (3 to 4 × width)	1.15
Long slot (more than 4 × width)	1.20

FACTOR "J" (Efficiency)	
95%	1.00
90%	1.15
85%	1.35
80%	1.50
75%	1.70
70%	1.90

Source: Vibrating Screen Manufacturers Association.

Screen efficiency = (% of undersize in feed which actually passes) / (% of undersize in feed)

The bed depth at the discharge end of a screen should be <4 × the size of the screen opening for 1,605 kg/m³ material and <3 × for 802 kg/m³ material. A formula is

$$DBD = (O \times C) / (5 \times T \times W) \qquad \text{(EQ 17.15)}$$

where:
- DBD = discharge-end bed depth in inches
- O = oversize in STPH
- C = bulk density in ft³/short ton
- T = rate of travel in ft/min (nominal 75 fpm for inclined screen at slope of 18° to 20° with flow rotation and nominal 45 fpm for horizontal screen)
- W = width of screening area in feet

First determine the width of the screen deck based on the bed depth calculation, then select the length of the screen.

For wet screening, apply water at a rate of 19 to 27 L/min per m³ of ore; use some of the water to prepare a slurry in the feed box for presentation to the screen; very little screening of fines takes place between spray bars where no water is washing over the material.

SORTING
by Douglas K. Maxwell, P.E.

All Sorting Techniques
- Lighting is critical; greatest differences are seen when particles are wet.
- Wash slimes from particles to expose all surfaces.
- Sorts are best when feed is closely sized; e.g., large:small = 2:1 or 3:1
- Typically only one surface per particle is evaluated; thus, liberation is critical to quality of sort. Expect some misplacement of particles.
- Sorting rate depends on particle size; bigger particles = higher tonnes/h.

TABLE 17.7 Commercial sorting systems

Property	Sensor	Separation	Particle Size, mm	Sorting Rate	Rate Units	Applications	Typical Use
Visual appearance	Human eye	Hand sorting	3 to 450	0.05 to 10	Tonnes per workershift	Gems, industrial minerals, coal, metal ores, tramp material, oversize	Rougher, cleaner, scavenger
Laser reflectance	Photo multiplier	Trajectory deflection	10 to 120	10 to 100	Tonnes per hour	Magnesite, limestone, wollastonite, feldspar, quartz, spodumene, gold ore, silver ore	Rougher, cleaner
Radioactive "Grade"	Scintillometers and camera	Trajectory deflection	10 to 120	10 to 100	Tonnes per hour	Uranium ore, gold ore	Rougher, scavenger
X-ray fluorescence	Photo multiplier	Trajectory deflection	3 to 75	3 to 40	Tonnes per hour	Diamonds	Cleaner

Hand Sorting
- Feed preparation—little preparation needed for tramp material or oversize removal.
- Feed presentation—particles should cover <20% of sorting surface.
- Sorting rates
  - Rate ≈ constant in particles per worker-hour. Tonnage depends on particle size.
  - Gem sorting takes more evaluation time per particle.

Laser Reflectance (Photometric)
- Feed preparation
  - Maximum particle size ≈120 mm; minimum particle size ≈10 mm technically but ≈20 mm for most economic operations.
  - One sorter can handle different size ranges at different times (i.e., campaigning).
- Feed presentation
  - Dust control is essential for sensing system.
  - Separate the particles; particles should cover <20% of sorting surface.
- Separation
  - Horizontal belt speed ≈4 m/s.
  - Reflectance measurement is after particles leave the belt and start free trajectory.
  - Air jets are down firing and are controlled by an electronic processor.
- Sorting rate—depends on particle size. Rule of thumb: tonnes/h = average particle size in mm (i.e., 60 tph for 40 mm to 80 mm particles).
- Sort characteristics
  - Sort is based on reflectance of scanned laser spot. Red (helium-neon) laser used most; blue (argon) lasers used occasionally. Any visible light laser can be adapted for use.
  - Expect 10%–20% misplacement of particles.

Radioactive
- Feed preparation
  - Maximum particle size is 120 mm.
  - Each sorter is designed to handle only one limited size range.
- Feed presentation: Feed is directed to channels in which particles are spread apart for separate evaluation.
- Separation
  - Belt moves at 4 m/s.
  - Scintillometers measure radiation of each particle in each channel and videocam measures each particle's length and width (i.e., size).
  - Down-firing air jets controlled by processor.
- Sorting rate approximates particle size (i.e., 60 tonnes/h for 40 mm to 80 mm particles).
- Sort characteristics
  - Particle grade = scintillation count divided by camera's measurement of size.
  - Expect 10%–20% misplacement of particles.

X-ray Fluorescence Diamond Sorters
- Feed presentation—a monolayer of material on an inclined chute in total darkness.
- Separation
  - Diamonds fluoresce with visible blue light when exposed to X-rays. Glow is detected by a bank of photomultipliers that trigger a down-firing air jet as material drops off end of chute.
  - Sorters used mostly to clean gravity concentrates, but in Russia for roughing also.
- Sorting rate of gravity concentrates is typically 2 to 5 tonnes/h; up to 40 tph for roughing of −20 mm feed.
- Sort characteristics
  - Every diamond must be collected so concentrates are low grade. Low grade is also caused by other minerals that fluoresce and are collected.
  - Diamond losses may be caused by diamonds that do not fluoresce brightly enough or quickly enough to be detected.
  - Sorters are enclosed machines and they offer a high degree of security.

GRAVITY CONCENTRATION
by D. Erik Spiller

The key to successful gravity concentration of minerals is careful attention to both liberation of the components to be separated and presentation of the feed material to the separating machine. There are many mechanical devices to separate minerals based on particle-specific gravity differences. Effective commercial separations can be made on materials as coarse as 1,200 mm (5 in.) to as fine as 0.01 mm (10 μm); the upper size is limited only by the size of the separating machine, and of course, liberation. The following guidelines cover gravity separation via the classical system definitions of stratification, flowing film, and shaking (Kelly and Spottiswood 1982); density systems, i.e., dense medium separations (DMS), are not included.

TABLE 17.8 Commercial characteristics of gravity concentration machines*

Type	Machine	Operating Size Range, mm	Water Requirements[†]	Capacity[‡]
Stratification	Jigs—conventional	0.10 → 100	High	Medium
	Jigs—circular	0.05 → 100	High	High
	Jigs—centrifugal	0.03 → 2.0	High	Medium
Flowing film	Sluice box	0.15 → 10.0	High	Medium
	Reichert cone	0.05 → 1.5	Low	High
	Pinched sluice	0.05 → 1.5	Low	Medium
	Strake	0.15 → 2.0	High	Low
	Spiral	0.03 → 2.0	Medium	Medium
Shaking	Shaking table	0.02 → 2.0	Medium	Medium
	Orbital	0.01 → 0.07	High	Low
	Centrifugal	0.01 → 0.15	High	Low
	Crossbelt	0.01 → 0.03	High	Low
Centrifugal	Spinning bowls	0.01 → 1.7	Very high	High
Air dry	Pneumatic jig	0.5 → 25	None	Medium
	Air table	0.15 → 6	None	Low

* Generalized—specific machine operating on site-specific materials may or may not perform to these characteristics.
† Relative.
‡ Relative per unit—multiple units always equate to high capacity.

Concentration becomes more expensive and machine capacity lower as particle size decreases. Gravity concentration is usually difficult below 0.074 mm (200 mesh) and very difficult below 0.038 mm (400 mesh). Flowing film separators such as sluices, Reichert cones, and spirals tend to work best on −1.0 mm (16 mesh) to +0.1 mm (about 150 mesh) liberated particles; although there are numerous applications where satisfactory recovery efficiencies extend to finer sizes. Jigs perform well on coarser materials up to 75 mm, and in practice their recoveries drop off significantly on particles finer than about 0.15 mm (100 mesh), although there are exceptions, especially when applied to high specific gravity differentials found with gold, tin, etc. Shaking tables and their variations can be effective down to 0.02 mm (20 μm) when treating narrowly sized feed materials.

Virtually all gravity-concentrating machines tend, at times, to confuse gravity concentration with size classification. It's the nature of the system.

Narrower feed size distributions produce better gravity concentration results.

Gravity-concentrating machines are most effective at producing either a concentrate or tailing. It takes multiple-unit operations, cleaners, and scavengers, to do both.

More than about 10 wt % slimes (roughly −0.01 mm particles) significantly reduces performance, except for machines designed for beneficiation of slimes (Mozley MGS and centrifugal-based machines). The reason for this is the increased viscosity of the fluid/pulp.

A consistent feed rate and feed pulp density are essential for shaking tables (all types), spiral concentrators, and Reichert cone concentrators. Sluices are a little less sensitive to feed rate. Jigs and centrifugal-based devices are relatively insensitive to feed rate and feed pulp density up to the point of overloading, and then performance drops dramatically.

The concentration criterion, CC, suggests if gravity concentration is practical (other than dense medium, heavy liquid, etc.).

$$CC = (D_h - D_f) / (D_l - D_f) \qquad \text{(EQ 17.16)}$$

where:
D_h = the density of the heavy particles
D_l = the density of the light particles
D_f = the density of the fluid (usually water at 1.0)

Gravity concentration will usually be successful if $CC > 2.5$. When $CC < 1.25$, gravity concentration is virtually impossible. When $2.5 > CC > 1.25$, gravity concentration is very difficult, but may be possible to some extent with narrow feed size classification and slow and careful feed presentation.

Complete component liberation is not necessary before introducing feed to gravity separation. However, there must be "effective liberation," which is the size at which a separation can be made that will have a positive economic impact on downstream processing.

Feeding dry material to a wet machine is discouraged because certain minerals, including metallic gold, are difficult to "wet" and they can "float off" and be lost to tailings. It is best to provide a feed mixing/wetting tank in the circuit prior to a gravity separation device. This same tank will provide surge protection to feed rate sensitive gravity separators.

Dry gravity separation can be effective, although it is never as efficient as wet separation. A word of caution on dry separators: "dry deposits" are often not really dry and, therefore, feed dispersion problems can impede performance at commercial scale.

For maximum sluice performance, barren coarse rock (+25 mm, or even +6 mm, if possible) should be removed from the feed. A constant feed rate at a constant feed pulp density (water addition) improves performance. Site-specific factors, including mineral associations, particle size, and relative abundance of "heavies," influence sluice performance; testing is always necessary. Each deposit/operation should determine "clean-up" intervals based on riffle loading and tonnes processed; shorter is often better. For all practical purposes, a strake performs the same as a sluice with respect to mineral recovery.

Installing free-gold gravity recovery equipment on the circulating load in a grind circuit, i.e., cyclone underflow, improves gold recovery. Treating as little as 10 wt % of the circulating load can be effective, although up to 30% may be optimal. Free-gold recovery can improve the amount of payable gold (at sale) by moving gold to a different saleable product. Capturing the gravity recoverable free gold increases security issues, but also makes mill metallurgical accounting easier and more accurate.

Collected spills should be returned to circuit gradually.

FLOTATION
by Deepak Malhotra

General
Three main groups of variables are (1) ore and mineral properties, (2) reagent combinations and flotation environment, and (3) flotation machine characteristics.

The objective of rougher flotation is to maximize recovery at as coarse a grind as possible while maintaining grade at an acceptable level. The objective of cleaner flotation is to produce a high grade at acceptable recovery. Liberated particles float fast; middlings float slowly; thus, regrind them. Very fine grinding does not increase recovery. Efficiency drops off below 15 to 25 μ.

Recovery and grade have an inverse relationship. Increased recovery may not mean increased profit, e.g., a finer grind costs money but may give higher recovery, and pulling froth harder may give higher recovery but result in a low-grade unsalable concentrate; economic recovery occurs when revenue from incremental recovery equals or exceeds the incremental cost of getting it.

Operating Controls

Controllable variables are
- Soluble impurities in process water
- Alkalinity or acidity of process water
- Type of grinding mill
- Type of grinding media
- Oxidation during grinding
- Pulp density during grinding
- Grinding time
- Chemicals added during grinding
- Grinding temperature
- Particle size
- Particle size in flotation
- Pulp density in flotation
- Temperature in flotation
- pH in flotation
- Circulating load in flotation
- Flotation time
- Impeller speed
- Froth height
- Degree and type of aeration
- Chemicals added
- Flow rate
- Pulp level control.

Noncontrollable variables are
- Nature of minerals and gangue
- Soluble constituents in ore
- Degree of oxidation of ore
- Grain size and mineral/gangue associations
- Hardness of minerals and gangue
- Relative time for different minerals in grinding circuit.

Plants treating >2,000 tpd can justify expensive instrumentation such as onstream analyzers for tailings; even smaller plants must have at least automatic samplers.

Maintain a constant feed volume and density for smooth operation with all cell types.

The order of reagent addition can be critical for industrial minerals or two-product systems such as Pb/Zn. Proper mixing and dilution of reagents is important for efficient operations. Xanthates in sulfide flotation are the cheapest collectors; they are also nonselective. Dithiophosphates are more selective and less pH sensitive than xanthates.

Reagents must be added at the proper points in the circuit for best performance. Water-soluble collectors are usually added after grind to the conditioner or pump sump. Non-water-soluble collectors should be added to the mill.

Reagents affect the froth in addition to recovery and grade; too much oil can kill froth; too much frother can create overfrothing conditions.

Sulfidization often increases recovery of oxide minerals; determine percentage of oxidized mineral first.

Pulp level in mechanical cells should be controlled only on the last cell of the bank by weirs and dart valves. Pulp level in column cells is regulated by tailings removal rate.

Froth level should be deep in roughers and shallow in scavengers. The last cell should be used as the control cell; it should have little or no desirable mineralization in the froth to ensure maximum recovery.

The higher the flotation air, the lower the retention time; the higher the feed rate, the lower the retention time.

To avoid short-circuiting, 8 to 10 cells in a row are needed in roughers. It can also be controlled by proper agitation and aeration. Sample rougher overflows to determine if flotation time is sufficient.

Rougher flotation usually floats <10% of feed weight; hence, a quick method of checking recovery would be to check tail and feed assays, i.e., $R = 100(1.0 - t/f)$. Cleaner flotation usually recovers over 50% of cleaner feed. Check the upgrading ratio of concentration (i.e., c/f, for flotation efficiency; a ratio <1:1 indicates poor selectivity).

Pulp density affects residence time and, hence, recovery. Roughers are typically 30%–40% solids and cleaners typically <20% solids. Increase pulp density with increasing number of cleaner cells.

Consider replacing mechanical cleaners with flotation columns. Reduce column cell air supply pressure to supplier's recommendation.

Recycling of streams can help in some situations and hinder in others.

Process water in industrial minerals can be acidic in one part of the plant and basic in another; keep them separate to avoid problems.

Flotation circuits can be open circuit (easiest to operate), co-current, or counter-current (hardest to operate even with small upsets).

Solids are building up in a thickener if solids in feed are > solids in underflow.

Slimes create problems of poor selectivity, high reagent consumption, and low concentrate grade. Add dispersant such as sodium silicate to grinding circuit; overdosing causes problems in thickening, filtration, and settling in the tailings pond.

Short-circuiting of material is common in hog trough type cells.

If cells sand up, the cause is probably worn impeller/diffusors. Replace them.

To check airflow rate in the cells, place an inverted 2-liter cylinder filled with water in the cell's pulp.

Testing

- Conduct a recovery versus grind P_{80} series.
- If flotation is poor and recovery low, test a feed sample in the lab without additional reagents; follow up tests with additional collector dosages.
- Investigate other reagents.
- Perform tests at various agitation speeds and compare recovery and grade. If higher agitation gives higher recovery with the same grade, increase airflow to cells.
- Calculations (Taggart 1956):

$$Ff = Cc + Tt \quad (EQ\ 17.17)$$

where:
F, C, T = weight of feed, concentrate, and tails
f, c, t = assays for feed, concentrate, tails

$$K = \text{ratio of concentration} = F/C = (c - t) / (f - t) \quad (EQ\ 17.18)$$

$$R = \%\ \text{recovery} = 100c(f - t) / f(c - t) \quad (EQ\ 17.19)$$

SOLID–LIQUID SEPARATIONS
by David S. Davies, P.E.

General

Most important variables are size distribution of solids and density of solids. Solids with relatively uniform particle size will normally separate well, even fine sizes. The separation of solids with broad size distribution is greatly affected by the amount of –44-µm (325-mesh) particles.

- Particle size range (microns or μm) versus unit operation: screening, 40 to large sizes; classification, 10 to 1,000+; hydrocyclones, 5 to 1,000+; thickening, 1 to 1,000; clarification, 1 to 1,000; centrifugation, 0.5 to 600; filtration, 0.1 to 600.
- To reduce filter cake moisture consider: surfactants, steam hooding, belt presses, hyperbaric filters, pressure filter, and ceramic filters.
- Data for selecting an S–L separator:
  - Process: what results are expected of the separator?
  - Feed: quantity of solids in feed; temperature; pH; viscosity; specific gravity (SG) of solution and solids; percent solids; size distribution of solids.
  - Separation rates: filtration rate in Buchner funnel (L/min/m^2); filter leaf rate; measure vacuum in mm Hg; time required to form a cake of 'x' thickness; at what rate do solids settle by gravity? What percent of total feed volume do the settled solids occupy after settling 'x' hours?
  - Are flocculants, filter aids, or anti-foaming agents needed? What's their impact?
  - Washing: is it needed? temperature; quantity of wash allowable.
  - Clarity of separated liquid? What percent solids is needed in underflow or cake?

Thickening

Feed is usually 5% solids or more.

TABLE 17.9 Theoretical concentration in solution from final counter-current decantation (CCD) thickener

Number of Stages	Wash Ratio		
	1	2	3
2	0.333	0.143	0.077
3	0.250	0.067	0.025
4	0.200	0.032	0.008
5	0.167	0.016	0.003

Concentration in feed solution = 1.000
WR = wash flow / feed solution flow
Leaching is assumed complete before CCD
Complete mixing of slurry and wash water in each stage
Feed solution flow = solution in each underflow
Wash water flow = each thickener overflow
No evaporation; no other solutions added
Wash water concentration = 0.000
If wash water has a concentration, that concentration is added to the concentration of every solution in the circuit.

TABLE 17.10 Typical settling data

Application	% Solids Feed	Underflow	Solids T/m^{2}d
High-Rate Thickeners and Clarifiers			
Alum sludge	0.5	2–4	0.1–0.25
Alumina, red mud washers	10–15	30–40	0.25–1.0
Coal, clean fines	1–7	25–40	3.5–16.0
Copper concentrate	15–30	60–75	25–50
Copper leach residue	5–15	45–50	6–33
Copper tailings	10–35	50–65	16–50
Gold ore, NaCN leached	10–33	40–60	12–50
Kaolin slurry	1–6	20–30	—
Lime-neutralized pond water			
Partially neutralized	2.7	30–40	4
2nd stage neutralized	2.5	13–20	2.5
Mg(OH)$_2$ from brine	9	25–50	20
Mg(OH)$_2$ from sea water	2–6	10–25	1–10
Magnetite (heavy media)	12–18	60–70	30–50
Metal hydroxides (effluent treatment)	0.1–1.0	2–10	—
Molybdenum tailings	20–30	50–70	30–50
Phosphate slimes	2–8	12–18	6–17
Sand tailings	2–10	30–40	—
Soda ash, primary feed	1–3	8–20	30–50
Taconite tailings	10–15	55–65	—
Uranium acid-leached ore	10–30	45–65	16–50
Uranium, yellowcake	1–10	15–40	1–10
Zinc concentrate	15–40	60–75	40–50
Zinc tailings	20–35	50–70	16–50
Conventional Thickeners			
Activated sludge (sewage treatment)	0.2	2–3	0.025
Alumina, red mud primary	3–4	10–25	0.2–0.5
Coal, refuse	0.5–6.0	20–40	1–2
Copper flotation concentrate	15–40	60–75	2–8
Lead flotation concentrate	15–40	60–85	3–10
Nickel flotation concentrate	15–40	60–85	2–8
TiO$_2$ pigment clarification	0.1–1.0	20–25	0.06
TiO$_2$ effluent clarification	0.1–1.0	45–50	0.01
Zinc flotation concentrate	15–40	55–70	2.6

Source: Enviro-Clear 1997.

When thickening:
- Determine thickener area (m^2/(tonne solids × day)) versus feed rate or percent solids in feed.
- Determine underflow percent solids versus solids residence time.
- Be sure flocculants are dilute enough and well mixed into thickener feed.
- Conduct a flocculant comparison study.
- Repair center wells and overflow weirs to stop short-circuits.
- Consider an additional CCD stage for reduced soluble loss.
- Raising the water temperature from 7.2°C (45°F) to 23.9°C (75°F) changes the viscosity of water so much that settling rates increase by about 50%. This implies that flocculant usage can be decreased during warm weather months.

- Conventional thickener: Easy to operate; high underflow densities; high throughput; feed enters feed well above settled solids.
- High-density thickener: Extended solids retention time; high torque requirements limit diameter to 60 m; may increase underflow density by 10–12 percentage points; feed enters in settled solids zone; always use flocculants.
- High rate thickener: 2 to 8 times capacity compared to conventional thickeners; efficient flocculation required; generally better overflow quality and lower underflow densities than conventional; tight control required; size usually limited to 35–40 m ø.
- Ultra high rate thickener: Deep settling/compaction zone; no rakes; internals recycle solution for feed dilution and flocculation; generally low overflow turbidity and high underflow density; diameter generally 33% to 50% of high rate thickener diameter for equal capacity.

Clarification
- Generally used for <5% solids slurries (solids are in "free settling" mode).
- Sizing is controlled by overflow rate, $m^3/(m^2 \times h)$.
- Determine bulk settling rate, overflow and underflow solids concentration versus residence time, and rheology of underflow.
- Conventional clarifier: For very slow settling particles without feed enhancement; sizing in range of 5 to 25 $m^3/(m^2 \times day)$; use with or without flocculant; subject to upset with feed or temperature change.
- Reactor clarifier: Internal solids recycle with draft tube arrangement; promotes improved settling properties of solids; flocculation required; good overflow quality.
- Inclined plate and tube clarifier: Compact; small flows; fast settling solids; limited underflow retention time; usually no rake mechanism; limited to nonsticking solids.
- Hopper clarifier: A rougher clarifier used ahead of polishing filters; flocculation required; fluidized solids bed operation; control difficult; no rake mechanism.

Hydraulic Classifiers and Sizers

These provide gravitational classification of slurry solids by particle size and density. An upward-moving column of liquid will separate solids into fractions that can be separately collected.

TABLE 17.11 Wet classification machines

Type of Classifier	Normal Mesh of Separation Range*	Normal Feed Tonnage Range	Max. Oversize in Feed	Normal Overflow, % Solids Range	Normal Feed Density, % Solids Range	Normal Sand Product, % Solids Range	Motor Range (hp)	Typical Applications
Non-mechanical:								
Cone classifier	28–325	2–100 tons/h	¼ in.	5–30	Not critical	35–60	None	For desliming and primary dewatering
Liquid cyclones	35 mesh to 5 μ	½–1,500 gal/min	14–325 mesh	5–30	10–60	55–70	Power for pressure head 5–60 lb/in.²	For medium or fine separations and closed-circuit grinding
Mechanical:								
Drag classifier	28–200	5–350 tons/h	1½ in.	5–30	Not critical	70–83	1–10	For desliming, conveying, and closed-circuit grinding
Rake and spiral classifiers	20–200	5–350 tons/h	1 in.	5–30	Not critical	75–83	½–25	Closed-circuit grinding, washing and dewatering, desliming, process feed control
Bowl classifier	100–325	5–200 tons/h	½ in.	5–25	Not critical	75–80	Bowl: 1–7½ Rake: 1–25	Closed-circuit grinding usually in secondary circuits
Bowl desilter	100–325	5–250 tons/h	½ in.	1–15	Not critical	75–83	Bowl: 1–10 Rake: 5–25	Recovery of fine sand, limestone, coal, and fine phosphate rock from large flow volumes
Hydro separator	100–325	5–700 tons/h	½ in.	1–20	Not critical	30–50	1–15	For fine separation where large feed volumes are involved and drainage not critical
Solid bowl centrifuge	200 mesh to 1 μ	10–600 gal/min	½ in.	1–40	5–50	10–70	15–150	For fine-size fractionating
Countercurrent classifier	35–100	1–600 tons/h	3 in.	5–30	Not critical	75–83	¼–25	Sand-slime separations, washing, closed-circuit grinding
Hydraulic:								
Jet sizer	8–150	2–100 tons/h	3⁄16 in.	1–10	30–60	40–60	1–2 for air pressure	Multiproduct unit for exceptionally clean sands fractionated into narrow size ranges. Min. 3 tons hydraulic water per ton sand
Super sorter	8–150	40–150 tons/h	⅜ in.	1–10	30–60	40–60	1 to operate pincer valves	Multiproduct unit for exceptionally clean sands fractionated into narrow size ranges. Min. 3 tons hydraulic water per ton sand
Siphon sizer	14–150	1–100 tons/h	1 in.	1–10	30–60	40–60	None	Two-product unit efficient for desliming and exceptionally clean sands, washing, closed-circuit grinding. Min. 2 tons hydraulic water per ton sand

* Size of screen retaining 1½% of the overflow solids.
Source: Perry 1980 (reprinted with permission of the McGraw-Hill Companies).

Hydrocyclones

The achievable separation is dictated by diameter of the unit, inlet and outlet dimensions, cone angle, feed pressure, slurry solids concentration, slurry viscosity, liquid density, solids density, particle shape factor, and particle size distribution.

The "separation" is the particle size of which 1% to 3% reports to overflow; larger sizes report to underflow. Determine separation of a cyclone as follows:
- Obtain base D_{50} from Figure 17.1
- Correct base D_{50} with factors from Figure 17.2 (feed concentration), Figure 17.3 (pressure drop between feed and overflow), and Figure 17.4 (SG of solids)
- Multiply base D_{50} by each of the three correction factors to get an actual D_{50}, and multiply the actual D_{50} by 2.2 to get the separation size. This is approximate.

Flow rate through cyclone versus pressure drop is shown in Figure 17.5. Multiple cyclones may be needed to handle the flow.

Centrifugation
- Can make finer separations than a hydrocyclone.
- Screening centrifuges: for solids dewatering and washing.
- Disk centrifuges: for separation of very fine solids from liquids, i.e., 1-µm range.
- Solid bowl scroll conveyor centrifuge: most common type; can process up to 265 L/min of slurry and make separations in the 2- to 5-µm range.

Vacuum Filtration
- Factors affecting filter selection: particle size distribution in feed; % solids in feed; temperature; corrosiveness of the slurry; cake moisture; cake washing requirements; wash availability; wash composition; altitude.
- To obtain reliable results the following need to be obtained: detailed process knowledge of the separation to be made; separation objectives; bench scale and/or pilot testing; representative samples.
- Cake buildup on a laboratory vacuum leaf filter is "rapid" if 0.1 to 10 cm/s; "medium" if 0.1 to 10 cm/min; and "slow" if 0.1 to 10 cm/h. Filtration is difficult if 0.3-cm cake thickness cannot be formed in <5 min.
- Gravity filters are used for free-draining solids. In the case of sand or mixed media filters, they are used to remove trace particulates.
- Types: precoat drum, drum, disk, horizontal belt, and pan.

MINERAL PROCESSING | 309

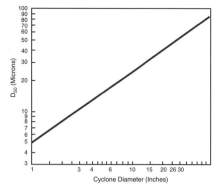

Source: Besendorfer 1996 (reprinted with permission of Chemical Engineering).

FIGURE 17.1 Cyclone diameter versus D_{50}

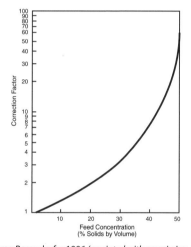

Source: Besendorfer 1996 (reprinted with permission of Chemical Engineering).

FIGURE 17.2 Correction for feed concentration

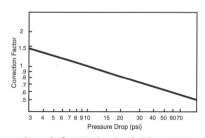

Source: Besendorfer 1996 (reprinted with permission of Chemical Engineering).

FIGURE 17.3 Correction for pressure drop

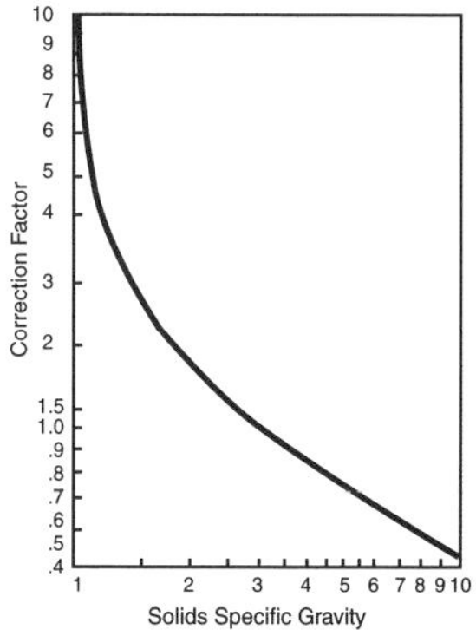

Source: Besendorfer 1996 (reprinted with permission of Chemical Engineering).

FIGURE 17.4 Correction for solids specific gravity

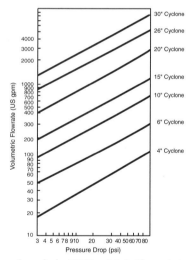

Source: Besendorfer 1996 (reprinted with permission of Chemical Engineering).

FIGURE 17.5 Volumetric flow rate, pressure drop versus U.S. gpm

TABLE 17.12 Vacuum filter characteristics

Filter Type	Solids Characteristics	Cake Thicknesses (mm)	Wash Application Possible
Precoat drum	Very fine, used for liquor clarification	<1	Yes, but limited by wash arc available
Rotary drum	Must suspend feed solids with mild agitation; can be fine	>5	Yes, but limited by wash arc available
Rotary disk	Must suspend feed solids with moderate agitation; good filter rate	>10	No
Ceramic disk	Must suspend feed solids with moderate agitation; must be discrete, nonblinding solids as cake heel remains below knife	>3	No
Horizontal belt	Broad range; feed flocculation applies; versatile unit	>10	Yes, excellent; can be applied in counter-current fashion
Pan type	Fast filtering solids; feed can be flocculated	>20	Yes, excellent; can be set up for counter-current wash

TABLE 17.13 Typical vacuum filtration results

Feed Solids	Feed (wt % solids)	Particle Size	Filter Type	Filter Cake Moisture (wt %)	Filtration Rate (kg dry solids/h/m²)
Alumina red mud	25–35	95+% –100 μ	Roller discharge drum	30–40	75–175
Alumina trihydrate seed	30–40	–40 μ	Disk	15–19	800–1,400
Alumina trihydrate product	40–50	30–60 μ	Horizontal belt scroll	12–14 13–16	2,000–5,000 750–1,500
Flotation coal	20–28	30%–35% –200 mesh	Disk	18–23 (surface)	350–450
Nonflotation coal	35–40	–28 mesh	Disk	16–26	250–500
Cement slurry	45–55	80%–90% –200 mesh	Drum disk	18–27	80–400
Copper concentrates	55–65	80%–95% –200 mesh	Disk drum	11–16	300–500
Cyanide slurry	40–50	70%–90% –200 mesh	Drum	24–30	80–400
Fluorspar	55–65	75%–90% –200 mesh	Continuous belt drum	8–13	250–1,000
Iron ore concentrate					
Magnetite	50–60	1,600–2,000 Blaine	Disk	8–10	625–1,500
Hematite	50–65	1,800–2,220 Blaine	Disk	9–11	300–1,000
Kaolin clay	30–35	70%–100% –2μ	Roller discharge drum	35–40	75–250

continues next page

TABLE 17.13 Typical vacuum filtration results (continued)

Feed Solids	Feed (wt % solids)	Particle Size	Filter Type	Filter Cake Moisture (wt %)	Filtration Rate (kg dry solids/h/m²)
Lead concentrates	60–70	80%–95% –325 mesh	Disk continuous belt drum	11–14	350–600
Magnesia precipitate	8–12	—	Drum	—	30–65 (MgO basis)
Digested phosphate rock	30–35	—	Pan	16–22	100–1,000
Sand	55–65	85%–95% +200 mesh	Horizontal belt, scroll	8–11	2,000–4,000
Tailings					
Coal	25–35	80% –200 mesh	Roller discharge drum	30–40	75–125
Iron ore	55–60	10%–15% –200 mesh	Horizontal belt	15–20	1,000–1,500
Uranium ore					
Acid leach	50–55	50%–60% –200 mesh	Continuous belt drum string horizontal belt	20–24	200–500
Carbonate leach	45–50	65%–80% –200 mesh	Horizontal belt	24–30	200–350
Uranium yellow cake	12–20	—	Drum	60–70	25–100
Zinc concentrates	50–65	80%–95% –325 mesh	Disk continuous belt drum	14–17	300–600

Source: Dahlstrom 1985.

Pressure Filtration

Use for fine solids separations that are not economic with vacuum units, to obtain lower cake moisture, and to provide more efficient solute removal from the cake.

TABLE 17.14 Pressure filtration equipment

Type	Attributes
Membrane (ultra-filtration)	Limited application; sensitive to both liquor and solids characteristics; produces good quality liquor and semi-concentrated pulp; membrane fouling control requires evaluation.
Leaf	Usually for clarification with use of precoat where liquor contains product for recovery; cleaning presents unique problems; batch or semi-continuous.
Horizontal plate	Inexpensive; single-platen types for low tonnages; labor intensive.
Tower press	Small footprint; multiple horizontal platens stacked vertically. Specific applications require sealing, feeding, cloth tracking and cake discharge impact determination; can be maintenance intensive; relatively complex operation.
Tube	Capable of high pressure; small footprint; cake discharge and re-sealing make operation labor intensive.
Hyperbaric (fabric or ceramic media)	An adaptation of a disk filter in a pressure vessel; efficacy of mechanical seals on drive and cake discharge are application specific. In situ acid cleaning of media may be required; maintenance intensive; expensive spares.
Plate and frame	Good for small tonnage applications; labor intensive.
Recessed plate	Good for large tonnages if automated, otherwise labor intensive.
Moving belt	Light duty with low driving force capability; used primarily for clarification; limited application.
Cross-flow	Used to concentrate solids; requires pumped or centrifugally induced velocity to preclude cake deposition on media; produced liquor rate and quality good; solids concentration to be discharged limited to pumpable slurry.
Filter/thickener	Used to concentrate difficult-to-dewater solids; limited application and use.

Flocculation
- Probably most used and most abused enhancement tool.
- Common problems: plant feed changes; feed variability; overflocculation; insufficient dilution of flocculant; poor flocculant/feed contact; high shear after flocculation.
- Flocculation benefits are higher settling rates and smaller equipment, but there are tradeoffs (i.e., liquor is bound within the floc structure and washing is made more difficult and less predictable). Overflocculation or improper selection of flocculant with respect to charge and molecular weight can dramatically increase slurry viscosity and negate any chance of reducing equipment size or increasing throughput.
- Effective flocculation can usually be maintained through a 5-stage CCD thickener circuit but not 6 stages or more.

Process and Engineering Considerations

Solid–liquid separation enhancements can be achieved by flocculation, improved upstream particle size control, solids densification, feed concentration alteration, use of admix, viscosity alteration, pH adjustment, and temperature adjustment.

The use of high rate and ultra high rate thickeners that require flocculation make it imperative to have controls and the ability to rapidly adjust operating conditions.

Design and control philosophy must be compatible with operating philosophy.

Samples need to be carefully prepared and evaluated. Sample aging effects, such as oxidation or other chemical or physical changes, can strongly affect results.

A small change in particle size can drastically affect results.

Combining two S–L separation techniques may provide optimal performance (e.g., filtration after thickening). Flocculation is particularly affected by multistage operations and the effects must be taken into account.

Slurry rheology must be considered because of its impact on efficient S–L separations, rake mechanisms, pumpability, pipeline design, and mixing. When pumps start to cavitate or slurries won't flow easily, rheology modifiers may be helpful. Determine rheology of the thickened solids versus underflow percentage of solids with portable units.

Measure viscosity in the field from ΔP measurements over a known pipeline length.

SOLVENT EXTRACTION (SX)
by John E. Litz

General
- Two types: (1) liquid–liquid SX with organic extractant in an organic diluent, and (2) ion pair SX in which recoverable value is soluble in both the aqueous and organic.
- Applications: Cu, Zn, Ni, Co, Mo, V, W, Re, Zr, Hf, Be, B, Li, rare earths, Ga, Ge, Nb, Ta, Pt, Pa, and U.
- Extraction: Mass transfer from aqueous solution to organic solution is governed by the laws of chemistry and by the practical aspects of mixing two immiscible liquids. Provides most purification because of solvent selectivity.
- Stripping: Provides most concentration. Recovers solvent in a form ready for recycle.
- Washing: May be needed (e.g., wash loaded organic to remove Cl^- if saline water is used in leach because electrowinning (EW) cannot tolerate much Cl^-.
- Regeneration: May be needed to put organic into chemical form needed for extraction, e.g., if an amine is NH_3 stripped, it will be in free-base form; must wash with weak acid to put it into hydrogen-bisulfate form.
- Extractant properties (favorable): High selectivity for element to be recovered; high stability; low toxicity; high flash point; high capacity; low solubility in aqueous; low specific gravity; high interfacial tension; noncorrosive; low viscosity; low cost.

- Extractant selection is based on literature reviews, batch tests, and continuous laboratory testing. General examples are amines for uranium, oximes for copper, methyl isobutyl ketone for hydrofluoric acid (HF) solutions of Ta and Nb, and alcohols for boric acid.
- Diluents: Most are aliphatic rather than aromatic.
- Solvent power of diluent can be exceeded by having too concentrated an extractant; molybdenum-amine complexes are notorious for this.
- Viscosity has significant effect on phase separation time.

TABLE 17.15 Effect of viscosity on separation time

Aliphatic Diluents (by Exxon)	Kinematic Viscosity, cSt at 25°C	Settling Time (s)
Aromatic 150	1.3	93
Escaid 100	1.8	94
Mining kerosene	2.1	114
Escaid 110	2.5	132
Isopar M	3.1	148

Mixers

- Have most impact on efficiency and operating cost of the overall SX circuit.
- Configuration: Cylindrical with baffles (eliminate vortexing) or rectangular without baffles. Discharge to settler should approach full width of mixer.
- Impellers: Radial gives more shear than axial at the same rpm and tip speed.
- Pump mixers: Preferred as they reduce capital and operating costs by eliminating need for pumping between mixing stages. May increase entrainment because of high-speed impeller shear. Place impeller at mixer bottom or above a draft tube. If many shutdowns are expected, placement above a draft tube allows the mixer to start in the dominant phase (the phase that most of the liquid surrounding the impeller is), making it the continuous phase rather than waiting for the mixed phase to invert to the desired continuous phase. Mixer head controlled by impeller speed and recirculation rate; thus, overdesign the impeller so that recirculation can be used to control head. If draft tube is used and aqueous phase has high SG, tube should have holes that allow recirculation from bottom of mixer.
- Pump efficiency ≈ proportional to impeller tip speed, e.g., in cylindrical tank with height = diameter.

TABLE 17.16 Pump efficiency versus impeller tip speed

m/min	1.0/0.8 Aqueous/Organic (A/O) Specific Gravity	1.2/0.8 A/O Specific Gravity
240	50%	40%
210	40%	
180	30%	

- Entrainment: Minimum when impeller diameter is one-third of the tank diameter; it also is related to tip speed. Tip speed should be 150 to 300 m/min.
- Residence time for transfer = time in mixer plus time in settler dispersion band.
- Improve time in mixer by installing a top baffle ≈80% of distance from bottom to overflow; baffle should have centered circular hole ≈50% of impeller diameter.
- Phase continuity

- If impeller starts in aqueous (dominant phase), aqueous is continuous. If it starts in organic, organic is continuous. Inverting continuity is difficult; a draft tube makes it easier because, at rest, the impeller can be in the dominant phase.
- Continuity is hard to maintain as mixer volume increases; e.g., for organic continuous, need to increase organic flow as percentage of total flow.
- Aqueous continuous (1) entrains less aqueous in organic phase, (2) can minimize transfer of strip solution to extraction circuit, (3) may take more horsepower for the same transfer rate.
- Organic continuous (1) entrains less organic in aqueous phase, (2) is good for raffinate and strip product stages.

Settlers
- Objective is to produce two phases with minimum entrainment.
- Settler sizing: done as part of laboratory test program to get approximate size; $\approx$81 L/min per m^2 is rule of thumb for oxime systems but can vary widely. Also, settlers must be designed larger to handle two problems that occur with cooler solutions: (1) flooding the settler with mixed phases that are slow to disengage, and (2) stabilizing the interface with solids or fungus, which slows phase disengagement.
- Concentration of extractant in diluent can reduce rate of phase separation if it is too high. Diluents with a higher solvent power may help. Alcohol or phosphate modifiers may help.
- Entry of the dispersion into the settler should be full width at the level of the dispersion band. A horizontal baffle directs the dispersion velocity horizontally.
- Minimize linear velocity of solutions through settler; maximum length/width = 4. L/W should decrease as throughput increases.
- Baffles are needed to reduce solution velocities. Picket fences work best. First fence is close to inlet to hold back the dispersion band. Usual picket is 50 mm wide. Fence is two rows of pickets, one picket width apart, with picket spacing in each row about one-third to one-quarter of a picket width apart. If considerable dispersion still occurs downstream, a second picket fence at the settler midpoint will help.
- Flow rate of aqueous and organic should be <9 cm/s. If flow is higher, increase the thickness of that phase in the settler.

Controls and Testing
- Improve clarification of pregnant leach solution.
- Check mixing efficiency (each stage): take 2 samples from mixer, allow one to separate immediately and mix the other for 10 minutes more, assume the extra mixing achieves 100% mixing efficiency. If the transfer in sample #1 is <80% of the transfer in sample #2, the mass transfer in that stage of mixing can be improved. Recirculation can improve efficiency. If adequate total mass transfer is being achieved, improved stage efficiency may allow elimination of one stage of mixing, thus reducing power cost.
- Check dispersion band thickness throughout settler's length to determine if settler is nearing maximum capacity.
- Check for organic entrainment; use centrifuge with Babcock bottles; >0.1 L/m^3 of aqueous is probably a problem for oxime systems; for other systems the acceptable loss is proportional to the cost of extractant.
- Maintain constant solution flow to extraction. Use interface level controls to check on flow rates.
- Use conductivity meter to check for phase continuity in mixers on a periodic basis.
- Monitor the mixer phase ratios on a periodic basis.
- A/O ratios are very system-dependent and need to be optimized for each system in a continuous pilot plant.

HEAP LEACHING
by Paul D. Chamberlin, P.E.

Symptoms of Percolation Problems
- Seepage from heap slope more than 1 m above pad
- Ponding on top of heap
- Washouts or slumping or cracks or irregular slope of heap's side slopes
- Ore has slid off the pad
- Perched water table within heap
- Very low application flow rates.

Typical Heap Operating Conditions

TABLE 17.17   Typical heap operation conditions

	Gold	Copper
Application rate, L/(h·m^2)	2.5–12.0	12.0–35
Crush size, mm	1.7 to ROM	6.0 to ROM
Agglomeration	From H$_2$O only to as much as 20 kg/T cement; from conveyors only to pug mills.	H$_2$SO$_4$ cure; gypsum from H$_2$SO$_4$ + Ca(OH)$_2$
Heap heights, m	6–100	6–120
Lift heights, m	6–12	Shorter, 3–8
Preg	0.5–2.0, g/T	<1 to 6 g/L
Barren	<0.03 g/tonne	0.14–1.8 g/L
Evaporation, % of applied	2–10	2–10
Overall heap slope	2:1	2:1
Agglomerates, dry T/m^3	1.28–1.92	

General Rules for Column Testing of Gold Ores
- Acquire core without drilling muds. If muds are needed, test them for preg robbing. Keep core moist in plastic sleeves. Get rock quality determination (RQD) and photograph of core.
- Primary leach time in commercial heap ≈1.5× leach time in 6-m-high column test and ≈ 2 to 4× leach time in 2-m-high column test.
- About the same gold extraction is obtained if the same number of tonnes of leaching solution is applied to a tonne of ore in either a column test or a commercial heap.
- 0.5% gold extraction/month is about the economic limit in the United States for 1.5 g/T ore.
- Cyanide consumption in a 96-h, 1.7-mm crush size bottle leach test is an approximation of commercial heap consumption.
- Gold extraction from a 96-h bottle leach test of –1.7-mm ore is typically 10% less than gold extraction after 60 days from column leached ore that is –12.7 mm; this may be a quick way to forecast commercial heap leach results.
- Typical heap detoxification time to reach 0.2 ppm CN$_w$ is 50% of leach time.

- Use 6-m-high column tests; 2-m-high column tests for scoping tests.
- To obtain "ultimate gold extraction," perform weekly regression analysis of column test data and extrapolate to 180–360 days. Continue tests until extraction remains constant.

General Rules for Column Testing Copper Ores
- Pretreat with H_2SO_4 (acid cure, 5 to 12 kg/T).
- Do not allow solution pH to rise above 3 to 4.
- Be aware of assaying problems for determination of "free acid" concentration.
- Crush ores to permit greater control of pH within the particles; generally <2.5 cm.
- Solution effectively fills all of the voids for rock sizes <0.3 mm and air is excluded; if rock is >1 to 2 mm, drainage is almost complete and voids are filled with air.
- Thin layer leaching avoids acid depletion.
- Column diameter should be 4 to 6 times the largest ore particle dimension.
- Solution irrigation rates ≈ 12 to 35 $L/(h\text{-}m^2)$.

Design Features
- Use counter-current heap leach; total flow to heap = 2 to 4 times preg flow.
- Use balls or covers on ponds; save NaCN, CaO, water, and bird kills.
- Use drip irrigation to minimize destruction of agglomerates by droplets; or use fine sprays; or cover the heap with porous plastic cloth to absorb the shock of droplets.
- Use preg ponds with dividers so pond repairs can be made without shutting down.
- Arrange crush circuit to allow bypass of secondary and tertiary crushing if both refractory and very leachable ores are processed.
- Scalp −25 mm fines from ore ahead of primary crusher; agglomerate only fines.
- Blend hard ore into saprolite/laterite to allow higher heap height.
- Spray CaO (1) into trucks hauling ROM ore to heap or (2) onto heap slope.
- Use less than 5% slope on heap's surface to minimize runoff and make reclamation easier.
- Heap stability: need containment berm? Remember friction angles can be very low for plastic on plastic or some ores on plastic; need benches to contain slumps and raveling of side slopes; use telescoping well casing with breakable welds when using submersible pumps inside the heap.
- Drainage piping under heap: use perforated and corrugated high-density polyethylene (HDPE); usually 100 mm ø on 6-m to 10-m centers and at 45° angle to fall line; it is connected to a header that discharges to sump or ditch.
- Surfactants have had variable success in past.
- Use gravity separation to recover gold from "fines fraction" if ore requires fine crushing; then recombine wet fines with dry "coarser fraction" ahead of agglomeration.

Pads
- Permeability of soil in a composite pad should be 10^{-6} cm/s or lower; geomembranes are theoretically 10^{-14} cm/s or lower but have a much higher permeability because of punctures and imperfect installation.
- Stackers can operate on up to 8% slopes if done carefully; i.e., first lift only.
- Minimum slope of pad should be >2% to account for settlement of material under it.
- Drainage/cushion layer of crushed gravel on top of pad is to permit good drainage from under the heap (prevent buildup of hydraulic head and high seepage rates) and to prevent puncturing the liner when placing the first lift of ore. A 300-mm-thick layer must be placed with small equipment; a 1-m-thick layer can be placed with small mine equipment. If Cat 777s are used, the thickness should be at least 1.5 m.
- HDPE is a crystalline material and sharp rocks may puncture it over time.

- Polyvinyl chloride (PVC) has high plasticity and will stretch over sharp rocks.
- Polypropylene is becoming more widely used.
- Be aware of friction angles, especially between synthetics and between clays and smooth geomembranes.

Miscellaneous

- High cyanide concentration makes $Cu(CN)_3^-$; it does not load on carbon very well.
- High Cu or Zn concentrations may give free-cyanide assays that are too high when using AgCl titration.
- Water balance techniques
    - Reduce evaporation by (1) covering ponds, (2) using drip irrigation rather than sprays, and (3) using pipe to conduct the preg solution from heap to pond rather than open ditches.
    - Enhance evaporation by (1) using sprays on top of the heap rather than emitters, (2) spraying barren solution on black plastic (up to 50% evaporation and heat solution to 35°C), (3) installing snow-making fogging machines on heaps (up to 30% evaporation), and (4) installing sprays in ponds.
- Causes of low recovery percentages during the first year of operation may be:
    - Not irrigating nearly all of heap's surface.
    - An overly optimistic forecast of the rate at which values will be extracted.
    - Different ore than forecast.
    - Poor blast hole sampling that gives wrong ore grade.

TABLE 17.18 Dissolved oxygen per liter of distilled water, milligrams

Pressure (mm Hg)	Temperature (°C)			
	5	20	35	50
760	12.7	9	6.9	5.6
700	11.8	8.4	6.4	5
650	11	7.7	6	4.8
600	10.1	7.2	5.5	4.3
550	9.3	6.7	5	4
500	8.5	6	4.6	3.7
450	7.8	5.5	4.2	3.3

Source: Liddel 1922.

CARBON PLANTS
by Vernon F. (Fred) Swanson, P.E.

Carbon-In-Leach/Carbon-In-Pulp (CIL/CIP)

- High slurry density makes carbon concentrate at top of tank.
- High viscosity traps carbon in pockets of pulp; poor contact with new solution.
- Only sampling throughout tanks, and testing, will show if carbon is uniformly suspended.
    - Sampling slurry/carbon pulp: rod to reach almost to agitator, 1-liter small neck bottle attached to rod, cork attached to cord, submerge bottle to desired depth and pull cork for sample. Pour pulp through 500- to 600-μm sieve and wash out bottle to get all solids. Wash slurry from carbon and dry carbon and weigh it to determine grams carbon per liter of pulp. Assay carbon for metal loading. Assay solution and solids for metallurgical balance.
- Carbon concentration, 10 to 25 g/L

TABLE 17.19 Carbon expansion characteristics

Bed, % Expansion	Temperature = 35°C Flow		Temperature = 25°C Flow		Temperature = 15°C Flow		Temperature = 5°C Flow	
	gpm/ft²	L/m²	gpm/ft²	L/m²	gpm/ft²	L/m²	gpm/ft²	L/m²
Carbon, –2.8 × 3.0 mm								
25	22.62	922	20.50	835	7.26	296	16.40	668
50	33.38	1,360	30.03	1,223	6.05	247	26.22	1,068
75	40.46	1,648	38.86	1,583	5.19	211	33.14	1,350
100	47.86	1,950	45.00	1,833	4.54	185	39.56	1,612
Carbon, –1.0 × 0.72 mm								
25	8.84	360	8.06	328	7.26	296	5.07	207
50	14.36	585	12.89	525	6.05	247	8.88	362
75	19.23	783	18.12	738	5.19	211	12.68	517
100	22.99	937	21.80	888	4.54	185	15.87	647
125					4.04	164	18.69	762
Carbon, –0.5 × 0.3 mm								
25	3.20	130	2.68	109	7.26	296	1.74	71
50	5.36	218	4.50	183	6.05	247	3.23	132
75	6.91	282	5.93	242	0.00	211	4.09	167
100	8.43	343	7.04	287	5.19	185	4.83	197
125	9.57	390	8.26	337	4.54	164	6.30	257

- Use high concentration for Ag or preg-robbing ores
- Low concentration in first tank for high loading. High concentration in middle tanks
- Residence time
- Step loading efficiencies
- Carbon inventory
- Carbon contamination: oil spills, flotation reagents. Cure contamination with regeneration.

Carbon-in-Column (CIC)
- Amount of carbon in each tank: should be equal (if not, check transfer facilities).
- Dead bed depth (no flow) ≈1.2 m.
- Number of stages: typically 4–6 adsorption tanks.
- Uniform distribution of solution throughout the tank requires >6,895 N/m² (1 psi) across the bed plate.
- Lower carbon attrition with recessed impeller, Hidrostall, or screw pumps: more with eductors.
- Carbon kinetics and loading tests are needed for proper design.
- If Ag/Au ratio has increased over design, a higher stripping frequency or higher tonnage of carbon per strip may be needed. High barrens (>0.001 oz/ST or 0.028 g/tonne) may mean high Ag/Au ratio.
- Carbon loads Au, Ag, Cu, Hg, and Ni.
- Reduce Cu in dore with cold cyanide strip.
- Remove Ni from carbon with hot acid wash (90°C); may require redesign of tank.
- Analyze carbon for metallics (Ni, Fe, Co, etc.) if regeneration isn't working well.

Acid Washing, 1% HCl, 30 Minutes, 1 BV at 2BV/h
- Removes $CaCO_3$; acid wash before stripping to enhance strip efficiency.
- Reduce wash frequency to minimum for good carbon activity: saves on costs.

Stripping (Elution)

- Zadra: atmospheric with or without alcohol; pressure stripping; IPS (integrated pressure stripping); and stripping that has pressurized EW and no heat exchangers (tough to remove cathodes from EW). Usually 1% NaCN and 3% NaOH. Are strip times too slow? Solution temperature and composition all right? Higher temperature will speed up strip, but will strip vessel handle pressure and heat exchangers handle extra heat?
- Anglo American Research Laboratory (AARL): (1) single pass of ≈10 BV, or (2) 10 BV with last 5 BV used in next strip. Usual sequence (all flows at ≈2 BV/h): 1 BV of hot 3%–5% NaCN; 4–5 BV of strip solution from previous strip and combine with first BV to make pregnant solution; 4–5 BV of fresh water (<200 ppm TDS), saved for next strip; 1 BV of cold fresh water to cool the carbon and transfer it. Preg is batched through EW.
- Usually height:diameter ratio of strip column = 6:1. Hotter solution temperature may allow lower ratio.

Regeneration (Most Problem-Prone; Check Mechanicals and Fabrication)

- O_2-free atmosphere with steam; $C + H_2O = CO + H_2$; endothermic; indirect heating.
- More than 600°C and less than 750°C; maybe 800°C if many organic contaminants are on carbon.
- Feed wet carbon (water makes the steam); add water spray at kiln discharge if needed.
- Check temperature profile along kiln's length.
- Two fans: (1) combustion gases (need to be reducing if they are allowed to enter kiln for heat recuperation) and (2) regeneration products.
- Regenerate only as needed, not necessarily every time.

BACTERIAL OXIDATION
by Andy Briggs, P.E.

TABLE 17.20 Bacteria

Bacteria	Oxidation of	Doubling time (h)	Approximate operating temperature (°C)
Mesophiles			
Thiobacillus ferrooxidans	Fe^{2+} and sulfide minerals	3.6–12.0	25–45
Thiobacillus thiooxidans	S^0 and soluble sulfide compounds	10–24	25–45
Leptospirillum ferrooxidans	Fe^{2+}	9–36	25–45
Moderate thermophiles	(diverse group) Fe^{2+}, S^0, sulfide minerals	1	45–55
Sulfobacillus thermosulfidooxidans			
Extreme thermophiles	Fe^{2+}, S^0, sulfide minerals	≈42	>50
Sulfolobus species			
Acidianus species			

- Typical cell formula (Mintek) $CH_{1.67}N_{0.2}P_{0.014}O_{0.27}$; Size—Mesophiles 0.5 × 1.0 μm (rod or comma shaped)
- Concentration: 10^6/mL in heap leach solutions; 10^8–10^9/mL in stirred tank solutions.

Toxicity

Toxicity is variable and not known with any accuracy. Minimum inhibitory concentrations are approximately (mg/L): SX reagents, 16; other organics, <24; Cl^-, 1,500–7,000 (inhibited), 19,000 toxic; NO_3, 200; SCN^-, 2; CNO^-, 10; metal CN^- complexes, 10; As^{3+}, 6,000; Hg, ~4.

Minerals Oxidized
- Sequence: FeS > Cu_2S > FeAsS > ZnS > CuS > FeS_2 > $CuFeS_2$ (passivates)
- Reactions:
  Pyrite: $4FeS_2 + 15O_2 + 2H_2O \rightarrow 2Fe_2(SO_4)_3 + 2H_2SO_4$
  Arsenopyrite: $2FeAsS + 7O_2 + 2H_2O + H_2SO_4 \rightarrow 2\,H_3AsO_4 + Fe_2(SO_4)_3$
  Pyrrhotite: $2FeS + 4.5\,O_2 + H_2SO_4 \rightarrow 2Fe^{3+} + 3SO_4^{2-} + H_2O$
  Chalcopyrite: $4CuFeS_2 + 2H_2SO_4 + 17O_2 \rightarrow 4Cu^{2+} + 4Fe^{3+} + 10SO_4^{2-} + 2H_2O$
  Chalcocite: $Cu_2S + H_2SO_4 + 2.5\,O_2 \rightarrow 2Cu^{2+} + 2SO_4^{2-} + H_2O$
  Covellite: $CuS + H_2SO_4 + 2O_2 \rightarrow Cu^{2+} + 2SO_4^{2-} + 2H^+$
  Bornite: $Cu_5FeS_4 + 4H_2SO_4 + 6O_2 \rightarrow FeSO_4 + 5CuSO_4 + 2S° + 4H_2O$
  Sphalerite: $ZnS + 2.5O_2 + 2H^+ \rightarrow Zn^{2+} + SO_4^{2-} + H_2O$
  Galena: $PbS + 2O_2 = Pb^{2+} + SO_4^{2-} \rightarrow PbSO_4$
  Siegenite: $(Co,Ni)_3S_4 + 7.5\,O_2 + H_2O \rightarrow xCo^{2+} + (3-x)Ni^{2+} + 4SO_4^{2-} + 2H^+$.

Calculations
- O_2 utilization efficiency: 30%–40% for stirred tanks, 20%–25% for heaps.
- Pressure: Stirred tanks hydrostatic head +15 kPa line loss; heaps need approximately 8 kPa + line losses (the heap is geometry-dependent).
- Gas held up in reactors is 7%–15% by volume.
- Blower power:

$$kW = 1.0195\,WT((P_2/P_1)^{0.283} - 1) \qquad \text{(EQ 17.20)}$$

where:
W = air flow in kg/s
T = inlet air temperature, K
P_1 = inlet pressure
P_2 = delivery pressure (P_1 and P_2 absolute [*not* gauge] pressure)

- Blower efficiency: 74%–78%.
- Agitator power (stirred tank reactors): approximately 27 W/Nm^3h air added (agitator-dependent).

TABLE 17.21 Heat of reaction, oxygen requirement, and H_2SO_4 demand

| | | Heat of reaction | | Oxygen requirement | | H_2SO_4 demand |
		Mineral kJ/kg	Sulfide kJ/kg S^{2-}	mol O_2/mol mineral	kg O_2/kg S^{2-}	(kg/kg mineral)
Pyrrhotite	FeS	−11,373	−31,245	2.25	2.25	0.557
Arsenopyrite	FeAsS	−9,415	−48,036	3.5	3.5	0.301
Pyrite	FeS_2	−12,884	−24,173	3.75	1.88	−0.408
Chalcopyrite	$CuFeS_2$	−9,593	−27,505	4.25	2.13	
Chalcocite	Cu_2S	−6,201	−30,811	2.5	2.5	
Covellite	CuS	−792	−24,756	2	2	
Pentlandite	$(Ni,Fe)_9S_8$	−10,174	−30,644	17.63	2.2	
Ankerite	$Ca(Fe,Mg)(CO_3)_2$	−219.2				0.979
Siderite	$FeCO_3$	−326.7		0.069*		1.267

* kg/kg mineral for $Fe^{2+} \rightarrow Fe^{3+}$.

Heat Balances
- Stirred tank reactors
  - Gains: Reactions, agitator power, air addition.
  - Losses: Evaporation (air leaves reactors at operating temperature, saturated in moisture); losses through tank walls; heat up of incoming slurry; Joule/Thomson effect of air expansion.
  - Warm cooling water ~4°C less than reactor temperature.

$$\text{Flow rate (m}^3/\text{h)} = \frac{\text{Heat load (kW)} \times 3.6}{\Delta T \times 4.186} \quad \text{(EQ 17.21)}$$

- Heaps
  - Gains—reactions, solar radiation (day)
  - Losses—evaporation (air addition leaves heap saturated in moisture); evaporation of irrigation solutions; convection from heap surfaces; radiation (night); heat up of irrigation solutions.
  - At 1% S^{2-} oxidized in the ore, the challenge is to maintain heap temperature; at 3% S^{2-} oxidized, the challenge is to remove the heat.
  - Temperature control in heaps by rest periods, irrigation rates, air addition rates, use of covers.

Typical Circuit Conditions
Slurry density, 15%–20% solids; residence time, 4–6 days; temperature, 40°C; pH range, ≈1.2–1.6; dissolved O_2, 2 ppm.

Nutrient Additions, kg/T of Concentrates
- Standard biological oxidation addition: N, 1.7; P, 0.3; K, 0.9
- Calculated from cell requirements: N, 0.11; P, 0.016; K, zero

TABLE 17.22 Gold recovery in stirred tank reactors—plant comparisons

Item	Unit	Fairview	Sao Bento*	Harbour Lights	Wiluna	Ashanti (Sansu)	Youanmi[†]	Tamburaque
Arsenopyrite	%	13	18	17	22	17	9.4	56
Pyrite	%	33	16	28	33	6	49	35
Pyrrhotite	%	NIL	19	NIL	NIL	13	NIL	NIL
Gold	g/t	109	30	86.6	92.9	76.4	64	23
S^{2-}	%	20	18.7	18.6	22	11.4	28	30
S^{2-} oxidized	%	89	—	87	95	94	32	86
Concentrate feed	t/day	35	—	40	158	960	120	60
Total reactor volume (aerated)	m³	913	1,374	880	3,391	18,278	3,000	1,336
Heat of reaction	MW	2	—	2.2	11.1	41.1	4	6.1
Oxygen demand	kg/h	540	—	612	3,024	10,872	987	1,656
Oxygen demand	kg/h/m³	0.59	—	0.69	0.89	0.59	0.33	1.24
S^{2-} oxidation rate	kg/m³/day	6.8	—	7.4	9.7	5.6	3.3	11.6
Specific power consumption	kWh/kgS^{2-}	1.9	1.8	1.9	1.5	1.9	2.3	
Gold leach	h				24	32		
CN consumption	kg/t cons				30	10	7.5	
Lime addition	kg/t cons				50	30	25	
Gold recovery	%				96	94	90–95	

* Sao Bento: pretreatment by bacterial oxidation, followed by pressure oxidation—throughput very elastic.
† Youanmi is Bactech technology.

- Additional parameters to check: $Fe^{2+}:Fe^{3+}$ ratio (separate from redox potential), oxygen uptake rate, presence of poisons in solution, and reagents (shale flask tests).
- In most cases, when the circuit is operating poorly, the concentrate feed to the circuit should be reduced or stopped until the parameters have stabilized.

TABLE 17.23 Bacterial oxidation control and troubleshooting

Control Parameter	Response Time	Comments
Dissolved oxygen (DO) level	Within minutes	If DO is high and same air addition, then oxidation rates have dropped. If DO is low, not enough air is being added.
Redox potential	Within a day	If redox potential is low, oxidation rate is low.
Cooling water addition	Within hours	If cooling water addition is low, oxidation rate drops. If temperature is high, not enough water is being added.
pH	Within a day	If pH is low, oxidation of pyrite has increased and more limestone is required. If pH is high, oxidation rate is decreasing.
Bacterial counts	A few days	Lower counts indicate reproduction rate lower than washout rate—slow feed down.
Froth	Within minutes	Excessive froth indicates bacteria under stress—watch other parameters. Froth should be made up of fine bubbles, not coarse.
Smell		A strong acid smell should be evolved from the top of the slurry.
Color		The color should change from dark gray (feed) to yellowish green (discharge). Final leach solution should be reddish (Fe^{3+}).

Solid–Liquid Separation

Usually performed in 3 CCD high-rate thickeners; thickener area required typically is 4.2–9.0 $m^2/t\cdot h$ (depends on particle size and level of oxidation); solids content in underflow ~33%; wash ratio typically 6 tonnes solution per tonne biooxidation residue solution; flocculant additions ≈80 g/t for the first stage and 25 g/t for the second and third stages.

Neutralization

A two-stage pH-controlled circuit for arsenic precipitate stability:

Stage 1: neutralization to pH 4–5:

$$H_2SO_4 + Ca(OH)_2 \rightarrow CaSO_4 \cdot 2H_2O$$
$$2Fe_2(SO_4)_3 + 2H_3AsO_4 + H_2SO_4 + 7CaCO_3 + 13H_2O \rightarrow 2FeAsO_4 + 2Fe(OH)_3 + 7CaSO_4 \cdot 2H_2O + 7CO_2$$

Stage 2: Neutralization to pH 6–8:

$$MSO_4 + Ca(OH)_2 + 2H_2O \rightarrow M(OH)_2 + CaSO_4 \cdot 2H_2O$$

- Number of stages is 4–6; residence time (h/stage) is 1.5–1.0
- Arsenic stability factors: Fe/As ratio, optimum ≥3:1; pH, optimum 4.5–5.0; temperature, higher increases stability; co-precipitation of other compounds (gypsum, Zn, Cu, Cd salts) increases stability.
- Products, depending on pH and Fe/As molar ratio, may be scorodite ($FeAsO_4 \cdot 2H_2O$), basic ferric arsenate ($FeAsO_4 \cdot xFe(OH)_3 \cdot CaHAsO_4$), amorphous calcium arsenate ($Ca_3(AsO_4)_2$), hydrated calcium arsenate ($Ca_2AsO_4 \cdot OH$).

Stirred Tank Reactors for Base Metals

- No commercial operations exist.
- Kasese (commissioned 1999) 240 tpd of pyrite concentrate for cobalt recovery.
- BioCOP® process under development by Billiton.
- MinBac process for chalcopyrite concentrates under development by Mintek/Bactech.

Bacteria Heap Leach Operations
- All commercial operations as of 1999 are only for Cu recovery from Cu_2S and CuS.
- Gold operations are at the development phase.

COAL CLEANING
by Gary F. Meenan
- Cleaning = separation of low specific gravity, low ash coal (product) from high specific gravity, high ash coal (refuse) to give the yield and quality required.
- Raw coal characteristics are determined by a specific gravity analysis on a dry weight basis at specific gravities of 1.3, 1.35, 1.4, 1.45, 1.5, 1.55, 1.6, 1.65, and 1.7. Each fraction, including float and sink, is weighed and assayed for ash (%) and sulfur (%). Higher specific gravity fractions have higher ash contents.

TABLE 17.24 Relative difficulty of separation for different values of the specific gravity distribution

Wt % Feed in the ±0.10 Specific Gravity Units	Degree of Difficulty
0 to 7	Simple
7 to 10	Moderately difficult
10 to 15	Difficult
15 to 20	Very difficult
20 to 25	Exceedingly difficult
Above 25	Formidable

Source: Bird 1928.

Coal cleaning performance criteria are yield, coal recovery, and distribution curves (probable error, E_p, and error area).
- Yield depends on ROM coal's specific gravity distribution; use ash balances on feed, product, and refuse streams.

$$Y_p \times 100\% = (A_r - A_f) / (A_r - A_p) \quad (EQ\ 17.22)$$

where:
Y_p = yield of product
A_p = % ash in product
A_f = % ash in feed
A_r = % ash in refuse

- Recovery is the amount of ash-free coal in product divided by the amount of ash-free coal in feed and can be expressed as

$$CR_p = (Y_p \times 100\%)(100 - A_p) / (100 - A_f) \quad (EQ\ 17.23)$$

where:
CR_p = ash-free coal recovery of product
CR_r = ash-free coal recovery of refuse

- Distribution curve—method for assessing performance of coal cleaning equipment.
  - Distribution curve = % recovery to product versus mean density of each specific gravity fraction.
  - Distribution curve is independent of ROM coal distribution by specific gravity; it depends on the cleaning equipment.
  - Construct distribution curve as follows:
    - Perform specific gravity analysis of product and refuse and get ash assay of feed as shown in following table.

- Determine product yield: Y_p = 100(77.65 − 36.61) / (77.65 − 5.88) = 57.19%
- Determine refuse yield: Y_r = 100% − 57.19% = 42.81%
- Multiply the weight percent of each specific gravity fraction of product by $Y_p/100$ for a table of adjusted weight percents.
- Multiply the weight percent of each specific gravity fraction of refuse by $Y_p/100$ for a table of adjusted weight percents.
- Add the adjusted weight percents of the product and refuse in each specific gravity fraction

TABLE 17.25 Example of coal feed data (ash content 36.61%)

Specific Gravity Fraction	Wt (%)	Ash (%)	Sulfur (%)	Cumulative Weight (%)	Cumulative Ash (%)	Cumulative Sulfur (%)
Product						
Float 1.30	74.26	4.02	1.17	74.26	4.02	1.17
1.30 × 1.35	15.70	8.41	2.19	89.96	4.79	1.35
1.35 × 1.40	6.59	13.40	2.97	96.55	5.37	1.46
1.40 × 1.45	2.58	18.44	3.31	99.13	5.71	1.51
1.45 × 1.50	0.62	22.52	3.78	99.75	5.82	1.52
1.50 × 1.55	0.15	27.04	3.98	99.90	5.85	1.53
1.55 × 1.60	0.05	32.99	4.62	99.95	5.86	1.53
1.60 × 1.70	0.03	37.19	4.24	99.98	5.87	1.53
1.70 sink	0.02	53.92	5.07	100.0	5.88	1.53
Refuse						
Float 1.30	0.50	4.04	1.11	0.50	4.04	1.11
1.30 × 1.35	0.23	8.33	2.32	0.73	5.39	1.49
1.35 × 1.40	0.24	14.20	2.58	0.97	7.57	1.76
1.40 × 1.45	0.53	19.68	2.89	1.50	11.85	2.16
1.45 × 1.50	1.61	23.85	3.27	3.11	18.06	2.73
1.50 × 1.55	1.58	28.42	3.49	4.69	21.55	2.99
1.55 × 1.60	1.15	33.55	3.19	5.84	23.91	3.03
1.60 × 1.70	2.02	39.21	2.98	7.86	27.85	3.02
1.70 sink	92.14	81.90	1.80	100.0	77.65	1.90

TABLE 17.26 Weight percent of feed

Specific Gravity Fraction	Wt % of Product	+	Wt % of Refuse	=	Wt % of Feed
Float 1.30	42.47		0.21		42.68
1.30 × 1.35	8.98		0.10		9.08
1.35 × 1.40	3.77		0.10		3.87
1.40 × 1.45	1.48		0.23		1.70
1.45 × 1.50	0.35		0.69		1.04
1.50 × 1.55	0.09		0.68		0.76
1.55 × 1.60	0.03		0.49		0.52
1.60 × 1.70	0.02		0.86		0.88
1.70 Sink	0.02		39.44		39.46

- For each SG fraction of product, divide weight percent of product by weight percent of feed for that fraction.
- For each SG fraction of refuse, divide weight percent of refuse by weight percent of feed for that fraction.

TABLE 17.27 Weight fraction to product

Specific Gravity Fraction	Wt % Product/Wt % Feed	= Wt fraction to Product
Float 1.30	42.47/42.68	0.995
1.30 × 1.35	8.98/9.08	0.989
1.35 × 1.40	3.77/3.87	0.974
1.40 × 1.45	1.48/1.70	0.871
1.45 × 1.50	0.35/1.04	0.337
1.50 × 1.55	0.09/0.76	0.118
1.55 × 1.60	0.03/0.52	0.058
1.60 × 1.70	0.02/0.88	0.023
1.70 Sink	0.02/39.46	0.001

- Plot weight fraction to product coal versus mean specific gravity of each specific gravity fraction. This is the distribution curve. The mean specific gravity of 1.3 float and 1.7 sink gravity fractions are ≈1.25 and ≈1.90.

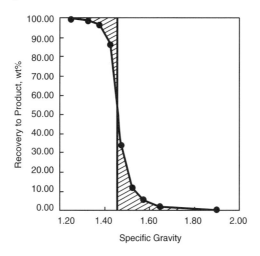

FIGURE 17.6 Distribution curve showing error area

- Performance criteria derived from distribution curve are probable error, E_p, and the error area. Sharp separations have small error and small area. E_p gives no information about recovery of low ash clean coal or the rejection of high ash impurities. Typical E_p for heavy media = 0.020; for hydrocyclones = 0.225.

$$E_p = (d_{25} - d_{75}) / 2 = (1.50 - 1.44) / 2 = 0.030 \quad \text{(EQ 17.24)}$$

where:
E_p = probable error
d_{25} = SG at 25% recovery to product
d_{75} = SG at 75% recovery to product

- Error area determined by planimeter is ≈20 for curve shown above. Typical error areas for heavy media are 10 and for hydrocyclones are as high as 125. Jigs and tables are between.

Operations and Troubleshooting
- Typical equipment for cleaning coarse coal (+9.5 mm) are heavy-media vessels and Baum jigs; for intermediate coal (9.5 mm × 0.6 mm), heavy-media cyclones, Batac jigs, Deister tables; for fine coal (–0.6 mm), mechanical flotation cells; for ultra-fine coal (–50 µm) that is high in clay, column flotation cells.
- Most new coal plants have a spiral circuit in them. The feed stream must be properly sized and constant. Optimal size range is 1.18 mm × 0.25 mm (16 × 60 mesh). Retreating the primary middlings in secondary spirals is more common and it minimizes coal losses. Consider retreating the secondary spiral middlings in a heavy-media cyclone circuit.
- Coarse and intermediate size bituminous coal are dewatered with screens and basket centrifuges to a final moisture content of ≈5%. The fines are dewatered with vacuum disc filters (≈25% moisture) and screen-bowl centrifuges (≈20%–21% moisture). Feed size of fines is typically 0.6 mm × 0 with 55% greater than 0.15 mm. Filtration rate (kg/h·m^2) increases with (1) the square root of pressure, (2) increasing feed solids content, and (3) quicker cake formation or filter cycle.
- Vacuum filter maintenance: (1) change sectors on a routine basis, (2) use high pressure sprays to minimize the buildup of solids internally and keep screen cloth clean, (3) provide a minimum of 5 cfm/ft^2 of air suction to maximize throughput and moisture removal, (4) maintain the slurry agitators to minimize the "donut" effect and avoid sanding out in the tub (air agitation with simple mushroom valves effectively keep the solids in suspension), (5) use a snap blow cake discharge mechanism to avoid sector buildup, (6) control tub level to avoid conditions where the minimum submergence level to maintain vacuum is exceeded, and (7) maintain a minimum vacuum of 10–12 in. of Hg.
- Limits on impurities are dictated by the end user or contract specifications. Product sulfur content can range from 0.2%–5.0%, the heating value can range from 8,000 Btu/lb for lignite to more than 13,000 Btu/lb for bituminous coals. The product ash content range can be from 5% to 13%. The moisture content also varies with the rank of coal. The total moisture of bituminous coal (plant product) ranges from 6 to 12%. Thermally dried bituminous coal is usually under 6%. Western or lignite coal can have +40% moisture. The metallurgical coal must have a sulfur content of 1% or less, high heating value (about 13,000 Btu/lb), an ash content of around 5%, and a moisture content of about 5%.
- Anthracite and bituminous coals are treated similarly. Lignite is not cleaned.

DUST COLLECTION
by Paul D. Chamberlin, P.E.

Fabric Dust Collectors
- Suggested air-to-cloth ratios (ft^3/ft^2):cement at 1.5 to 2; coal at 2 to 2.5; limestone at 2 to 2.5; sand at 3 to 3.5. Conversions are 35.31 ft^3/m^3 and 10.76 ft^2/m^2.
- Floating dust can usually be made to flow at 61 m/min. Total air flow = 61 m/min times total square meters of all openings in the housing on the dust creating machine. Velocity in ducts should be about 1,066 m/min for light dusts to 1,827 m/min for heavy dusts.

TABLE 17.28 Fabric bag materials

	Maximum Long-Term Operating Temperature (°C)	Tensile Strength	Resistance		
			Abrasion	Acid	Alkali
Cotton	82	G	F	P	E
Polypropylene	93			E	E
Nylon	93	E	E	P	E
Acrylic	121		G	G	F
Polyester	135	E	G	G	F
Nomex felt	218	E	G	F	E
Teflon, Gore-Tex	218			E	E
Fiberglass	260	E	P	F	P
P84, polyimide	260			E	
Nextel 312, ceramic	500				

E = excellent; G = good; F = fair; P = poor
Source: Bergmann 1981 (reprinted with permission of Chemical Engineering).

Suggested Dust Collection Equipment versus Particle Size

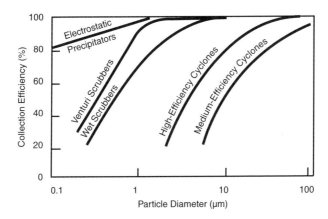

Source: Rossano 1969 (reprinted with permission of the McGraw-Hill Companies).

FIGURE 17.7 Collection efficiency versus particle size

Power Requirements versus Particle Size for Various Collectors

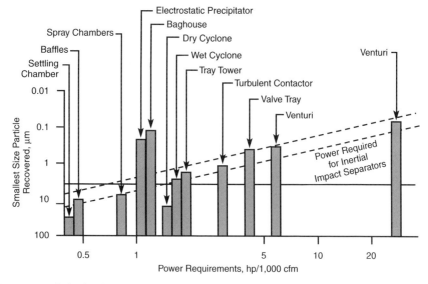

Source: Agarwal, Flood, and Giberti 1974 *(reprinted with permission of the Journal of Metals).*

FIGURE 17.8 Power requirements versus particle size

Troubleshooting

- Electrostatic precipitators (ESPs): Keep flush water below 1,000 mg/L $CaSO_4$ and 15 mg/L CaF_2 to prevent scaling. Adjust pH to control $CaCO_3$. Oils in gas or flush water lower efficiency of ESP. Adequate grounding is needed; check for galvanic corrosion between dissimilar metals. Keep flush water below its adiabatic saturation temperature.
- Needlepoint felts are better than woven fabrics for pulse jet filters.
- Pleated paper cartridge filters are best for control rooms; dirty air inlet typically is <23 g/m³ and pH must be 6 to 8.

Dust Control

- Generally, dust becomes a problem when moisture content is <4%.
- Pressure drop for venturi scrubbers: ≈25.4 cm water gauge removes ≈100% of 2-μ particles; 50 cm, 0.8 μ; 90 cm, 0.6 μ; 140 cm, 0.4 μ; 216 cm, 0.2 μ.
- Cooling hot gases from 650°C to 290°C using injected water sprays increases needed baghouse capacity by 25% (it is a straight line relationship). Water injection must be far enough upstream to be sure it evaporates before reaching the baghouse. Also beware of fan erosion, scale buildup, metal distortion, refractory failures, and nozzle failures.
- Gas velocity to avoid material buildup in the bottom of ducts should range from 1,066 m/min to 1,827 m/min.
- Chemical control of dust is by (1) wet suppression with treated or untreated water, (2) foam suppression, and (3) crusting of storage piles.
- Wet suppression generally requires 4 to 8 L of treated water (0.05%–2.0% wetting agent) per tonne when wetting ore on a conveyor; without a wetting agent, water use is 3 to 4 times more.
- Foam suppression usually requires 0.8 to 1.6 L of treated water per tonne of ore; it does a better job of reducing respirable dust than wet suppression.
- Crusting agents are applied by spraying or with a hydroseeder.

- Mechanical control of dust is by (1) dust collectors, (2) hoods and ducts, and (3) tight control of conveyor systems (<150 m/min, transfer chutes that drop ore onto receiving belt with same velocity and direction, skirt board seals, install two belt cleaners in series, install slider beds or cradles at loading point to avoid belt sag).

ELECTROWINNING (EW) AND ELECTROREFINING (ER)
by Douglas J. Robinson

Background
- Primary cathode reaction: $M^{+z} + ze^- = M°$
- Primary anode reaction in EW: $H_2O = 2H^+ + 0.5\,O_2(gas) + 2e^-$
- Or, in the case of an EW chloride electrolyte: $2Cl^- = Cl_2 + 2e^-$
- Primary anode reaction in ER: $M° = M^{z+} + ze^-$
- Electromotive series (typical voltages); most active metals dissolve first and most noble metals deposit first:

Noble							Active
Au^{+3}/Au	Ag^+/Ag	Fe^{+3}/Fe^{+2}	Cu^{+2}/Cu	H^+/H_2	Pb^{+2}/Pb	Ni^{+2}/Ni	Zn^{+2}/Zn
1.35v	0.79	0.55	0.335	0.00	−0.126	−0.25	−0.76

$$\text{Faraday's Law: } W = I \times t \times M \times CE / (z \times F) \qquad \text{(EQ 17.25)}$$

where:
- W = grams metal dissolved or deposited
- I = current in amps
- t = time in seconds
- M = molecular weight of the metal in grams
- CE = current efficiency as a decimal fraction
- z = change of valence in the deposition or dissolution reaction, i.e., moles electrons per mole metal
- F = faraday = 96,500 coulombs per mole electrons

coulombs = amps times seconds

For example, 1,000 amps will produce about 28.5 kg of $Cu°$ from Cu^{++} per day per cell at 100% CE.

TABLE 17.29 Typical electrolytic plant operating data

Item	Cu	Zn	Ni	Pb	Ag
Range of plant production, stpd	9–450	270–820	36–180	91–540	0.1–1.6
Cathode current density, amps/m²	215–377	377–807	215–269	161–215	54–215
Cell voltage, EW	1.96	3.2	3.8		
Cell voltage, ER	0.28–0.48			0.09–0.15	0.13
kWh/kg produced, EW	2.1	3.5	3.8		
kWh/kg produced, ER	0.3–0.5			0.33–0.52	0.5
Cathode material	SS, Cu	Al	SS, Ni	Pb	SS
Cathode size, m²	1.86–2.32	1.86–3.25	1.86–2.32	1.49–2.32	0.28–0.56
Anode material	Rolled Ca-Sn-Pb	Cast Ag-Pb	Sb-Pb	Gr	Ag
Metal salt	$CuSO_4$	$ZnSO_4$	$NiSO_4$, $NiCl_2$	$PbSiF_6$	$AgNO_3$
Acid concentration, g/L	180	160	pH 3	100	pH 3
Reagents, 5–15 mg/L	Guar	Glue		ECA, glue	Glue

- There is no adequate method of making a "first principles" calculation of current density that is general for all metals. Current density is selected from operating experience. The practical current density will be <40% of the limiting current density to avoid the onset of powdery deposits. Mass transfer theory says that an approximation of the limiting current density will be I_L (in amps/m^2) = 18.6 × [M$^+$] in g/L; i.e., for a 40 g/L Cu^{++} solution, $I_L \approx$ 744 amps/m^2.
- Deposition generally should be under activation or charge transfer control. The cathode over-voltage typically exceeds 30 mV. Temperatures often exceed 40°C.
- A surface-active agent is generally required to control crystal growth if the electrolyte temperature exceeds 45°C. Concentrations in the mg/L range are suitable.
- The submerged area of the cathodes should be as large as possible to maximize the current carried by each electrode (this minimizes the total number of cathodes and the overall capital cost needed to meet design production).
- The submerged anode dimensions should be 2.5 to 3.8 cm less than the cathode on each side and at the bottom when the anode-cathode spacing is about 3.8 cm. If the anode is too wide or long, the cathode edges will be rough. Anode edges radiate current cylindrically, with good throwing power at angles up to 45°. Therefore, cathode overlap should approximately equal anode-cathode spacing.
- Solution resistance depends primarily on electrode spacing: R (in ohms) = $r(L/A)$ where r = resistivity of the electrolyte in ohm-cm, A = submerged cathode area in cm^2, and L = electrode spacing in cm. Total plant voltage = (number of cells × cell voltage) + bus bar losses of 1–2 volts maximum.
- The number of electrodes per cell is a compromise. Typical values for any metal are 20–30 cathodes/cell at a production rate of 20 tpd, 50–60 at 100–300 tpd, and 70–90 for larger plants. An increase in the number of cathodes/cell decreases operating labor costs whereas a decrease in the number of cathodes/cell decreases capital cost by reducing both the rectifier current and the cross-sectional area of bus bars.
- Cathode width is determined by the width of cathode material that can be purchased and by the door opening size on the melting furnace. The total number of cathodes = total current/(current density × submerged cathode area). The number of cells = total cathodes/(cathodes/cell). Cells should be adjacent with their tops all at the same elevation and the ends of all cells in perfect alignment. Each row of cells should have many cells in it and the number of rows should be an even number.
- Backup rectifiers and transformers should be installed.
- Bus bars should not be operated >40°C above ambient air. Short circuits lead to overheating of the intercell bus bar, which can lead to two warped electrodes in the adjacent cells. Their current-carrying capacity, ampacity, depends on the material, i.e., copper ampacity is 3 times that of aluminum. The ampacity also depends on bus bar shape; a perfect shape would be hollow bars with coolant running through them to keep the bus bars cool. Typical shapes are rectangular and the ampacity varies with orientation which affects cooling, e.g.:
  - A bar 6 mm wide × 100 mm tall (vertical orientation) can carry 155 amps/cm^2.
  - A bar 100 mm wide × 6 mm tall (horizontal orientation) can carry 77 amps/cm^2.
  - Four bars 6 mm wide × 100 mm tall at 6 mm spacing can carry 109 amps/cm^2.
- Pull cathodes in gangs of every second or third cathode so that at least one surface of each anode is always active. Pull as large a gang as possible to minimize crane trips. Wash electrolyte from cathodes before it dries and crystallizes; once crystallized, it can never be totally washed off. Remove deposits from blanks by flexing and high frequency, low amplitude hammering. High impact hammering can deform and damage the blanks.
- The reactive film on anodes (PbO$_2$ on Pb alloy anodes and RuO$_2$ or other Pt group oxide on titanium anodes) must be maintained while repairing or bypassing cells to prevent chemical contamination of the plated cathode metal. This is done with properly designed shorting frames or by draining the electrolyte. Ramping the current down

(15 min) and then back up (45 min) is another way but it loses production and can be uneconomic. Power failures can ruin the reactive film and contaminate cathodes; such problems can be avoided with backup generators.
- Electrode envelopes (diaphragms):
  - Can obtain up to 15 g/L metal concentration change if diaphragms are put around cathodes and a solution that has been purified by hydrometallurgical methods is fed into the cathode bag and allowed to permeate through the diaphragm into the common anode chamber, e.g., nickel refining.
  - Can get up to 90 g/L acid change across diaphragms that are put around anodes when high pH electrolyte is fed into the cathode chamber and some of it flows through the diaphragms into the individual anode chambers; e.g., Co EW.
  - Chlorine gas from chloride electrolysis can be totally captured by enclosing the anodes in diaphragm bags.
- Control acid mist generated at the anodes in sulfate electrolysis by enclosing anodes in a bag; fitting anodes with hoods; covering each cell with hydrophobic beads, balls, blankets; providing cross-flow ventilation to carry the mist outside the building or to a scrubber; and by using alternative anode reactions such as the ferrous-ferric couple.
- Uniform electrolyte flow is essential for (1) each cell, (2) the flow of current between each pair of electrodes, and (3) the electrolyte concentration throughout the inter-electrode gaps. Items 2 and 3 are achieved by creating a uniform reservoir of electrolyte beneath the electrodes and then allowing the natural convective flow generated by the flow current to the electrodes to pump the electrolyte into the gap. Some plants incorporate a flow distributor in each cell; others use a 0.3-m-deep space below the electrodes to create the reservoir naturally. A sufficient flow is generally created with a change in metal ion concentration of 3–5 g/L between the feed and the spent electrolytes. The flow rate of spent electrolyte overflowing each cell can be determined by use of a rectangular weir plate in the overflow box.
- Uniform electrode spacing is essential and is established by insulating-spacing capblocks on the intercell walls and by insulators on the anode body.
- Electrode bars and hanger bars. The connection between the electrode blade and bar must be metallurgically sound to minimize voltage drop and ensure uniform current distribution. In Zn EW, the cathode blade is Al alloy and the bar is Al° fitted with a copper contact tip and Al° lifting hooks. In copper, the blade is stainless steel (SS) and bars are either a slotted solid Cu° or a stainless-steel tube electroplated with a 3 mm thick coating of Cu°. Some metal processes use Ti cathode blades welded to a copper-cored Ti bar.
- Materials of construction:
  - New cells are generally of cast polymer concrete, a mixture of 12%–15% vinyl ester resin with at least two sizes of silica sand grains, 0.3 mm and 6 mm.
  - Buildings are reinforced concrete.
  - Floors are concrete coated with a 12-mm-thick layer of polymer concrete.
  - Steel columns are coated with epoxy paint. Sometimes an SS or fiber-reinforced plastic (FRP) sheathing is attached to the inside of the columns for increased corrosion protection.

Checklist
- Tidy plant appearance
- Cells in alignment—tops, spacing, front and back walls
- Electrodes in perfect straight line rows and with uniform spacing
- Electrodes flat and straight
- Electrode contacts cleaned regularly
- Bus bars cool to the touch
- Salt spills promptly cleaned up

- Working platforms unobstructed
- Flow rates to all cells uniform
- Spent solution concentrations on target
- Additive flow rates and concentrations properly set
- Every pair of anode-cathode surfaces identical with every other pair
- Handrails maintained in perfect condition (that will ensure that everything else is under control).

PYROMETALLURGY
by Terry P. McNulty, P.E.

Physical Chemistry
- 1 calorie (small) = 4.184 Joules; mol = gram molecular weight
- Heat capacity = heat energy needed to raise temperature of 'x' weight by 1.0°C.
- Specific heat of water = 1 calorie/g°C = 4.184 Joules/g°C = 1 Btu/lb°F
- Specific heat of many minerals and rocks ≈ 0.25 cal/g°C

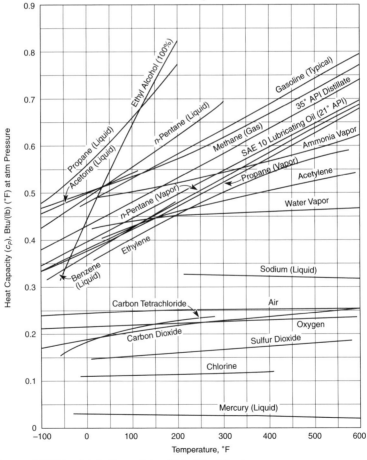

Source: Foust et al. 1980 (reprinted with permission of John Wiley & Sons).

FIGURE 17.9 Heat capacities of gases and liquids

TABLE 17.30 Ratio of heat capacities at 1 atm (C_p/C_v)

Compound	Formula	Temperature (°C)	Ratio of Specific Heats ($\gamma = C_p/C_v$)
Acetylene	C_2H_2	15	1.26
		−71	1.31
Air	...	925	1.36
		17	1.403
		−78	1.408
		−118	1.415
Ammonia	NH_3	15	1.310
Argon	Ar	15	1.668
		−180	1.76
		0–100	1.67
Benzene	C_6H_6	90	1.10
Carbon dioxide	CO_2	15	1.304
		−75	1.37
		−180	1.41
Chlorine	Cl_2	15	1.355
Dichlorodifluormethane	CCl_2F_2	25	1.139
Ethane	C_2H_6	100	1.19
		15	1.22
		−82	1.28
Ethyl alcohol	C_2H_6O	90	1.13
Ethyl ether	$C_4H_{10}O$	35	1.08
		80	1.086
Ethylene	C_2H_4	100	1.18
		15	1.255
		−91	1.35
Helium	He	−180	1.660
Hexane (n−)	C_6H_{14}	80	1.08
Hydrogen	H_2	15	1.410
		−76	1.453
		−181	1.597
Methane	CH_4	600	1.113
		300	1.16
		15	1.31
		−80	1.34
		−115	1.41
Methyl alcohol	CH_4O	77	1.203
Nitrogen	N_2	15	1.404
		−181	1.47
Oxygen	O_2	15	1.401
		−76	1.415
		−181	1.45
Pentane (n−)	C_5H_{12}	86	1.086
Sulfur dioxide	SO_2	15	1.29

Source: Foust et al. 1980 *(reprinted with permission of John Wiley & Sons).*

TABLE 17.31 Heat capacities of solids

Metals	Heat Capacity, Btu/lb °F
Aluminum	0.374 (0°C), 0.405 (100°C)
Copper	0.164 (0°C), 0.169 (100°C)
Iron, cast	0.214 (20–100°C)
Lead	0.0535 (0°C), 0.0575 (100°C)
Nickel	0.0186 (0°C), 0.206 (100°C)
Steel	0.12
Tin	0.096 (0°C), 0.104 (100°C)
Zinc	0.164 (0°C), 0.172 (100°C)

Miscellaneous	Heat Capacity, Btu/lb °F
Alumina	0.2 (100°C), 0.274 (1,500°C)
Asbestos	0.25
Brickwork	About 0.2
Carbon (mean values)	0.168 (26–76°C)
	0.314 (40–892°C)
	0.387 (56–1,450°C)
Cellulose	0.32
Cement, portland clinker	0.186
Charcoal (wood)	0.242
Chrome brick	0.17
Clay	0.224
Coal	0.26 to 0.37
Coke (mean values)	0.265 (21–400°C)
	0.359 (21–800°C)
	0.403 (21–1,300°C)
Concrete	0.156 (70–312° F), 0.219 (72–1,472° F)
Fireclay brick	0.198 (100°C), 0.298 (1,500°C)
Fluorspar	0.21 (30°C)
Glass (crown)	0.16 to 0.20
(flint)	0.117
(pyrex)	0.20
(silicate)	0.188 to 0.204 (0–100°C)
	0.24 to 0.26 (0–700°C)
wool	0.157
Graphite	0.165 (26–76°C), 0.390 (56–1,450°C)
Gypsum	0.259 (16–46°C)
Limestone	0.217
Magnesia	0.234 (100°C), 0.188 (1,500°C)
Magnesite brick	0.222 (100°C), 0.195 (1,500°C)
Marble	0.21 (18°C)
Quartz	0.17 (0°C), 0.28 (350°C)
Sand	0.191
Stone	About 0.2
Wood (oak)	0.570
Most woods vary between	0.45 and 0.65

Source: Foust et al. 1980 (reprinted with permission of John Wiley & Sons).

TABLE 17.32 Saturated steam tables

Temp., °F	Absolute Pressure lb/in.²	Volume (ft³/lb) Liquid	Volume (ft³/lb) Vapor	Enthalpy (Btu/lb) Liquid	Enthalpy (Btu/lb) Vapor	Entropy Btu/(lb)(°R) Liquid	Entropy Btu/(lb)(°R) Vapor
32	0.08854	0.01602	3,306	0.00	1,075.8	0.0000	2.1877
35	.09995	0.01602	2,947	3.02	1,077.1	0.0061	2.1770
40	.12170	0.01602	2,444	8.05	1,079.3	0.0162	2.1597
45	.14752	0.01602	2,036.4	13.06	1,081.5	0.0262	2.1429
50	.17811	0.01603	1,703.2	18.07	1,083.7	0.0361	2.1264
60	.2563	0.01604	1,206.7	28.06	1,088.0	0.0555	2.0948
70	.3631	0.01606	867.9	38.04	1,092.3	0.0745	2.0647
80	.5069	0.01608	633.1	48.02	1,096.6	0.0932	2.0360
90	.6982	0.01610	468.0	57.99	1,100.9	0.1115	2.0087
100	.9492	0.01613	350.4	67.97	1,105.2	0.1295	1.9826
110	1.2748	0.01617	265.4	77.94	1,109.5	0.1471	1.9577
120	1.6924	0.01620	203.27	87.92	1,113.7	0.1645	1.9339
130	2.2225	0.01625	157.34	97.90	1,117.9	0.1816	1.9112
140	2.8886	0.01629	123.01	107.89	1,122.0	0.1984	1.8894
150	3.718	0.01634	97.07	117.89	1,126.1	0.2149	1.8685
160	4.741	0.01639	77.29	127.89	1,130.2	0.2311	1.8485
170	5.992	0.01645	62.06	137.90	1,134.2	0.2472	1.8293
180	7.510	0.01651	50.23	147.92	1,138.1	0.2630	1.8109
190	9.339	0.01657	40.96	157.95	1,142.0	0.2785	1.7932
200	11.526	0.01663	33.64	167.99	1,145.9	0.2938	1.7762
210	14.123	0.01670	27.82	178.05	1,149.7	0.3090	1.7598
212	14.696	0.01672	26.80	180.07	1,150.4	0.3120	1.7566
220	17.186	0.01677	23.15	188.13	1,153.4	0.3239	1.7440
230	20.780	0.01684	19.382	198.23	1,157.0	0.3387	1.7288
240	24.969	0.01692	16.323	208.34	1,160.5	0.3531	1.7140
250	29.825	0.01700	13.821	216.48	1,164.0	0.3675	1.6998
260	35.429	0.01709	11.763	228.64	1,167.3	0.3817	1.6860
270	41.858	0.01717	10.061	238.84	1,170.6	0.3958	1.6727
280	49.203	0.01726	8.645	249.06	1,173.8	0.4096	1.6597
290	57.556	0.01735	7.461	259.31	1,176.8	0.4234	1.6472
300	67.013	0.01745	6.466	269.59	1,179.7	0.4369	1.6350
310	77.68	0.01755	5.626	279.92	1,182.5	0.4504	1.6231
320	89.66	0.01765	4.914	290.28	1,185.2	0.4637	1.6115
330	103.06	0.01776	4.307	300.68	1,187.7	0.4769	1.6002
340	118.01	0.01787	3.788	311.13	1,190.1	0.4900	1.5891
350	134.63	0.01799	3.342	321.63	1,192.3	0.5029	1.5783
360	153.04	0.01811	2.957	332.18	1,194.4	0.5158	1.5677
370	173.37	0.01823	2.625	342.79	1,196.3	0.5286	1.5573
380	195.77	0.01836	2.335	353.45	1,198.1	0.5413	1.5471
390	220.37	0.01850	2.0836	364.17	1,199.6	0.5539	1.5371
400	247.31	0.01864	1.8633	374.97	1,201.0	0.5664	1.5272
410	276.75	0.01878	1.6700	385.83	1,202.1	0.5788	1.5174
420	308.83	0.01894	1.5000	396.77	1,203.1	0.5912	1.5078
430	343.72	0.01910	1.3499	407.79	1,203.8	0.6035	1.4982
440	381.59	0.01926	1.2171	418.90	1,204.3	0.6158	1.4887

continues next page

TABLE 17.32 Saturated steam tables (continued)

Temp., °F	Absolute Pressure lb/in.²	Volume (ft³/lb)		Enthalpy (Btu/lb)		Entropy Btu/(lb)(°R)	
		Liquid	Vapor	Liquid	Vapor	Liquid	Vapor
450	422.6	0.0194	1.0993	430.1	1,204.6	0.6280	1.4793
460	466.9	0.0196	0.9944	441.4	1,204.6	0.6402	1.4700
470	514.7	0.0198	0.9009	452.8	1,204.3	0.6523	1.4606
480	566.1	0.0200	0.8172	464.4	1,203.7	0.6645	1.4513
490	621.4	0.0202	0.7423	476.0	1,202.8	0.6766	1.4419
500	680.8	0.0204	0.6749	487.8	1,201.7	0.6887	1.4325
520	812.4	0.0209	0.5594	511.9	1,198.2	0.7130	1.4136
540	962.5	0.0215	0.4649	536.6	1,193.2	0.7374	1.3942
560	1,133.1	0.0221	0.3868	562.2	1,186.4	0.7621	1.3742
580	1,325.8	0.0228	0.3217	588.9	1,177.3	0.7872	1.3532
600	1,542.9	0.0236	0.2668	617.0	1,165.5	0.8131	1.3307
620	1,786.6	0.0247	0.2201	646.7	1,150.3	0.8398	1.3062
640	2,059.7	0.0260	0.1798	678.6	1,130.5	0.8679	1.2789
660	2,365.4	0.0278	0.1442	714.2	1,104.4	0.8987	1.2472
680	2,708.1	0.0305	0.1115	757.3	1,067.2	0.9351	1.2071
700	3,093.7	0.0369	0.0761	823.3	995.4	0.9905	1.1389
705.4	3,206.2	0.0503	0.0503	902.7	902.7	1.0580	1.0580

Source: Keenan and Keyes 1936 (reprinted with permission of John Wiley & Sons).

Enthalpy (heat content), heat of formation, and free energy are all in terms of kJ/mol.

Free-energy diagrams (Ellingham or Richardson diagrams) consolidate a wealth of thermodynamic information on a single page. They are useful for rapidly estimating thermodynamic data. Figures 17.10 and 17.11 respectively are for oxides and sulfides. Some useful features of the diagrams are

- Changes of state for elements and their oxides are shown on the diagram (see legend). Temperatures (°C) at which changes occur are read from the x-axis.
- The standard Gibbs free energy of formation ($\Delta G°_f$ per mole of gas) for the oxidation reactions is read directly from the y-axis at the temperature of interest.
- Standard entropy of reaction is determined by measuring the slope of the line for the reaction of interest; the slope is equal to the negative entropy ($\Delta S°$).
- Knowing the standard free energy and standard entropy of reaction, calculate the standard enthalpy of reaction ($\Delta H°$) by means of the Gibbs-Helmholtz equation:

$$\Delta G° = \Delta H° - T\Delta S° \tag{EQ 17.26}$$

This approximation is valid up to the first-phase transition.

- Equilibrium oxygen partial pressure (P_{O_2}) and corresponding CO/CO_2 and H_2/H_2O ratios can be determined for specific oxidation reactions and temperatures by first drawing a line from the appropriate symbol (O, C, or H on the vertical line to the left of the diagram) to the point on the desired reaction line that corresponds to the temperature of interest, and then extending the construction line to the appropriate scale to determine the P_{O_2}, CO/CO_2, and H_2/H_2O ratios. One can determine whether the reaction will spontaneously proceed under a non-equilibrium gas composition by applying the van't Hoff reaction isotherm:

$$\Delta G = \Delta G° + RT \ln K \tag{EQ 17.27}$$

where K = dimensionless equilibrium constant defined as ratio of the molar concentrations of reactants and products.

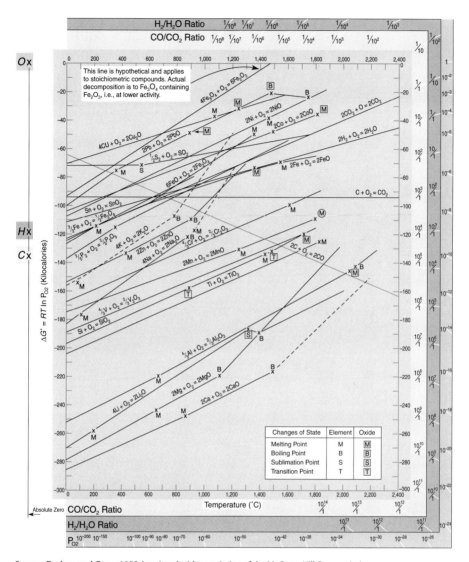

Source: Darken and Gurry 1953 *(reprinted with permission of the McGraw-Hill Companies).*

FIGURE 17.10 Free energy diagram for oxides

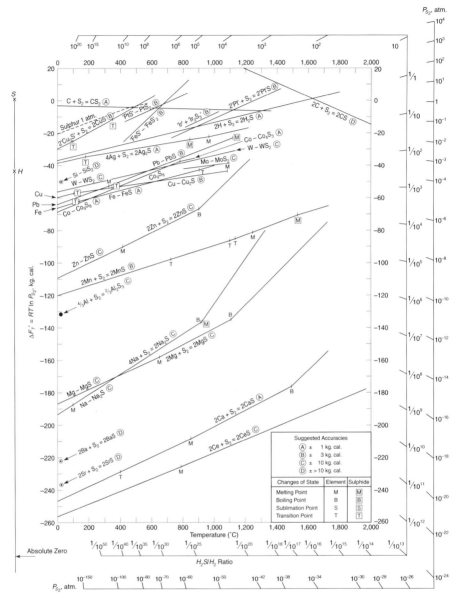

Source: Richardson and Jeffes 1952.

FIGURE 17.11 Free energy of formation for sulfides

Troubleshooting
- Look for obvious hot spots (build up the area with refractories, check on chemistry of reactions (a change in feed composition?).
- Burners: Has net heat content of fuel changed? Check exhaust gas analysis to see if correct amount of air is flowing through burner. Is flame being "thrown" too far?

Drying
- Reduce filter cake moisture (filter aids, air blow, hot water/vacuum evaporation in plate and frame filters, pressure filtration, beltpress filters, Ceramec disc filters).
- Use infrared or induction heat to eliminate natural gas (NO_X) and to reduce dusting.
- Heat required to evaporate 1 g of water ≈ 666 to 1,221 calories = 2,787 to 5,109 J.

Sintering
- Used to prepare an agglomerated calcine for blast furnaces.
- Feed size ≈ 12.7 mm or less.
- Feed ≈ 3 cm deep for ignition and then up to 0.3 m deep on traveling grate for sintering.
- Grates travel 25 to 120 cm/min.
- Capacity generally between 2.9 and 16.6 tonnes/m² of grate area per day.
- Pb industry generally uses updraft machines; Zn industry generally uses downdraft.
- Monitor: fines in product, sulfur burning rate in T/m²/day, SO_2 concentration in exhaust gas, changes in chemical analysis of feed material and final sinter product, clogging of grates.

Fluid Bed Reactors
- A temperature profile across the bed should be nearly constant; ±3°C in circulating bed fluid reactors (CBFRs). This allows more precise roasting and, possibly, 5% better gold extraction.
- Autogenous roasting requires about 4% sulfide sulfur in the feed.
- 700°C needed to burn off organic carbon.
- Dead roasted material should contain <0.1% sulfur.
- Feed should typically be <6.3 mm.
- Organic carbon in calcine should be 40% of the amount in the feed.
- CBFRs are replacing many stationary bed reactors.
- Questions: Dry or slurry feed? Is preheating of the roasting air needed? Should heat from the calcine be recovered? Should feed be preheated? Is direct injection of additional fuel needed? Is cooling of the reactor needed by integrated cooling surfaces, water injection, or recycling of calcine? Should the calcine be cooled by a fluid bed cooler or by direct quenching? Should heat be recovered in a waste heat boiler or by direct/indirect gas cooling? Prefer a baghouse or an electrostatic precipitator?

Rotary Kilns
- Sizing and retention times are calculated best by supplier's computer models. Data needed about the feed material are feed rate, bulk density, and angle of repose. Data needed about the kiln are inside diameter, length, slope, dam geometry, and rpm.
- The U.S. Bureau of Mines formula, simplified by Allis Chalmers and shown next, varies by as much as 40% from actual operations and is best for lightly loaded kilns; i.e., 5% loading rather than a more typical 15% to 20% loading (loading = percent of cross-sectional area of kiln's interior occupied by bed of material).

$$L/T = V_{avg} = SDN / 1.77\beta^{.5} \qquad \text{(EQ 17.28)}$$

where:
- L = kiln length in m
- T = total retention time in min
- V_{avg} = average axial solids velocity in m/min
- S = kiln slope in degrees
- D = kiln diameter in m
- N = kiln rotation rate in rpm
- β = dynamic angle of repose in degrees

Flash Smelter

- Slag contains ≈ 2.5% Cu (about 10× the Cu in reverb furnace slag). It is reprocessed by flotation to yield a tailing with ≈ 0.3% to 0.5% Cu (0.1% to 0.3% Ni). High slag assays may be from "pushing" capacity. Metal assays alone do not allow complete diagnosis of problems; optical microscopy is needed for characterization of losses.
- Matte grade ≈ 60% to 70% Cu.
- Fusion is often incomplete because (1) lack of understanding of slags and buildup of magnetite on furnace bottom resulting in less retention time, (2) flux is too coarse, (3) inadequate blending of furnace feed.
- Failure to obtain and analyze frequent, concurrent, and representative samples of feed, slag, metal, and dust results in a poor understanding of impurity distribution.
- Sulfur distribution (typical) is 97.4% to H_2SO_4, 0.5% to atmosphere from the acid plant, 0.2% to atmosphere from dryer ahead of smelter, 0.3% to atmosphere from flash smelter, 0.8% to atmosphere from converting and anode casting operations, 0.8% in slag flotation tailings.

Emissions from Typical Pyrometallurgical Operations

TABLE 17.33   Emissions from typical pyrometallurgical operations

Metal	Type of Operation	Particulates kg/MT	Particulates lb/ton	Sulfur Oxides kg/MT	Sulfur Oxides lb/ton	Carbon Monoxide kg/MT	Carbon Monoxide lb/ton	Gaseous Hydrofluoric Acid kg/MT	Gaseous Hydrofluoric Acid lb/ton
Copper	Roasting	22.5	45	30	60	—	—	—	—
	Reverb, furnace smelting	10	20	160	320	—	—	—	—
	Converting	30	60	435	870	—	—	—	—
Lead	Sintering (updraft)	106.5	213	275	550	—	—	—	—
	Blast furnace	180.5	361	22.5	45	—	—	—	—
	Dross reverb furnace	10	20	Negligible	Negligible	—	—	—	—
Zinc	Roasting (multiple hearth)	60	120	550	1,100	—	—	—	—
	Sintering	45	90	Included in above		—	—	—	—
Pig iron	Blast furnace	82.5	165	—	—	875	1,750	—	—
	Sintering	10	20	—	—	—	—	—	—
Steel	Open hearth furnace (no oxygen lance)	4.15	8.3	—	—	—	—	0.05	0.1
	Open hearth furnace (oxygen lance)	8.7	17.4	—	—	—	—	0.05	0.1
	Basic oxygen furnace	25.5	51	—	—	69.5	139	Negligible	Negligible
	Electric arc furnace (no oxygen lance)	4.6	9.2	—	—	9	18	0.006	0.012
	Electric arc furnace (oxygen lance)	5.5	11	—	—	9	18	0.006	0.012

Source: Sittig 1975 (reprinted with permission of Noyes Data Corporation).

INDUSTRIAL MINERALS
by Edwin H. Bentzen III

General

Industrial minerals are sold on the basis of their physical and chemical properties. They can be sold into more than one market or for more than one end-product use, often at two extremely different prices based only on the change in one chemical or physical characteristic. A product may sell in one market but not in another because of the differing needs of the buyer. There are many tests used to assess the quality of an end product; some are written and some are not.

The following abbreviated table shows some of the tests used for a few industrial minerals. It is not all inclusive. These tests often force the use of certain unit operations in an effort to make a salable product.

TABLE 17.34 Common industrial mineral tests

Mineral/Commodity	Major Markets	Chemical Analysis – Major Value	Chemical Analysis – Major Impurities	Moisture Criteria	Bacteria Count	Particle Size Distribution	Particle Shape	Mineralogical/Chemical Composition	Optical Factors (4)	Oil Absorption	Odor/Gas Absorption	Rheology Properties (5)	Specific Gravity	Pyro Factors; Other
Chemical Industries														
Alumina	C	Al$_2$O$_3$	Fe, Si	Y		N	N	Y						(1)
Beryllium	C	Be		N		N	N	Y						
Borates	C, F, A	B	As	N		Y	N	Y						
Bromine	C	Br	Cl, I, H$_2$O, organics			N	N	Y						
Chromium	C	Cr	Fe, Mg, S	N		Y	N	Y						
Fluorine	C	F		N		N	N	Y						
Iodine	C	I	Br, Cl			N	N	Y						
Lime	C, F, A	Ca	Mg, Fe	N		S	N	Y						
Lithium	C	Li		N		N	N	Y						
Phosphate	C, F, A	PO$_4$	U, V, Ni	N		N	N	Y						
Potash	C, F, A	K$_2$O	SiO$_2$, NaCl	Y		Y	N	Y						
Rare Earths	C			N		Y	N	Y						
Salt	C	NaCl	As, organics	Y		Y	N	Y						
Soda ash	C, A	Na$_2$CO$_3$	Fe, organics, other carbonates, pyrite	Y		Y	N	Y						
Strontium minerals	C	Sr	Ba, Fe, F	N		Y	N	Y						
Sodium sulfate	C, F, A	Na$_2$SO$_4$	Fe, organics, other sulfates	Y		Y	N	Y						
Sulfur	C, F, A	S	Organics	N		S	Y	Y						
Zircon	C	Zr	Hf, Th, U, Fe	N		Y	N	Y						
Fillers														
Calcium carbonate	Fl, P		Dolomite, iron oxides	Y	N	Y	S	S	Y	N	N	Y		
Common clays	Fl, CO		Quartz	Y	N	Y	Y	N	N	N	N	Y		(2)
Diatomite	Fl, CO, FA, P		Clays, quartz	Y	N	Y	Y	Y	N	N	N	N		
Perlite, expanded	Fl		Obsidian	Y	N	Y	N	Y	Y	S	N	N		
Iron oxides	P		Quartz	Y	N	Y	N	Y	N	N	N	Y		
Kaolin	Fl, CO, FA, P		Iron oxides, quartz	S	Y	Y	Y	Y	Y	N	N	Y		
Mica	Fl, CO		Quartz, feldspar	Y	N	Y	Y	S	Y	S	N	Y		
Natural zeolites	Fl		Quartz, clays	Y	N	Y	N	Y	N	Y	Y	N		
Talc	Fl, CO, P		Asbestos serpentine	N	Y	Y	Y	Y	Y	S	S	Y		

continues next page

TABLE 17.34 Common industrial mineral tests (continued)

Mineral/Commodity	Major Markets	Chemical Analysis - Major Value	Chemical Analysis - Major Impurities	Moisture Criteria	Bacteria Count	Particle Size Distribution	Particle Shape	Mineralogical/Chemical Composition	Optical Factors (4)	Oil Absorption	Odor/Gas Absorption	Rheology Properties (5)	Specific Gravity	Pyro Factors; Other
Fillers (continued)														
Titanium minerals	FI, P		Zircon, garnet, quartz, iron oxides, manganese	N	N	S	N	S	N	N	N	N		
Tripoli	FI			Y	N	Y	Y	N	N	N	N	S		
Ceramics														
Asbestos	I			Y	Y	Y	Y	Y	S	S	N	S		
Chromite	R	Cr	Olivine, serpentine	Y	Y	S	Y	N	N	N	N		(3)
Feldspar	CE, G, I, FL	Al₂O₃, alkali	Mica, quartz, iron minerals	Y	Y	N	S	Y	S	N	N		
Graphite	R, I	Carbon, LOI	Quartz	Y	Y	Y	Y	N	S	N	S		
Kyanite	CE, R	Al₂O₃, alkali	Quartz	Y	Y	Y	Y	N	N	N	N		
Magnesite	CE, R	Mg, Ca, LOI	CaCO₃, quartz, Fe	Y	Y	N	Y	S	N	N	N		
Nepheline syenite	CE, R, G	Al₂O₃, Fe, alkali	Iron minerals	Y	Y	N	Y	N	N	N	N		
Olivine	R, FL	LOI	Serpentine, talc	Y	Y	S	Y	N	N	N	N		
Perlite, expanded	I	LOI	Obsidian	Y	Y	N	Y	S	S	S	S		
Pyrophyllite	CE, R	Al₂O₃, alkali	Quartz	Y	Y	Y	Y	Y	N	N	N		
Silica sand	CE, R, G, I, FL	SiO₂	Clay, refractory heavy minerals	Y	Y	Y	S	S	N	N	N		
Silimanite	CE, R	Al₂O₃, alkali	Quartz	Y	Y	Y	Y	Y	N	N	N		
Vermiculite	I	LOI	Quartz, feldspar	Y	Y	Y	S	N	S	S	S		
Well Drilling														
Barite	W	BA	Quartz	Y	Y	N	Y	N	N	N	Y	Y	
Bentonite	W, AD			Y	Y	S	Y	N	Y	S	Y	N	
Corundum	AB	Al₂O₃	Quartz	N	Y	S	N	N	N	N	N	S	
Diamonds	AB	C		N	Y	S	N	N	N	N	N	N	
Diatomite	AD, E, AB	Si	Quartz	S	Y	Y	N	S	Y	S	S	N	
Emery	AB		Quartz	N	Y	S	N	N	N	N	N	S	
Fracture sand	W, AB	SiO₂	Clay, feldspar	N	Y	Y	N	N	N	N	N	N	
Garnet	AB		Quartz	N	Y	S	N	N	N	N	N	Y	
Limestone	E	CaCO₃	Quartz	S	S	N	N	S	S	S	S	N	
Silica sand	AB	SiO₂	Clay, feldspar	S	S	S	N	N	N	N	N	N	
Zeolite	AD, E, AB			Y	Y	Y	Y	N	Y	Y	S	N	
Construction														
Crushed rock			Sand, silt, clay	S	Y	S	S	N	N	N	N		
Dimension stone			Pyrite, iron carbonates	N	Y	S	S	Y	N	N	N		
Gypsum			Anhydrite, quartz	Y	Y	N	Y	Y	N	N	Y		
Pumice, scoria			Quartz	Y	Y	S	S	S	S	N	N		
Sand and gravel			Clay, organics	Y	Y	S	Y	N	N	N	Y		

Legend:
Y = yes; N = no; S = sometimes.
A = agriculture; AB = abrasives; AD = adsorbents; C = chemicals; CE = ceramics; CO = coatings.
E = environmental; F = fertilizer; FA = filter aid; FI = fillers; FL = flux; G = glass; I = insulators.
P = pigments; R = refractories; W = well drilling.
(1) = Polished stone value, load capacity, thermal expansion; (2) = Swelling, adsorption, load capacity; (3) = Load capacity, expansion, pyrometallurgical tests; (4) = Optical factors = opacity, color, brightness, reflectance; (5) = Rheology properties = swelling, loading, adsorbing.

GOLD LEACHING
by Paul D. Chamberlin, P.E.

- $4Au + 8NaCN + O_2 + 2H_2O = 4NaAu(CN)_2 + 4NaOH$
- Install dissolved oxygen probes and aerate to >2 ppm dissolved oxygen.
- Dissolved oxygen in leach solution (up to 6–7 ppm naturally) is enough to dissolve Au, but may not be enough to react with other oxygen consumers.
- If preg robbers are present, take cyanide out of grinding circuit.
- Test CIL tails for gravity concentration and regrinding of pyrite.
- Conduct a recovery versus grind P_{80} series.
- As little as 100 mg/L of NaCN is enough to dissolve gold ($\approx 10\times$ more for Ag).
- $Cu(CN)_x$ and $Zn(CN)_x$ can titrate as free cyanide; be careful.

TABLE 17.35   Solubility of minerals in cyanide

Mineral	% Dissolved in 24 h	Source	Mineral	% Dissolved in 24 h	Source
Calaverite, $AuTe_2$	Readily soluble	1	Hydrozincite, $3ZnCO_3 \cdot 2H_2O$	35.1	3
Argentite, Ag_2S	Readily soluble	2	Franklinite, $(Fe,Mn,Zn)O \cdot (Fe,Mn)_2O_3$	20.2	3
Cerargyrite, AgCl	Readily soluble	2	Sphalerite, ZnS	18.4	3
Proustite, Ag_3AsS_3	Sparingly soluble	2	Gelamine, $H_2Zn_2SiO_4$	13.4	3
Pyrargyrite, Ag_3SbS_3	Sparingly soluble	2	Willemite, Zn_2SiO_4	13.1	3
Azurite, $2CuCO_3 \cdot Cu(OH)_2$	94.5	3	Pyrrhotite, FeS	Readily soluble	4
Malachite, $CuCO_3 \cdot Cu(OH)_2$	90.2	3	Pyrite, FeS_2	Sparingly soluble	4
Chalcocite, Cu_2S	90.2	3	Hematite, Fe_2O_3	Sparingly soluble	4
Cuprite, Cu_2O	85.5	3	Magnetite, Fe_3O_4	Nearly insoluble	4
Bornite, $FeS \cdot 2Cu_2S \cdot CuS$	70	3	Siderite, $FeCO_3$	Nearly insoluble	4
Enargite, $3CuS \cdot As_2S_5$	65.8	3	Orpiment, As_2S_3	73	4
Tetrahedrite, $4Cu_2S \cdot Sb_2S_3$	21.9	3	Realgar, AsS	9.4	4
Chrysocolla, $CuSiO_3$	11.8	3	Arsenopyrite, FeAsS	0.9	4
Chalcopyrite, $CuFeS_2$	5.6	3	Stibnite, Sb_2S_3	21.1	4
Smithsonite, $ZnCO_3$	40.2	3	Galena, PbS	Soluble at high pH	5
Zincite, ZnO	35.2	3			

Sources: (1) Johnstone 1933; (2) Leaver, Woolf, and Karchmer 1931; (3) Leaver and Woolf 1931; (4) Hedley and Tabachnik 1968; (5) Lemmon 1940.

- Merrill-Crowe recovery of gold and silver
  - Preg solution: <1 ppm total suspended solids after clarification and <1 ppm dissolved oxygen after deaeration. Eliminate surge capacity between clarification and Zn precipitation.
  - Add NaCN into Zn cone to prevent formation of ZnO_2.
  - Free NaCN concentration should be $\approx 0.1-0.5$ g/L entering the precipitate filter presses.
  - Zinc consumption should be ≈ 1 g Zn per gram Ag or 3 g per gram Au.
  - Pb acetate (1–10 ppm Pb^{++}) combines with sulfides and reduces NaCN consumption as SCN^-.
  - An Hg:Au ratio of >5 makes a slimy, hard-to-filter precipitate.
  - Keep filter press feed pump submerged in barren solution.
  - Optimal pH is 11 to 12.

- Excess zinc will collect in the filter presses; at the end of a press run, stop adding zinc through the Zn cone and use up the zinc in the presses until barrens rise.
- More than 200 ppm Cu in preg retards Au precipitation because it consumes most of the free cyanide; nickel acts similarly.
- Sulfide ions react with zinc and free cyanide, thus inhibiting Au precipitation; arsenic and antimony are also deleterious.
- Good barren assays should be <0.034–0.100 g Au/tonne solution.
- Use very fine, high-grade zinc dust, not shavings.
- Use antiscalants to minimize lime scaling in the clarifiers and filter presses.

VAT LEACHING
by Louis W. Cope, P.E.

General
Vat leaching is a more controlled, contained version of heap leaching. It is applicable to gold, copper, and other leachable mineral ores. Whereas a heap is contained only on its bottom side, a vat of ore is normally contained from below and on all four sides; i.e., a tub (vat). A vat leaching operation can have any number of vats, with piping for flexible movement of solutions. Solution flow through the ore can be upward or downward. Loading and excavating a vat can be with conveyors, bucket-wheel excavators, clamshells, mobile front-end loaders, or by sluicing.

The Ecovat is a continuous vat in which ore and solution are fed at one end while the spent ore and metal-loaded solution are continuously removed from the vat.

Advantages   Ore in a vat can be totally immersed for better contact by the leaching solution. Like heap leaching, it is applicable to ores that do not require fine grinding for metal extraction. Vat leaching operations can be put under a roof or even within a building for use in cold, rainy, high evaporation climates, or where there is a shortage of water.

Disadvantages   Vat leaching is applicable only to fast-leaching ores. Vat leaching is constrained by time the same as on-off heap leaching. Loading of ore and removal of spent ore must be maintained on a schedule.

TABLE 17.36   Comparisons of leaching by vat, heap, and agitation

	Vat	Heap	Agitation
Capital cost	1	0.8	2.5
Operating cost	1	0.9	3.0
Treatment time, days	2–3	90–300+	<3
Gold recovery (%)	85	70	90+

Order-of-magnitude comparisons.

MISCELLANEOUS
by Paul D. Chamberlin, P.E.

- Percent solids of a pulp = $[100(SG_p - 1)SG_s]/(SG_s - 1)SG_p$

  where:
  SG_p = specific gravity of pulp
  SG_s = specific gravity of solids

- Fans raise pressure ≈3%; blowers raise pressure up to ≈40 psig; compressors are for higher pressures.
- Vacuum pumps: reciprocating piston types are used down to 1 torr (1 mm Hg; 760 mm Hg per 14.7 psi); rotary piston down to 0.001 torr; two-lobe rotary down to .0001 torr; steam-jet ejectors—1-stage down to 100 torr, 3-stage to 1 torr, 5-stage to .05 torr.

- In-leakage of air to evacuated equipment:

$$1 \leq P < 10 \text{ torr}; W = 0.026 P^{0.34} V^{0.6} \quad \text{(EQ 17.29a)}$$

$$10 \leq P < 100 \text{ torr}; W = 0.032 P^{0.26} V^{0.6} \quad \text{(EQ 17.29b)}$$

$$100 \leq P < 760 \text{ torr}; W = 0.106 V^{0.6} \quad \text{(EQ 17.29c)}$$

where:
- P = pressure in equipment in torr
- V = volume in ft^3
- W = equipment leakage in lb/h (does not include seals)

Double W to account for static seals; $2W$ plus 5 lb/h for each rotary seal; $2W$ plus 2 lb/h for each mechanical seal and o-ring.
- Better fragmentation in mine blasting usually gives many benefits in processing (i.e., better sampling, fewer boulders to rehandle at crushing plant, more fractures and faster leaching rate, lower crushing and grinding costs). Therefore, insert cone-shaped plastic plugs in blast holes so that stemming isn't blown out of the hole and the blast energy is retained in the rock.
- Locate conveyor belt scales 6 m to 9 m from loading point. Avoid locating scale near drive pulley (belt tension is too high). Locate scale >12 m from tangent points on belt curves. Align (all in same plane) at least 2 and maybe 4 idlers either side of the scale. Calibration methods are (1) material test, which is the best—pass known weight of material over the scale; (2) static weights hung from scale, (3) test chains that roll on the moving conveyor, (4) electronic, which tests only the electronics and does not weigh anything—use only when the scale cannot be easily accessed.
- One ton (2,000 lb) of refrigeration equals the removal of 12,000 Btu/h of heat. The following heat transfer coefficients in Btu/(h)(ft^2)(°F) are approximate: water to liquid, 150; condensers, 150; liquid to liquid, 50; liquid to gas, 5; gas to gas, 5.
- Feeders and bins: The size of the bin opening must be smaller than feeder to ensure flow. For belt feeders, (1) is bin behind feeder rather than on top of it?, and (2) is downstream bin opening wider than upstream? For screw feeders, increasing pitch of screw gives more uniform withdrawal. For rotary valves, add short pipe between bin bottom and feeder to eliminate stagnant zone in bin. Bin discharge width >2 to 3 times the largest particle size (if few large particles). Downstream bin wall should be vertical; other bin walls should be greater than 50°.

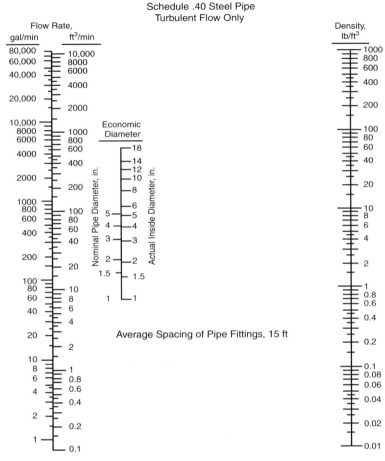

FIGURE 17.12 Economic pipe diameter

TABLE 17.37 Feeder capacities and effect of material characteristics

Type of Feeder	Capacity Range (tph)	Size			Flowability			Abrasiveness			Path of Travel		
		Fine	Granular	Large Lumps	Very Free	Free	Sluggish	Nonabrasive	Mildly Abrasive	Very Abrasive	Horizontal	Inclined	Combined
Apron (pan)	57 to 688		•	•		•		•	•	•	•	•	•
Belt	40 to 1,200	•	•		•	•		•	•		•	•	
Flight	36 to 182	•	•	•		•	•	•	•		•	•	
Screw	10 to 187	•	•			•	•	•	•		•		
Reciprocating	82 to 450		•	•	•	•		•	•	•	•		
Vibratory, electrical	25 to 1,000	•	•	•	•	•	•	•	•	•	•		
Vibratory, mechanical	35 to 1,550	•	•		•	•	•	•	•	•	•		
Rotary table, stationary plow	12 to 57	•	•			•	•	•	•		•		
Rotary plow, stationary table	300 to 3,500	•	•			•	•	•	•	•	•		
Rotary (vane)	1.5 to 117	•				•		•			•		

Source: Anon. 1979 (reproduced with permission of Rock Products).

TABLE 17.38 Precipitating metals

Metal	Precipitant	pH Range	% Removal*	Metal	Precipitant	pH Range	% Removal*
Sb	$Fe_2(SO_4)_3$	8–3	96		alum	9–8	2
	$FeCl_3$	8–3	96		CaO	11–10	5
As^{+5}	$Fe_2(SO_4)_3$	9–4	99	Cu	$Fe_2(SO_4)_3$	10.5–8.5	98
	$FeCl_3$	9–4	99	Pb	$FeSO_4$/filter	8.5–9.0	N/A
	CaO	12–10	98		$Fe_2(SO_4)_3$	10–6	98
	alum	9.5–7.0	85		CaO	12–9	99
As^{+3}	$Fe_2(SO_4)_3$	9.5–5.0	96	Hg	CaO	12.5–11.0	70
	$FeCl_3$	9.5–5.0	96	Ni	$Fe_2(SO_4)_3$	11.0–9.5	90
	CaO	12–10	80		$FeSO_4$	9.0–8.5	?
	alum	9.5–5.5	15	Se	$Fe_2(SO_4)_3$	?–5.5	85
Be	$Fe_2(SO_4)_3$				$FeCl_3$	7.0–5.5	85
	CaO/filter	11.5			alum	7.0–5.5	25
Cd	$Fe_2(SO_4)_3$	11–9	98		CaO	11.5–10.0	40
	$FeCl_3$	11–9	98	Ag	$Fe_2(SO_4)_3$/filter	9.5	85
	alum	9.0–8.5	50		$FeCl_3$	6.2	
Cr^{+3}	$Fe_2(SO_4)_3$	10–6	92		alum	8–7	78
	alum	9.5–8.0	90		CaO	12–11	90
	CaO	12–11	98	Zn	$Fe_2(SO_4)_3$	9.5	90
Cr^{+6}	$FeSO_4$	10–6	92				

* Full dosage at 5–10 times stoichiometric.
Source: Rice 1999 (reprinted with permission).

REFERENCES

Agarwal, J.C., H.W. Flood, and R.A. Giberti. 1974. Preliminary economic control systems in metallurgical plants. *Journal of Metals*. 26(12):9.

Anon. 1979. Dry solids handling. *Engineering & Mining Journal*. June:128.

Austin, L.G. 1990. A Mill Power Equation for SAG Mills. *Minerals & Metallurgical Processing*, Feb: 57–62.

Bergmann, L. 1981. Baghouse filter fabrics. *Chemical Engineering*. October 19:177.

Besendorfer, C. 1996. Exert the force of hydrocyclones. *Chemical Engineering*. September:108–114.

Bird, B.M. 1928. Interpretation of float-and-sink data. In *Proceedings of Second International Conference on Bituminous Coal*: 82–111.

Dahlstrom, D.A. 1985. Thickening, filtering, drying. In *SME Mineral Processing Handbook*. Edited by N.L. Weiss. New York: Society of Mining Engineers of the American Institute of Mining, Metallurgical, and Petroleum Engineers, Inc. 9–27.

Darken, L.S., and R.W. Gurry. 1953. *Physical Chemistry of Metals*. New York: McGraw-Hill.

Enviro-Clear Company, Inc. 1997. Clarifier/thickener, thickening rate data. In company advertising brochure. High Bridge, NJ: Enviro-Clear Company, Inc. CC-3.

Foust, A.S., L.A. Wenzel, C.W. Clump, L. Maus, and L.B. Andersen. 1980. *Principles of Unit Operations*. New York: John Wiley & Sons.

Hedley, N., and H. Tabachnik. 1968. *Chemistry of Cyanidation*. Parsippany, NJ: American Cyanamid Co.

Johnstone, W.E. 1933. Tellurides are soluble in cyanide. *Engineering & Mining Journal*. August.

Keenan, J.H., and F.G. Keyes. 1936. *Thermodynamic Properties of Steam*. New York: John Wiley & Sons.

Kelly, E.G., and D.J. Spottiswood. 1982. *Introduction to Mineral Processing*. New York: John Wiley & Sons.

Kjos, D.M. 1986. Semiautogenous mill liners: designs, alloys, and maintenance procedures. In *Minerals and Metallurgical Processing*. 3(2):80–87.

Klumpar, I.V. 1987a. Air Classification, Part I—Equipment and Selection. *Powder and Bulk Engineering,* Aug: 42–58.

Klumpar, I.V. 1987b. Air Classification, Part II—Performance and Dynamics. *Powder and Bulk Engineering,* Sep: 12–16.

Leaver, E.S., and J.A. Woolf. 1931. *Copper and Zinc Cyanidation*. Technical Paper 494. Washington, DC: U.S. Bureau of Mines.

Leaver, E.S., J.A. Woolf, and N.K. Karchmer. 1931. Oxygen as an aid in the dissolution of Ag by cyanide from various silver minerals. In *Report of Investigation 3064*. Washington, DC: U.S. Bureau of Mines.

Lemmon, R.J. 1940. Reaction of minerals in the cyanidation of gold ores. *Chemical Engineering and Mining Review.* March: 227–229.

Liddel, D.M. 1922. *Handbook of Chemical Engineering*. 1st ed., Volumes 1 and 2. New York: McGraw-Hill.

MacPherson, A.R. 1989. Autogenous grinding 1987—update. *CIM Bulletin*. January: 75–82.

Matthews, C.W. 1985. Screening. In *SME Mineral Processing Handbook*. Edited by N.L. Weiss. New York: Society of Mining Engineers of the American Institute of Mining, Metallurgical, and Petroleum Engineers, Inc. 3E-1–3E-41.

Ottergren, C., and J. Steer. 1996. Crusher selection and the design of crushing. *Engineering & Mining Journal*. May: WW-30.

Perry, J.H. 1969. *Chemical Engineers Handbook*. 4th ed. New York: McGraw-Hill. 3-191; 5-29.

Perry, R.H. 1980. *Chemical Engineers Handbook*, 6th ed. New York: McGraw-Hill. 19-91.

Rice, R.J. 1999. Bonfield, Ontario, Canada. Personal communication.

Richardson, F.D., and J.H.E. Jeffes. 1952. The thermodynamics of substances of interest in iron and steel making. *Journal of Iron & Steel Institute*. 171:165–175.

Rossano, A.T. 1969. *Air Pollution Control Guidebook for Management*. New York: ERA Inc., Environmental Services Division. 143.

Rowland, C.A., and D.M. Kjos. 1978. Rod and ball mills. In *Mineral Processing Plant Design*. Edited by A. Mular and R. Bhappu. New York: Society of Mining Engineers of the American Institute of Mining, Metallurgical, and Petroleum Engineers, Inc. 239–278.

Scott, J.W., and D.J. Barratt. 1987. Testwork, selection, and design for grinding circuits: an engineering company viewpoint. Paper presented at Canadian Institute of Mining, Metallurgy, and Petroleum (CIM) District 6 meeting, October 30–31, 1987, Vancouver, BC.

Sittig, M. 1975. *Environmental Sources and Emissions Handbook*. Westwood, NJ: Noyes Data Corporation.

Taggart, A.F. 1956. *Handbook of Mineral Dressing*. 6th ed. New York: John Wiley & Sons.

Vibrating Screen Manufacturers Association. Undated. Selection of screen size and type. Chapter 5 in *VSMA Vibrating Screen Handbook*. Stamford, CT.

CHAPTER 18

Site Structures and Hydrology

Brett F. Flint, P.E.

FACILITY LAYOUT

Although potential facility layouts are practically infinite in variety, they can be placed in general categories. For example, surface impoundments may be categorized as follows:

- Ring dikes—typically used on flat terrain, and all sides of the impoundment are enclosed by embankments or dikes.
- Cross-valley—the typical layout of a water-storage reservoir; constructed between the natural walls of a valley or other topographic depression. Placing the facility near the head of the drainage will reduce inflow from runoff or the need for large diversion structures. Placing the facility further down the drainage path may result in greater storage capacity for a given embankment height or fill volume.
- Sidehill—generally used on slopes with a grade of 10% or less. Uses a natural slope as one side of the impoundment with dikes or embankments around the other three sides.
- Valley-bottom—this layout is a combination of cross-valley and sidehill layouts and is generally used when upstream flows are large and would require significant diversion structures.

Figure 18.1 shows typical impoundment layouts. Similar layout considerations are applicable to dry stack facilities such as waste rock storage facilities and leach facilities, although the potential for combinations and variations of layouts increases. Consider liner installation, leachate recovery, and delivery systems in the layouts.

Table 18.1 shows typical siting considerations for tailings storage facilities.

TABLE 18.1 Factors for siting tailings facilities

Parameter	Effects
Location and elevation relative to mill	Length of tailings and return-water pipelines Capital and operating cost for pumps
Topography	Embankment layout Embankment fill requirements Diversion feasibility
Hydrology and catchment area	Long-term water accumulation Flood-handling requirements
Geology	Availability of natural borrow types and quantities Seepage losses Foundation stability
Groundwater	Rate and direction of seepage movement Contamination potential Moisture content of borrow materials

Source: Vick 1983.

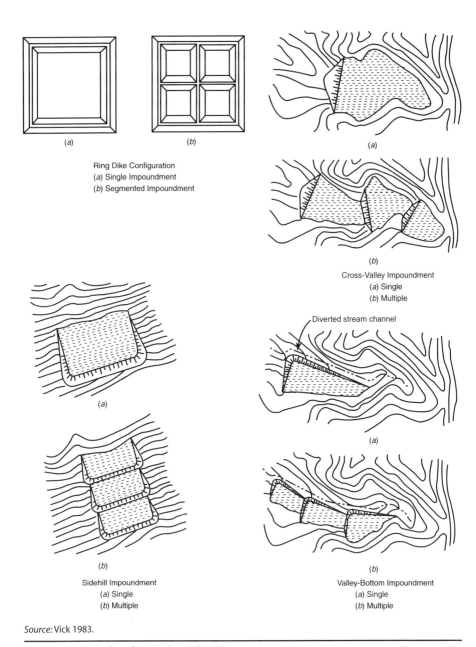

Source: Vick 1983.

FIGURE 18.1 Surface impoundment layouts

EARTHWORKS

Generally, any soil or rock material can be used for construction of embankments and other fills, as long as the material is clean and is properly placed and compacted. Rock and gravel will generally be stable at steeper slopes; use sands and silts with caution in applications where erosion may be possible. Sands and silts exposed to erosive forces should be protected. Silts are also susceptible to frost heave in cold climates.

Embankments are generally constructed with downstream slopes of 2.5 to 3 H: 1 V (horizontal to vertical) and upstream slopes of 2 to 2.5 H: 1 V. Actual slopes should be based on the material used and service conditions of the fill. Flatter slopes will offer more stability and provide better access for maintenance and reclamation.

Embankments may be constructed as a homogeneous fill for small facilities that will not continuously impound water. A membrane structure is sometimes used for small impoundments. This consists of a homogeneous embankment with a membrane of fine-grained soils, or a synthetic liner on the upstream face.

Larger embankments intended for long-term storage of water or other fluids are generally constructed as a zoned fill. Semi-pervious erosion-resistant materials are used for the upstream and downstream portions of the embankment. An impervious core of clay or silt provides resistance to water seepage; and drain structures constructed of free-draining materials control seepage within the embankment to maintain stability. Figure 18.2 shows a typical zoned earth-fill embankment.

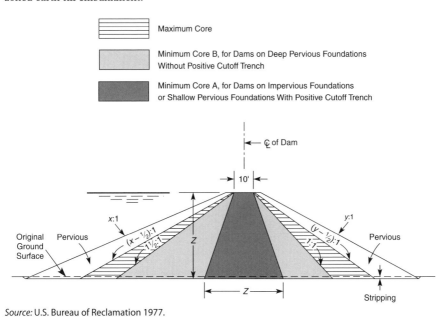

Source: U.S. Bureau of Reclamation 1977.

FIGURE 18.2   Zoned earth-fill embankment

LINER SYSTEMS

Liner systems are used to control and contain water, process solutions, or other liquids; to collect and conduct leachate; and to protect environmental resources.

Liner systems may be as simple as a layer of fine-grained soil placed and compacted for low permeability or they can include several elements or layers such as a geomembrane, a geotextile, a geocomposite, a geonet, or other elements. Liner systems may be used in conjunction with leachate collection and recovery systems, leak detection systems, or various types of drainage systems.

Soil Liners

Compacted soil liners are generally constructed using fine-grained soils such as silts and clays; however, any soil that can provide a low permeability may be used. Use Darcy's law to calculate movement of a liquid through a soil as follows:

$$Q = KIA \qquad (EQ\ 18.1)$$

where:
- Q = flow rate
- K = coefficient of permeability (depends on soil and liquid properties and is generally determined by laboratory or in situ testing)
- I = hydraulic gradient
- A = cross-sectional flow area, perpendicular to direction of flow

This equation is valid for any units that are consistent for each parameter. The coefficient of permeability is commonly reported in meters per second (m/s), centimeters per second (cm/s), or feet per second (ft/s). Values of K for various soil types can be estimated from Figure 18.3.

Values in Figure 18.3 assume that the fluid is generally clean water. When fluids other than water are used, or temperatures significantly affect the viscosity, the coefficient of permeability can be separated into two components: intrinsic, which is related to soil properties, and hydraulic, which is related to fluid properties. Relevant relationships are

$$k = Cd^2 \qquad (EQ\ 18.2)$$

$$K = k\mu/\gamma \qquad (EQ\ 18.3)$$

where:
- k = intrinsic permeability (typically cm/s)
- C = constant shape factor that is dependent on the density, grain size distribution, and other soil properties (typically 1/s-cm)
- d = mean grain size or a defined grain size such as d_{10}, d_{40}, d_{50}, etc. (cm)
- K = hydraulic conductivity (cm^2)
- γ = specific gravity of the fluid (gr/cm^3)
- μ = viscosity of fluid (gr-s/cm^2)

Other units can be used in these equations as long as unit consistency is maintained.

General values for the coefficient C have been difficult to determine. A proposed relationship for clean sands is

$$k = 100 d_{10}^2 \qquad (EQ\ 18.4)$$

Research suggests that for a similar relationship based on the d_{10} grain size, C would vary as shown in Table 18.2.

TABLE 18.2 Range of values for coefficient C

Soil	d_{10} Particle Size (cm)	C (1/s-cm)
Coarse gravel	0.082	16
Sandy gravel	0.020	40
Fine gravel	0.030	8
Silty gravel	0.006	11
Coarse sand	0.011	1
Medium sand	0.002	7
Fine sand	0.003	1
Silt	0.0006	42

Source: Lambe and Whitman 1969 (reprinted with permission of John Wiley & Sons).

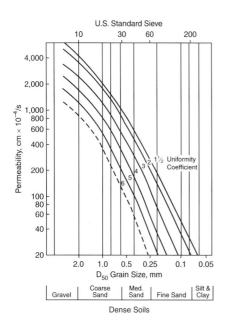
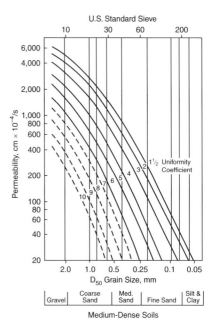
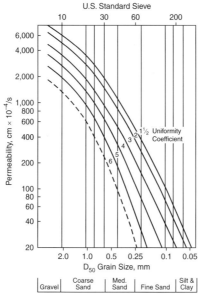

The hydraulic conductivity of materials to be dewatered can be estimated on the basis of particle size, uniformity coefficient, and density when information from pumping tests is not available. In these graphs, the D_{50} grain size represents the diameter of the 50-percent-retained size. To convert permeability in cm × 10^{-4}/s to hydraulic conductivity in gpd/ft², multiply by 2.13.

Source: Driscoll 1986 *(reprinted with permission of Moretrench American Corporation)*.

FIGURE 18.3 Hydraulic conductivity for various soils

Composite Liners

Composite liners, in general, consist of a compacted soil liner covered by a synthetic liner. In some facilities, the liner system may be extended to include additional layers of synthetic or soil liner materials to provide a multilayered system. A composite liner system takes advantage of the low permeability of a synthetic material; the compacted soil liner provides a suitable subgrade and limits potential seepage from defects or minor punctures in the synthetic liner.

Geomembranes

Geomembranes are often used as primary liner systems or as part of composite liner systems. They are also used as aquitards in embankments, as channel revetment, and in other applications for solution control and containment. These materials are manufactured from synthetic polymers and their properties are variable. Contact the manufacturers to obtain up-to-date properties.

Thickness Determination Geomembrane thickness selection should be based on material properties and on potential deformation that could be experienced during the service life of the material. Figure 18.4 shows the model used in conjunction with the thickness determination equation.

Summation of forces: $\Sigma F_x = 0$, gives $F \cos \beta = T_U + T_L$ or $(\sigma_{allow} \, t) \cos \beta = (p \tan \delta_U + p \tan \delta_L)x$, which leads to

$$t = (p/\cos \beta)(x/\sigma_{allow})(\tan \delta_U + \tan \delta_L) \qquad (EQ\ 18.5)$$

where:
- t = thickness of the geomembrane (in.)
- ΔH = deformation that mobilizes stresses (in.)
- F = force in the geomembrane (lb/ft²)
- σ_{allow} = allowable liner stress (lb/ft²)
- T_U = shear force on top of the geomembrane (zero if the upper surface of the geomembrane is exposed to water or similar fluid; lb/ft²)
- T_L = shear force below the geomembrane (lb/ft²)
- T = $p \tan \delta$ (lb/ft²)
- p = pressure applied to the surface of the geomembrane (lb/ft²)
- δ = angle of shearing resistance between the liner and adjacent material (degrees)
- x = distance of mobilized liner deformation (in.)
- β = slope angle (degrees)

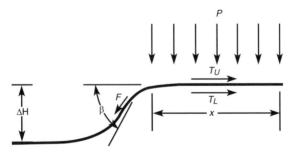

Source: Koerner 1994 (reprinted with permission of Prentice Hall).

FIGURE 18.4 Model used for calculating liner thickness

Anchor Trench Design The most common method for anchoring synthetic liner materials is the use of an anchor trench or liner runout. Figure 18.5 shows a model that can be used to determine the length of runout or the depth of the anchor trench for a synthetic liner system. This model uses imaginary frictionless pulleys to simplify the calculation. Using the pulleys in the model will furnish a conservative result.

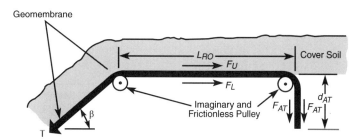

Source: Koerner 1994 (reprinted with permission of Prentice Hall).

FIGURE 18.5 Design model for calculating anchor trench depth and liner runout

F_U is the friction force above the geomembrane and is generally taken to be zero because the soil cover is likely to move with the geomembrane as it deforms. The following design equation can then be developed.

$$\sigma_{allow}(t) = q \tan \delta (L_{RO}) + 2(1 - \sin \varphi)(\gamma H_{ave}) \tan \delta (d_{AT}) \qquad \text{(EQ 18.6)}$$

There are two unknowns in this equation—either the anchor trench depth or the length of runout must be assumed. If the anchor trench depth is taken as zero, the equation gives a conservative result for runout only. A more accurate result for runout is achieved by removing the frictionless pulley from the model, which gives

$$\sigma_{allow}(t) \cos \beta = q \tan \delta (L_{RO}) + \sigma_{allow}(t) \sin \beta \tan \delta \qquad \text{(EQ 18.7)}$$

where:
- q = surcharge pressure (γ times the depth of cover soil) (lb)
- γ = unit weight of the cover soil (lb/ft^3)
- L_{RO} = length of runout (ft)
- φ = internal shear angle of the anchor trench fill soil (degrees)
- H_{ave} = average depth of the anchor trench (may require an estimate) (ft)
- d_{AT} = depth of the anchor trench (ft)

Geotextiles

Geotextiles are used as drainage pathways in ponds and dams to improve subgrades under liner systems, and to prevent migration of fine-grained soils into drainage systems. Table 18.3 provides a typical range of geotextile properties; additional detailed properties may be obtained from manufacturers. Because geotextiles are compressible, the thickness under a given load is taken into consideration. Parameters used in fluid flow are defined as permittivity for cross-plane flow, and transmissivity for in-plane flow.

TABLE 18.3 Typical range of geotextile properties

	Standard Units	SI Units
Physical properties		
Specific gravity	0.9 to 1.7	
Mass per unit area	4–20 oz/yd^2	130–700 g/m^2
Thickness	10–300 mils	0.25–7.5 mm
Stiffness	nil–22 lb-mils	nil–25,000 mg-cm
Mechanical properties		
Compressibility	nil to high	
Tensile strength (grab)	100–1000 lb	0.45–4.5 kN
Tensile strength (wide width)	50–1,000 lb/in.	9–180 kN/m
Confined tensile strength	100–1,000 lb/in.	18–180 kN/m
Seam strength	50%–100% of tensile	
Fatigue strength	50%–100% of tensile	
Burst strength	50–750 lb/in.^{2}	350–5,200 kPa
Tear strength	20–300 lb	90–1,300 N
Impact strength	10–150 ft-lb	14–200 J
Puncture strength	10–100 lb	45–450 N
Friction behavior	60%–100% of soil friction	
Pullout behavior	50%–100% of geotextile strength	
Hydraulic properties		
Porosity (nonwovens)	50%–95%	
Percent open area (wovens)	1%–36%	
Apparent opening size (sieve size)	#10–#200	
Permittivity	0.02–2.2 s^{-1}	
Permittivity under load	0.01–3.0 s^{-1}	
Transmissivity	0.1 to 20 × 10^{-3} ft^3/min-ft	0.01 to 2.0 × 10^{-3} m^3/min-m
Soil retention: turbidity curtains	m.b.e.*	
Soil retention: silt fences	m.b.e.	
Endurance properties		
Installation damage	0%–70% of fabric strength	
Creep response	g.n.p.† if <40% strength is being used	
Confined creep response	g.n.p. if <50% strength is being used	
Stress relaxation	g.n.p. if <40% strength is being used	
Abrasion	50%–100% of geotextile strength	
Long-term clogging	m.b.e. for critical conditions	
Gradient ratio clogging	m.b.e. for critical conditions	
Hydraulic conductivity ratio	0.3 to 0.6 appear acceptable	
Degradation properties		
Temperature degradation	High temperature accelerates degradation	
Oxidative degradation	m.b.e. for long service lifetimes	
Hydrolysis degradation	m.b.e. for long service lifetimes	
Chemical degradation	g.n.p. unless aggressive chemicals	
Radioactive degradation	g.n.p.	
Biological degradation	g.n.p.	
Sunlight (UV) degradation	Major problem unless protected	
Synergistic effects	m.b.e.	
General aging	Track record to date is excellent	

* m.b.e.: must be evaluated.
† g.n.p.: generally no problem.
Source: Koerner 1994 *(reprinted with permission of Prentice Hall).*

$$\Psi = k_n/t \qquad \text{(EQ 18.8)}$$

$$\theta = k_p t \qquad \text{(EQ 18.9)}$$

where:
- Ψ = permittivity
- θ = transmissivity
- k_n = cross-plane permeability coefficient for geotextile
- k_p = in-plane permeability coefficient for geotextile
- t = thickness of the geotextile under given normal load

Units for these equations can be any combination as long as consistency is maintained. Permittivity is generally in units of 1/s and transmissivity in generally in units of ft³/min-ft or m³/min-m.

Geonet

Geonet is often used to provide a flow path in liner systems. Planer flow is a function of the manufactured characteristics of the geonet. The concept of transmissivity is applicable and compression under load should be considered in design calculations. Applicable equations are

$$q = kiA \qquad \text{(EQ 18.10)}$$

$$\theta = q/iW \qquad \text{(EQ 18.11)}$$

where:
- q = volumetric flow rate
- k = coefficient of permeability
- i = hydraulic gradient
- W = width of flow area
- A = flow cross-sectional area (= $W \cdot t$ where t is thickness under load)
- θ = transmissivity

Units for transmissivity in the previous equation are similar to those above for geotextiles. The range of geonet properties is large and properties often change as manufacturing methods and materials change; consult the manufacturer for properties.

Interface Friction

Table 18.4 shows frictional values for the interfaces of soils to geomembranes, geomembranes to geotextiles, and soils to geotextiles.

TABLE 18.4 Interface frictional values

	Soil Types		
Geomembrane	Concrete Sand ($\varphi = 30°$)*	Ottawa Sand ($\varphi = 28°$)	Mica Schist Sand ($\varphi = 26°$)
EPDM-R†	24° (0.77)	20° (0.68)	24° (0.91)
Polyvinyl chloride (PVC)			
Rough	27° (0.88)	—	25° (0.96)
Smooth	25° (0.81)	—	21° (0.79)
CSPE-R‡	25° (0.81)	21° (0.72)	23° (0.87)
High-density polyethylene (HDPE)	18° (0.56)	18° (0.61)	17° (0.63)
Geomembrane-to-geotextile friction angles			

continues next page

TABLE 18.4 Interface frictional values (continued)

	Geomembrane				
		PVC			
Geotextile	EPDM-R†	Rough	Smooth	CSPE-R‡	HDPE
Nonwoven, needle punched	23°	23°	21°	15°	8°
Nonwoven, heat bonded	18°	20°	18°	21°	11°
Woven, monofilament	17°	11°	10°	9°	6°
Woven, slit film	21°	28°	24°	13°	10°

Soil-to-geotextile friction angles

	Soil Types		
	Concrete Sand	Ottawa Sand	Mica Schist Sand
Geotextile	($\varphi = 30°$)	($\varphi = 28°$)	($\varphi = 26°$)
Nonwoven, needle punched	30° (1.00)	26° (0.92)	25° (0.96)
Nonwoven, heat bonded	26° (0.84)	—	—
Woven, monofilament	26° (0.84)	—	—
Woven, slit film	24° (0.77)	24° (0.84)	23° (0.87)

* Efficiency values in parentheses are based on the relationship $E = (\tan \delta)/(\tan \varphi)$.
† Synthetic elastomer based on ethylene, propylene, and nonconjugated diene—reinforced.
‡ Chlorosulfonated polyethylene—reinforced.
Source: Martin, Koerner, and Whitty 1984.

Liner Leakage

In practice there is no such thing as a perfect liner; however, with modern materials and techniques, and with a high level of quality control during installation, potential leakage through a liner system can be reduced to small levels. Leakage rates may be required for the design of leak detection and secondary leachate collection and recovery systems. Estimates of leakage rates for typical liners at various levels of overall quality control are shown in Table 18.5.

TABLE 18.5 Leakage rates for various liners

Type of Liner	Overall Quality of Liner	Assumed Values of Key Parameters	Rate of Flow (gal/acre/day)*
Compacted soil	Poor	$k_s = 1 \times 10^{-6}$ cm/s	1,200
Geomembrane	Poor	30 holes/acre; a = 0.1 cm²	10,000
Composite	Poor	$k_s = 1 \times 10^{-6}$ cm/s 30 holes/acre; a = 0.1 cm²	100
Compacted soil	Good	$k_s = 1 \times 10^{-7}$ cm/s	120
Geomembrane	Good	1 hole/acre; a = 1 cm²	3,300
Composite	Good	$k_s = 1 \times 10^{-7}$ cm/s 1 hole/acre; a = 1 cm²	0.8
Compacted soil	Excellent	$k_s = 1 \times 10^{-8}$ cm/s	12
Geomembrane	Excellent	1 hole/acre; a = 0.1 cm²	330
Composite	Excellent	$k_s = 1 \times 10^{-8}$ cm/s 1 hole/acre; a = 0.1 cm²	0.1

* L = gal × 3.785.
Source: U.S. Environmental Protection Agency 1991.

PIPES

Pipes of various sizes, types, and materials are used to convey liquid and slurry to and from ponds, leach pads, tailings storage facilities and other impoundments and as part of leachate collection, leak detection, and drainage systems. Physical design considerations are presented below.

Pipe Thickness

For pipe made of plastic or other synthetic materials, the standard dimension ratio (SDR) is used as a measure of internal and exterior strength and other pipe properties.

SITE STRUCTURES AND HYDROLOGY

$$SDR = D/t \quad \text{(EQ 18.12)}$$

where:
- SDR = standard dimension ratio (dimensionless)
- D = outside pipe diameter (units of length)
- t = minimum pipe wall thickness (units of length same as for D)

Wall thickness requirements for pipe may be determined by

$$t = pD/2f_a \quad \text{(EQ 18.13)}$$

where:
- t = pipe wall thickness (units of length)
- D = outside diameter of pipe (units of length consistent with those for t)
- p = internal pressure in pipe (force per unit area in units consistent with units for t)
- f_a = allowable stress in pipe material (force per unit area in units consistent with units for t)

Earth Pressure on Buried Pipe

In applications such as leachate collection systems in heap leach pads, drainage systems in tailings facilities, and embankment drains, pipe deflection should be considered in selecting pipe size, material, and wall thickness. Buried pipe may be classified as ditch conduit or projecting conduit as illustrated in Figure 18.6.

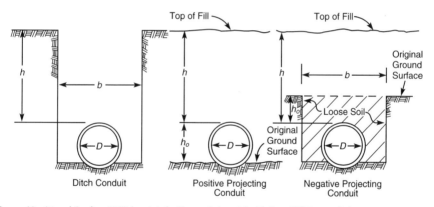

Source: Merritt and Gardner 1983 (reprinted with permission of the McGraw-Hill Companies).

FIGURE 18.6   Pipe backfill types

For a ditch conduit, the load on a ridged pipe is computed from

$$W = C_D whb \quad \text{(EQ 18.14)}$$

And the load on a flexible pipe is

$$W = C_D whD \quad \text{(EQ 18.15)}$$

For a positive projecting conduit the load is

$$W = C_p whD \quad \text{(EQ 18.16)}$$

And for a negative projecting conduit:

$$W = C_N whD \quad \text{(EQ 18.17)}$$

where:
- W = load on the pipe due to earth pressure (lb/linear ft)
- C_D = ditch conduit load coefficient
- C_P = load coefficient for positive projecting conduit (typically 1.0 for flexible pipe and 1.5 for ridged pipe in the absence of site-specific data)
- C_N = load coefficient for negative projecting conduit (use 0.9 if site-specific data is not available)
- w = unit weight of fill (lb/ft³)
- h = height of fill above top of pipe (ft)
- b = width of trench at top of pipe (ft)
- D = outside diameter of pipe (ft)

C_D is calculated from the equilibrium of vertical forces acting on the backfill above the pipe as

$$C_D = (1 - e^{-kh/b})/k(b/h) \qquad \text{(EQ 18.18)}$$

where:
- $k = 2K_a \tan \theta$
- K_a = coefficient of active earth pressure [= $(1 - \sin \phi) / (1 + \sin \phi)$]
- θ = angle of friction between the fill and the adjacent soil (generally equal to or less than the angle of internal friction [ϕ] of the fill)
- $e = 2.1718$

See Table 2.5 in Chapter 2, which covers material properties, for typical strength characteristics of soils.

Any system of units may be used for Equations 18.14 through 18.18 as long as dimensional consistency is maintained.

Deflection resulting from earth pressure can be found from

$$\Delta X = (D_L KW)/(EI/r^3 + 0.061 E_S) \qquad \text{(EQ 18.19)}$$

where:
- ΔX = vertical deflection of pipe (in.)
- D_L = deflection lag factor (varies from 1.0 to 1.5; typically taken as 1.5 in the absence of specific data)
- K = bedding constant (varies from 0.0 to 0.8; typically taken as 0.1 in the absence of specific data)
- W = earth pressure due to backfill (psi)
- E = modulus of elasticity of pipe material (psi)
- I = moment of inertia of pipe wall, in.⁴ per in. (in.³ = $t^3/12$; where t = average wall thickness of pipe [in.])
- r = mean radius of pipe (in.)
- E_S = modulus of soil reaction (psi)

See Table 2.6 in Chapter 2 for typical values of modulus of soil reaction.

HYDROLOGY AND HYDRAULICS

See the fluid mechanics section in Chapter 4, which covers physical science and engineering, for more information.

Water Balance

A water balance is generally completed for any impoundment structure that will contain solutions as part of mine operations. The water balance may be calculated from

$$S = I - Q - E \qquad (EQ\ 18.20)$$

where:
- S = change in storage for the facility
- I = inflow into the facility from all sources (operational inputs, direct precipitation, runoff, secondary inputs, etc.)
- Q = outflow (removal of solutions for process, recycle, treatment, other uses, etc.)
- E = evaporation from the facility (free-water surface and wet tailings or other materials)

Storm Water Runoff

Precipitation events are important to the overall water balance and for operational considerations. Runoff data is required to design the size of diversion channels and storage ponds. For minor hydraulic structures and small drainage basins (up to five square miles) the peak discharge may be calculated using the rational equation:

$$Q = CIA \qquad (EQ\ 18.21)$$

where:
- Q = peak discharge (ft^3/s)
- C = runoff coefficient based on percentage of precipitation that is direct runoff
- I = rainfall intensity (in./h)
- A = size of the drainage area (acres)

The runoff coefficient is dependent on local soil conditions, type, and amount of vegetative cover and the topographic slopes within the drainage basin. Table 18.6 gives some common runoff coefficients.

TABLE 18.6 Runoff coefficients for the rational method

Description of Area	Runoff Coefficients
Residential	
Single-family areas	0.30–0.50
Suburban	0.25–0.40
Industrial	
Light areas	0.50–0.80
Heavy areas	0.60–0.90
Unimproved areas	0.10–0.30
Lawns; sandy soil	
Flat, ≤2%	0.05–0.10
Average, 2%–7%	0.10–0.15
Steep, >7%	0.15–0.20
Rural Areas (clay and silt loam)	
Woodland	
Flat 0%–5%	0.30
Rolling 5%–10%	0.35
Hilly 10%–30%	0.50
Pasture	
Flat 0%–5%	0.30
Rolling 5%–10%	0.36
Hilly 10%–30%	0.42
Cultivated	
Flat 0%–5%	0.50
Rolling 5%–10%	0.60
Hilly 10%–30%	0.72

Source: Warner 1992.

The rainfall intensity is a function of storm duration or time of concentration and storm frequency. A common method used to calculate intensity is the Steel formula:

$$I = K/(t + b) \quad \text{(EQ 18.22)}$$

where:
- I = rainfall intensity (in./h)
- t = duration of storm (min) (time of concentration)
- K and b = factors dependent on storm frequency and region

Use Figure 18.7 and Table 18.7 to obtain values for the Steel formula for the United States.

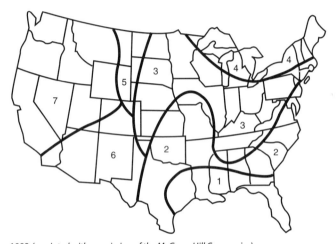

Source: Nelson 1983 (reprinted with permission of the McGraw-Hill Companies).

FIGURE 18.7 Regions for use with the Steel formula

TABLE 18.7 Coefficients for use in the Steel formula

Frequency (years)	Coefficients	Region						
		1	2	3	4	5	6	7
2	K	206	140	106	70	70	68	32
	b	30	21	17	13	16	14	11
4	K	247	190	131	97	81	75	48
	b	29	25	19	16	13	12	12
10	K	300	230	170	111	111	122	60
	b	36	29	23	16	17	23	13
25	K	327	260	230	170	130	155	67
	b	33	32	30	27	17	26	10
50	K	315	350	250	187	187	160	65
	b	28	38	27	24	25	21	8
100	K	367	375	290	220	240	210	77
	b	33	36	31	28	29	26	10

Source: Nelson 1983 (reprinted with permission of the McGraw-Hill Companies).

SITE STRUCTURES AND HYDROLOGY

For large drainage areas, there are several methods for determining runoff. Generally, these methods use a unit hydrograph for a given basin and runoff is found by multiplying the unit hydrograph by the rainfall amount from the design storm. For basins with no existing unit hydrograph, an estimate may be made using the Soil Conservation Service's synthetic unit hydrograph. The following equations are used to estimate key parameters:

$$t_r = 0.133 t_c \qquad \text{(EQ 18.23)}$$

$$t_p = 0.5 t_r + t_l \qquad \text{(EQ 18.24)}$$

$$Q_p = 0.756 A_d / t_p \qquad \text{(EQ 18.25)}$$

where:
- t_r = storm duration (h)
- t_c = time of concentration (h)
- t_p = time to peak flow (h)
- t_l = lag time from the centroid of the distribution to peak discharge, may be estimated as $t_l = 0.6 t_c$
- Q_p = peak discharge (cfs)
- A_d = drainage area (acres)

Q_p and t_p provide one point on a hydrograph. Additional points may be found using Figure 18.8. Using time as the independent variable, arbitrary values of time are selected and the ratio t/t_p computed. The figure is then used to find values of Q_t/Q_p and a unit hydrograph is constructed.

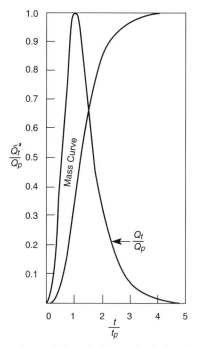

Source: Lindeburg 1992 (reprinted with permission of Professional Publications, Inc.).

FIGURE 18.8 Graph for use in estimating a unit hydrograph

Sediment Ponds

A common form of sedimentation control is sediment or settling ponds. Sediment removal is accomplished by reducing the flow velocity to a point where sediment particles will settle. Sediment ponds are often sized by rules of thumb, such as 0.5 acre-in. of storage per acre of disturbed area. Ponds that use this type of empirical design generally are effective at removing the course sediment faction. However, more detailed design and consideration of a number of factors are required to provide more complete settling and sediment removal.

The overflow velocity of a sediment pond may be calculated from

$$V_0 = Q_0/A \qquad \text{(EQ 18.26)}$$

where:
- V_0 = overflow velocity
- Q_0 = pond outflow
- A = surface area of the pond

If the critical settling velocity of any size sediment particle is greater than the overflow velocity, that particle and particles of a larger size will settle out in the pond. Settling velocity of an ideal particle in a solution of low solids concentration is given by one of the following three equations, depending on the Reynolds number.

Stokes' law:

$$V_s = g/18v \, (S_s - 1) \, D^2 \, ; \, R_e < 1 \qquad \text{(EQ 18.27)}$$

Transition region:

$$V_s = [2.32 \, (S_s - 1) \, D^{1.6v - 0.6}]^{0.714} \, ; \, 1 < R_e < 1{,}000 \qquad \text{(EQ 18.28)}$$

Newton's law:

$$V_s = 1.82g[(S_s - 1) \, D]^{0.5} \, ; \, 1{,}000 < R_e < 25{,}000 \qquad \text{(EQ 18.29)}$$

where:
- V_s = critical settling velocity (cm/s)
- g = acceleration due to gravity (981 cm/s^2)
- D = diameter of a spherical (ideal) particle (cm)
- S_s = specific gravity of particle
- v = kinematic viscosity of water (cm^2/s)
- R_e = Reynolds number

Sediment pond surface area and free depth (depth above accumulated sediments) is determined from the size of sediment particle to be settled. The total depth of the pond is based on the amount of sediment to be stored. Design should allow for equipment access so accumulated sediments can be removed periodically.

Emergency Spillways

Design of reservoirs and other impoundments should include a spillway to pass flows from extreme storms or other events and to protect the structure from damage from these flows. Generally spillways are designed as overflow spillways in the crest of the dam or embankment and flow is calculated from

$$Q = C_s L H^{3/2} \qquad \text{(EQ 18.30)}$$

where:
- Q = flow (ft^3/s)
- C_s = spillway coefficient

L = width of the spillway crest (ft)
H = total head at the spillway crest (ft; generally the static head; if the velocity head at the spillway is significant then $H = H_s + V^2/2g$)
H_s = static head at the spillway (ft)
V = velocity of flow at the spillway (ft/s)
g = acceleration of gravity (ft/s^2)

For a sharp crested (ogee) spillway the coefficient C_s is generally between 3 and 4. A value of 3.97 may be used in the absence of site-specific data.

In some cases, chute spillways may be cut into the abutment of a dam or embankment, or an overflow spillway may be constructed alongside the facility. These may be designed using open-channel flow theory.

INSPECTION AND MAINTENANCE

Any earthen or rock structure that involves cut or fill earth slopes, liner systems, pipe systems, or channels should be inspected on a regular basis by a qualified professional. Damage or irregularities identified should be investigated fully and corrected in a timely manner. A checklist for inspection of an earth-fill dam is given in Figure 18.9 on page 368.

REFERENCES

Driscoll, F.G. 1986. *Groundwater and Wells*. 2nd ed. St. Paul, MN: Johnson Division.
Koerner, R.M. 1994. *Designing With Geosynthetics*. 3rd ed. Englewood Cliffs, NJ: Prentice Hall.
Lambe, T.W., and R.V. Whitman. 1969. *Soil Mechanics*. New York: John Wiley & Sons.
Lindeburg, M.R. 1992. *Civil Engineering Reference Manual*. 6th ed. Belmont, CA: Professional Publications, Inc.
Martin, J.P., R.M. Koerner, and J.E. Whitty. 1984. Experimental friction evaluation of slippage between geomembranes, geotextiles, and soils. *Proceedings of the International Conference on Geomembranes*. Denver: 191–196.
Merritt, F.S., and W.S. Gardner. 1983. Geotechnical engineering. In *Standard Handbook for Civil Engineers*. 3rd ed. Edited by F.S. Merritt. New York: McGraw-Hill Companies. 7-1–7-103.
Nelson, S.B. 1983. Water engineering. In *Standard Handbook for Civil Engineers*. 3rd ed. Edited by F.S. Merritt. New York: McGraw-Hill Companies. 21-1–21-143.
U.S. Bureau of Reclamation. 1977. *Design of Small Dams*. Washington, DC: Government Printing Office.
U.S. Environmental Protection Agency (EPA). 1991. *Design and Construction of RCRA/CERCLA Final Covers*. Seminar Publication of the Office of Research and Development. EPA 625/4-91/025. Washington, DC: EPA.
U.S. Department of the Interior. 1980. *Safety Evaluation of Existing Dams*. A Water Resources Technical Publication. Denver, CO: Water and Power Resources Service.
Vick, S.G. 1983. *Planning, Design, and Analysis of Tailings Dams*. New York: John Wiley & Sons.
Warner, R.C. 1992. Design and management of water and sediment control systems. In *SME Mining Engineering Handbook*. 2nd ed., Vol 1. Edited by H.L. Hartman. Littleton, CO: Society for Mining, Metallurgy, and Exploration, Inc. 1159.

Dam

Upstream Face
Slope Protection _____
Erosion-beaching _____
Vegetative Growth _____
Settlement _____
Debris _____
Burrows or Burrowing Animals _____
Unusual Conditions _____

Downstream Face
Signs of Movement _____
Seepage or Wet Areas _____
Vegetative Growth _____
Channelization _____
Condition of Slope Protection _____
Burrows or Burrowing Animals _____
Unusual Conditions _____

Abutments
Seepage _____
Cracks, Joints, and Bedding Planes _____
Channelization _____
Slides _____
Vegetation _____
Signs of Movement _____

Crest
Surface Cracking _____
Durability _____
Settlement _____
Lateral Movement (Alignment) _____
Camber _____

Seepage and Drainage Summation
Location(s) _____
Estimated Flow(s) _____
Color (Staining) _____
Erosion of Outfall _____
Toe Drain and Relief Wells _____

Measurement
Method _____
Amount _____
Change in Flow _____
Clearness of Flow _____
   Color _____
   Fines _____
   Condition of
     Measurement Devices _____
   Records _____

Other
_____
_____

Performance Instruments
Piezometer Well
   Well _____
   Frostfloor _____
   Ventilation _____
   Gauges _____
   Piping _____
   Security _____
Surface Settlement Points _____
Crossarm Devices
   (Deviation, Station, and Offset) _____
Reservoir-level Gauge _____
Ice-prevention System _____
Other
_____
_____

Spillway

Approach Channel
Vegetation (Trees, Willow, etc.) _____
Debris _____
Slides Above Channel _____
Channel Side Slope Stability _____
Log Boom _____
Slope Protection _____

Control Structures (Observed Operation)
Apron
   Surface Condition _____
   General Condition of Concrete _____
   Movement _____
   Settlement _____
   Joints _____
   Cracks _____
Crest
   Surface Condition _____
   General Condition of Concrete _____
   Cracks or Areas of Distress _____
   Signs of Movement _____
Walls
   Surface Condition _____
   General Condition of Concrete _____
   Movement (Offsets) _____
   Cracks or Areas of Distress _____
   Settlement _____
   Joints _____
   Drains _____
   Backfill _____
Gates
   Condition _____
   Hoist Equipment _____
   Control Equipment _____

Source: Water and Power Resources Service, U.S. Department of the Interior 1980.

FIGURE 18.9 Dam inspection checklist

CHAPTER 19

Placer Mining

Louis W. Cope, P.E.

Placer ore bodies are alluvial deposits that can contain the economic metals or minerals of gold, tin, magnetite, titanium, tungsten, zircon, garnet, diamonds, and semiprecious gemstones, to name the most common valuable constituents. These deposits are usually formed by deposition in rivers, on shorelines, or colluvially by in-place weathering.

Mining of the deposits can vary from small operations ranging from 1–5 yd^3/h by hand or mechanized digging with small machines. Operations ranging from 5–200 yd^3/h can be land-based, usually skid-mounted, or barge-mounted units with separate digging machines. These latter items can be bulldozers, front-end loaders, backhoe excavators, draglines, cutter-head, or suction machines. When the digging and recovery machinery are on the same barge with the recovery section, such as on a 200–2,000 yd^3/h floating dredge, the digging is either by a bucket ladder or bucket-wheel excavator.

The ore is usually treated in two steps. First the muck is slurried before it is sceened in a rotating trommel. Recovery of target metals or minerals is normally by a combination of sluices, jigs, centrifugal concentrators, vibrating tables, or all four.

PLACER ORE BODIES

TABLE 19.1 Grain size of material in placer deposits

Names of Particles	Average Diameter (mm)
Boulders	Greater than 256 mm (10 in.)
Cobbles	64 mm to 256 mm (2½ in. to 10 in.)
Pebbles	4 mm to 64 mm (³⁄₁₆ in. to 2½ in.)
Gravel	Greater than 2 mm
Sand	2 mm to $\frac{1}{16}$ mm
Silt	$\frac{1}{16}$ mm to $\frac{1}{256}$ mm
Clay	Less than $\frac{1}{256}$ mm

Source: Wells 1969.

370 | SME MINING REFERENCE HANDBOOK

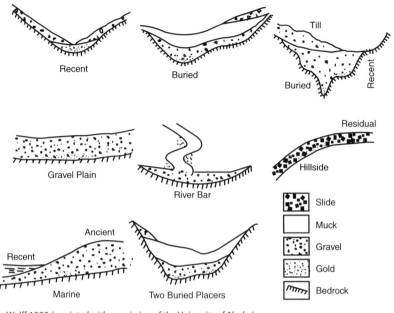

Source: Wolff 1980 (reprinted with permission of the University of Alaska).

FIGURE 19.1   Cross sections of types of placers

TABLE 19.2   Common heavy minerals in placers

Mineral	Specific Gravity	
	From	To
Magnetite	—	5.2
Ilmenite	5.6	5.7
Garnet	3.2	4.3
Zircon	4.2	4.7
Hematite	4.9	5.3
Chromite	4.3	4.6
Epidote	3.2	3.5
Olivine	—	3.3
Limonite	3.6	4.0
Rutile	—	4.2
Pyroxene	—	3.3
Monazite	4.9	5.3
Platinum group metals	14	19

NOTE: This is a partial list, arranged in approximate order of frequency.
Source: Wells 1969.

TABLE 19.3 Approximate swell of earth and gravel when disturbed (%)

Ordinary gravel	20 to 30
Cemented gravel	40
Sand and gravel	15
Gravel and clay	35
Loam	20
Dense clay	50

Source: Wells 1969.

PLACER EVALUATION

TABLE 19.4 Prospecting methods

Metallic Minerals—Au, Sn, Pt
Deposits other than beach sand, thawed ground:
A. Depth to about 15 ft
1. Dry: Hand-dug pits
Power digger, square or cylindrical bucket
Occasionally placer churn drill, cased hole
2. Wet: Hand-dug pits with caisson or lagged lining
Placer churn drill, cased hole
B. Depth 15 to 200 ft
1. Dry or wet: Placer churn drill, cased hole (rarely open hole)
Becker percussion drill
Deposits other than beach sand, frozen ground:
A. Depths to 200 ft or more
Placer churn drill, open hole
Hand-dug shafts, usually little or no lagging
Becker percussion drill—not yet proved to 1969
Beach sand deposits (at or above present sea level)
A. Depth to about 15 ft—greater depth rarely encountered
1. Above water level: Hand auger
Occasionally placer churn drill
2. To below water level: Hand auger with pipe casing
Placer churn drill

Source: Daily 1973.

TABLE 19.5 Effect of a single gold particle on a sample

Size of Drill Hole or Channel (in.)	Size of Gold Particle and Effect on Sample ($/yd³)		
	20-Mesh (6.57 Mg)	40-Mesh (0.91 Mg)	60-Mesh (0.27 Mg)
7½ diameter	0.58	0.08	0.025
5¼ diameter	1.18	0.16	0.05
3 diameter	3.60	0.50	0.14
3 × 6	1.42	0.02	0.06
6 × 6	0.71	0.10	0.03
6 × 12	0.35	0.05	0.015
12 × 12	0.18	0.025	0.0075
16-in. pan*	1.18	0.16	0.05

* 180 pans/yd³.
NOTE: Values shown are those that would result from one gold particle in a 1-ft sample increment or drive, and are based on gold weights determined by the author, with gold at $35/oz.
Source: Wells 1973.

TABLE 19.6 Character of gold versus distance from source

5 miles	Rough nuggety
8 miles	Small nuggety, water-worn
11 miles	Fine granular
25 miles	Fine scaly

Source: Wells 1969.

TABLE 19.7 Classification of gold particles

Coarse gold (nuggets), which remains on a 10-mesh screen (openings 1/16 in.).

Medium gold (small nuggets), which passes 10-mesh and remains on a 20-mesh screen (openings 1/32 in.). Value about 1 cent apiece.

Fine gold, which passes 20-mesh and remains on a 40-mesh screen (openings 1/64 in.). Value about 1/3 cent per color.

Very fine gold, which passes a 40-mesh screen.

Source: Boericke 1933.

TABLE 19.8 Sample methods: advantages and disadvantages

Bulk Sampling

Advantages: The advantages of bulk sampling include a good view of the gravel in place, and knowledge of the amount of force required to excavate it. Some of the same digging machines used in bulk sampling are used in production. The large samples obtained minimize "nugget effect" and errors.

Disadvantages: The fact that pits or trenches may not reach bedrock in all cases is a serious disadvantage. It is often difficult to accurately measure a pit due to curves in the excavation, if by a backhoe, and sloughing from the sides. If groundwater is encountered, sample integrity is lost. Larger samples require larger transport and processing equipment, and the situation becomes more complicated. As unit costs of each sample are high, the exploration cost to delineate reserves is very expensive.

Drill Sampling

Advantages: Historically, drilling has been the sample method of choice and has been proven by operation in most cases. Drilling gives greater coverage in developing gravel grade and reserves at less unit cost, the profile of bedrock, and indications of enriched and barren areas.

Disadvantages: Drilling gives a small sample which aggravates the nugget effect. At times, boulders will necessitate abandonment of uncompleted drill holes. Vibration from driving casing or from a reverse circulation hammer-drill may cause gold to migrate downward into the gravel. Drill rigs require at least minimal access roads and drill pads.

Source: Cope 1992a.

PLACER OPERATIONS

Placer ore is unconsolidated material, usually found on the surface. It is excavated by many means including pick and shovel, dragline, front-end loader, backhoe, bulldozer, bucket ladder, bucket wheel, hydraulicking, and suction. After it is excavated, the target valuable minerals are recovered using various processing methods.

TABLE 19.9   Methods of concentration

Mineral	Method of Concentration	
	Primary	Secondary
Gold	Sluice box with riffles	Amalgamation
	Placer jig	Wet panning, dry blowing
Platinum	As above	Magnetic
		Hindered settling
		Wet vibrating table
		Air-deck table
Tin	Palong sluice	Willoughby concentrator
	Placer jig	Lanchute
		Electro
Diamond	Diamond pan	Grease table
	Harz jig	Plietz jig
	Heavy-media separation	Magnetic—high-tension electro
	Placer jig	Optical—X-ray
	X-ray (coarse fractions)	Chemical
		Hand picking (always terminal)
Beach-sand minerals		
Magnetite	Magnetic	Magnetic
All other	Humphrey spiral	Magnetic—high-tension electro
	Cannon concentrator	Wet vibrating tables
	Others—not jigs	Dry vibrating tables

Source: Daily 1973.

TABLE 19.10   Water requirements

Mining Method	Water Requirements
Rocker boxes	4 to 5 gpm, 50 to 100 gal/yd^3
Small scale hand mining	170 to 225 gpm for steep 12-in.-wide sluice
Ground sluicing	22,000 to 162,000 gal/yd^3
Hydraulicking	2,000 to 32,000 gal/yd^3
Stationary washing plants	650 to 2,000 gal/yd^3
Land-based mobile plants	480 to 3,200 gal/yd^3
Floating dragline-fed dredges	570 to 2,500 gal/yd^3
Bucket-line dredges	3,500 to 10,000 gal/yd^3

Source: Abstracted from Wells 1969.

TABLE 19.11 Water requirements of a sluice

Width of Sluice Box (in.)	Depth of Flow (in.)	Grade (%)	Water Flow		
			Cubic Feet per Minute	Equivalent Miner's Inches	Cubic Yards Gravel per 24 Hours
10–12	6–7	4.1	45	30	67–135
12–14	10	6.2	100	66	150–300

Source: Boericke 1933.

TABLE 19.12 Treatment methods: pros and cons

Method	Advantages	Disadvantages
Sluice	▪ Simple ▪ Inexpensive ▪ Construction materials available everywhere ▪ No power required	▪ Reduced gold recovery ▪ Time-consuming to clean up ▪ Low capacity
Fabric	▪ Simple ▪ Inexpensive ▪ No power required ▪ Improved fine gold recovery ▪ Best used in conjunction with a normal sluice	▪ Low capacity
Amalgamation	▪ Simple ▪ Inexpensive ▪ No power required ▪ Recovers very fine gold ▪ Best used for final cleanup	▪ Low capacity ▪ Cost of makeup of lost mercury ▪ Legal and moral constraints ▪ Not all gold will amalgamate
Thin film	▪ No moving parts in separating machine ▪ Recovers fine gold	▪ Requires screening to fine size ▪ High headroom required ▪ Pump and piping wear
Jig	▪ Feed-size forgiving ▪ Feed-rate forgiving ▪ Requires little operator attention ▪ Recovers fine gold	▪ Complicated machine ▪ Several adjustments
Vibrating table	▪ Can produce high-grade gold concentrate ▪ Takes jig concentrate as produced ▪ Best for cleanup ▪ Separation clearly visible	▪ Requires screening to fine size ▪ Requires considerable operator attention
Bowl	▪ Recovers fine gold ▪ Requires little operator attention ▪ Long history of use in cleanup ▪ Longer run gives better concentrate	▪ Requires screening to relatively fine size ▪ Has to be stopped to be cleaned up ▪ Requires absolutely clean water (for water injection models)
Pan	▪ Minimum equipment required ▪ Simple ▪ Inexpensive ▪ Often used in last stage of cleanup ▪ Best used for prospecting	▪ Very low capacity ▪ Hard work

Source: Cope 1992b.

PLACER MINING | 375

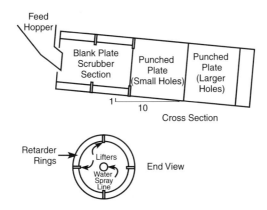

Source: Cope 1992b.

FIGURE 19.2 Elements of a scrubbing/screening trommel

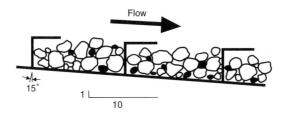

Source: Cope 1992b.

FIGURE 19.3 Cross section of a sluice with Hungarian riffles

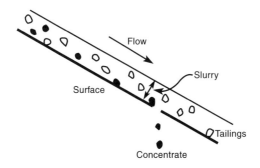

Source: Cope 1992b.

FIGURE 19.4 How spirals and cones work

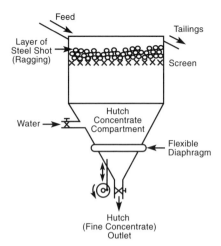

Source: Cope 1992b.

FIGURE 19.5   Components of a jig

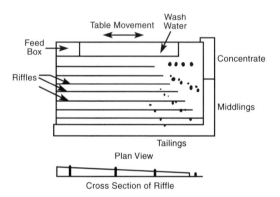

Source: Cope 1992b.

FIGURE 19.6   How a vibrating table separates minerals

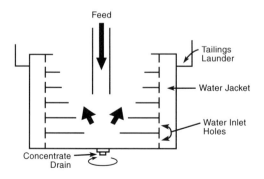

Source: Cope 1992b.

FIGURE 19.7   Cross section of a centrifugal separator

COSTS

TABLE 19.13 Bucketline dredge capital costs*

Bucket Capacity		Cost US $ (millions)
ft³	m³	
10	0.283	9
15	0.425	16
20	0.566	23
30	0.850	50

* Not including transportation, insurance, or duties.
Source: McLean 1992.

The budgetary capital costs for small placer treatment operations of less than 200 yd³/h are approximately \$1,500/yd³ of capacity per hour adjusted to year 2001. Because of the variation in locations and gravel conditions, this figure does not include transportation, insurance, duties, or the digging machine. Operating costs should be in the \$2.50 to \$4.00/yd³ range (Cope 1976).

GRAVEL

TABLE 19.14 Useful gravel facts

	One pan (heaping) equals 1/10 to 1/2 ft³ of gravel
	One cubic yard of gravel weighs about 3,000 lb
	One person can pan 1/2 to 1 yd³ of gravel per day
	One person can shovel up to 7 yd³ of gravel per day
Source: Wolff 1980.	

REFERENCES

Boericke, W.F. 1933. *Prospecting and Operating Small Gold Placers.* New York: John Wiley & Sons.

Cope, L.W. 1976. "Doodlebug" Placer Gold Mining. Lecture in Placer Exploration and Mining Short Course, October 25–29, Reno, NV.

Cope, L.W. 1992a. Samples, bulk versus drill. In *Practical Placer Mining.* Edited by L.W. Cope and L.R. Rice. Littleton, CO: Society for Mining, Metallurgy, and Exploration, Inc. (SME). 19–23.

Cope, L.W. 1992b. Placer processing. In *Practical Placer Mining.* Edited by L.W. Cope and L.R. Rice. Littleton, CO: SME. 43–56.

Daily, A.F. 1973. Placer mining. In *SME Mining Engineering Handbook,* Vol. 2. Edited by A.B. Cummins and I.A. Given. New York: Society of Mining Engineers of the American Institute of Mining, Metallurgical, and Petroleum Engineers, Inc. 17-151–17-179.

McLean, C.A. 1992. Placer mining costs. In *SME Mining Engineeering Handbook*. 2nd ed., Vol. 2. Edited by H.L. Hartman. Littleton, CO: SME. 1471–1473.

Wells, J.H. 1969. *Placer Examination, Principles and Practice.* Technical Bulletin 4, Bureau of Land Management. Washington, DC: U.S. Government Printing Office.

Wells, J.H. 1973. Special exploration techniques—placer deposits. In *SME Mining Engineering Handbook,* Vol. 1. Edited by A.B. Cummins and I.A. Given. New York: Society of Mining Engineers of the American Institute of Mining, Metallurgical, and Petroleum Engineers, Inc. 5-44–5-53.

Wolff, E. 1980. *Handbook for the Alaskan Prospector.* 2nd ed. Ann Arbor, MI: Edwards Brothers, Inc.

CHAPTER 20

In Situ Leaching

Paul D. Chamberlin, P.E.

IN SITU LEACHING
by Ray V. Huff

In situ leaching generally means leaching undisturbed ore rather than in-place rubble. It does not imply heap or dump leaching. Commercial operations have been used for uranium, copper, boron, trona, nahcolite, and potash. Potential operations are possible for precious metals, manganese, and some industrial minerals. Sulfur is not leached in situ; it is melted.

Ore and Deposit Characteristics
Many deposits are formed by precipitation of values from either percolating meteoric or upwelling fluids. Thus, the deposit has or had both porosity and permeability.

TABLE 20.1 Typical ranges of porosity and permeability

Ore type	Porosity (%)	Permeability (md [millidarcies])
Fractured, crystalline	5 to 10	3 to >30
Sandstones	5 to 30	10 to >1,000
Siltstones	<Sandstones	<Sandstones

NOTE: Shear zones have higher values than fractured crystalline ores.

The deposit must have permeability and values must be in flow channels (not encapsulated). Ideally, the deposit should have an ore zone beneath the water table, uniformly high permeability, low porosity, be bounded by low permeability formation(s), and have values lining the flow channels.

Lixiviants
- Uranium—carbonate/bicarbonate with an oxidant (O_2 or peroxide)
- Oxide copper—dilute sulfuric acid
- Sulfide copper—sulfuric acid with oxidant, often $Fe_2(SO_4)_3$ or $FeCl_3$
- Boron—hydrochloric acid
- Precious metals—complexing agent (CN^- or Cl^-) and oxidant (dissolved O_2 or OCl^-)
- Manganese—reducing acid (water with SO_2)

NOTE: gangue minerals may dictate the nature and concentration of lixiviant.

Wells and Well Fields

Completed wells at 300 m to 600 m deep cost about US$ 328/m.

Shallow wells (<300 m) can be substantially less costly when low-cost materials of construction can be used.

An inverted 5-spot well pattern, consisting of one injection well and four production wells, is common. This provides wells at the perimeter of well field, which maximizes the recovery of pregnant solution, although the solution grade is somewhat diluted. Other well patterns are line drive and inverted 4-spot, 7-spot, and 9-spot.

Well spacing is optimized by iterative designs considering cost of wells and predicted values recovered per 5-spot unit.

Approximate flow rate in the 5-spot pattern is

$$q = (1.035 \times 10^{-4} kh(\Delta p))/\mu[\ln(d/r_w) - 0.619] \qquad (EQ\ 20.1)$$

where:
- q = flow rate in gpm
- k = permeability in millidarcies
- h = vertical ore interval open to flow in feet
- Δp = applied differential pressure in psig
- μ = viscosity of the fluid in centipoise
- d = spacing between injection and production wells in feet
- r_w = well bore radius in feet

Materials of construction are fiber-reinforced plastic (FRP) for high pressure and corrosive environments, other plastic materials for lower pressure and less corrosive environments, and stainless steels or other metallics as required for pumps or valves.

Operating Considerations

- Annual plant production rate depends on the grade of pregnant solution and the number of producing wells.
- Well flow rate depends mostly on permeability of the ore zone, the differential pressure between injection and production wells, the spacing of the wells, and the thickness of ore interval open to flow.
- Injection pressure is limited by rock parting pressure.
- Δp is related to injection pressure and drawdown of fluid in production wells.
- Equations describing flow from various well patterns can be found in water flood manuals published by the Society of Petroleum Engineers.

Troubleshooting

- Look for reduced flow rates from production wells over time.
- Look for increasing injection pressures over time.
- Look for unexpected decreases in pregnant solution grade over time.
- Data collection would include fluid injection rates and pressures for each well, fluid production rates from each well, and pregnant solution grades from each well.
- Diagnosing most troubles often requires down-hole wireline measurements.

IN SITU LEACHING OF URANIUM
by Joseph R. Stano

Deposit Characteristics

- Roll-type sandstone deposits (redox fronts)
- Ore beds usually unconsolidated or loosely cemented with calcium carbonate
- Typical depths from 45 to 137 m; maximum 304 to 609 m

- Typical ore grades between 0.04% and 0.2% U_3O_8
- Predominant ore minerals are uraninite, coffinite, and sometimes carnotite; radioactive disequilibrium is relatively common
- Common accessory minerals are quartz, feldspars, clays, calcite, pyrite, and carbonaceous material
- Less common minerals are jordisite, vanadates, selenates, arsenates, humates, hydrogen sulfide, and methane
- Typical hydrology is in confined aquifers; leach zone preferably below water table
- Typically 10%–30% open porosity with permeability of 50–1,000 millidarcies
- Typical reserves: 10,000,000–30,000,000 lb U_3O_8; isolated pods less than 1,000,000 lb

Wells and Well Fields
- Wells—modified water well designs; rotary-drilled; constructed with plastic casings (polyvinyl chloride [PVC] and FRP) cemented back to the surface; single or multiple completions; most common casing diameter is nominal 4 in.; 6 in. sometimes used in production wells and for deep completions
- Well completions—injection and production wells perforated in the selected ore zone(s) and usually lined with stainless well screens
- Typical well field spacings—50 to 200 ft (15.2–61.0 m)
- Predominant well field configurations—5-spots and line drives
- Monitor wells—completed in upper and sometimes lower aquifers with provisions for water level measurements and water sampling
- Typical single well yields—5 to 10 gpm (19 to 38 L/min)

In Situ Leaching Process

Most operations are continuous—injection of fortified barren leach solution; flow through ore controlled by induced pressure gradients and well spacings; dissolution of uranium; withdrawal of pregnant liquor from recovery wells with submersible pumps; and chemical restoration of groundwater on completion of leaching.

The typical concentration of uranium in the pregnant leach solution is 35 to 200 ppm U_3O_8, with peaks as high as 1,000 ppm.

Injected leach solutions have low concentrations of oxidant (50–500 ppm) and lixiviant (500–2000 ppm). Oxidation of the uranium minerals is the rate-controlling step.

Dissolved oxygen is the most common oxidant; hydrogen peroxide is used in shallow deposits (<76 m) and to accelerate the leach.

The primary lixiviant is a variable mixture of sodium carbonate and bicarbonate adjusted with carbon dioxide to regulate the pH between 7 and 9. Ammonium carbonate was used in early projects. Leaching has been conducted successfully with pH levels as high as 10. Lower pH levels tend to dissolve more calcium carbonate mineral, which is present in most roll-front deposits.

Sulfide minerals (such as pyrite and jordisite) in the ore tend to react with the oxidant to produce acid and lower the pH of the pregnant leach solution.

Oxidation of sulfide minerals can mobilize molybdenum, arsenic, and polythionates. Radium is generally mobilized. Radon gas is entrained in leach solutions and provisions are made for in-plant ventilation.

Hydrogen sulfide is sometimes present in leach liquor.

Principal leaching reactions are

$$2UO_2 + O_2 = 2UO_3$$
$$UO_3 + 2HCO_3^- = UO_2(CO_3)_2^{-2} + H_2O$$
$$UO_3 + CO_3^{-2} + 2HCO_3^- = UO_2(CO_3)_3^{-4} + H_2O$$
$$2CaCO_3 + CO_2 + H_2O = 2Ca(HCO_3)_2$$
$$FeS_2 + O_2 = Fe^{+2} + SO_4^{-2}$$

Plant Recovery Process
- Pregnant solution is pumped from the well field(s) to a plant feed tank and clarified with sand filters and cartridge filters.
- Uranium is extracted with anion exchange resins (capacity 16 to 80 kg U_3O_8 per m^3).
- Ion exchange (IX) equipment can be either fixed bed or moving bed systems.
- Resins are stripped with alkaline or acidic chloride brines (concentrates U_3O_8 5 to 20 times).
- Eluate is neutralized with acid.
- Uranium is precipitated with ammonia, sodium hydroxide, or hydrogen peroxide to produce a slurry of ammonium diuranate, sodium diuranate, or uranyl peroxide.
- Uranium precipitate is dewatered by thickening and centrifugation (or alternately filtration) to produce a moist yellow cake.
- Cake is dried or calcined to a yellow cake product that is actually yellow, gray, or black depending on calcining temperature. The product contains from 71%–95% U_3O_8.
- Barren leach solution is replenished with leach chemicals and reinjected.

Principal recovery reactions are:

Na diuranate: $8Na^+ + 2UO_2(CO_3)_3^{-4} + 6NaOH = Na_2U_2O_{7(solid)} + 6Na_2CO_3 + 3H_2O$

NH_4 diuranate: $2UO_2^{+2} + 6NH_4OH = (NH_4)_2U_2O_{7(solid)} + 4(NH_4)^+ + 3H_2O$

Uranyl peroxide: $UO_2^{+2} + H_2O_2 + 2H_2O = UO_4 2H_2O_{(solid)} + 2H^+$

Environmental Concerns
- Dispose of excess leach solution by solar evaporation or reverse osmosis.
- Dispose of (1) scale residues and plugged filter elements (both radioactive); (2) by-product precipitates such as molybdenum sludges; (3) surplus IX brines; and (4) groundwater restoration residues.
- Restore well fields by groundwater flushing with concurrent aboveground separation of dissolved solids.
- Accomplish aboveground radium removal with selective ion exchange resins or barium chloride precipitation.
- Impound separated brines.
- Fix uranium and molybdenum underground.

Operating Problems and Remedies
- Plugged wells: (1) usually caused by scale buildup on well screens, relieved by acidification with hydrochloric acid; (2) bacterial plugging can be relieved by injecting hypochlorite; and (3) severe cases may also require stimulation by surging and jetting.
- Induced formation fracturing: generally self-healing after pump pressure is relieved.
- IX resin fouling: (1) perform an acid wash to remove calcium scale on resin beads; (2) oxidize hydrogen sulfide, polythionates, and thiomolybdates in the pregnant leach solution so they will not poison the resin; beware that too much oxidant can ruin resins; (3) pre-extract selenium in a sacrificial resin column; (4) condition resins with chemicals; and (5) replace irreversibly fouled resins. Note that Mo can poison resin.
- Premature declines and abrupt drops in preg grade: (1) check assays; (2) check injection chemical concentrations; (3) check injection pH; and (4) check for plugged wells.
- Gas blockage: temporarily curtail or reduce input of oxidant.
- Freeze protection of well field surface piping: install fast drainage devices and blow out with compressed carbon dioxide.

CHAPTER 21

Maintenance and Inventory

Marcus A. Wiley, P.E.

MAINTENANCE

Maintenance Theory

Equipment availability is key to an optimum return on mine investment. The proper role of maintenance is to provide the lowest cost in maintenance labor and materials, and to minimize production losses. Maintenance is a service function and must be coordinated with management and operations personnel. As shown in Figure 21.1, increased maintenance beyond a certain point may ultimately increase overall mine costs. The goal of maintenance should be to achieve the lowest overall cost.

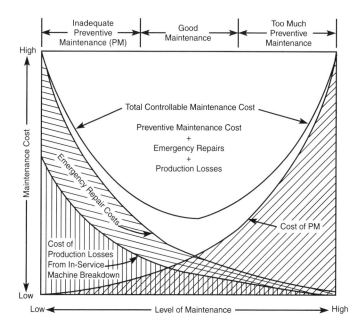

Source: Charlton 1989 (reprinted with permission of the Canadian Institute of Mining, Metallurgy, and Petroleum).

FIGURE 21.1 Maintenance cost versus level of maintenance

Mines are unique working places and maintenance often is performed outside of a shop and where working conditions are not favorable. Proper maintenance at mines involves judiciously combining the management of people, equipment, and resources with methods for anticipating problems before they occur (Bise 1992).

Definitions (Bise 1992)

Area maintenance is the responsibility of a front-line supervisor for all maintenance within a given geographical area.

Backlog is the total employee-hours required to complete maintenance work. A backlog index is used to determine the effectiveness of the maintenance department.

Work order is the formal document used to communicate maintenance requirements for planning, scheduling, and controlling work to be performed.

Overhaul is restorative maintenance taken before equipment fails.

Preventive maintenance is performed to prolong the life of equipment and to avoid premature failure. It includes equipment inspection, lubrication, adjustment, and cleaning.

Predictive maintenance is a special preventive maintenance that involves nondestructive testing techniques to predict wear rate, state of deterioration, or imminent equipment failure.

Repair is restorative maintenance performed after equipment has failed.

Scheduled maintenance is planned in advance to prevent premature equipment failure.

Unscheduled maintenance is done after equipment fails and requires repair before it can be used.

Mine Maintenance Organization

The 11 steps that follow can build an effective maintenance system (Herbaty 1983):

1. Establish goals and objectives, policies, and procedures.
2. Establish permissible variance from these guidelines.
3. Measure performance to establish guidelines.
4. Compare performance measurement information to guidelines.
5. Isolate and identify deviations beyond tolerances.
6. Determine basic causes for deviations.
7. Determine corrective action.
8. Plan method of implementing corrective action.
9. Schedule plan for implementing corrective action.
10. Implement corrective action.
11. Follow up to ensure completion of corrective action and to prevent overswing.

Categories of maintenance department work are (Sneddon 1973)
- Daily and routine maintenance
- Planned preventive maintenance
- Standing work orders (SWOs)
- Cost improvement orders
- Safety work
- Capitalized construction
- Production work done by maintenance personnel
- Contract work for out-of-plant customers
- Manufacturing items for plant stock.

A list of all equipment to be maintained should be assembled. The list should show equipment most critical to the operation (see Table 21.1). A small percentage of the equipment is generally responsible for a large percentage of production losses from downtime. A mine maintenance program that focuses on the few critical items should result in substantial savings (Bise 1992).

TABLE 21.1 Criticality determination

Item Number	Machine Number	Cumulative Maintenance Cost $	Cumulative Maintenance Cost %	Annual Cumulative Cost (%)	Annual Cumulative Number of Items (%)
1	001	xx.xx	30.0	30.0	0.1
2	008	xx.xx	29.0	59.0	0.2
3	109	xx.xx	10.0	69.0	0.3
4	110	xx.xx	5.0	74.0	0.4
5	005	xx.xx	3.0	77.0	0.5
6	854	xx.xx	1.0	78.0	0.6
7	562	xx.xx	1.0	79.0	0.7
8	687	xx.xx	0.4	79.4	0.8
9	295	xx.xx	0.3	79.7	0.9
10	346	xx.xx	0.3	80.0	1.0
.	.	.	.	.	.
.	.	.	.	.	.
.	.	.	.	.	.
1,000	272	xx.xx	0.1	100.0	100.0

Source: Herbaty 1983 (reprinted with permission of Noyes Data Corporation).

An overall maintenance system is diagrammed in Figure 21.2. In the figure, jobs of unscheduled maintenance work requests (MWRs), emergencies, and preventive maintenance through SWOs are initiated at (1); assignments are made at (2); at (3) work is performed, labor reported, and stock materials obtained; jobs requiring planning are identified and planned at (4); approved schedule is assigned to supervisor at (5); at (6) information is provided so work may be evaluated, controlled, or converted into additional work; other sources of planned work are sent to planning at (7); decision-making reports are prepared at (8) (Tomlingson 1994).

Preventive Maintenance

Preventive maintenance includes equipment inspection, lubrication, adjustments, and cleaning. The goal is to keep equipment running effectively and to avoid downtime. Success of a preventive maintenance program can be evaluated by (1) increase in equipment operating time, (2) increased capability to do more planned maintenance, (3) increase in product output, and (4) decrease in breakdown maintenance. Figure 21.3 indicates the important steps in organizing and operating a preventive maintenance program.

Figure 21.4 shows the relationship between equipment deterioration over time with cost and downtime.

Predictive Maintenance

Predictive maintenance is specialized and it strengthens a preventive program. Figure 21.5 shows preventive and predictive maintenance techniques.

Figure 21.6 illustrates how to start a predictive maintenance program.

Lubrication and Equipment Analysis

Oil and grease are used to lubricate, cool, and to keep machinery parts clean thereby reducing heat build-up and wear from excessive friction between moving parts. A sample of engine oil can be analyzed for wear metals, contaminants, and condition of the oil in comparison to new oil. This will help determine the condition of the equipment and help evaluate the success of the maintenance program. Table 21.2 is a typical oil sampling schedule.

386 | SME MINING REFERENCE HANDBOOK

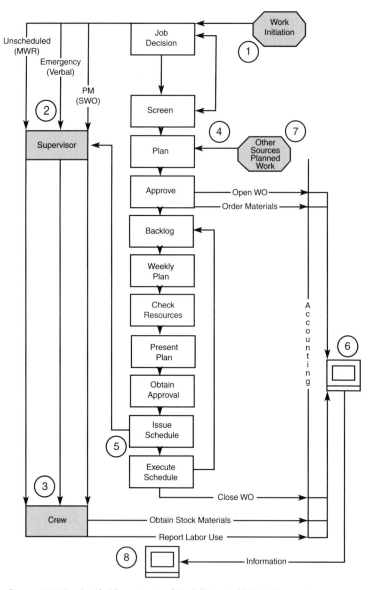

Source: Tomlingson 1994 *(reprinted with permission of Kendall/Hunt Publishing Company).*

FIGURE 21.2   Maintenance management system

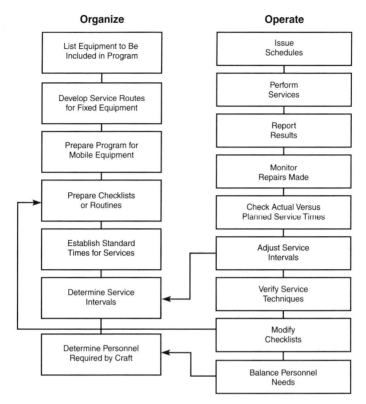

Source: Tomlingson 1998 *(reprinted with permission of Kendall/Hunt Publishing Company).*

FIGURE 21.3 Preventive maintenance program

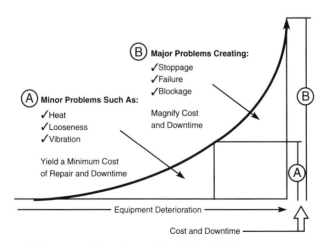

Source: Tomlingson 1998 *(reprinted with permission of Kendall/Hunt Publishing Company).*

FIGURE 21.4 Equipment deterioration versus cost and downtime

388 | SME MINING REFERENCE HANDBOOK

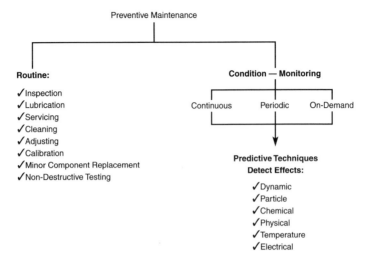

Source: Tomlingson 1998 *(reprinted with permission of Kendall/Hunt Publishing Company).*

FIGURE 21.5 Preventive and predictive maintenance techniques

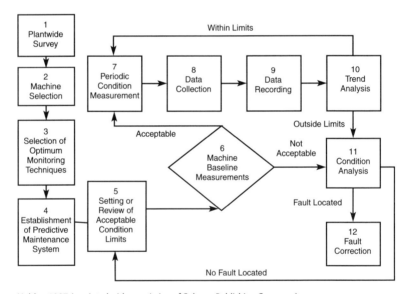

Source: Nolden 1987 *(reprinted with permission of Cahners Publishing Company).*

FIGURE 21.6 Essential steps in building a predictive maintenance program

TABLE 21.2 Typical oil sampling intervals

Diesel engines	Monthly or at 500 h
Natural gas engines	Monthly or at 500 h
Gas turbines	Monthly or at 500 h
Steam turbines	Bimonthly
Air compressors	Monthly or at 500 h
Air conditioning compressors	Beginning, middle, and end of season
Gears and bearings	Bimonthly
Hydraulic systems	Bimonthly

Source: Dunn 1987 *(reprinted with permission of Cahners Publishing Company).*

Other techniques are used to help determine when equipment needs maintenance prior to a breakdown thus preventing unscheduled repairs. Some of these methods are (Bise 1992)

- *Vibration analysis* is the measurement of machinery vibration at various intervals compared to baseline data taken when placing equipment in service. Increased vibration is an indication of deterioration of components that can be replaced during a scheduled maintenance procedure rather than waiting for failure.
- *Shock-pulse analysis* measures lubrication and mechanical condition in ball and roller bearings that, when damaged, generate abnormal high-frequency energy.
- *Ferrography or wear-particle analysis* is a technique that analyzes metal particulates after separation from oil samples. Information on particle shape, composition, size, and quantities can provide information on types of wear within a piece of equipment.
- *Thermography or infrared inspection* uses a hand-held scanner to measure the temperatures of various machinery components to predict potential failure through overheating.

RECORDKEEPING

Gathering and managing information on equipment and personnel utilization, costs, inventory control, repair history, and equipment specifications is critical to the success of a maintenance program.

Reporting

Records and reports that should be maintained and distributed to individuals having a need for the information are as follows (Sneddon 1973):

- Equipment data sheet
- Work order form
- Daily maintenance work schedule
- Employee's daily time card
- Supply requisition
- Standing work orders
- Monthly equipment cost report
- Backlog report
- Personnel performance report
- Equipment history records.

A sample work-order form and instructions for filling it out are provided as Figure 21.7 and Table 21.3, respectively. Work that may require a standing work order includes lubrication, preventive maintenance, janitor services, and delivery of parts and materials.

The Anaconda Company		WORK ORDER				W. O. Number
Priority	Equipment Name		Plant Charge	Acct. No.	Acct. Sub	Equipment No.

Category	Location	Available	Required	Requested By	Date

Description of and Reason for Work

Review After Estimate	☐ Yes ☐ No	Authorized By	Date

Job Seq.	Skill Code	Type of Work	Description of Work	Estimated Personnel and Hours

Materials

Quan.	Description	Loc.	Maj.	Min.	Sub	Price	Unit	Total Cost

Estimated Job Cost →	Personnel-Hours	Labor $	Material $	Other $	Total $
Estimated By			Date		Ckd.

Source: Sneddon 1973.

FIGURE 21.7 Sample work-order form

INVENTORY

A key component of a maintenance department is having the proper inventory of spare parts on hand to carry out required maintenance work. Inventory control may need as much management attention as the actual servicing of equipment itself.

An inventory control system maintains inventory, purchases spare parts, and arranges for spare part manufacture and component rebuilding to ensure that parts will be in stock when needed. An inventory control system should provide information on (Tomlingson 1994)

- Stocked parts and substitutes
- Stocked parts related to types of units and components
- Quantities on hand and quantities set aside for identified work

TABLE 21.3 Instructions for preparing a work-order form

1. Priority—urgency of work to be done:
   "E"—emergency work, work around the clock until done
   "1"—high-priority work, may require overtime
   "2"—work that should be completed within 3 days
   "3"—routine work
   "P"—preventive maintenance.
2. Equipment name—name of piece of equipment or building to be worked on
3. Plant charge—three-digit code number designating the appropriate cost center
4. Account number—two-digit account number within the cost center
5. Account sub—sequential series of numbers for each piece of numbered equipment; enter "00" if the work does not pertain to a numbered item
6. Equipment number—number assigned to equipment to be worked on, or a sequential series of numbers if not a numbered item
7. Category—category of work as shown in text above
8. Location—location within plant
9. Available—date equipment is available for work to begin
10. Required—date by which work must be completed
11. Requested by—signature of requester
12. Date—date work request was made
13. Description of and reason for work—a complete description of work to be done, including dimensions, etc., and the reasons or need. The information in this space should be complete enough to enable the work to be planned and accomplished without unnecessary delays to determine what is to be done. Drawings and any other pertinent data should be listed, or sketches attached.
14. Review after estimate—the person responsible for costs can designate whether he or she wishes to have the work order fully estimated before granting approval
15. Authorized by—signature of person responsible for costs, indicating approval to have work accomplished
16. Date—the date authority to proceed was granted
17. Job sequence—the chronological order of steps required to complete work
18. Skill code—the craft group code of the particular craft required for each sequence of work
19. Type of work—the name of the craft required (e.g., "mach" for machinist)
20. Description of work—a short description of the work to be done in each sequence
21. Estimated personnel and hours—estimated time required to complete sequence (shown as 2 × 8 for 2 individuals, 8 h)
22. Materials—all material not readily available is listed so as to ensure its availability when required
23. Estimated job cost—shows the total estimated labor dollars and material dollars and, when completed, is the budget for that particular work order.

Source: Sneddon 1973.

- Number on order and reorder points and order quantities
- Unit costs
- Resupply sources
- Costs of issued parts
- Turnover rate
- Number of issue days on hand for critical parts
- Reconciliation of records and physical counts
- Stocking cost
- Value of stock by code and user area
- Status actions for restocking.

An excessive amount of inventory unnecessarily ties up capital resources; however, an insufficient level will not allow an operation to continue efficiently and without interruption. A minimum stock level is the average usage of an item times the lead time necessary to receive it when ordered. This does not take into account the cost to purchase or store an item; nor does it provide a safety factor quantity. Once minimum stock level is reached a new order should be placed. The optimum number of items to be ordered can be determined as follows (adapted from Brown 1973 and Moore and Jaedicke 1967):

$$Q_o = [(2 \cdot C_i \cdot U_i) / C_s]^{1/2} \qquad (EQ\ 21.1)$$

where:
- Q_o = optimum number of items to order
- C_i = cost of item
- U_i = usage or frequency of item that needs replacing
- C_s = cost to store or warehouse item

REFERENCES

Bise, C.J. 1992. Equipment maintenance. In *SME Mining Engineering Handbook*. 2nd ed., Vol. 1. Edited by H.L. Hartman. Littleton, CO: Society for Mining, Metallurgy, and Exploration, Inc. (SME) 1260–1269.

Brown W.L. Sr. 1973. Warehouse function and inventory control. In *SME Mining Engineering Handbook*. Vol. 2. Edited by A.B. Cummins and I.A. Given. New York: Society of Mining Engineers of the American Institute of Mining, Metallurgical, and Petroleum Engineers, Inc. 25-12–25-23.

Charlton, D.J. 1989. Maintenance systems. In *Operation and Maintenance in Mineral Processing Plants*. Edited by P.G. Claridge. Montreal: The Canadian Institute of Mining, Metallurgy, and Petroleum.

Dunn, R.L. 1987. Advanced maintenance technologies. *Plant Engineering* 41(12):80–87.

Herbaty, F. 1983. *Cost-Effective Maintenance Management*. Park Ridge, NJ: Noyes Publications.

Moore, C.L., and R.K. Jaedicke. 1967. *Managerial Accounting*. Cincinnati, OH: South-Western Publishing Company.

Nolden, C. 1987. Predictive maintenance. *Plant Engineering* 41(41):38–43.

Sneddon, G.L. 1973. Reports and records. In *SME Mining Engineering Handbook*. Vol. 2. Edited by A.B. Cummins and I.A. Given. New York: Society of Mining Engineers of the American Institute of Mining, Metallurgical, and Petroleum Engineers, Inc. 24-53–24-63.

Tomlingson, P.D. 1994. *Mine Maintenance Management*. 9th ed. Dubuque, IA: Kendall/Hunt Publishing Company.

Tomlingson, P.D. 1998. *Equipment Management*. Dubuque, IA: Kendall/Hunt Publishing Company.

CHAPTER 22

Environment and Reclamation

Conrad H. Parrish, P.E. and Russell F. Price, P.E.

Environmental laws governing mining have been in place in the United States for more than a hundred years. State and federal statutes apply to mining claims, exploration activities, mining and beneficiation operations, water and air quality, historic sites, and endangered species.

MAJOR FEDERAL LAWS

The Clean Air Act (CAA) 42 *U.S. Code* (U.S.C.) § 7401 et seq (1970) is the comprehensive federal law that regulates air emissions from area, stationary, and mobile sources. This law authorizes the U.S. Environmental Protection Agency (EPA) to establish National Ambient Air Quality Standards (NAAQSs) that protect public health and the environment.

The Clean Water Act (CWA) 33 U.S.C. § 1251 et seq (1977) is an amendment to the Federal Water Pollution Control Act (FWPCA) of 1972, which set the basic structure for regulating discharges of pollutants to U.S. waters. The law gave EPA the authority to set effluent standards on an industry basis (technology-based) and continued the requirements to set water quality standards for all contaminants in surface waters. The CWA makes it unlawful for any person to discharge any pollutant from a point source into navigable waters unless a permit (National Pollutant Discharge Elimination System [NPDES]) is obtained under the act. The CWA allows the EPA to delegate many permitting, administrative, and enforcement aspects of the law to state governments. In states with the authority to implement CWA programs, EPA retains oversight responsibilities.

In 1972, amendments to the FWPCA added what is commonly called Section 404 authority (33 U.S.C. § 1344) to the program. The Secretary of the Army, acting through the Chief of Engineers, is authorized to issue permits, after notice and opportunity for public hearings, for the discharge of dredged or fill material into U.S. waters at specified disposal sites. Selection of such sites must be in accordance with guidelines developed by the EPA in conjunction with the Secretary of the Army; these guidelines are known as the 404(b)(1) guidelines. The discharge of all other pollutants into U.S. waters is regulated under Section 402 of the Act, which supersedes the Section 13 permitting authority.

The Comprehensive Environmental Response, Compensation, and Liability Act (CERCLA) (42 U.S.C. § 9601) (1980) is commonly known as Superfund. This law created a tax on the chemical and petroleum industries and provided broad federal authority to respond directly to releases or threatened releases of hazardous substances that may endanger public health or the environment. The Superfund Amendments and Re-authorization Act of 1986 (SARA) amended CERCLA and made several important changes and additions, which stressed the importance of permanent remedies and innovative treatment technologies in cleaning up hazardous waste sites, provided new enforcement authorities, increased state involvement, focused on human health problems posed by hazardous waste, encouraged greater citizen participation in deciding how sites should be cleaned up, and increased the size of the trust fund to $8.5 billion.

The Endangered Species Act (ESA) 7 U.S.C. § 136; 16 U.S.C. § 460 et seq (1973), provides a program for the conservation of threatened and endangered plants and animals and the habitats in which they are found. The U.S. Fish and Wildlife Service (FWS) maintains a list of 632 endangered species (326 are plants) and 190 threatened species (78 are plants). Species include birds, insects, fish, reptiles, mammals, crustaceans, flowers, grasses, and trees. Anyone can petition FWS to include a species on the list. The law prohibits any action, administrative or real, that results in a "taking" of a listed species, or adversely affects habitat.

Executive Order 12898 (59 *Fed. Reg.* 7629; February 16, 1996) is a federal action to address environmental justice in minority populations and low-income populations. It was put in place to ensure that federal agencies conduct programs and activities and institute policies that substantially affect human health or the environment in a manner that ensures that such actions do not have the effect of excluding persons (including populations) from participating in, denying persons the benefits of, or subjecting persons to discrimination because of race, color, or national origin. It requires agencies to identify, research, collect, analyze, and consult minority and low-income populations to prevent a disproportional risk to such populations from programs that might have a high or adverse human health risk or environmental effect.

The National Historic Preservation Act (NHPA) of 1966 (16 U.S.C. § 470f) established a policy of the federal government to foster conditions under which a modern society and prehistoric and historic resources can exist in productive harmony, and fulfill the social, economic, and other requirements of present and future generations. Under Section 106 of NHPA, federal agencies must consider the impact of their programs and projects on places of historic value. They must incorporate ways to protect and enhance historic resources through land use, funding, and licensing actions.

The National Environmental Policy Act (NEPA) (42 U.S.C. § 4321) (1970) and the Environmental Quality Improvement Act (42 U.S.C. § 4371 et seq) (1970) provide the basic national charter for environmental protection. They establish policy, set goals, and provide a means for carrying out the policy. NEPA is administered by the Council on Environmental Quality. Federal agencies must use a systematic, interdisciplinary approach when making a decision that has an impact on the human environment. They must prepare environmental impact statements (EISs) on federal actions that significantly affect the quality of the human environment. Federal agencies must also obtain comments and views from other federal, state, and local agencies, along with the public. The EIS is intended to develop, study, and describe alternatives to the recommended decision with clear reasons as to why the preferred alternative was selected.

The Resource Conservation and Recovery Act (RCRA) 42 U.S.C. § 6901 et seq (1976) gave the EPA the authority to control hazardous waste from "cradle-to-grave." This includes the generation, transportation, treatment, storage, and disposal of hazardous waste. RCRA also set forth a framework for the management of nonhazardous wastes.

The Surface Mining Control and Reclamation Act (SMCRA) of 1977 (30 U.S.C. § 1201 et seq) applies to active and abandoned surface and underground coal mines on private, state, federal, and Native American lands. SMCRA has two major provisions: (1) the regulation of active coal mines and (2) the reclamation of abandoned mines (Abandoned Mine Lands or AML). SMCRA is administered through grants to states under the concept of primacy, except for those states that choose not to exercise primacy and on Native American lands. Coal mines on federal lands are regulated under cooperative agreements with the primacy states. Abandoned mines are reclaimed through the Abandoned Mine Land Fund, which is based on fees on active coal mining. The AML program also responds to emergencies that result from hazards of past coal mining practices, such as landslides, coal fires, subsidence from underground mines, and unstable highwalls.

CITATIONS AND REGULATIONS

TABLE 22.1 Citations to selected federal statutes and regulations

Popular Name of Statute [state delegation (Y)es, (N)o]	Statutory Cite	Selected Citations in the *U.S. Code of Federal Regulations* (C.F.R.)
Atomic Energy Act (N)	42 U.S.C. §§ 2011–2296	10 C.F.R. Parts 0–171, 760–763, 962; 40 C.F.R. Parts 190–192
CERCLA (Superfund) (N)	42 U.S.C. §§ 9601–9675	40 C.F.R. Parts 300–311
Clean Air Act (Y)	42 U.S.C. §§ 7401–7642	40 C.F.R. Parts 50–87
Clean Water Act (Y)	33 U.S.C. §§ 1251–1387	40 C.F.R. Parts 104–149, 401–471
Endangered Species Act (N)	16 U.S.C. §§ 1531–1544	7 C.F.R. Parts 355–356; 50 C.F.R. Parts 17, 20–23, 81, 217–227, 401–453
Federal Land Policy and Management Act (N)	43 U.S.C. §§ 1701–1784	36 C.F.R. Parts 200–297; 43 C.F.R. passim
Forest and Range Land Renewable Resources Planning Act (N)	16 U.S.C. §§ 1600–1687	36 C.F.R. Parts 200–251
Geothermal Energy Research, Development, & Demonstration Act (N)	30 U.S.C. §§ 1101–1164	10 C.F.R. Part 790
Global Climate Protection Act (N)	15 U.S.C. § 2901	
Low-Level Radiation Waste Policy (N)	42 U.S.C. §§ 2021b–2021j	
Marine Protection, Research, and Sanctuaries Act (N)	33 U.S.C. §§ 1401–1445	40 C.F.R. Parts 220–229; 50 C.F.R. Parts 215–229
Migratory Bird Treaty Act (N)	16 U.S.C. § 701	50 C.F.R. Part 10
Mineral Resources Research Act (N)	30 U.S.C. §§ 1221–1230	
Mining in the Parks Act (N)	16 U.S.C. §§ 1901–1912	36 C.F.R. Part 9
Multiple-Use Sustained Yield Act (N)	16 U.S.C. §§ 528–53115	43 C.F.R. Parts 23, 3701–3740
National Climate Program Act (N)	15 U.S.C. §§ 2901–2908	
National Environmental Policy Act (N)	42 U.S.C. §§ 4321–4370b	40 C.F.R. Parts 1500–1517 (each agency has separate regulations)
National Historic Preservation Act (Y)	16 U.S.C. § 470	36 C.F.R. Part 800
Noise Control Act (N)	42 U.S.C. §§ 4901–4918	40 C.F.R. Parts 201–211
Nuclear Waste Policy Act (N)	42 U.S.C. §§ 10101–10270	10 C.F.R. Parts 51, 60, 72
Oil Pollution Act (N)	33 U.S.C. §§ 2701–2761	33 C.F.R. Parts 151–159
Pollution Prevention Act (N)	42 U.S.C. §§ 13101–13190	
Refuse Act of 1899 (N)	33 U.S.C. § 407	
Resource Conservation and Recovery Act (N)	42 U.S.C. §§ 6901–6991i	40 C.F.R. Parts 240–279, 280-2 (underground storage tanks)
Safe Drinking Water Act (N)	42 U.S.C. §§ 300f–300j–11	40 C.F.R. Parts 141–149
SARA (Community Right to Know) (N)	42 U.S.C. §§ 11001–11050	40 C.F.R. Parts 350–374
Soil and Water Resource Conservation Act (N)	16 U.S.C. §§ 2001–2009	
Surface Mining Control and Reclamation Act (Y)	30 U.S.C. §§ 1201–1328	30 C.F.R. Parts 700–955; 43 C.F.R. Parts 3400–3408
Toxic Substances Control Act (N)	15 U.S.C. §§ 2601–2654	40 C.F.R. Parts 700–799, 761 (PCBs)
Uranium Mill Tailings Radiation Control Act (N)	42 U.S.C. §§ 7901–7942	40 C.F.R. Part 192

Source: Adapted from Gilbert 1997 *(shown with permission of Imperial College Press).*

TABLE 22.2 Environmental regulations in selected Latin American and Asia-Pacific rim countries

Country	Status of Environmental Regulations and Standards	Regulatory/Permitting Agencies
Bolivia	In transition—new Environmental Law No. 1333 passed in 1992; regulations and standards being drafted by Council for Sustainable Development	Ministry of Sustainable Development Law and Environment; Ministry of Human Development; and Ministry of Economic Development
Chile	Rapidly evolving—government drafting environmental policy to combine economic growth and environmental protection	Servicio Nacional de Geología y Minera (SERNAGEOMIN); Superintendencia de Servicios Sanitarios; plus several other federal and regional ministries and agencies.
Mexico	Being formulated by representatives of Ministry of Social Development (SEDESOL), CAMIMEX and SEMIP	SEDESOL, composed of National Ecology Institute and Office of Attorney General for Environment
Peru	Legislation revised 1993—evolving	Consejo Nacional de Medio Ambiente (CONAMA); Dirección General de Asuntos Ambientales
Venezuela	Regulations covering all aspects (air, water, emissions, and effluents) of submission of environmental studies	Regional offices of the Ministry of Environment
China	Protection Law (1989); Law on the Prevention and Control of Water Pollution (1987); Law on the Prevention and Control of Air Pollution (1984); regulations on the Prevention and Control of Noise Pollution (1989)	EPA
India	Mining Act (1957 et seq); Water (Prevention and Control of Pollution) Act (1974 et seq); Air (Prevention and Control of Pollution) Act (1981 et seq); Environmental Act (1986); Forest Act (1980 et seq)	EPA
Indonesia	Mining Law (1967) Article 30; Environmental Management Act (1982); Regulation 29/1986 covers environmental impact act, 11/1974 regulates water pollution and quality, 20/1990 covers management of polluted waters	Ministry for Population and Environment
Malaysia	Mining Enactment FMS Cap. 147; Environmental Quality Act	Department of Mines; Ministry of Science, Technology and Environment
Papua New Guinea	Environmental Planning Act (1978); Environmental Contaminants Act (1978); Conservation Areas Act (1978); Water Resources Act (1982); National Parks Act (1982)	Department of Environment and Conservation
Thailand	National Reserved Forest Act (1964); Improvement and Conservation of National Environmental Quality Act (1975 et seq)	Ministry of Agriculture and Cooperatives, National Environmental Board

Source: Adapted from Malhotra 1997 (shown with permission of Imperial College Press).

PERMITS AND APPROVALS

TABLE 22.3 Environmental permits and approvals required for projects

Federal Permits or Approvals*	
Department of the Interior (DOI) Bureau of Land Management (BLM) and/or U.S. Department of Agriculture (USDA) Forest Service	**EPA**
(a) Environmental Impact Statement	(a) Hazardous waste generator identification number
(b) Plan of Operations	(b) Storm Water Pollution Prevention Plan (may be delegated to state)
(c) Archeological clearance (Section 106 process)	(c) Spill Prevention, Control and Countermeasures plan
(d) Rights-of-way for utility corridors, etc.	(d) NPDES permit (may be delegated to state)
(e) Mineral material sale (borrow areas)	(e) Clean Air Act Title 5 permit
DOI FWS	(f) Toxic release inventory reporting
(a) Endangered Species Act compliance	**U.S. Army Corps of Engineers**
(b) Bald Eagle Protection Act compliance	(a) Clean Water Act Section 404 permit
Department of Labor, Mine Safety and Health Administration (MSHA)	**Department of Justice, Bureau of Alcohol, Tobacco, and Firearms**
(a) MSHA mine identification number	(a) Permit to purchase and store explosives
(b) Fire, evacuation, and rescue plans	(b) Explosives inventory and use reports
(c) Notice of start of operations	
(d) Miner training plan	

State Permits or Approvals†	
Mining Authority	**State Engineer**
(a) Plan of operations or reclamation plan	(a) Water rights
(b) Environmental impact report (or equivalent)	(b) Dam safety permit
Water Quality Control Authority	**Air Pollution Control Authority**
(a) NPDES permit	(a) Authority to construct
(b) Section 401 (CWA) certification	(b) Permit to operate
(c) Groundwater protection permit	(c) Air toxics emission inventory plan (California only)
(d) Waste disposal permit (leach pads, tailings, etc.)	
State Historic Preservation Office	**State Lands Commission** (if on state-owned lands)
(a) Archaeological clearance	(a) Mining lease or permits
State Occupational Health and Safety Authority	(b) Water well lease
(a) Notice of mine opening	**Fish and Game Authority**
(b) Injury and illness prevention program	(a) Stream alteration permit
(c) Hazardous materials communication plan	(b) State Endangered Species Act compliance
	(c) Artificial industrial pond permit

Local Permits or Approvals‡	
(a) Conditional use permit	(e) Acutely hazardous materials registration
(b) Mining reclamation plan	(f) Domestic water system permit
(c) Hazardous materials business plan	(g) Sewage disposal permit
(d) Risk management prevention program	(h) Water well permits

* Not all permits will be required for every project; depends on location, physical and environmental setting, and project design.
† State permit requirements vary; some permits may be delegated to local or regional authorities, especially in California.
‡ Local permits and approvals vary widely, this list is not exhaustive and local investigation is always required.
Source: Adapted from Struhsacker 1997 (shown with permission of Imperial College Press).

AIR QUALITY

TABLE 22.4 Summary of national ambient air quality standards

Pollutant	Averaging Time	(µg/m³)
CO_2	8 h	10,000
	1 h	40,000
Pb	Calendar year	1.5
NO_2	Annual	100
Ozone	8 h	157
PM10 (particulate matter smaller than 10 µm in diameter)	Annual	50
	24 h	150
PM2.5 (particulate matter smaller than 2 µm in diameter)	Annual	15
	24 h	65
SO_2	Annual	80
	24 h	365
	3 h	1,300

Source: Adapted from Struhsacker 1997 (shown with permission of Imperial College Press).

TABLE 22.5 Prevention of significant deterioration incremental standards* (µg/m³)

Pollutant	Class 1	Class 2	Class 3
SO_2			
Annual	2	20	40
24 h	5	91	128
3 h	25	512	700
TSP			
Annual	5	19	37
24 h	10	37	75
NO_2	2.5	25	50

* Short-term increments not to be exceeded more than once per year.
Source: Adapted from Struhsacker 1997 (shown with permission of Imperial College Press).

TABLE 22.6 Significant emission rates

Pollutant	Emissions (tons per year)
CO	100
NO_x	40
SO_2	40
Particulate matter (TSP)	25
Particulate matter (PM10)	15
Ozone	40 of volatile organic compounds (VOCs)
Pb	0.6
Asbestos	0.007
Be	0.0004
Hg	0.1
Vinyl chloride	1
Fluorides	3
Sulfuric acid mist	7
H_2S	10
Reduced sulfur compounds (including H_2S)	10

Source: Adapted from Struhsacker 1997 (shown with permission of Imperial College Press).

TABLE 22.7 Mining fugitive emission controls, effectiveness, and costs

Mining Activity Emission Control Technique	Control Effectiveness	Cost Factors L = low M = moderate H = high
Topsoil removal		
Prewatering	50%	L
Topsoil or overburden stockpile		
Wind breaks	50%	L
Rapid revegetation	75%	L
Mulch	85%	L
Chemical dust suppressant	85%	M
Blasting		
Reduce blasting needed	Function of reduction	L
Prevent overshooting	Function of reduction	L
Overburden removal		
Prewatering	50%	L–M
Overburden shaping		
Leave ridges	Function of soil ridge roughness	L
Establish wind breaks	Function of the height and wind speed	L
Rapid revegetation	85%	L
Minimize spoil pile area	Function of area reduced	L
Product removal, truck/shovel or front-end loader		
Minimize fall distance	Function of distance reduced	L
Product dumping, end or bottom dump		
Spray dumped material	50%–85%	L
Product storage		
Keep storage pile wet	50%–85%	L–M
Enclose with a structure	Up to 100%	H
Haul roads		
Limit speeds	Function of the speed reduction	L
Chemical stabilization	85%	M
Restrict off-road use	100%	L
Road maintenance		
Remove loose debris (grading)	Function of material removed	L
Chemical stabilization	85%	M
Disturbed areas		
Rapid revegetation	75%	L–M
Mulch	85%	L–M
Chemical dust suppressant	85%	M
Crushers and screens		
Baghouse	99% control of captured dust	M–H
Water sprays	50%–75%	L–M
Conveyor belts		
Full covering	100% control	H
Water sprays	50%–75% control	L–M
Transfer points		
Enclose and vent to a baghouse	99% control of captured dust	M–H
Water sprays	50%–75%	L–M

Source: Lowrie 1997 (shown with permission of Imperial College Press).

WATER QUALITY

TABLE 22.8 Typical water quality standards and goals (μg/L unless otherwise indicated)

Inorganic Constituent	State or EPA Drinking Water Standards Maximum Contaminant Levels (MCLs) State Levels Unless Indicated			Health Advisories or Suggested No-Adverse-Response Levels (SNARLs) (Cancer Risk Not Considered)		EPA—National Ambient Water Quality Criteria Based on:		EPA—Criteria to Protect Freshwater Aquatic Life	
	Primary MCL	Secondary MCL*	EPA MCL Goal	EPA	NAS†	Public Health Effects	Taste and Odor or Welfare	4-d Average	1-h Average
Aluminum	1,000					5,000 (7-day)		87	750
Ammonium sulfonate				1,500‡					
Antimony						146			
Arsenic	50		100‡					190	360
Asbestos†§			7,100,000 fibers/L						
Barium	1,000			1,500	4,700	1,000			
Beryllium			1,500‡						
Bromide					2,300				
Cadmium	10		5‡	5	5	10		55	1.4
Chloride		250,000**					250,000	230,000	860,000
Chlorine								11	19
Chromium (III)						170,000		98	820
Chromium (VI)	50††		120‡,††	120††		50		11	16
Copper		1,000	1,300‡				1,000	5.4	7.5
Cyanide	200‡‡			154		200		5.2	22
Fluoride	4,000	2,000	4,000						
Iodide					1,190				
Iron		300							
Lead	50§§/5‡		zero‡	20‡		50		0.99	25
Manganese		50					300		
Mercury	2		3‡	1.1		0.144	50	0.012	2.4
Nickel				150		13.4		73	653
Nitrate	45,000***		10,000†††,‡	10,000†††		10,000			
Nitrite			1,000†††,‡	1,000†††					
pH (S.U.)						5–9			
Selenium	10		45‡			10		5	19
Silver	50					50			
Specific conductivity (EC)		900 μmhos/cm‡‡‡							
Strontium				8,400 (7-day)					
Sulfate		250,000**					250,000		
Thallium						13			
Total dissolved solids (TDS)		500,000§§§							
Uranium	20 pCi/L				35				
Zinc		5,000					5,000	49	54

* Other secondary drinking water standards include: color = 15 color units; foaming agents = 0.5 mg/L; odor = 3 (threshold odor number).
† National Academy of Sciences.
‡ Proposed.
§ Limited to fibers longer than 10 Fm.
** Recommended level: upper limit = 500 mg/L; short-term limit = 600 mg/L.
†† Measured as total chromium.
‡‡ EPA. May 1986. *Quality Criteria for Water*, plus updates. The form of cyanide is not stated; however, the standard has frequently been applied as a free cyanide standard.
§§ EPA value.
*** As nitrate (NO_3).
††† As nitrogen.
‡‡‡ Recommended level: upper limit = 1,600 μmhos/cm; short-term limit = 2,200 μmhos/cm.
§§§ Recommended level: upper limit = 1,000 mg/L; short-term limit = 1,500 mg/L.
Source: Adapted from Marshack 1989.

ENVIRONMENT AND RECLAMATION | 401

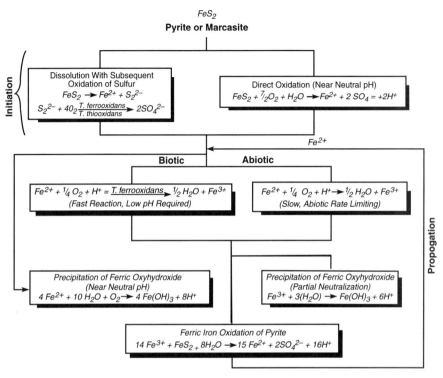

Source: Schmiermund and Drozd 1997 (shown with permission of Imperial College Press).

FIGURE 22.1 Flow chart of major reactions and pathways involved in weathering of pyrite

TABLE 22.9 Potentially mobile chemical species in tailings liquids

Chemical Groups	Mine Waste Type						Spent Heap Leach Ore
	Mine Tailings			Waste Rock			
	Flotation	Concentrate	Undifferentiated	Acid-Generating	Nonacid-Generating		
Cations and metal cations	Calcium, ammonia, transition metals, lead, mercury, and barium	Calcium, transition metals, lead, mercury, and barium	Calcium, transition metals, lead, mercury, and barium	Calcium, ammonia, transition metals, lead, mercury, and barium	Calcium, ammonia		Calcium, ammonia, transition metals and metals that form cyanide complexes, e.g., mercury
Anions	Nitrate, sulfate	Nitrate, chloride, sulfate	Nitrate, chloride, sulfate	Nitrate, sulfate	Nitrate, sulfate		Nitrate, sulfate
Amphoteric species	Arsenic, antimony, chromium, molybdenum, selenium	Arsenic, antimony, chromium, molybdenum, selenium	Arsenic, antimony, chromium, molybdenum, selenium	Arsenic, antimony, chromium, selenium	Arsenic, antimony, chromium, molybdenum, selenium		Arsenic, antimony, molybdenum, selenium
Cyanide complexes (where cyanide is used as a process reagent)	—	Free cyanide (CN and HCN) Weak metal cyanide complexes (e.g., zinc, copper, nickel cyanides). Strongly complexed cyanides (e.g., iron, cobalt complexes).	—	—	—		—

Source: Hutchison and Ellison 1992 (reprinted with permission of the California Mining Association).

TABLE 22.10 Summary evaluation of kinetic prediction techniques

Kinetic Test	Use	Advantages	Disadvantages	Correlation With Field Data — Tailings	Correlation With Field Data — Waste Rock
British Columbia Research confirmation	Frequently used for base metal and gold mines in Canada	▪ Relatively simple to use ▪ Relatively low cost ▪ Allows assessment of potential for biological leaching	▪ Bacteria may not be acclimated to sample ▪ May be difficult to interpret if pH does not strongly rise or fall ▪ Does not simulate initial acid rock drainage (ARD) production step ▪ May take several weeks for sample pH to stabilize	6 of 8	4 of 4
Shake flasks	Infrequently used for base metal mines in Canada	▪ Simulates acid production and consumption reactions ▪ Assesses effect of several controlling conditions on acid production	▪ May be difficult to interpret results ▪ Moderate cost ▪ Involves large amount of data to compile and interpret ▪ Requires long time to complete testing ▪ Not practical for large number of samples ▪ Limited in considering different waste particle sizes	8 of 8	1 of 4
Soxhlet extraction	Infrequently used in coalfields of Appalachia, United States	▪ Relatively simple to use ▪ Results are available in relatively short time period ▪ Allows assessment of interaction between acid production and consumption reactions	▪ Still in the development stages ▪ Results are difficult to interpret ▪ Relationship of results to actual site-specific conditions is not clear	8 of 8	3 of 4
Humidity cell	Relatively common in coalfields of Appalachia, United States	▪ Simulates acid production and consumption reactions kinetically ▪ Can approximately simulate wet and dry weather cycles	▪ May be difficult to interpret results ▪ Moderate cost ▪ Involves large amount of data to compile and interpret ▪ Requires long time to complete testing ▪ Not practical for large number of samples ▪ Limited in considering different waste particle sizes	8 of 8	3 of 4
Columns	Gradually being used more extensively on gold mine projects in the western United States	▪ Simulates acid production and consumption reactions ▪ Can simulate effect of different rock types within the same column ▪ Can simulate wet and dry weather cycles ▪ Can consider different waste particle sizes within column size constraints	▪ May be difficult to interpret results ▪ Relatively high costs ▪ Involves large amount of data to compile and interpret ▪ Requires long test period ▪ Not practical for large number of samples ▪ May have problems with uneven leachate application and channelization		
Pilot-scale waste piles	Infrequently used	▪ Most realistic simulation of waste and site-specific climatic conditions ▪ Can model relevant natural conditions in a single test	▪ High cost ▪ Involves large amount of data to compile and interpret ▪ Very long test period ▪ Not practical for more than a very small number of individual tests		

Source: Fergusen 1985.

TABLE 22.11 Summary evaluation of static prediction techniques

Prediction Procedure	Simplicity of Test*	Time Required	Special Equipment Requirement†	Approximate Cost per Sample‡ ($)	Ease of Interpretation	Correlation of Test Results With Field Observations
Static Tests						
B.C. Research initial	Yes	2 days§	None	75–200	Easy	Two conservative** errors in evaluating tailings.
Sobek	Yes	4 hours§	None	50–130	Easy	Three conservative errors in evaluating tailings.
Sobek modified	Yes	1 day§	None	50–130	Easy	One conservative error in evaluating tailings and one unconservative†† error in evaluating waste rock.
Alkaline production	Yes	4 hours§	None	50–130	Moderate	Three conservative errors in evaluating tailings.
Net acid production	Yes	4 hours§	None	40–80	Easy	One conservative error in evaluating tailings.
Other Tests						
Hydrogen peroxide test	Yes	4 hours§	None	80–120	Moderate	Comparison not performed. However, the test method is known to provide a poor estimate of pyrite content.

* This rating assumes that the test is carried out by a trained laboratory technician or technologist.
† This assumes that the tests are carried out in a chemical laboratory equipped to undertake environmental testing.
‡ Costs will depend on the laboratory used, the number of samples tested at the same time, and on the number and costs of the sulfur species assays required. Costs are estimates as of March 1989.
§ Additional time may have to be allowed for sulfur assay turnaround.
** Conservative means the test predicted the potential for the waste to be acid-generating when none was actually observed.
†† Unconservative means the test does not predict acid generation that is known to have occurred.
Source: CANMET 1989.

TABLE 22.12 Federal hazardous waste classification criteria*

Waste Characteristic Category	Waste Classification Testing Criteria
Ignitability	- Liquids with a flashpoint of less than 60°C (140°F) - Nonliquids capable of causing fire through friction or absorption of moisture, occur as spontaneous combustion, and burn vigorously enough to create a hazard - Ignitable compressed gas - Oxidizes as defined in 49 C.F.R. 173.151
Corrosivity	- Is a liquid and corrodes steel at a rate greater than 6.35 mm per year - pH ≥ 12.5 - pH ≤ 2
Reactivity	- Normally unstable and readily undergoes violent change - Reacts violently with water - Forms a potential explosive mixture with water - Cyanide or sulfide bearing which, when exposed to pH conditions between 2 and 12.5, can generate toxic gases in quantities sufficient to endanger human health - Readily capable of explosive decomposition at standard temperature and pressure or when subjected to a strong initiating source.
Toxicity	Wastes with concentrations that exceed the following standards† when subjected to toxicity characteristic leaching procedure (TCLP)‡

Constituent	Maximum Concentration (mg/L)
Arsenic	5.0
Barium	100.0
Cadmium	1.0
Chromium	5.0
Lead	5.0
Mercury	0.2
Selenium	1.0
Silver	5.0

* Wastes are also judged to be hazardous if they are a listed waste in 40 C.F.R. 261.30. Mining related wastes that appear on these lists include: acid from primary copper and zinc production; and several smelter wastes.
† The list of constituents also includes a range of organic compounds. As these usually are not associated with mining activities, they are not included in the table.
‡ The extraction test procedure requires a buffered acetic acid (pH 5.0) at a liquid-to-solid ratio of 20 to 1. EPA intends to replace this test with a modified batch test procedure designated EPA 1312.
Source: 40 C.F.R. 261.

TABLE 22.13 Federal Bevill-excluded wastes*

Waste Type Excluded from Subtitle C	Definition of Waste Type
All Mining Extraction and Beneficiation Waste[†]	Wastes that involve "crushing; grinding; washing; dissolution; crystallization; filtration; sorting; sizing; drying; sintering; pelletizing; briquetting; calcining to remove water and/or carbon dioxide; roasting, autoclaving; and/or chlorination in preparation for leaching (except where the roasting [and/or autoclaving and/or chlorination]/leaching sequence produces a final or intermediate product that does not undergo further beneficiation or processing); gravity concentration; magnetic separation; electrostatic separation; flotation; ion exchange; solvent extraction; electrowinning; precipitation; amalgamation; and heap, dump, vat, tank, and in situ leaching."
Selected Mineral Processing Waste[‡]	The EPA has indicated that only the following 18 mineral processing wastes fall under the Bevill Exclusion. • Red and brown muds from bauxite refining • Treated residue from roasting/leaching of chrome ore • Gasifier ash from coal gasification • Process wastewater from coal gasification • Slag from primary copper processing • Calcium sulfate wastewater treatment plant sludge from primary copper processing • Slag tailings from primary copper processing • Slag from elemental phosphorus production • Iron blast furnace air pollution control dust/sludge • Iron blast furnace slag • Basic oxygen furnace and open-hearth furnace air pollution control dust/sludge from carbon steel production • Basic oxygen furnace and open-hearth furnace slag from carbon steel production • Fluorogypsum from hydrofluoric acid production • Process wastewater from hydrofluoric acid production • Slag from primary lead processing • Process wastewater from primary magnesium processing by the anhydrous process • Chloride process waste solids from titanium tetrachloride production • Slag from primary zinc processing

* As of August 1992.
† C.F.R. 261.4[b][7].
‡ *Fed. Reg.*, Volume 56, No. 114 (1991).

TREATMENT

TABLE 22.14 Common ionic forms of chemicals found in mining operations

	Ion	Environment	Ionic Charge		
			Neutral	Positive	Negative
Metal cations	Barium	All*		Ba^{+2}	
	Beryllium	All*	$Be(OH)_2(s)$	BE_2OH^{+3} $BE_3(OH)^{+3}$	
	Cadmium	All*	$CdCO_3(s)$ CdS	Cd^{+2} $Cd(OH)^+$	
	Chromium III	All*	$Cr_2O_3(s)$ $Cr(OH)_3(s)$		
	Cobalt	Oxidizing	$Co(OH)_2(ss)$ $Co_3O_4(s)$ $Co_2O_3(s)$	Co^{+2} Co_2OH^{+3}	$CoCl_4^{-2}$ $Co(CN)_5^{-3}$
		Reducing	$Co(s)$ $CoS(ss)$		$CuCl_2^{-2}$ $CuCl_5^{-3}$
	Copper	Oxidizing	$Cu_2O(s)$ $Cu(OH)_2(s)$ $CuCO_3(s)$	Cu^{+2} $Cu(OH)^+$ $Cu_2(OH)_2^{+2}$	
		Reducing	$Cu(s)$ $CuS(s)$		
	Lead	Oxidizing	$PbO_2(s)\ PbSO_4(ss)$ $PbO(s)\ PbCO_3(ss)$	Pb^{+2} $PbOH^+$	
		Reducing	$Pb(s)\ PbS(s)$		
	Mercury	Oxidizing	$HgO(ss)\ HgCL_2(aq)$ $Hg_2CL_2(ss)$	Hg^{+2} $HgOH^+$	
		Reducing	$Hg(l)\ HgS(s)$		
	Nickel	Oxidizing	$Ni(OH)_2(ss)$ $Ni_3O_4(s)$	Ni^{+2} $NiOH^+$	
		Reducing	$Ni(s)\ NiS(s)$		
	Zinc	Oxidizing	$Zn(OH)_2(s)$ $ZnCO_3$	Zn^{+2} $Zn_2(OH)^+$	
		Reducing	$ZnS(s)$		
Major anions	Antimony	Oxidizing	$Sb_2O_3(s)$		SbO_3^{-3}
		Neutral	$Sb_2O_3(s)$		
		Reducing	$SbO(s)$		
	Arsenic	Oxidizing	$H_3AsO_4\ (aq)$		H_2AsO_4 $HAsO_4^{-2}$
		Neutral	$As_2O_3\ (aq)\ H_3AsO_3$		$HAsO_4^{-2}$
		Reducing	$AsO(s)$		$H_3AsO_3^{-1}$ $H_2AsO_3^{-2}$ $HAsO_3^{-2}$ AsO_3^{2-}
	Boron	All*			$B_4O_7^{-2}$
	Chloride	All*			Cl^-
	Chromium VI	Oxidizing			$HCrO_4^{-1}$ CrO_4^{-2}
	Nitrate	All*			NO_{3-}
	Sulfate	Neutral and oxidizing			SO_4^{-2}

continues next page

TABLE 22.14 Common ionic forms of chemicals found in mining operations (continued)

	Ion	Environment	Ionic Charge		
			Neutral	Positive	Negative
Cyanides	Cyanide, free	All*			CN^{-1}
	Cyanide, weakly complexed	All*			$Zn(CN)_4^{-2}$ $Cd(CN)_3^{-1}$
	Cyanide, strongly complexed	All*			$Fe(CN)_6^{-3}$ $Fe(CN)_6^{-4}$

* All includes oxidizing, neutral, or reducing environments.

NOTES: l = liquid phase; s = solid phase; ss = suspended solid; aq = aqueous phase.

Source: Hutchison and Ellison 1992 (reprinted with permission of the California Mining Association).

TABLE 22.15 Attenuation mechanisms of chemical species

	Chemical Species	Physical				Physiochemical		Chemical					Biological	
		Filtration	Dispersion	Dilution	Volatilization	Adsorption	Fixation	Precipitation	Hydrolysis	Complexation	Oxidation	Reduction	Bacterial Reaction	Cellular Uptake
Very Predictably Attenuated	Asbestos	●	○	○*				○						
	Chromium III	●	○	○				●	●		○	○		○
	Lead		○			●	○	○	○	○				
	Mercury	●	○		○	●		○			○	○		○
	Beryllium	●				○		●	●					○
	Cyanide, free		○	○	●	●		●	●	○	○		●	○
	Arsenic		○	○		●	○	●	○		○	○		
Strongly Attenuated	Barium		○	○		●	○	●						
	Copper		○	○		●	●		○	○	○	○		
	Nickel		○	○		○	○	○	○					○
	Zinc		○	○		●	○	●	○					○
	Cadmium		○	○		○		○	○					○
	Cobalt		○	○		●	○		○					
Attenuation Uncertain	Antimony		○	○		○	○	○	○		○			
	Cyanide, strongly complexed		○	○		○		●	○					○
	Cyanide, weakly complexed		○	○	○	○		○	●	○	○		●	○
	Sulfate		○	○				●						
	Nitrate		○	○							○	○	●	●
	Chromium VI†		○	○				○	○			●		○
	Boron		○	○				○						

* With sufficient water flow, Mg will gradually leach out, thus changing the nature of the asbestos with loss of its fibrous structure.
† Because of the high solubility of Cr^{+6} it is not easily attenuated unless reduced to Cr^{+3}.

NOTES: ● denotes primary attenuation mechanism; ○ denotes secondary mechanism.

Source: Hutchison and Ellison 1992 (reprinted with permission of the California Mining Association).

TABLE 22.16 Examples of the nature and importance of vadose zone properties relative to attenuation

		Soil/Rock Physical Properties								Soil/Rock/Pore Water Chemistry												Pore Water Volume	
		Rock			Soil																		
											pH			Eh		Cation Exchange Capacity		Cation Availability		Anion Availability		Organic Content		Nutrient Content			
	Inorganic Chemical Elements and Minerals	Highly Fractured	Microfractured	Organic Content	Clay	Silt	Sand	Gravel	"Fractured"*	Ph <6	6 < pH < 8	pH >8	Oxidizing	Reducing	High	Low	Hydrous Metal Oxides	High	Low	High	Low	High	Low	High	Low	Below Field Capacity	At Field Capacity

(table body continues with symbol entries per row for Asbestos, Beryllium, Chromium III, Barium, Cadmium, Cobalt, Copper, Lead, Mercury, Nickel, Zinc, Antimony, Arsenic, Boron, Chromium VI, Chloride, Nitrate, Sulfate, Cyanide free, Cyanide weakly complexed, Cyanide strongly complexed)

Importance of property to attenuation of chemical

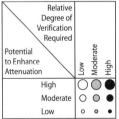

* "Fractured" soil may relate to soil conditions such as very deep desiccation cracks or active fault areas that could disrupt normal sedimentary patterns. Fractured limestone could be an exception to this general pattern, since limestone tends to neutralize acidic solutions promoting precipitation and adsorption mechanisms.

† Gravel conditions could potentially assist attenuation of free and weak acid dissociable cyanide by volatilization, if the pH is below about 9.5.

Source: Hutchison and Ellison 1992 *(reprinted with permission of the California Mining Association).*

TABLE 22.17 Federal recommendations for maximum trace element loadings for agricultural lands

CEC range	Metal Loading* by CEC†		
	<5	>5 to <15	>15
Cd	5	10	20
Cu	125	250	500
Ni	50	100	200
Pb	500	1,000	2,000
Zn	250	500	1,000

* Loading in kg/ha (kg/ha × 0.9 = lb/ac).
† CEC = cation exchange capacity in meq/100 g.
Source: Anon. 1977.

TABLE 22.18 General ranges for cation exchange capacities of soil clays, soil organic matter, and several soil types

	CEC (milliequivalents/100 g)
Soil type:	
Sand	2–7
Sandy loam	2–18
Loam	8–22
Silt loam	9–27
Clay loam	4–32
Clay	5–150
Clay minerals:	
Chlorite	10–400
Illite	10–40
Kaolinite	3–15
Montmorillonite	80–150
Oxides and oxyhydroxides	2–6
Saponite	80–120
Vermiculite	100–150
Soil organic matter	>200

Source: Dragun 1988.

REFERENCES

Anon. 1977. *Municipal Sludge Management: Environmental Factors*. EPA Technical Bulletin. 430/9-7-004. Denver, CO: U.S. Environmental Protection Agency.

CANMET (Canadian Centre for Mineral and Energy Technology). 1989. *Investigation of Prediction Techniques for Acid Mine Drainage*. MEND Project 1.16.1a. DSS File No. 30SW.23440-7-9187. Ottawa, Ontario: Energy, Mines and Resources Canada.

Dragun, J. 1988. *The Chemistry of Hazardous Materials*. Silver Spring, MD: Hazardous Materials Control Research Institute.

Ferguson, K.D. 1985. Methods to predict acid mine drainage. Paper presented at the International Symposium on Biohydrometallurgy, August 22–24, Vancouver, British Columbia.

Gilbert, A.J., chapter ed. 1997. The legal bases of federal environmental control of mining. In *Mining Environmental Handbook*. Edited by J.J. Marcus. London: Imperial College Press. 38–98.

Hutchison, I.P.G. and R.D. Ellison, eds. 1992. *Mine Waste Management*. Sacramento, CA: California Mining Association.

Lowrie, R.L., chapter ed. 1997. Technologies for environmental protection. In *Mining Environmental Handbook*. Edited by J.J. Marcus. London: Imperial College Press. 190–282.

Malhotra, D., chapter ed. 1997. International environmental control of mining. In *Mining Environmental Handbook*. Edited by J.J. Marcus. London: Imperial College Press. 659–680.

Marshack, Jon B. 1989. A Compilation of Water Quality Goals. Sacramento, CA: Regional Water Quality Control Board, Central Valley Region.

Schmiermund, R.L., and M.A. Drozd, chapter eds. 1997. Acid mine drainage and other mining-influenced waters (MIW). In *Mining Environmental Handbook*. Edited by J.J. Marcus. London: Imperial College Press. 599–617.

Struhsacker, D.W., chapter ed. 1997. Environmental permitting. In *Mining Environmental Handbook*. Edited by J.J. Marcus. London: Imperial College Press. 283–411.

CHAPTER 23

Health and Safety

Kelvin K. Wu, P.E. and William J. Francart, P.E.

LEGISLATIVE HISTORY

1910—Public Law 61-179
Bureau of Mines created in the U.S. Department of the Interior. Federal safety and health role established for research and investigation of accidents.

1941—Public Law 77-49
Right of entry given to federal inspectors to conduct inspections and investigations in coal mines to obtain information. No safety or health regulations mandated.

1947—Public Law 80-328
Federal safety standards adopted for bituminous coal and lignite mines. Provisions made for federal inspectors to notify mine operators and state mine agencies of violations. No enforcement provisions. Expired after 1 year.

1952—Public Law 82-552
Federal Coal Mine Safety Act—Emphasis put on preventing major coal mine disasters. Annual inspections required at underground coal mines. Mandatory safety standards established for underground coal mines, with more stringent standards for "gassy" mines. Federal inspectors given authority to issue notices of violation, as well as orders of withdrawal in situations of imminent danger. In addition, orders of withdrawal mandated where less serious violations were not properly corrected. Enforcement of federal standards by state inspectors allowed under a state plan system. Anthracite mines covered, but all surface coal mines and any mine employing fewer than 15 people underground exempted.

1961—Public Law 87-300
Authorized study of causes and prevention of injuries and health hazards in metal and nonmetal mines. Federal officials given right of entry to collect information.

1966—Public Law 89-376
Extended coverage of 1952 law to small underground coal mines. Provided for orders of withdrawal in cases of repeated unwarrantable failures to comply with safety standards. Education and training programs expanded.

1966—Public Law 89-577

Federal Metal and Nonmetallic Mine Safety Act of 1966—Set up procedures for developing safety and health standards for metal and nonmetal mines. Standards could be advisory or mandatory. Annual inspections required for underground mines. Federal inspectors given authority to issue notices of violation and orders of withdrawal. Enforcement of federal standards by state inspectors allowed under state plan system. Education and training programs expanded.

1969—Public Law 91-173

Federal Coal Mine Health and Safety Act of 1969—Enforcement powers in coal mines increased vastly. Surface mines covered. Four annual inspections required for each underground coal mine. Stricter standards for gassy mines abolished, but additional inspections required in these mines. Miners given the right to request a federal inspection. State enforcement plans discontinued. Mandatory fines assessed for all violations. Criminal penalties established for knowing and willful violations. Safety standards for all coal mines strengthened and health standards adopted. Procedures incorporated for developing new health and safety standards. Training grant program instituted. Benefits provided to miners disabled by black lung disease.

1973—Secretarial Order 2953

Mining Enforcement and Safety Administration (MESA) created as a new U.S. Department of the Interior agency by administrative action. New agency assumed safety and health enforcement functions formerly carried out by the Bureau of Mines.

1977—Public Law 95-164

Federal Mine Safety and Health Amendments Act of 1977—Placed coal, metal, and nonmetal mines under a single law with enforcement provisions similar to the 1969 Act. (Separate safety and health standards retained.) Moved enforcement agency to U.S. Department of Labor and renamed it Mine Safety and Health Administration (MSHA). Established requirement for four annual inspections at all underground mines and two annual inspections at all surface mines. Advisory standards for metal and nonmetal mines eliminated. State enforcement plans in metal and nonmetal sector discontinued. Contained provisions for mandatory miner training. Mine rescue teams required for all underground mines. Increased involvement of miners and their representatives in health and safety activities.

HEALTH AND SAFETY REGULATIONS

TABLE 23.1 Health and safety topics in 30 C.F.R. Parts 1 to 199

Part	Subject
	Subchapter B—Testing, Evaluation, and Approval of Mining Products
5	Fees for testing, evaluation, and approval of mining products
7	Testing by applicant or third party
15	Requirements for approval of explosives and sheathed explosive units
18	Electric motor-driven mine equipment and accessories
19	Electric cap lamps
20	Electric mine lamps other than standard cap lamps
21	Flame safety lamps
22	Portable methane detectors
23	Telephones and signaling devices
24	Single-shot blasting units
26	Lighting equipment for illuminating underground workings
27	Methane-monitoring systems
28	Fuses for use with direct current in providing short-circuit protection for trailing cables in coal mines

continues next page

HEALTH AND SAFETY

TABLE 23.1 Health and safety topics in 30 C.F.R. Parts 1 to 199 (continued)

Part	Subject
29	Portable coal dust/rock dust analyzers, and continuous duty, warning light, portable methane detectors for use in coal mines
33	Dust collectors for use in connection with rock drilling in coal mines
35	Fire-resistant hydraulic fluids
Subchapter G—Filing and Other Administrative Requirements	
36	Approval requirements for permissible mobile diesel-powered transportation equipment
40	Representative of miners
41	Notification of legal identity
43	Procedures for processing hazardous conditions complaints
44	Rules of practice for petitions for modification of mandatory safety standards
45	Independent contractors
Subchapter H—Education and Training	
47	National Mine Health and Safety Academy
48	Training and retraining of miners
49	Mine rescue teams
Subchapter I—Accidents, Injuries, Illnesses, Employment, and Production in Mines	
50	Notification, investigation, reports and records of accidents, injuries, illnesses, employment, and coal production in mines
Subchapter K—Metal and Nonmetal Mine Safety and Health	
56	Safety and health standards—surface metal and nonmetal mines
57	Safety and health standards—underground metal and nonmetal mines
58	Health standards for metal and nonmetal mines
Subchapter M—Uniform Mine Health Regulations	
62	Occupational noise exposure
Subchapter O—Coal Mine Safety and Health	
70	Mandatory health standards—underground coal mines
71	Mandatory health standards—surface coal mines and surface work areas of underground coal mines
72	Health standards for coal mines
74	Coal mine dust personal sampler units
75	Mandatory safety standards—underground coal mines
77	Mandatory safety standards—surface coal mines and surface work areas of underground coal mines
90	Mandatory health standards—coal miners who have evidence of the development of pneumoconiosis
Subchapter P—Civil Penalties for Violations of the Federal Mine Saftey and Health Act of 1977	
100	Criteria and procedures for proposed assessment of civil penalties
Subchapter Q—Pattern of Violations	
104	Pattern of violations

Source: 30 C.F.R. Parts 1–199.

MINE SAFETY AND HEALTH ADMINISTRATION'S MISSION

The mission of MSHA is to administer the provisions of the Federal Mine Safety and Health Amendments Act of 1977 (Mine Act) and to enforce compliance with mandatory safety and health standards as a means to eliminate fatal accidents; to reduce the frequency and severity of nonfatal accidents; to minimize health hazards; and to promote improved safety and health conditions in the nation's mines. MSHA carries out the mandates of the Mine Act at all mining and mineral processing operations in the United States, regardless of size, number of employees, commodity mined, or method of extraction.

STATUTORY FUNCTIONS

The Mine Act provides that MSHA inspectors shall inspect each surface mine at least two times a year and each underground mine at least four times a year (seasonal or intermittent operations are inspected less frequently) to determine whether an imminent danger exists and whether there is compliance with health and safety standards or with any citation, order, or decision issued under the Mine Act.

MSHA activities include

- Investigating mine accidents, complaints of retaliatory discrimination filed by miners, hazardous condition complaints, knowing or willful (criminal) violations committed by agents or mine operators, and petitions for modification of mandatory safety standards
- Developing improved mandatory safety and health standards
- Assessing and collecting civil monetary penalties for violations of mine safety and health standards
- Reviewing for approval mining plans and education and training programs from mine operators
- Maintaining the National Mine Health and Safety Academy to train inspectors, technical support personnel, and mining industry personnel
- Approving and certifying the design of certain mining products
- Providing technical assistance to operators in meeting the requirements of the Mine Act
- Providing assistance to mine operators in improving their education and training programs
- Cooperating with states in the development of mine safety and health programs
- Making grants to states in which mining takes place
- Overseeing rescue and recovery operations.

ORGANIZATIONAL STRUCTURE

MSHA was created in 1978, when the 1977 Mine Act transferred the federal mine safety program from the U.S. Department of the Interior to the U.S. Department of Labor.

MSHA is headed by an assistant secretary of labor who administers a broad regulatory program to reduce injuries and illnesses in mining. Enforcement of safety and health rules and other responsibilities are carried out by two functional entities:

- The Coal Mine Safety and Health activity conducts its mine inspection, investigation, and training programs through 11 district offices and a system of subordinate offices in the nation's coal mining regions.
- The Metal and Nonmetal Mine Safety and Health activity administers its programs for all non-coal mines through six district offices in mining areas throughout the United States.

Other entities that have important roles include

- The Office of Standards, Regulations and Variances coordinates the development and issuance of safety and health rules and revision of existing rules, continually involving the public in the process.
- The Office of Assessments administers civil penalty assessments against mine operators for failing to comply with health or safety requirements.
- The Technical Support Directorate supplies engineering and technical aid, approves equipment and materials for safe mining use, and assists in mine emergencies and accident investigations. Technical Support operates major facilities in Pennsylvania and West Virginia.

- The Educational Policy and Development Office administers the agency's training programs. From the National Mine Health and Safety Academy, training is conducted on a variety of mine health and safety topics for specialists from government, industry, and labor. Through the Educational Field Services Division, training-related assistance is offered to mine operators throughout the country.
- The Office of Program Evaluation Information Resources (PEIR) conducts internal reviews, evaluates the effectiveness of agency programs, and conducts follow-up reviews to ensure that appropriate corrective actions have been taken. Another PEIR function is to collect, analyze, and publish data obtained from mine operators on the prevalence of work-related injuries and illnesses in the mining industry. PEIR is also responsible for support and training for all MSHA automated information systems, data communications networks, and automated data processing equipment. National mine injury and illness data is compiled, analyzed, and distributed to the mining community and public by specialists of the Division of Mining Information Systems (DMIS).

ENFORCEMENT

MSHA has a number of important tools for reducing injuries and illnesses in the nation's mines. Among them are various enforcement actions that MSHA can use to help ensure that dangerous conditions or practices are prevented or corrected. These are codified under 30 C.F.R. Parts 1 to 199.

Civil Penalties Imposed for Violations

MSHA inspectors must issue a citation or order for each violation of a health or safety standard they encounter. Each issuance entails a civil penalty. These fines may range up to $55,000 per violation. MSHA's Office of Assessments sets the penalties. Most non-serious violations that are corrected promptly are assessed a flat $55 penalty, except that mining operations found to have an excessive history of safety and health violations are not eligible for the $55 penalty. Most other violations are assessed according to a formula that considers six factors: (1) history of previous violations; (2) size of the operator's business; (3) any negligence by the operator; (4) gravity of the violation; (5) the operator's good faith in trying to correct the violation promptly; and (6) effect of the penalty on the operator's ability to stay in business. These factors are determined from the inspector's findings, MSHA records, and information supplied by the operator. In some cases (often involving fatalities or serious injuries), the formula would not yield an appropriate penalty. In these cases, MSHA may waive the formula and make a special assessment. Civil penalties are assessed against the mine operator. However, agents of corporate operators may individually be fined for violations they knowingly caused or permitted. Individual miners can be fined for violating smoking prohibitions.

"Significant and Substantial"

Several provisions of the act concern significant and substantial (S&S) violations. An S&S violation is one that is reasonably likely to result in a reasonably serious injury or illness. In writing each citation, the MSHA inspector determines whether the violation is S&S. These violations are not eligible for the flat $55 penalty.

Orders of Withdrawal

In several situations, the law provides that MSHA may order miners withdrawn from a mine or part of a mine. Some of the most frequent reasons for orders of withdrawal are (1) imminent danger to the miners; (2) failure to correct a violation within the time allowed; and (3) failure to secure an area during an accident investigation.

Unwarrantable Failures

If an MSHA inspector finds an S&S violation resulting from an "unwarrantable failure" by the operator to comply with a standard, the inspector incorporates that finding into the citation. If another violation, also resulting from an unwarrantable failure, is found within 90 days, MSHA issues a withdrawal order until it is corrected. Thereafter, any violation similar to the one that led to this withdrawal order will trigger another withdrawal order. This applies until an inspection of the mine discloses no similar violations.

Pattern of Violations

If MSHA determines that a mine has a pattern of S&S violations, the law and regulations provide that the agency shall notify the operator, who is then given an opportunity to improve compliance. Thereafter, if a mine is notified that it has a pattern of violations, any S&S violation found within 90 days would automatically trigger a withdrawal order. Each additional S&S violation would mean another withdrawal order until the mine had a "clean" inspection with no S&S violations.

Discrimination Protection

The law prohibits discrimination against miners, their representatives, or job applicants for exercising their safety and health rights. MSHA investigates all complaints of discrimination. If evidence of discrimination is found, the U.S. Department of Labor can take the miner's case before the independent Federal Mine Safety and Health Review Commission. In some cases, miners who have been fired can get their job back temporarily while a discrimination complaint is being adjudicated.

Criminal Penalties

The Mine Act provides for criminal sanctions against mine operators who willfully violate safety and health standards. MSHA initially investigates possible willful violations; if evidence of such a violation is found, the agency turns its findings over to the Department of Justice for prosecution.

Appeals

Before any citation or order is assessed, the operator or miners' representative can confer with an MSHA supervisor about any disagreement with the inspector's findings. If the disagreement cannot be resolved on this level, the operator is entitled to a hearing before an administrative law judge with the Federal Mine Safety and Health Review Commission. An operator or miners' representative who disagrees with any other enforcement action by MSHA also is entitled to a hearing. The administrative law judge's decision can be appealed to the commissioners and thereafter to the U.S. federal court system.

TECHNICAL ASSISTANCE

MSHA gives direct assistance for mines around the country from its Technology Center in Pittsburgh, Pennsylvania. Specialists work with mining companies and local MSHA inspectors to gather information on safety and health problems and to offer engineering or other types of solutions. MSHA provides mining companies with help in overcoming health hazards such as harmful dusts, liquids, vapors, and gases, or physical agents such as noise, ionizing radiation, and heat stress. Specialists also help with ventilation and electrical systems, roof support and ground control methods, mine waste facilities, equipment use, and many other aspects of the mining environment.

In addition, the staff of the Technology Center works to keep the mining industry current with state-of-the-art information on mine safety and health issues. MSHA regularly sponsors lectures, seminars, and training classes, and publishes reports on the latest scientific and technical information.

For more information on MSHA technical support programs, contact:
- MSHA's Directorate of Technical Support at (703) 235-1580
- MSHA's Pittsburgh Safety and Health Technology Center at (412) 386-6902
- MSHA's Approval and Certification Center at (304) 547-2044
- MSHA's Office of Information and Public Affairs at (703) 235-1452

For state mining agencies see Chapter 25 on mining-related Web sites or check MSHA's Web site at http://www.msha.gov.

TESTING PRODUCTS USED IN MINING

MSHA investigates and tests a wide range of mining equipment, components, instruments, and materials to ensure that they meet government standards for safe design and construction. This work helps to ensure that the various products will not contribute to an explosion, fire, electrical failure, vehicle crash, or other kind of accident.

The extensive list of mining products that MSHA must approve includes equipment such as multi-ton loading scoops, electrical cable and splice kits, panic bar designs, fire-resistant hydraulic fluids, vehicle braking systems, conveyor belts, diverse kinds of electrical equipment, hoses, explosives, mine illumination systems, and monitoring devices such as methane gas detectors. These products are used in coal, metal, and nonmetal mines.

MSHA conducts these tests and investigations at the Approval and Certification Center near Wheeling, West Virginia. Laboratories, a test track, explosion galleries, and offices for administrative work and records keeping make up the center.

MANDATORY TRAINING

The Federal Mine Safety and Health Amendments Act of 1977 recognizes training as an important tool for preventing accidents and avoiding unsafe and unhealthy working conditions. The act authorizes MSHA to "expand programs of education and training of operators ... and miners ...."

MSHA requires that each U.S. mine operator have an approved plan for miner training. This plan must include
- Forty hours of basic safety and health training for new miners who have no underground mining experience before they begin work underground
- Twenty-four hours of basic safety and health training for new miners who have no surface mining experience before they begin work at surface mining operations
- Eight hours of refresher safety and health training for all miners each year
- Safety-related task training for miners assigned to new jobs.

For more information on MSHA training programs, contact:
- Director, Educational Policy and Development, MSHA Headquarters, Arlington, Virginia, (703) 235-1400
- Superintendent, National Mine Health and Safety Academy, Beckley, West Virginia, (304) 256-3200
- Educational Field Services
  Eastern Operations   Western Operations
  Beckley, West Virginia   Denver, Colorado
  (304) 256-3223   (303) 231-5434
- Office of Information and Public Affairs, MSHA Headquarters, Arlington, Virginia, (703) 235-1452.

HISTORICAL DATA ON MINE DISASTERS IN THE UNITED STATES

TABLE 23.2   Accidents in the United States with five or more fatalities since 1970

Year	Day	Mine	Location	Type	Deaths
1992	12/07	No. 3 Mine, Southmountain Coal Co.	Wise Co., Norton, VA	Explosion	8
1989	09/13	William Station No. 9 Mine, Pyro Mining Co.	Union Co., Wheatcroft, KY	Explosion	10
1986	02/06	Loveridge No. 22, Consolidation Coal Co.	Marion Co., Fairview, WV	Suffocation (surface stockpile)	5
1984	12/19	Wilberg Mine, Emery Mining Corp.	Emery Co., Orangeville, UT	Fire	27
1983	06/21	McClure No. 1 Mine, Clinchfield Coal Co.	Dickinson Co., McClure, VA	Explosion	7
1982	01/20	No. 1 Mine, RFH Coal Co.	Floyd Co., Craynor, KY	Explosion	7
1981	12/08	No. 21 Mine, Grundy Mining Co.	Marion Co., Whitwell, TN	Explosion	13
1981	12/07	No. 11 Mine, Adkins Coal Co.	Knott Co., Kite, KY	Explosion	8
1981	03/15	Dutch Creek No. 1, Mid-Continent Resources, Inc.	Pitkin Co., Redstone, CO	Explosion	15
1980	11/07	Ferrell No. 17, Westmoreland Coal Co.	Boone Co., Uneeda, WV	Explosion	5
1979	06/08	Belle Isle Mine, Cargill, Inc. (salt)	St. Mary Parish, Franklin, LA	Explosion	5
1978	04/04	Moss No. 3 Portal A, Clinchfield Coal Co.	Dickinson Co., Duty, VA	Suffocation (oxygen deficiency)	5
1977	03/01	Porter Tunnel, Kocher Coal Co.	Schuykill Co., Tower City, PA	Inundation	9
1976	03/9–11	Scotia Mine, Blue Diamond Coal Co.	Letcher Co., Oven Fork, KY	Explosion	26
1972	12/16	Itmann No. 3 Mine, Itmann Coal Co.	Wyoming Co., Itmann, WV	Explosion	5
1972	07/22	Blacksville No. 1, Consolidation Coal Co.	Monongalia Co., Blacksville, WV	Fire	9
1972	05/02	Sunshine Mine, Sunshine Mining Co. (silver)	Shoshone Co., Kellog, ID	Fire	91
1971	04/12	Barnett Complex, Ozark-Mahoning Co. (fluorspar)	Pope Co., Rosiclair, IL	Hydrogen sulfide gas	7
1970	12/30	Nos. 15 and 16 Mines, Finley Coal Co.	Leslie Co., Hyden, KY	Explosion	38

Source: Compiled from records and documents of U.S. Bureau of Mines, MESA, and MSHA.

REFERENCES

30 C.F.R. Parts 1 to 199. 2001. Mineral resources. In *Code of Federal Regulations*. Washington, DC: Office of the Federal Register, National Archives and Records Administration.

CHAPTER 24

Bonding and Liabilities

Brent C. Bailey, P.E.

BONDING (FINANCIAL ASSURANCE)

Closure is the general term used to describe all the activities involved in decommissioning a mining operation and/or processing facility including reclamation, revegetation, removing equipment and structures, removal of chemicals and reagents, removal of hazardous wastes, remediation of any releases of hazardous substances to the environment, and postclosure monitoring. Closure refers to decommissioning an operation in accordance with its reclamation plan (or closure plan) to meet the prescribed land uses, to protect human health and the environment, and to minimize the need for further maintenance.

A mine operation generally must guarantee that it will perform closure or reclamation in accordance with its closure plan or reclamation plan before beginning operations. This guarantee is often provided with a bond from a surety company, but many regulating entities allow the use of cash deposits, certificates of deposit, irrevocable letters of credit, trust funds, or other security instruments.

The amount of a bond should equal the cost of successfully completing reclamation of the disturbed lands. The amount can be determined by applying the general principles of cost estimating. The following methodology demonstrates a general approach to developing reclamation cost estimates (State of California 1998).

- Describe the tasks to be performed:
    - Reclamation—may include establishing final slopes on all cuts and fills, removal of haul/access roads, constructing drainage/erosion controls, decompacting stockpile areas, topsoil replacement and distribution, finish grading, demolition and disposal of building foundations and other underground structures (i.e., storage tanks and septic tanks).
    - Revegetation—may include soil preparation and amendment, mulching, installation of irrigation systems, collection of custom seeds and plants, nursery services, hydroseeding, seeding and planting, plant protection, and revegetation maintenance.
    - Removal of equipment and structures—may include dismantling and removing of plant equipment and buildings. Although there will be a salvage value to the equipment and buildings, separate costs for removal of each major piece of equipment and building should be estimated. Similarly, the salvage value for each major piece of equipment and building should be estimated and deducted from the corresponding cost of removal to provide a net cost (or salvage value).
    - Miscellaneous costs—may include cleanup of boneyard areas, well closures, remediation of any contaminated soil, disposal of chemical or hazardous substances, and establishing access restrictions.

- In identifying the various reclamation tasks, it is important to include the tasks of reclaiming mineral processing facilities. Such facilities may require additional special considerations such as detoxification of process solutions, pumping costs (electricity), water treatment costs, removal of tanks and lined ponds, and disposal of sludge. For example, reclaiming a cyanide leaching operation will require detoxification of the leached ore. Column detoxification tests can be used to determine the amount of fresh water or detoxifying agent required to adequately detoxify all cyanide in the pond solutions and in the ore that had been processed with cyanide solutions. Consideration must be given to future draindown solutions and the possibility of reducing these with a cover and applying treatment if they do not meet water quality requirements (Posey et al. 1997).
- Monitoring
- Identify the equipment, labor, materials, and miscellaneous items necessary to complete the proposed tasks, and determine the unit costs for these items.
- Calculate the necessary quantities required for each of the tasks (i.e., cubic yards of material to be moved, hours of labor). This can be determined from production rates of equipment and labor, and from standard seed application rates, plant densities, or hydroseeding application rates.
- Determine the total costs for equipment, labor, and materials from the quantities and the unit costs for a particular task.
- Determine the total base cost of reclamation by adding the costs of the individual tasks.
- Add charges for mobilization, supervision, profit and overhead, and contingencies.

Mobilization costs are attributed to moving personnel, equipment, office trailers, and support facilities to the project site. These costs will vary depending on the site location and the cost of the reclamation work. They will normally vary between 1% to 5% of the total direct cost of reclamation (State of California 1998).

Supervision refers to management of the reclamation programs and can range from 2% to 7% for a $100,000,000 project to a $10,000 project, respectively (State of California 1998).

Many regulating entities require that the reclamation cost estimate be prepared as if the site had been abandoned by an operator and a third-party contractor was performing the reclamation (U.S. Department of Interior 1997).

Profit and overhead is provided in the case of a third-party contractor performing the reclamation. It can range from 3% to 14% for a $100,000,000 project to a $250,000 project, respectively (State of California 1998).

Contingencies are supplied to cover uncertainties in the cost estimate. Table 24.1 provides a sliding scale.

TABLE 24.1 Costs covering contingencies

Total Direct Cost	Contingency (%)
0 to $500,000	10
$500,000 to $5 million	7
$5 million to $50 million	4
Greater than $50 million	2

Source: OSM 1987.

LIABILITIES

Liabilities are the aspects, elements, or components of an operation or facility that impose an obligation to do or refrain from doing something and deduct from the value of a property (Gifis 1975). Environmental liability relates to (1) permitting (i.e., can the permits be obtained in a timely manner? or, for an operating facility, are the permits in place and is the facility in compliance?) and (2) contingent liabilities associated with environmental claims; i.e., claims under an environmental statute, such as a hazardous waste cleanup statute (Garver and Butler 1993).

Environmental Due Diligence

Future liability and financial risk can be minimized through environmental due diligence, which is the process of investigating and determining the existence of potential environmental liabilities or risk involved in the transfer of property ownership. The following is an abbreviated outline of an approach to environmental due diligence (Garver and Butler 1993).

- Identify constraints on the operation or expansion.
  - Identify all current permits imposed by regulating authorities and determine the expiration or renewal dates. For each permit:
    - Determine how long the renewal process is expected to take and when the application for renewal should be submitted.
    - Identify any substantive criteria that will affect renewal.
  - Determine the terms and conditions of the permits.
    - Determine what limits, if any, the permit places on operations or expansions. If there are limits, determine if the permit can be amended to accommodate any new operating plans.
    - Determine whether permits can be transferred.
    - If the operation or facility is exempt from any current applicable permitting requirements, determine whether the exemption will remain applicable if the facility is transferred to a new owner.
    - Review applicable land use plans to confirm that contemplated uses are consistent.
  - Determine the "regulatory environment" in which key operations exist.
    - Identify regulatory agencies and determine status of relationship with the seller.
    - Determine nature and extent of opposition to the project and assess the likelihood of appeal, citizen litigation, or other actions that could delay permitting or hinder operations.
  - Review the status of any performance bonds and liability policies. Determine how the transfer of authority to operate the facility may affect bonding.
  - Review company files and interview key employees.
  - Review regulatory agency records and interview key employees in regulatory agencies. (Caution: This must be done in cooperation with the target company.)
  - Review permit regulations, guidance, and correspondence.
- Identify significant contingent liabilities.
  - Identify potential significant contingent liabilities.
    - Define a "threshold of materiality" or a "risk tolerance" for the buyer.
    - Develop an understanding of the history of the company, facilities, or properties.
    - Define or identify potential contingent liabilities for the companies or properties to be acquired.
  - Identify potential sources of significant environmental liabilities.
    - Review laws defining liabilities and remediation for releases of hazardous substances.
    - Determine potential unfunded or underfunded reclamation expenses.
    - Review waste disposal practices for solid and hazardous wastes regulated under solid waste disposal statutes.

- Determine presence of other hazardous or toxic substances that may trigger liability or cleanup obligations.
- Determine presence and removal and cleanup obligations associated with any underground storage tanks.
- Determine applicability of other environmental regulatory programs that might lead to financial liability for compliance or penalties.
- If diligence discloses any historic or ongoing violations of permits or regulations, determine the potential for enforcement or citizen legal action.
- Determine whether existing permits and approvals include terms that create remediation or closure obligations that extend beyond applicable law.
– Review environmental indemnities in prior transactions.
– Review applicable insurance policies.
– Identify sources of information to review for environmental due diligence.
  - Target company files and interviews.
  - Perform site inspection; collect data onsite where necessary.
  - Review regulatory agency records (if possible).
  - Review all information relating to threatened or pending litigation with environmental claims.
  - Review title information for evidence of possible environmental problems.

REFERENCES

Garver, P.J., and J. Butler. 1993. Environmental due diligence in the acquisition of natural resource properties. Due diligence, landmen's section. *Proceedings of the 39th Annual Rocky Mountain Mineral Law Institute.*

Gifis, S.H. 1975. *Law Dictionary.* New York: Barron's Educational Series, Inc.

OSM (Office of Surface Mining). 1987. *Handbook for Calculation of Reclamation Bond Amounts.* Unpublished report. Washington, DC: OSM.

Posey, H.H., J.T. Doerfer, C. Kamnikar, B. Keffelew, A. Moore, and A.C. Sorenson. 1997. *Guidelines for the Characterization, Monitoring, Reclamation and Closure of Cyanide Leaching Projects (Draft).* Denver, CO: Colorado Department of Natural Resources, Division of Minerals and Geology.

State of California, State Mining and Geology Board. 1998. *Financial Assurance Guidelines.* Guidelines revised and readopted January 16, 1997. Revised bond forms added June 10, 1998. Available online at <http://www.consrv.ca.gov/smgb/index.htm>

U.S. Department of Interior, Bureau of Land Management. 1997. *Federal Register.* Amendment of 43 C.F.R. subpart 3809, 62(40):9093–9103.

CHAPTER 25

Web Sites Related to Mining

Russell F. Price, P.E.

This chapter provides some of the more relevant Web site addresses (Universal Resource Locators or URLs) for entities related to or of interest to the mining industry. The world of the Internet is dynamic, with new Web sites being created and older ones disappearing. For this reason, the sites included here cannot be complete for all mining-related entities. Some of the sites could fit into more than one category, but were put into the one where the fit seemed best. Individual mining companies are not included.

PROFESSIONAL SOCIETIES, INSTITUTES, COUNCILS, ASSOCIATIONS, FOUNDATIONS, BOARDS, AND COMMISSIONS

American Association of State Geologists: www.kgs.ukans.edu/AASG/AASG.html
American Ceramic Society: www.acers.org
American Coal Foundation: www.acf-coal.org
American Concrete Institute: www.aci-int.org
American Geological Institute: www.agiweb.org
American Geophysical Union: http://earth.agu.org
American Institute of Hydrology: www.aihydro.org
American Institute of Mining, Metallurgical, and Petroleum Engineers (AIME): www.aimeny.org
American Institute of Professional Geologists: www.aipg.org
American Iron and Steel Institute: www.steel.org
American Rock Mechanics Association: www.armarocks.org
American Society for Testing and Materials (ASTM): www.astm.org
ASFE (Associated Soil and Foundation Engineers): www.asfe.org
Association for Women Geoscientists: www.awg.org
Association of Engineering Geologists (AEG): www.aegweb.org
Clay Minerals Society: http://cms.lanl.gov
Cobalt Development Institute: www.cobaltdevinstitute.com
Deep Foundations Institute: www.dfi.org
Earthquake Engineering Research Institute: www.eeri.org
Edison Electric Institute: www.eei.org
Electric Power Research Institute: www.epri.com
Energy & Mineral Law Foundation: www.emlf.org
(The) Environmental and Engineering Geophysical Society: www.eegs.org
Friends of Mineralogy: www.indiana.edu/~minerals

Gems, Rocks & Minerals: www.gemsrocks.com
GEOindex: www.geoindex.com/geoindex
Geo-Institute: www.geoinstitute.org
Geological Society of America: www.geosociety.org
Geosynthetic Institute: www.geosynthetic-institute.org
Gold Institute: www.goldinstitute.com
Greening Earth Society: www.greeningearthsociety.org
Illinois Clean Coal Institute: www.icci.org
International Center for Aggregates Research: www.ce.utexas.edu/org/icar
International Energy Agency: www.iea.org
International Ground Water Modeling Center: www.mines.edu/research/igwmc
International Lead Zinc Research Organization: www.ilzro.org
International Society of Explosive Engineers: www.isee.org
International Society of Soil Mechanics and Geotechnical Engineering: www.issmge.org
Interstate Mining Compact Commission: www.imcc.isa.us
Iron & Steel Society: www.issource.org
I&SM (Iron & Steelmaker): www.iss.org
Kentucky Coal Council–Education: www.coaleducation.org
Kentucky Mining Institute: www.miningusa.com/kmi
Lignite Energy Council: www.lignite.com
Maguire Energy Institute: maguireenergy.cox.smu.edu
Mineral Economics and Management Society: www.minecon.com
Minerals and Geotechnical Logging Society: www.mgls.org
Minerals & Metallurgical Processing: www.smenet.org
(The) Minerals, Metals & Materials Society (TMS): www.tms.org
Mining Industry Council of Missouri: www.momic.com
Mississippi Valley Coal Trade & Transport Council: www.coalcouncil.org/index.html
Multidisciplinary Center for Earthquake Engineering Research: http://mceer.buffalo.edu/default.asp
(The) National Academies: www.national-academies.org
National Association of Geoscience Teachers: www.nagt.org
National Council for Geo-Engineering and Construction: www.geocouncil.org
National Council of Examiners for Engineering and Surveying: www.ncees.org
National Geotechnical Experimentation Sites: www.geocouncil.org/nges/nges.html
National Mine Land Reclamation Center: www.nrcce.wvu.edu/nmlrc
National Research Center for Coal & Energy: www.nrcce.wvu.edu
North American Geosynthetics Society: www.nagsigs.org
North Carolina Coal Institute: www.nccoal.org
Nuclear Energy Institute: www.nei.org
Pennsylvania Mining and Mineral Resources Research Institute: www.research.psu.edu/iro/html/pmmrri.html
Precast/Prestressed Concrete Institute: www.pci.org
Rocky Mountain Association of Geologists: www.rmag.org
Rocky Mountain Mineral Law Foundation: www.rmmlf.org
Salt Institute: www.saltinstitute.org
Seismological Society of America: www.seismosoc.org
SEPM (Society for Sedimentary Geology): www.sepm.org/sepm.html
Silver Institute: www.silverinstitute.org
Society for Mining, Metallurgy, and Exploration Inc (SME): www.smenet.org

Society for Organic Petrology: www.tsop.org
Society of Economic Geologists: www.segweb.org
Society of Exploration Geophysicists: www.seg.org
Society of Independent Professional Earth Scientists: www.sipes.org
Society of Petroleum Engineers: www.spe.org
(The) Soft Earth: www.wsu.edu/~geology/pages/S_earth.htm
Soil Science Society of America: www.soils.org
Solution Mining Research Institute: www.solutionmining.org
United States Universities Council on Geotechnical Engineering Research: www.usucger.org
Western Coal Council: www.westcoal.org
Western Interstate Energy Board: www.westgov.org/wieb

REFERENCE, INFORMATION, AND PUBLICATIONS

Aggregates and Roadbuilding magazine: www.rocktoroad.com
ASM International–The Materials Information Society: www.asm-intl.org
Ceramics and Minerals: www.ceramics.com
Chemical Engineering: www.che.com
Coal Age: www.coalage.com
COALDaily: www.fieldston.com
Coalfields: www.consumersref.com/coal
Coal Information Network: www.coalinfo.com
Coal Week International: www.mhenergy.com/demos/coal
Copper News: www.coppernews.com
Copper Page: www.copper.org
Electronic Journal of Geotechnical Engineering: www.ejge.com
Energy Argus Inc: www.energyargus.com
Energy Market Report: www.econ.com
Engineering & Mining Journal (E&MJ): www.e-mj.com
Engineering News-Record (ENR): www.enr.com
FedStats (Federal Government Statistics): www.fedstats.gov
Geotimes: www.geotimes.org
GInfoServer: www.geo.uni-bonn.de/members/haack/gisinfo.html
Gold and Silver Mines.com: www.goldandsilvermines.com
Goldsheet Mining Directory: www.goldsheetlinks.com
Infomine: www.infomine.com
International California Mining Journal: www.icmj.com
International Coal Report: www.ftenergy.com
(The) Internet Geotechnical Engineering Magazine: http://geotech.civen.okstate.edu/magazine/index.htm
Journal of Metals (JOM): www.tms.org/jom.html
Key to Metals: www.key-to-metals.com
Marine Sand and Gravel Information Service: www.sandandgravel.com
Metal World: www.metalworld.com
Mine Depot Inc: www.minedepot.com
MineNet–White Pages: www.microserve.net/~doug/whitepg.html
MineNetwork: www.minenet.com
Mine-On-Line: www.mine-on-line.com
Mineral Information Institute: www.mii.org
Miner's News: www.minersnews.com

Minesite.com: www.minesite.com
Mining Business Digest: www.mining.com
Mining Engineering: www.smenet.org
Mining magazine: www.mininginformation.com
Mining Record: www.miningrecord.com
Mining Research Database: http://sac.uky.edu/~skgang0/research.html
Mining Voice: www.nma.org
MyPlant Inc: www.myplant.com
National Information Service for Earthquake Engineering: www.eerc.berkeley.edu/software_and_data
Northern Miner: www.northernminer.com
Physical Properties of Earth Materials: http://geoweb.tamu.edu/tectono/ppem
Pit & Quarry: www.pitandquarry.com
Platts: www.platts.com
Quarry World: www.quarryworld.com
Raj's Mining Research Database: http://sac.uky.edu/~ganguli/research.html
Resource Data International: www.resdata.com
Robert Kranz's Reference Database: www.armarocks.org/resources/bookstore/kranz_database/kranz_database.html
Rock & Dirt: www.rockanddirt.com
Skillings Mining Review: www.skillings.net
Tunnel Business magazine: www.tunnelingonline.com
W.M. Keck Earth Sciences and Mining Research Information Center: http://keck.library.unr.edu
World Tunnelling: www.worldtunnelling.com
World Wide Web Virtual Library—Environment: http://earthsystems.org/virtuallibrary/vlhome.html
WWW Virtual Library—Geotechnical Engineering: http://geotech.civen.okstate.edu/wwwVL
Yahoo (Geotechnical Engineering): http://dir.yahoo.com/science/earth_sciences/geotechnical_engineering
Yahoo (Mining and Mineral Exploration): yahoo.com/business_and_economy/business_to_business/mining_and_mineral_exploration
Yesresources: www.yesresources.com

ASSOCIATIONS

Alaska Miners Association: www.alaskaminers.org
Aluminum Association Inc: www.aluminum.org
American Coal Ash Association: www.ACAA-USA.org
American Iron Ore Association: www.aioa.org
American Zinc Association: www.zinc.org
Brick Industry Association: www.brickinfo.org
California Cast Metals Association: www.foundryccma.org
California Mining Association: www.calmining.org
Coal Operators and Associates Inc: www.miningusa.com/coa
Colorado Mining Association: www.coloradomining.org
China Clay Producers Association (Macon, Georgia): www.kaolin.com
Idaho Mining Association: www.idahomining.org
International Association of Foundation Drilling: www.adsc-iafd.com

International Copper Association: www.copperinfo.com
International Zinc Association: www.iza.com
Iron Mining Association of Minnesota: www.taconite.org
Kentucky Coal Association: www.kentuckycoal.org
National Aggregates Association: www.nationalaggregates.org
National Mining Association (NMA): www.nma.org
National Stone Association: www.aggregates.org
Nevada Mining Association: www.nevadamining.org
Nickel Industry Producers Environmental Research Association: www.nipera.org
Northwest Mining Association: www.nwma.org
Ohio Aggregates & Industrial Minerals Association: www.oaima.org
Ohio Coal Association: www.ohiocoal.com
Ohio Mining and Reclamation Association: www.omra.org
Pocahontas Coal Association: www.pocahontascoal.org
Steel Manufacturers Association: www.steelnet.org
Utah Mining Association: www.utahmining.org
West Virginia Coal Association: www.wvcoal.com
West Virginia Mining & Reclamation Association: www.wvmra.com
Wyoming Mining Association: www.wma-minelife.com

ECONOMICS

AME Mineral Economics (Australia): www.ame.com.au
American Metal Market: www.amm.com
American Stock Exchange (AMEX): www.amex.com
Bloomberg: www.bloomberg.com
Bloomsbury Minerals Economics (United Kingdom): www.bloomsburyminerals.com
Board of Trade Clearing Corporation: www.botcc.com
Bridge/Commodity Research Bureau: www.crbindex.com
Chicago Board of Trade (CBOT): www.cbot.com
Chicago Stock Exchange: www.chicagostockex.com
Coal Trading Association: www.coaltrade.org
Denver Gold Group: www.denvergold.org
Dow Jones: www.dj.com
Economic Insight Inc: www.econ.com
Financial Times Energy: www.ftenergy.com
International Organization of Securities Commissions: www.iosco.org
London Metal Exchange: www.lme.co.uk
Marshall and Swift: www.marshallswift.com
Metals Economics Group: www.metalseconomics.com
MineMarket.com: www.minemarket.com
NASDAQ: www.nasdaq.com
New York Board of Trade (NYBOT): www.nybot.com
New York Mercantile Exchange (NYMEX/COMEX): www.nymex.com
New York Stock Exchange (NYSE): www.nyse.com
OTC Bulletin Board: www.otcbb.com
Quadrem: www.quadrem.com
Qualisteam: www.qualisteam.com
Reuters: www.reuters.com
Standard & Poor's (S&P): www.standardandpoors.com

TXU Energy Trading: www.txu.com
Wall Street Journal: www.wsj.com
Western Mine Engineering Inc: www.westernmine.com
World Mine Cost Data Exchange Inc: www.minecost.com

COMPUTING APPLICATIONS

(The) Computer Oriented Geological Society: www.csn.net/~tbrez/cogs
Engineering Software Center: www.engsoftwarecenter.com
Geotechnical and Geoenvironmental Software Directory: www.ggsd.com
Gibbs Associates: www.earthsciswinfo.com
Hearne Scientific Software (Australia): www.hearne.com.au
Macintosh Geological Software: http://geowww.geo.tcu.edu/faculty/geosoftware.html
Mining Internet Services Inc: www.miningusa.com
Richard B. Winston's Home Page: www.mindspring.com/~rbwinston/rbwinsto.htm
RockWare: www.rockware.com
Scientific Software Group: www.scisoftware.com

MACHINERY AND EQUIPMENT TRADING

A.M. King Industries Inc: www.amking.com
Compressed Air: www.ingersoll-rand.com/compair
Construction & Aggregates Mining Machinery: www.gate.net/~cagmm
Conveyor Services Directory: www.mining-services.com
Hendrikx Equipment: www.hendrikx-equipment.com/English/english.html
IronOx: www.ironox.com
Machinery Trader.com: www.machinerytrader.com
Mininggear.com: www.mininggear.com
Mining-Technology: www.mining-technology.com/index/html
Ritchie Bros. Auctioneers: www.rbauction.com
World Mining Equipment: www.wme.com

TRADE UNIONS

AFL-CIO: http://www.aflcio.org/home.htm
United Mine Workers of America (UMWA): www.umwa.org

GOVERNMENT

Federal Agencies

U.S. Advisory Council on Historic Preservation: www.achp.gov
U.S. Army Corps of Engineers: www.usace.army.mil
U.S. Bureau of Labor Statistics: www.bls.gov
U.S. Commodity Futures Trading Commission: www.cftc.gov
U.S. Council of Economic Advisers: www.whitehouse.gov/WH/EOP/CEA/html
U.S. Council on Environmental Quality: www.whitehouse.gov/CEQ
U.S. Department of Commerce: www.doc.gov
    National Oceanic and Atmospheric Administration (NOAA): www.noaa.gov
    National Geophysical Data Center: www.ngdc.noaa.gov
U.S. Department of Energy (DOE): www.doe.gov
    Energy Information Administration (EIA): www.eia.doe.gov
    Fossil Energy: www.fe.doe.gov

U.S. Department of Interior (DOI): www.doi.gov
    Bureau of Land Management (BLM): www.blm.gov
    Bureau of Reclamation (Technical Service Center): www.usbr.gov/tsc
    Fish and Wildlife Service (F&WS): www.fws.gov
    Geological Survey (USGS): www.usgs.gov
        Minerals Statistics and Information: http://minerals.usgs.gov/minerals
        National Atlas of the United States of America: www.nationalatlas.gov
        State Minerals Information: http://minerals.usgs.gov/minerals/pubs/state
    Minerals Management Service (MMS): www.mms.gov
    Office of Surface Mining (OSM): www.osmre.gov
U.S. Department of Transportation: www.dot.gov
    Surface Transportation Board: www.stb.dot.gov
U.S. Environmental Protection Agency (EPA): www.epa.gov
U.S. Forest Service: www.fs.fed.us
U.S. Government Printing Office (GPO): www.access.gpo.gov/su_docs
U.S. Internal Revenue Service (IRS): www.irs.ustreas.gov
    Internal Revenue Code–Title 26: tns-www.lcs.mit.edu/uscode/TITLE_26/toc.html
U.S. Mine Safety and Health Administration (MSHA): www.msha.gov
U.S. National Institute for Occupational Safety and Health (NIOSH): www.cdc.gov/niosh/mining
U.S. National Institute of Standards and Technology (NIST): www.nist.gov
    Universal Coordinated Time: www.boulder.nist.gov
U.S. National Science Foundation: www.nsf.gov
U.S. National Technical Information Service (NTIS): www.ntis.gov
U.S. Nuclear Regulatory Commission: www.nrc.gov
U.S. Securities and Exchange Commission (SEC): www.sec.gov

State Agencies
Alabama Surface Mining Commission: www.surface-mining.state.al.us
Alaska Division of Mining and Water Management: www.dnr.state.ak.us/land
Arizona Office of State Mine Inspector: www.state.az.us
Arkansas Department of Environmental Control: www.adeq.state.ar.us/mining
California Division of Mines and Geology: www.consrv.ca.gov
Colorado Office of Active and Inactive Mines: www.dnr.state.co.us
Connecticut Department of Environmental Protection: www.dep.state.ct.us
Delaware Geologic Survey: www.udel.edu/dgs/minres.html
Florida Mine Reclamation Section: http://www.dep.state.fl.us/water/mines/office.htm
Georgia Surface Mining Unit: www.dnr.state.ga.us/dnr/environ
Hawaii Department of Lands and Natural Resources: www.state.hi.us/dlnr/lmd
Idaho Bureau of Lands, Range, and Minerals: http://www2.state.id.us/lands/bureau/lands.htm
Illinois Office of Mines and Minerals: www.state.il.us
Indiana Department of Natural Resources: www.state.in.us/dnr/reclamation
Iowa Division of Soil Conservation: www2.state.ia.us/agriculture/soilconservation.html
Kansas Department of Health and Environment: www.kdhe.state.ks.us/mining
Kentucky Department of Surface Mining, Reclamation and Enforcement: www.nr.state.ky.us/nrepc/dsmre
Louisiana Surface Mining Section: www.dnr.state.la.us/CONS/conserin/surfmine.ssi
Maine Bureau of Land and Water Quality: http://janus.state.me.us/dep/blwq

Maryland Department of the Environment: http://www.mde.state.md.us/wma/minebur/index.htm

Massachusetts Department of Environmental Management: www.state.ma.us

Michigan Department of Natural Resources: www.dnr.state.mi.us

Minnesota Minerals Division: www.dnr.state.mn.us/lands_and_minerals

Mississippi Mining and Reclamation Division: www.deq.state.ms.us

Missouri Department of Natural Resources: www.dnr.state.mo.us/deq/lrp/homelrp.htm

Montana Department of Environmental Quality: www.deq.state.mt.us

Nebraska Conservation and Survey Division-University of Nebraska: Lincoln: http://csd.unl.edu/csd.html

Nevada Division of Minerals: http://minerals.state.nv.us

New Hampshire Department of Environmental Services: www.des.state.nh.us

New Jersey Geologic Survey: www.state.nj.us/dep/njgs

New Mexico Mining and Minerals Division: www.emnrd.state.nm.us/mining

New York Division of Mineral Resources: www.dec.state.ny.us/website/dmn/index.html

North Carolina Land Quality Section: www.dlr.enr.state.nc.us/mining.html

North Dakota Public Service Commission: www.psc.state.nd.us

Ohio Division of Mines and Reclamation: www.dnr.state.oh.us

Oklahoma Department of Mines: http://www.state.ok.us

Oregon Department of Geology and Mineral Industries: http://sarvis.dogami.state.or.us/homepage

Pennsylvania Office of Mineral Resource Management: www.dep.state.pa.us/dep/deputate/mines

Rhode Island Department of Environmental Management: www.state.ri.us/dem/org/natres.htm

South Carolina Geologic Survey: www.dnr.state.sc.us/geology/geohome.html

South Dakota Minerals and Mining Program: http://www.state.sd.us/denr/des/mining/mineprog.htm

Tennessee Bureau of Environment: www.state.tn.us/environment

Texas Surface Mining and Reclamation: www.rrc.state.tx.us/division/sm

Utah Division of Oil, Gas, and Mining: http://dogm.nr.state.ut.us

Vermont Agency of Natural Resources: www.anr.state.vt.us/geology

Virginia Department of Mines, Minerals, and Energy: www.mme.state.va.us

Washington Division of Geology and Earth Resources: http://www.wa.gov/dnr/htdocs/ger/ger.html

West Virginia Department of Environmental Protection: www.dep.state.wv.us/mr

Wisconsin Bureau of Waste Management: http://www.dnr.state.wi.us/org/aw/wm

Wyoming Department of Environmental Quality: http://deq.state.wy.us/lqd.htm

INTERNATIONAL

Africa

Chamber of Mines of South Africa: www.bullion.org.za

MBendi (South Africa): www.mbendi.co.za

Asia

Asian Journal of Mining: www.asianmining.com

Chamber of Mining Engineers of Turkey: www.mining-eng.org.tr

China Coal Information Network: www.coalinfo.net.cn

China Coal Research Institute: www.ccri.ac.cn/mkzy/english/english.html

Chinese Coal Association: www.chinatone.com/huangye/mulu/kchy_e.htm
Japan Coal: www.jcoal.or.jp
Mineral Resources Authority of Mongolia: www.mram.mn
Mining India: www.miningindia.com

Australia and New Zealand
Association of Mining & Exploration Companies: www.amec.asn.au
Australasian Institute of Mining & Metallurgy: www.ausimm.com.au
Australian Centre for Mining Environmental Research: www.acmer.com.au
Australian Coal Association Research program: www.acarp.com.au
Australian Coal Industry: www.anzlink.com/Support/Sectors/Coal/coalcont.htm
Australian Institute of Geoscientists: www.aig.asn.au
Australian Mineral Foundation: www.amf.com.au/amf
Australian Mines & Metals Association Inc: www.amma.org.au
Chamber of Mines and Energy of Western Australia Inc: www.mineralswa.asn.au
Industry Science Resources: www.isr.gov.au/resources/coal/index.html
Institution of Engineering and Mining Surveyors Australia Inc: www.home.aone.net.au/iemsaust
Journal Mining and Exploration Australia and New Guinea: www.reflections.com.au/MiningandExploration/index.html
Julius Kruttschnitt Mineral Research Centre: www.jkmrc.uq.edu.au
Mining & Petroleum InfoPage: www.oberon.com.au/Mining_InfoPage
MinMetAustralia: www.minmet.com.au
New Zealand Minerals Industry Association: www.minerals.co.nz
NSW Minerals Council: www.nswmin.com.au
OzGold Database Technology: www.comcen.com.au/~ozgold/main.html
QTHERM: www.dynamics.com.au/qtherm
Queensland Mining Council: www.qmc.com.au
UIC—Uranium & Nuclear Power Information Centre, Australia: www.uic.com.au

Canada
Association Des Prospecteurs Du Quebec: www.apq-inc.qc.ca
B.C. & Yukon Chamber of Mines: www.bc-mining-house.com
CAMESE (Canadian Association of Mining Equipment and Services for Export): www.camese.org
Canadian Diamond Drilling Association: www.canadiandrilling.com
Canadian Institute of Mining, Metallurgy and Petroleum (CIM): www.cim.org
Canadian Mining Hall of Fame: www.halloffame.mining.ca
Canadian Mining Industry Research Organization (CAMIRO): www.camiro.org
Canadian Mining Journal: www.canadianminingjournal.com
Canadian Venture Exchange (CDNX): www.cdnx.ca
Coal Association of Canada: www.coal.ca
International Council on Metals and the Environment: www.icme.com
International Development Research Centre: www.idrc.ca/mpri/projects.html
McGill Mining Research: www.minmet.mcgill.ca/minres.htm
Mining Association of Canada: www.mining.ca
Natural Resources Canada: www.nrcan.gc.ca
NWT Chamber of Mines: www.miningnorth.com
Ontario Mining Association: www.oma.on.ca
Ontario Ministry of Northern Development and Mines: www.gov.on.ca/MNDM

Prospectors and Developers Association of Canada: www.pdac.ca
Uranium Mining Research–Environment Canada: www.mb.ec.gc.ca/pollution/e00s02.en.html
Yukon Chamber of Mines: web1.yukon.net/business/whitehorse/ycmines

Europe

Bismuth Institute (Belgium): www.bismuth.be
Euriscoal (Belgium): www.euriscoal.com
European Salt Producers' Association (France): www.eu-salt.com/sommaire.htm
German Brown Coal Association: www.braunkohle.de
German Coal Importers Association: www.verein-kohlenimporteure.de
IEA Coal Industry Advisory Board (France): www.iea.org/ciab
IISI–Worldsteel (Belgium): www.worldsteel.org
Institute of Coal and Coal-Chemical SB RAS (Russia): www.kemsc.ru
Institute of Coal Research of SB RAS (Russia): www-bras.nsc.ru/eng/sbras/copan/coal
International Association for Hydraulic Engineering and Research (Netherlands): www.iahr.nl
International Coal Encyclopedia (Ireland): www.coalservices.com
International Copper Study Group (Portugal): www.icsg.org
International Institute for Infrastructural, Hydraulic and Environmental Engineering (Netherlands): www.ihe.nl
International Nickel Study Group (Netherlands): www.insg.org
International Organization for Standardization (ISO)(Switzerland): www.iso.ch
International Society for Rock Mechanics (Portugal): http://www-ext.lnec.pt/ISRM
International Tungsten Industry Association (Belgium): www.itia.org.uk
Links for Mineralogists (Germany): www.uni-wuerzburg.de/mineralogie/links.html
Minerals and Energy-Raw Materials Report (Sweden): www.tandf.no/minerals
Rocas y Minerales (Spain): www.tsai.es/pymes/rocas
Rock Mechanics and Rock Engineering (Austria): http://link.springer.de/link/service/journals/00603
ROSUGAL (Russia): www.kemsc.ru/coalind/rosugol/koi-8/invest/inen.htm
Selenium-Tellurium Development Association (Belgium): www.stda.be
WWW–Server for Ecological Modelling (Germany): http://dino.wiz.uni-kassel.de/ecobas.html

Latin America

Construccion Pan Americana: www.cpa-mpa.com
Dirección General de Promoción Minera (Mexico): www.secofi-cgm.gob.mx
Editec (Chile): www.editec.cl
La Camara de Mineria del Ecuador: www.cme.org.ec
National Society of Mining Petroleum and Energy (Peru): www.snmpe.org.pe
Panorama Minero (Argentina): www.panoramaminero.com.ar

United Kingdom

Aggregates Advisory Service: www.planning.detr.gov.uk/aas/index.htm
Association of Geotechnical & Geoenvironmental Specialists: www.ags.org.uk
British Geological Survey: www.bgs.ac.uk
Coal Authority: www.coal.gov.uk
Coal International: www.tbarratt.force9.co.uk
(The) Concrete Society: www.concrete.org.uk

Department of Trade and Industry: www.dti.gov.uk
(The) Geological Society: www.geolsoc.org.uk
Geopages: www.geopages.co.uk
IEA Clean Coal Centre: www.iea-coal.org.uk
Industrial Minerals Information Ltd: www.mineralnet.co.uk
Institute of Materials: www.instmat.co.uk
Institute of Quarrying: www.inst-of-quarrying.org/iq
Institution of Mining and Metallurgy: www.imm.org.uk
International Association of Hydrogeologists: www.iah.org
McCloskey Coal Information Services: www.mccloskeycoal.com
Mining Journal Ltd: www.mining-journal.com
Palladian Publications Ltd: www.worldcoal.com
Simpson Spence & Young: www.ssyonline.com
Solid Fuel Association: www.solidfuel.co.uk
Spectron Global Coal Limited: www.globalcoal.com
World Coal Institute: www.wci-coal.com
World Nuclear Association: www.world-nuclear.org

Index

Note: *f* indicates figure; *t* indicates table.

A

Abandoned Mine Land Fund 394
Abrasives classification 173*t*–174*t*
Accelerated cost recovery system (ACRS) 166, 166*t*
Acceleration conversion factors 127*t*
Activity times and floats 125*t*
Activity-oriented graph 123*f*
Activity-oriented network 124*f*
Aerial photography scale 65
Aggregates
  grading requirements 175*t*
  uses and specifications 174*t*
Air
  composition 259*t*
  density at different altitudes 259*t*–260*t*
  in-leakage to evacuated equipment 345
Air quality
  ambient standards 398*t*
  mining fugitive emission controls, effectiveness, and costs 399*t*
  prevention of significant deterioration incremental standards 398*t*
  significant emission rates 398*t*
Airflow 263
  friction factors for mine airways 265*t*
  friction loss 264
  measurement points for fixed traversing in airways 264*f*
  quantity 263
  shock loss 266*t*
  velocity head 263
Airway resistance 95
Albite feldspar 32*t*
Algebra
  addition matrix 106
  associative law 105
  basic equations 105–106
  basic laws 105
  binomial formula for positive integer 105
  commutative law 105
  determinants 106
  distributive law 105
  equation of circle 106
  equation of straight line 105
  factorial *n* 105
  identity matrix 106
  inverse matrix 106
  matrices 106
  multiplication matrix 106
  proportion 105
  quadratic equation 105
  radius of circle 106
  special products and factors 105
  transpose matrix 106
Altitude correction 218
Aluminum specifications 175*t*
Alunite 32*t*
Amblygonite 32*t*
Analysis of retreat mining pillar stability 250*f*, 252*t*–253*t*
Andalusite 32*t*
Angle conversion factors 127*t*
Anglesite 32*t*
Anglo American Research Laboratory (AARL) stripping (elution) 319
Anhydrite 32*t*
Anorthite feldspar 32*t*
Anthracite 10
Antlerite 33*t*
Apatite 33*t*
Apparent dip 61
  determination of strike and dip from two apparent dips 63–64, 64*f*, 65*f*
Aragonite 33*t*
Area
  channel section 1 77
  channel section 2 77
  circular section 75
  conversion factors 127*t*–128*t*
  "H" section 1 76
  "H" section 2 76
  pipe section 75
  rectangular section 74
  of triangle 103
Area maintenance 384
Argentite 33*t*
Arithmetic mean 115
ARMPS. *See* Analysis of retreat mining pillar stability
Arsenic 33*t*
Arsenopyrite 33*t*
Atacamite 33*t*
Attenuation
  mechanisms of chemical species 407*t*
  vadose zone properties relative to 308*t*
Autunite 33*t*
Available energy conversion factors 130*t*

436 | INDEX

Average speed 217, 218t
Azurite 34t

B

Backhoes 222
Bacterial oxidation
  bacteria 319t
  calculations 320
  control and troubleshooting 322t
  gold recovery in stirred tank reactors 321t
  H_2SO_4 demand 320t
  heach leap operations 323
  heat balances 321
  heat of reaction 320t
  minerals oxidized 320
  neutralization 322
  nutrient additions 321–322
  oxygen requirement 320t
  solid-liquid separation 322
  stirred tank reactors for base metals 322
  toxicity 319
  typical circuit conditions 321
Barite 34t, 176
Barometric pressure at different altitudes 259t–260t
Beam analysis 79
  bending stresses 87
  deflection 88, 89f
  shear stresses 88
Belt conveyors 232
  angle of repose 235f
  angle of surcharge 235f
  belt width for given lump size 234f
  cross-sectional area of 20° troughed belt 233t
  cross-sectional area of 35° troughed belt 234t
  flowability 235f
  locating scales 345
  power requirements 235, 236f, 236t, 237f
  production capacity 232, 233t, 234t
  recommended belt speeds 233t
  weight per unit length of belt and idlers 236t
Belt friction 78
Benefit cost ratio (BCR) 168
Bentonite 176
Bernoulli equation 93
Beryl 34t
Best efficiency point (BEP) 278
Bins 345
Biotite mica 34t
Bismuth 34t
Bismuthinite 34t
Bituminous coal 10
Blasting. *See also* Explosives
  air concussion 211
  better fragmentation 345
  burden ratio 206–207, 207t
  charge diameter and detonation velocity 206f
  controlled 209–210
  design method based on burden ratios 206–207
  design method based on known powder factor 207–208
  determining amount of rock broken per borehole 208
  explosives per unit weight/volume of rock 208–209
  ground vibrations 211
  peak particle velocity (PPV) 211–213, 212f
  powder factor 207–209
  powder factor for underground blasting 209, 210f
  powder factors for surface coal mines 209t
  powder factors for surface metal mines 209t
  safe vibration level criteria 213, 213f
  scaled distance (scale factor) 211, 212f
  site constants 212t
  vector sum 213
  vibration attenuation 211–213
  water-resistant explosives to build out of water in borehole 211
Bonding 419–420
Book value 166
Borax 34t
Bornite 34t
Bournonite 35t
Brown coal 10
Brucite 35t
Bulk modulus 85
Bulldozers 228
  blades 228–229
  cycle time and production 229
  ripping 229
Buoyant force 92
Burden ratio 206–207, 207t

C

Calaverite 35t
Calcite 35t
Capacitance 100
Capacitive resistance 101
Capacitors 101
Capital structure 169t
Capitalized costs 166
Carbon plants
  acid washing 318
  carbon expansion characteristics 318t
  carbon-in-column (CIC) 318
  carbon-in-leach (CIL) 317–318
  carbon-in-pulp (CIP) 317–318
  regeneration 319
  stripping (elution) 319
Carnalite 35t
Carnotite 35t
Cash flow
  components for developing 170t
  factors for consideration 170t
Cassiterite 35t
Celestite 36t
Centrifugation 308
Centroid
  of an area 73
  channel section 1 77
  channel section 2 78
  circular section 75
  "H" section 1 76
  "H" section 2 76
  of a line 72–73
  of a mass 73

pipe section 75
rectangular section 74
of a volume 73
CERCLA. *See* Comprehensive Environmental Response, Compensation, and Liability Act
Cerussite 36*t*
Chalcanthite 36*t*
Chalcocite 36*t*
Chalcopyrite 36*t*
Channel section 1 formulas 77
Channel section 2 formulas 77–78
Chemical elements 69*t*–71*t*
Chi-square probability function 117, 120*t*
Chlorite 36*t*
Chromite 36*t*
Chromium 176
Chrysoberyl 36*t*
Chrysocolla 37*t*
Cinnabar 37*t*
Circular section formulas 75
Circulating load (CL) 287
Clamshells 222
Clarification 293, 306
    classifier types 293*t*
    performance measurements for air classifiers 294
Clays 176, 176*t*, 177*t*
Clean Air Act 393
Clean Water Act 393
Closure 419–420
Coal 10–11
    ash content 54*t*
    classifications by rank 51*t*
    fusion temperatures 54*t*
    heating values 51*t*
    petrographic and physical properties for various U.S. seams 52*t*–53*t*
    proximate analyses 51*t*
    rating of coking coals for blending 178*t*
    significance of characteristics for combustion performance 177*t*
    specific gravity 11
    strength properties 243*t*
    sulfur content and forms 53*t*–54*t*
    ultimate analyses 51*t*
Coal cleaning 323
    coal feed data 324*t*
    distribution curves 323–326, 325*f*
    operations 326
    recovery 323
    relative difficulty of separation 323*t*
    troubleshooting 326
    weight fraction to product 325*t*
    yield 323
Coalification 10
Cobalt 178
Cobaltite 37*t*
Coefficient of heat transfer
    conversion factors 130*t*
    selected 345
Coefficient of permeability 354
Column buckling 88, 90*f*
Combinations 114–115
Comprehensive Environmental Response, Compensation, and Liability Act 393–394

Compressed air
    air consumption multiplier by number of rock drills 284*t*
    air consumption multipliers for operation of rock drills 284*t*
    air requirements of representative drilling machines 286*t*
    factors for correcting capacity of single-stage compressors 284*t*
    pipe diameter 283
    and pressure 344
    pressure loss in hose 285*t*
Concentration criterion (CC) 301
Conceptual studies 1, 2*t*–4*t*
Conduction 99
Confidence intervals 116–117
Conservation of work and energy 82
Constants 140*t*
Consumer price index 169*t*
Continuity equation 92
Continuous miners 239
Controlled splitting 267
Convection 99
Coordinate relations 103
Copper 37*t*, 178
    heap leaching 316
Cordierite 37*t*
Corundum 37*t*
Costing
    cost reports 171*t*
    monthly income statement (full absorption costing basis) 172*t*
    monthly income statement (variable costing basis) 172*t*
    six-tenths rule 171
Costs
    accelerated cost recovery system (ACRS) 166, 166*t*
    benefit cost ratio (BCR) 168
    bucketline dredge capital costs 377*t*
    capitalized 166
    contingencies 420, 420*t*
    depletion 167
    deterioration vs. cost and downtime 387*f*
    equipment operating costs 220–222
    maintenance and repair 221, 221*t*, 383*f*
    mining fugitive emission costs 399*t*
    mobilization 420
    placer mining 377, 377*t*
    reclamation 419–420
Counter-current decantation (CCD)
    thickener 304*t*, 322
Couples 72
Covellite 37*t*
CPM. *See* Critical path method
Crawler-mounted loaders
    number of trucks 227
    truck cycle time and productions 226
    truck dump time 226
    truck load time 226
    truck loading 225–226
    truck spot time 226
    trucks 225
Cristobalite 37*t*
Critical path method 119
Crocoite 38*t*

Crushers and crushing 287, 288–289
  abrasion indices 289t
  circulating load (CL) 287
  cone/gyratory 288
  horizontal shaft impactors 288
  jaw 287
  maintenance 288
  Mining Machinery Developments, Ltd. (MMD) 288
  overall reduction ratio 287
  rolls 288
  types 287–288
  vertical shaft impactor (VSI) 288
  waterflush cone 288
  work index 289t
Cryolite 38t
Cuprite 38t
Current 100
Cyanide solubility of minerals 343t

D

Darcy's law 354
Decommissioning 419–420
Density 218–219
Density of heat flow rate conversion factors 130t
Depletion allowance 167
Depletion percentages for minerals 167
Depreciation 166, 166t
Derivatives
  of common functions 113
  definitions 112
  inflection points 113
  maxima 113
  minima 113
Dew point temperature 260t–261t
Diamonds 38t
  industrial 179t
Diaspore 38t
Diatomite 179t
Dip 60
  determining 61–62, 62f
  determining from two apparent dips 63–64, 64f, 65f
Dip slip 60, 61f
Dolomite 38t
  size analyses 182t
  typical analyses 181t
Draglines 222, 223t
Drawbar pull 215
Dry grinding systems 292
Ducts 270
  loss coefficients for area changes 271f
  loss coefficients for elbows 272f
Due diligence studies 4
  checklist 4–7
  economics 6–7
  metallurgical/processing 6
  mine engineering and planning 5
  reserves 4–5
Dust collectors and collection
  dust control 328–329
  fabric 326, 327t
  power requirements vs. particle size 328f
  suggested equipment vs. particle size 327f
  troubleshooting 328

Dynamic head 275–276
Dynamic viscosity conversion factors 137t
Dynamics
  basic kinematics 80–81
  conservation of work and energy 82
  impact 81
  impulse and momentum 81
  kinetic energy 82
  Newton's Second Law of Motion 81
  plane circular motion 80–81
  potential energy 82
  straight-line motion 80
  work 82

E

Earthworks (embankments) 353, 353f
Effective tax rate 168
Elastic strain energy 88
Electrical power 100, 281. *See also* Power
  allowable ampacities 281t–282t
  dissipation 100
  formulas 284t
  power factor 281, 283t
Electricity
  capacitance 100
  capacitive resistance 101
  capacitors 101
  conversion factors 128t–129t
  current 100
  electromotive force 100
  inductors 101
  Kirchoff's Laws 102
  resistance 100
  resistors 100
  time constant 101
  voltage drop 100
Electromotive force 100
Electrowinning and electrorefining 329–331
  checklist 331–332
  typical plant operating data 329t
Enargite 38t
Endangered Species Act 394
Energy 82
  conversion factors 129t
  divided by area time (conversion factors) 129t
Entropy conversion factors 131t
Environmental due diligence 421–422
Environmental Quality Improvement Act 394
Environmental regulations
  air quality 398–399
  ambient air quality standards 398t
  attenuation mechanisms of chemical species 407t
  cation exchange capacities of soil clays, soil organic matter, and several soil types 409t
  common ionic forms of chemicals found in mining operations 406t–407t
  federal Bevill-excluded wastes 405t
  federal hazardous waste classification criteria 404t
  kinetic prediction techniques 402t
  Latin American and Asia-Pacific rim laws 396t
  mining fugitive emission controls, effectiveness, and costs 399t

permits and approvals required 397*t*
potentially mobile chemical species in
    tailings liquids 401*t*
prevention of significant deterioration
    incremental standards (air
    quality) 398*t*
pyrite weathering reactions and
    pathways 401*f*
significant emission rates (air quality) 398*t*
static prediction techniques 403*t*
trace element loadings for agricultural
    lands 409*t*
treatment 406–409
typical water quality standards and
    goals 400*t*
U.S. laws 393–394, 395*t*
vadose zone properties relative to
    attenuation 308*t*
water quality 400–405
EPA. *See* U.S. Environmental Protection Agency
Epidote 38*t*
Epsomite 39*t*
Equilibrium 72
Equipment analysis 385–389, 389*t*
Euler buckling load 88, 90*f*
Event-oriented graph 123*f*
Event-oriented network 124*f*
Excavation, loading, and transport equipmen*t*
    See also specific types of equipment
    altitude correction 218
    average speed 217, 218*t*
    density 218–219
    drawbar pull 215
    fill factor 219
    fuel and power consumption 221–222
    grade resistance 216–217
    haulage cycle time 220
    haulage production 220
    load cycle time 219
    loading production 219–220
    maintenance and repair costs 221, 221*t*
    maximum speed 217, 218*t*
    operating costs 220–222
    operating efficiency 219
    performance factors 215–218
    production 219–220
    production calculations 218–220
    resistance 216–217, 217*t*
    rim pull 215
    rolling resistance 216–217, 217*t*
    swell 218–219
    tire life 221, 221*t*
    total resistance 216–217
    traction 215–216, 216*t*
    travel time 220
Executive Order 12898 394
Explosives. *See also* Blasting
    characteristics 204*t*
    classification by fume characteristics 204,
        205*t*
    descriptive classification 203*f*
    detonation pressure 204–205
    loading factor 205, 205*t*
    permissible 204
    water-resistant 211

F

F-(variance ratio) distribution function 119,
    122*t*
Fans
    horsepower requirements 266
    laws 266*t*
    raising pressure 344
Faults
    classification 60
    terminology 60, 61*f*
Feasibility studies 1
    as "bankable documents" 1
    checklist of minimum reporting
        requirements 2*t*–4*t*
Federal Coal Mine Health and Safety Act 412
Federal Coal Mine Safety Act 411
Federal Metal and Nonmetallic Mine Safety
    Act 412
Federal Mine Safety and Health Amendments
    Act of 1977 412, 417
Federal Water Pollution Control Act 393
Feeders 345
    capacities and effect of material
        characteristics 347*t*
Fill factor 219
First Law of Thermodynamics 97–98
Flocculation 312
Flotation 301–302
    operating controls 302–303
    testing 303
Flow networks 95
Fluid mechanics
    airway resistance 95
    basic concepts 90–91
    Bernoulli equation 93
    buoyant force 92
    continuity equation 92
    flow networks 95
    fluid momentum 92–93
    friction losses 93–94, 94*f*
    Froude number 91
    hydraulic diameter or radius 94
    Manning equation and coefficients 95, 96*t*
    mass density 90
    Moody diagram 93, 94*f*
    open channel flow 95, 96*t*
    pressure 91–92
    pressure prism 92
    Reynolds number 91
    specific gravity 90
    specific volume 90
    specific weight 90
    viscosity 91
Fluid momentum 92–93
Fluorite 39*t*
Fluorspar 179
Footwall 60
Force 71
    components 71
    conversion factors 129*t*–130*t*
    divided by length (conversion factors) 130*t*
Fosterite 39*t*
404(b)(1) guidelines 393
Fourier's Law of Conduction 99
Franklinite 39*t*
Free-energy diagrams 336, 337*f*, 338*f*

G

Friction 78
Friction head 275
Friction losses 93–94, 94f, 276t
Froude number 91
Fuel consumption conversion factors 131t

G

Galena 39t
Garnierite 39t
Gases
  conveying velocities 273t
  explosibility diagram 269f
  ideal gas law 97, 98t
  laws 263
  properties 268t
Geologic data collection 145t–146t
  equipment 147t
Geologic map symbols 56f
Geologic time scale 55f
Geometric mean 115
Geometry
  circle of radius r 110
  cone 110
  cube 111
  cylinder 110
  ellipse 110
  frustum of cone 111
  frustum of pyramid 111
  general quadrilateral 109
  general triangle 109
  parallelogram 109
  planar areas by approximation 111–112
  pyramid 111
  rectangle 109
  rectangular prism 111
  right triangle 109
  sphere 110
  trapezoid 109
Gibbsite 39t
Glauberite 39t
Goethite 40t
Gold 40t, 180
  bacterial oxidation 321t
  character vs. distance from source 372t
  classification of particles 372t
  effect of single particle on sample 372t
  heap leaching 315–316
  leaching 343–344, 343t
  Merrill-Crowe recovery 343–344
Grade resistance 216–217
Graphite 40t, 180
Gravel 185
Gravity concentration 300
  commercial characteristics of machines 300t
  concentration criterion (CC) 301
Greek alphabet 143t
Greenockite 40t
Grinding 289
  ball and liner wear 291
  ball diameter 291
  critical-speed rules of thumb 290t
  dry grinding systems 292
  high-pressure grinding rolls (HPGRs) 292–293
  rod and ball mills 290–291, 290t
  rod diameter 291
  SAG mills 292
  stirred media mills 293
  torque table 292t
  types of mills 290–293
Ground control and support 243
  pillars 248–253
  RMR system guidelines for excavation and support 245t
  rock bolting 246–248
  rock mass rating (RMR) system 244t–245t
  rock support selection chart for 20 ft coal entries 246f
  safety factors 246
  slope stability 256–257
  strength properties of rock and coal 243t
  subsidence 254–256
  support recommendations for mine drifts 247f
Gy's formula for minimum sampling error 158t
Gypsum 40t

H

"H" section 1 formulas 76
"H" section 2 formulas 76–77
Hade 60
Halite 40t
Hanging wall 60
Harmonic mean 116
Haul roads 200
  adhesion coefficients 200t
  gradients 199t
  number of lanes 196
  off-highway truck specifications 195t
  rolling resistance coefficients 200, 200t
  safe distance between trucks 197
  sight distance 199, 199t
  stopping distance 195, 196f, 197f, 198t
  subbase design (English units) 201f
  subbase design (SI units) 202f
  super elevation 199–200, 199t, 200t
  width for roads with sharp curves 198t
  width for straight regular grade roads 198t
Haulage cycle time 220
Haulage production 220
Hazen-Williams Equation 275–276
Health and safety. See also Mine Safety and Health Administration
  Federal Coal Mine Health and Safety Act 412
  Federal Coal Mine Safety Act 411
  Federal Metal and Nonmetallic Mine Safety Act 412
  Federal Mine Safety and Health Amendments Act of 1977 412, 417
  ground control and support 246
  legislative history 411–412
  Mining Enforcement and Safety Administration 412
  regulations 412t–413t
  U.S. mine disasters since 1970 418t
Heap leaching 317
  and bacterial oxidation 323
  comparison with vat leaching and agitation 344t
  copper ores 316
  design features 316

dissolved oxygen per liter of distilled
  water 317t
 gold ores 315–316
 operating conditions 315t
 pads 316–317
 symptoms of percolation problems 315
Heat capacities
 conversion factors 131t
 gases and liquids 332f
 ratio at 1 atm 333t
 solids 334t
Heat conversion factors
 available energy 130t
 capacity 131t
 density 130t
 density of heat flow rate 130t
 flow rate 131t
 fuel consumption 131t
 heat capacity and entropy 131t
 specific heat capacity and specific
  entropy 131t
 thermal conductivity 131t–132t
 thermal diffusivity 132t
 thermal insulance 132t
 thermal resistance 132t
 thermal resistivity 132t
Heat transfer. *See also* Thermodynamics
 coefficient of 130t, 345
 conduction 99
 convection 99
 conversion factors 130t
 Fourier's Law of Conduction 99
 radiation 99
Heave 60
Hematite 41t
High-pressure grinding rolls (HPGRs) 292–293
Hoists 238
 capacity 238
 cycle time 238
 drum 238
 friction (Koepe) 238
 power requirements 239
 ropes 238, 239t
 skips 238
Hydraulic classifiers and sizers 306
Hydraulic diameter or radius 94
Hydrocyclones 308, 309f
Hydrographs 365, 365f
Hydrology
 emergency spillways 366–367
 rainfall intensity 363–364
 runoff coefficients 363t
 sediment ponds 366
 Steel formula 364f, 364t
 storm water runoff 363–365
 unit hydrographs 365, 365f
 water balance 362–363

I

Ideal gas law 97, 98t
Igneous rocks
 classification 57f
 properties 21t–23t
Ilmenite 41t
Impact 81
Impoundments 351, 352f

Impulse 81
In situ leaching 379
 lixiviants 379
 operating considerations 380
 ore and deposit characteristics 379, 379t
 troubleshooting 380
 of uranium 380–382
 wells and well fields 380
Industrial minerals 341
 common tests 341t–342t
Inflation 168, 168f
Institute of Makers of Explosives 204
Integrals 113–114
Interest
 continuous interest lump sum end-of-period
  dollar values 161
 discrete interest end-of-period dollar
  values 161
 formulas 161
 single payment compound amount
  factor 162t
 single payment present worth factor 163t
 uniform series compound amount
  factor 164t
 uniform series present worth factor 165t
International System of Units 127
 conversion factors 127t–139t
 prefixes 143t
 selected unit equivalencies and
  approximations 141t–142t
Iron ore 180

J

Jamesonite 41t

K

Kaolin
 definitions 176t
 quality factors 177t
Kaolinite 41t
Kernite 41t
Kinematic viscosity 91
 conversion factors 138t
Kinematics 80–81
Kinetic energy 82
Kyanite 41t

L

Lascas grades and properties 185t
Law of Compound or Joint Probability 115
Law of Total Probability 115
Lazulite 41t
Lazurite 41t
Leaching. *See* Heap leaching, In situ leaching,
  Vat leaching
Lead 180
Left lateral 60, 61f
Length conversion factors 132t–133t
Lepidolite mica 41t
Leucite 41t
LHD units. *See* Load-haul-dump units
Liabilities 420–421
Light conversion factors 133t
Lignite 10
Lime fluxes 181t

Limestone
    physical and chemical specifications for glass-grade 181t
    size analyses 182t
    typical analyses 181t
Limiting friction 78
Limonite 42t
Liner systems 353
    anchor trenches 357, 357f
    composite 356
    geomembranes 356–357, 356f, 357f
    geonet 359
    geotextiles 357–359, 358t
    interface friction 359, 359t–360t
    leakage 360, 360t
    soil 354, 354t, 355f
Load cycle time 219
Load factors 14t
Loaders. *See* Crawler-mounted loaders, Wheel loaders
Load-haul-dump units 240
Loading. *See* Excavation, loading, and transport equipment
Longwall systems 240

M

Magnesite 42t
Magnetism conversion factors 128t–129t
Magnetite 42t
Maintenance 383–384
    area 384
    backlog 384
    categories 384
    cost vs. level 383f
    criticality determination 384, 385t
    deterioration vs. cost and downtime 387f
    ferrography or wear-particle analysis 389
    inventory of spare parts 390–392
    lubrication and equipment analysis 385–389, 389t
    organizing 384
    overhaul 384
    predictive 384, 385, 388f
    preventive 384, 385, 387f, 388f
    record keeping and reporting 389
    repair 384
    scheduled 384
    shock-pulse analysis 389
    system 385, 386f
    thermography or infrared inspection 389
    unscheduled 384
    vibration analysis 389
    work order 384
    work-order form 390f, 391t
Malachite 42t
Manganese 182
Manganite 42t
Manning equation and coefficients 95, 96t
Marcasite 42t
Mass and moment of inertia conversion factors 133t–134t
Mass density 90
Mass divided by area conversion factors 134t
Mass divided by length conversion factors 134t
Mass divided by time conversion factors 134t
Mass divided by volume conversion factors 134t–135t
Materials
    properties 12t–13t
    strength properties 90t
Mathematics. *See also* Algebra, Geometry, Probability and statistics, Trigonometry
    coordinate relations 103
    derivatives 112–113
    integrals 113–114
    regression analysis 104–105
    vectors 104
Matrices 106
Maximum speed 217, 218t
Means 115–116
Measures 140t
Mercury 182
Merrill-Crowe recovery 343–344
Metals, precipitating 348t
Metamorphic rocks
    classification 59f
    properties 23t–25t
Method of joints 79
Method of sections 79
Mica
    biotite 34t
    lepidolite 41t
    muscovite 43t
    phlogopite 44t
    typical chemical analysis and physical properties 183t
Microcline feldspar 42t
Millerite 43t
Mine Safety and Health Administration 204. *See also* Health and safety
    appeals 416
    Approval and Certification Center 417
    approved explosion-resistant seals 273
    civil penalties 415
    Coal Mine Safety and Health 414
    criminal penalties 416
    discrimination protection 416
    Educational Policy and Development Office 415
    enforcement 415–416
    founding 412
    functions 414
    mandatory miner training 417
    Metal and Nonmetal Mine Safety and Health 414
    mission 413
    Office of Assessments 414
    Office of Program Evaluation Information Resources 415
    Office of Standards, Regulations and Variances 414
    orders of withdrawal 415
    organizational structure 414–415
    as part of Department of Labor 412, 414
    pattern of violations 416
    product testing 417
    significant and substantial (S&S) violations 415
    Technical Support Directorate 414

Technology Center (technical
    assistance) 416–417
  unwarrantable failures 416
Mine trucks 240–241
Minerals 10
  Mohs hardness scale 60
  properties 32t–49t
  properties of major fillers 50t
  solubility in cyanide 343t
Mining Enforcement and Safety
    Administration 412
Mohr-Coulomb failure criterion 87
Mohr's circle, 86, 86f
Mohs hardness scale 60
Molybdenite 43t
Molybdenum 183
Moment of force or torque conversion
    factors 135t
Moment of inertia 74
  channel section 1 77
  channel section 2 78
  circular section 75
  conversion factors 133t–134t
  "H" section 1 76
  "H" section 2 77
  pipe section 76
  rectangular section
Moments 72
Momentum 81
Monthly income statements
  full absorption costing basis 172t
  variable costing basis 172t
Moody diagram 93, 94f
MSHA. See Mine Safety and Health
    Administration
Muscovite mica 43t
  typical chemical analysis and physical
    properties 183t

N

National Environmental Policy Act 394
National Historic Preservation Act 394
Natural splitting 267
Natural ventilating pressure 267
Net positive suction head 277, 277t
Net slip 60, 61f
Newton's Second Law of Motion 81
Niccolite 43t
Nickel 184t
Niter 43t
Normal (Gaussian) distribution 117, 118t
Normal fault 60, 61f

O

Opal 43t
Open channel flow 95, 96t
Orpiment 43t
Orthoclase feldspar 44t

P

Parallel flow 267
Peak particle velocity (PPV) 211–213, 212f
Peat 10
Pentlandite 44t
Percentage depletion 167
Perlite 179t

Permafrost 9
Permeability conversion factors 135t
Permutations 114–115
PERT. See Program evaluation and review
    technique (PERT)
Phenacite 44t
Phlogopite mica 44t
Phosphate rock 184
Physical constants 140t
Pillars
  analysis of retreat mining pillar
    stability 250f, 252t–253t
  distribution of abutment stress 251f
  load schematics 251f
  loading conditions 250f
  satisfactory cases for coal mines 252t–253t
  section layout parameters 250f
  tributary area method 248
Pipes and piping 360
  compressed air pipe diameter 283
  earth pressure on buried pipe 361–362, 361f
  economic diameters 346f
  equivalent lengths for vent pipe and
    tubing 272t
  equivalent lengths of pipe for pump
    fittings 276t
  section formulas 75–76
  thickness 360–361
Placer mining 369
  bucketline dredge capital costs 377t
  bulk sampling advantages and
    disadvantages 372t
  centrifugal separators 376f
  character of gold vs. distance from source 372t
  classification of gold particles 372t
  common heavy minerals in placers 370t
  concentration methods 373t
  costs 377, 377t
  drill sampling advantages and
    disadvantages 372t
  effect of single gold particle on sample 372t
  grain size in placer deposits 369t
  gravel 377t
  jig 376f
  operations 373–376
  prospecting methods 371t
  scrubbing/screening trommel 375f
  sluice water requirements 374t
  sluice with Hungarian riffles 375f
  spirals and cones 375f
  swell of earth and gravel when
    disturbed 371t
  treatment methods 374t
  types of placers 370f
  vibrating tables 376f
  water requirements 373t
Planes
  circular motion 80–81
  determining intersection of 62–63, 63f
  determining strike and dip 61–62, 62f
  equation of plane passing through
    points 104
  friction 78
  general equation 104
  strain 86
  stress 85

Plate tectonics 66*f*
Platinum 44*t*
Platinum group metals 184
Points
   distance between 103
   slope of line joining 103
Poisson's ratio 84
Polyhalite 44*t*
Potash product specifications 185*t*
Potential energy 82
Powder factors 207–209
   for surface coal mines 209*t*
   for surface metal mines 209*t*
   for underground blasting 209, 210*f*
Power 100. *See also* Compressed air, Electrical power
   belt conveyor requirements 235, 236*f*, 236*t*, 237*f*
   conversion factors 135*t*
   dissipation 100
   excavation, loading, and transport equipment consumption 221–222
   hoist requirements 239
   pumping power formulas 278
   ventilation fan horsepower requirements 266
Precipitating metals 348*t*
Predictive maintenance 384, 385, 388*f*
Prefeasibility studies 1, 2*t*–4*t*
Present value ratio (PVR) 168
Pressure 91–92
   conversion factors 135*t*–136*t*
   and fans, blowers, and compressors 344
   prism 92
Pressure filtration 311, 311*t*
Preventive maintenance 384, 385, 387*f*, 388*f*
Probability and statistics
   arithmetic mean 115
   chi-square probability function 117, 120*t*
   combinations 114–115
   confidence intervals 116–117
   F-(variance ratio) distribution function 119, 122*t*
   geometric mean 115
   harmonic mean 116
   Law of Compound or Joint Probability 115
   Law of Total Probability 115
   means 115–116
   normal (Gaussian) distribution 117, 118*t*
   permutations 114–115
   Probability of an Event 115
   standard deviation 116
   Student's t-distribution 119, 121*t*
   weighted arithmetic mean 115
Probability of an Event 115
Profitability index 168
Program evaluation and review technique (PERT) 125
Project evaluation
   conceptual studies 1, 2*t*–4*t*
   defined 1
   due diligence studies 4–7
   feasibility studies 1, 2*t*–4*t*
   prefeasibility studies 1, 2*t*–4*t*
Proustite 44*t*
Psilomelane 44*t*

Psychrometric chart 262*f*, 263
Pumice 185
Pumps and pumping 275
   affinity laws 279
   allowable operating range 278
   best efficiency point (BEP) 278
   calculating dynamic head 275–276
   centrifugal pumps 275
   efficiency values 278
   equivalent lengths of pipe for fittings 276*t*
   flow rate formula 278
   friction head 275
   friction loss 276*t*
   Hazen-Williams Equation 275–276
   net positive suction head 277, 277*t*
   parallel pumps 277
   power formulas 278
   pump characteristic curves 277–278, 277*f*
   series pumps 277
   system head curve 278
   vacuum 344
Pyragyrite 44*t*
Pyrite 44*t*
   weathering reactions and pathways 401*f*
Pyrolusite 45*t*
Pyrometallurgy 332, 336
   drying 339
   emissions from typical operations 340*t*
   flash smelter 340
   fluid bed reactors 339
   free-energy diagrams 336, 337*f*, 338*f*
   heat capacities of gases and liquids 332*f*
   heat capacities of solids 334*t*
   ratio of heat capacities at 1 atm 333*t*
   rotary kilns 339
   saturated steam tables 335*t*–336*t*
   sintering 339
   troubleshooting 338
Pyrophyllite 45*t*
Pyrrhotite 45*t*

Q

Quartz 45*t*
   crystal grades and properties 185*t*

R

Radiation 99
Radiology conversion factors 136*t*
Radius of gyration 74
   channel section 1 77
   channel section 2 78
   circular section 75
   "H" section 1 76
   "H" section 2 77
   pipe section 76
   rectangular section 75
Rail haulage 229
   acceleration 231
   adhesion 231, 231*t*
   braking 231–232
   cycle time and production 232
   frictional resistance 229, 230*t*
   track 232
   tractive effort 230
   tractive resistance 229–230
Rainfall intensity 363–364

INDEX

Rake 60, 61*f*
Realgar 45*t*
Reclamation 419–420
Rectangular section formulas 74–75
Regression analysis 104–105
Resistance 100, 216–217, 217*t*
Resistors 100
Resource Conservation and Recovery Act 394
Reverse fault 60, 61*f*
Reynolds number 91
Rhodochrosite 45*t*
Right lateral 60
Rim pull 215
Rock bolting
    bolt, hole and resin cartridge combinations 249*t*
    common coal mine roof bolt bar capacity 249*t*
    empirical rules 246–248
    maximum and minimum bolt spacing 248
    minimum bolt length 246
    roof bolt characteristics 249*t*
Rock mass rating (RMR) system 244*t*–245*t*
    guidelines for excavation and support 245*t*
Rocks 9
    classification charts 57*f*, 58*f*, 59*f*
    cohesion 17*t*
    compressive strength 10*f*
    engineering properties 30*t*
    friction angle 17*t*
    properties 21*t*–31*t*
    strength properties 243*t*
    unit weight 17*t*
Rod and ball mills 290–291, 290*t*
Rolling resistance 216–217, 217*t*
Rutile 45*t*

S

Safety. *See* Health and safety, Mine Safety and Health Administration
SAG mills 292
Sampling
    accuracy and precision 148, 149*f*
    alternate shoveling 152, 154*f*
    analysis 155, 158*t*
    augur procedure 152
    bulk sampling 151
    channel sampling 150, 150*f*
    coal sample sizes 148*t*
    coning and quartering 152, 154*f*
    conveyor belt procedure 151–152
    core sampling 151, 151*f*
    cuttings sampling 151
    density 146–149
    drill hole spacing in coalfields 148*t*
    error 153, 158*t*
    fractional shoveling 152, 155*f*
    frequency 146–149
    geologic data collection and equipment 145*t*–146*t*, 147*t*
    grab sampling 150
    in situ 149–151
    increments for a coal shipment up to 1,000 tonnes 149*t*
    non-in situ 149–150, 151–152
    sample preparation 153, 156*f*, 156*t*, 157*f*
    sample reduction diagram 152, 153*f*
    sample splitting 152
    size 146–149
    stockpile procedure 151, 152*f*
    theory 145
    wireline core drilling sizes 147*t*
Sand 185
Scheelite 45*t*
Scrapers 227
    cycle time and production 228
    load time 227, 227*t*
    loading and spreading 227–228
    pusher dozers and cycle time 228, 228*t*
    spread time 227, 227*t*
Screens and screening
    calculating screen area 296–298, 296*t*–297*t*
    construction materials 294
    international standard and U.S. sieve sizes 295*t*
    screen types 294–295
    troubleshooting 295–296
Second moment of area 74
    conversion factors 127*t*–128*t*
Section 404 393
Section modulus
    channel section 1 77
    channel section 2 78
    circular section 75
    "H" section 1 76
    "H" section 2 77
    pipe section 76
    rectangular section 75
Sediment ponds 366
Sedimentary rocks
    classification 58*f*
    properties 25*t*–29*t*
Seismic refraction surveys 65, 67*f*
Serpentine 46*t*
Shear modulus 84
Shovels 222, 224*t*
Shuttle cars 240
SI. *See* International System of Units
Siderite 46*t*
Silver 46*t*, 186
    Merrill-Crowe recovery 343–344
Site structures
    anchor trenches 357, 357*f*
    coefficient of permeability 354
    composite liners 356
    cross-valley 351
    earthworks (embankments) 353, 353*f*
    facility layout 351
    geomembranes 356–357, 356*f*, 357*f*
    geonet 359
    geotextiles 357–359, 358*t*
    impoundments 351, 352*f*
    inspection and maintenance 367, 368*f*
    interface friction 359, 359*t*–360*t*
    liner leakage 360, 360*t*
    liner systems 353–360
    pipes 360–362
    ring dikes 351
    sidehill 351
    siting of tailings facilities 351*t*
    soil liners 354, 354*t*, 355*f*
    valley-bottom 351

Six-tenths rule 171
Slope stability 256
  planar failure 256, 257f
Slushers 240
Smithsonite 46t
Soda ash 186
Soda niter 46t
Sodalite 46t
Soils 9
  cohesion 17t
  engineering properties and uses 19t–20t
  friction angle 17t
  modulus values 18t
  particle sizes 14t
  permeability 354, 354t, 355f
  strength characteristics 18t
  Unified Soil Classification System 15t–16t
  unit weight 17t
Solid-liquid separations 303–304
  and bacterial oxidation 322
  centrifugation 308
  clarification 306
  concentration from counter-current decantation (CCD) thickener 304t, 322
  flocculation 312
  hydraulic classifiers and sizers 306
  hydrocyclones 308, 309f
  pressure filtration 311, 311t
  process and engineering considerations 312
  thickening 304–306, 304t, 305t
  typical settling data 305t
  vacuum filtration 308, 310t–311t
  wet classification machines 307t
Solids
  heat capacities 334t
  percentage of pulp 344
Solvent extraction 312–313
  controls and testing 314–315
  effect of viscosity on separation time 313t
  mixers 313
  pump efficiency vs. impeller tip speed 313–314, 313t
  settlers 314
Sorting
  commercial systems 298t
  hand 298
  laser reflectance (photometric) 299
  radioactive 299
  techniques 298
  x-ray fluorescence diamond 299
Specific gravity 90
Specific heat capacity and specific entropy conversion factors 131t
Specific volume 90
Specific weight 90
Sphalerite 46t
Spillways 366–367
Spinel 46t
Standard deviation 116
Stannite 47t
Statistics. *See* Probability and statistics
Staurolite 47t
Steel formula 364f, 364t
Stephanite 47t
Stibnite 47t
Stirred media mills 293
Stone, crushed 186t–187t
Storm water runoff 363–365
Straight line depreciation 166
Straight-line motion 80
Strain 83
  plane 86
  relations with stress 84–85
  uniaxial 85
Stress 83
  maximum normal criterion 87
  maximum shear criterion 87
  Mohr-Coulomb failure criterion 87
  Mohr's circle 86, 86f
  plane 85
  relations with strain 84–85
  torsional 86
  transformation in two dimensions 86, 86f
  uniaxial 85
Strike 60
  determining 61–62, 62f
  determining from two apparent dips 63–64, 64f, 65f
Strike slip 60, 61f
Strike-slip fault 60, 61f
Strontianite 47t
Student's *t*-distribution 119, 121t
Subsidence
  minimum cover for total extraction under water bodies 256t
  protective zones 255f
  residual subsidence duration over longwall mines 254t
  sized pillars for protecting surface structures 255f
  typical angles of draw 254t
Sulfur 47t, 187
Superfund Amendments and Re-authorization Act 393
Surface Mining Control and Reclamation Act 394
Swell 218–219
  percentages 14t
SX. *See* Solvent extraction
Sylvanite 47t
Sylvite 47t

T

Tailings
  facilities siting 351t
  potentially mobile chemical species in tailings liquids 401t
Talc 47t
  properties 187t–188t
Temperature
  conversion factors 136t–137t
  at different altitudes 259t–260t
  interval conversion factors 137t
Tennantite 48t
Tetrahedrite 48t
Thermal conductivity conversion factors 131t–132t
Thermal diffusivity conversion factors 132t
Thermal insulance conversion factors 132t
Thermal resistance conversion factors 132t
Thermal resistivity conversion factors 132t

Thermodynamics. *See also* Heat transfer
  First Law of 97–98
  ideal gas law 97, 98*t*
  nomenclature 96
Thickening 304–306, 304*t*, 305*t*
Throw 60
Time 140*t*
  conversion factors 137*t*
Time-distance plots 65, 67*f*
Tin 188
Titanium
  composition 189*t*
  specifications 189*t*
Topaz 48*t*
Torsional stress 86
Total resistance 216–217
Tourmaline 48*t*
Traction 215–216, 216*t*
Transport. *See* Excavation, loading, and transport equipment
Trigonometry
  addition and subtraction of two circular functions 108
  addition formulas 108
  definitions 106–107
  half-angle formulas 108
  multiple-angle formulas 108
  negative angle formulas 107
  periodic properties 107
  products of sines and cosines 108
Truss analysis 79
Tubing 270–272
  equivalent lengths for vent pipe and tubing 272*t*
Turquoise 48*t*

U

Unified Soil Classification System 15*t*–16*t*
Units. *See* International System of Units
Uraninite 48*t*
Uranium
  deposit characteristics 380–381
  in situ leaching environmental concerns 382
  in situ leaching operating problems and remedies 382
  in situ leaching process 381
  plant recovery process 382
  wells and well fields 381
U.S. Bureau of Mines 412
  mine gas explosibility diagram 269*f*
U.S. Department of Labor 412
U.S. Environmental Protection Agency 393
U.S. Fish and Wildlife Service 394

Vacuum filtration 308, 310*t*–311*t*
Vacuum pumps 344
Vanadinite 48*t*
Vanadium specifications 190*t*
Vat leaching 344
  comparison with heap leaching and agitation 344*t*
Vectors 104
Velocity conversion factors 137*t*

Ventilation
  air composition 259*t*
  air density at different altitudes 259*t*–260*t*
  airflow 263–266
  barometric pressure at different altitudes 259*t*–260*t*
  capture velocities for mineral dusts 273*t*
  controlled splitting 267
  conveying velocities of dusts and gases 273*t*
  dew point temperature 260*t*–261*t*
  ducts 270–272
  explosion-resistant seals 273
  fan horsepower requirements 266
  fan laws 266*t*
  friction factors for mine airways 265*t*
  friction loss of airflow 264
  gas laws 263
  measurement points for fixed traversing in airways 264*f*
  mine gas explosibility diagram 269*f*
  mine gas properties 268*t*
  natural splitting 267
  natural ventilating pressure 267
  parallel flow 267
  psychrometric chart 262*f*, 263
  quantity of air flow 263
  regulatory requirements 270*t*
  shock loss 266*t*
  temperature at different altitudes 259*t*–260*t*
  tubing 270–272
  velocity head of airflow 263
Viscosity 91
  conversion factors 137*t*–138*t*
Void percentages 14*t*
Voltage drop 100
Volume
  conversion factors 138*t*–139*t*
  divided by time conversion factors 139*t*

W

Water balance 362–363
Water quality
  federal Bevill-excluded wastes 405*t*
  federal hazardous waste classification criteria 404*t*
  kinetic prediction techniques 402*t*
  potentially mobile chemical species in tailings liquids 401*t*
  pyrite weathering reactions and pathways 401*f*
  static prediction techniques 403*t*
  typical standards and goals 400*t*
Wavellite 48*t*
Websites
  Africa 430
  Asia 430–431
  associations 426–427
  Australia 431
  Canada 431–432
  computing applications 428
  economics 427–428
  Europe 432
  federal agencies 428–429
  information 425–426
Websites, *continued*
  Latin America 432

machinery and equipment trading 428
New Zealand 431
professional societies and other
    organizations 423–425
publications 425–426
reference 425–426
state agencies 429–430
trade unions 428
United Kingdom 432–433
Weighted arithmetic mean 115
Wheel loaders 222–225, 225*t*
Willemite 49*t*
Witherite 49*t*
Wolframite 49*t*
Work 82
Wulfenite 49*t*

Y

Young's Modulus 84

Z

Zadra stripping (elution) 319
Zeolite classification 190*t*
Zinc specifications 191*t*
Zincite 49*t*
Zircon 49*t*